LIMNOLOGY OF TUNDRA PONDS

US/IBP SYNTHESIS SERIES

This volume is a contribution to the International Biological Program. The United States effort was sponsored by the National Academy of Sciences through the National Committee for the IBP. The lead federal agency in providing support for IBP has been the National Science Foundation.

Views expressed in this volume do not necessarily represent those of the National Academy of Sciences or the National Science Foundation.

Volume

1 MAN IN THE ANDES: A Multidisciplinary Study of High-Altitude Quechua / *Paul T. Baker and Michael A. Little*

2 CHILE-CALIFORNIA MEDITERRANEAN SCRUB ATLAS: A Comparative Analysis / *Norman J. W. Thrower and David E. Bradbury*

3 CONVERGENT EVOLUTION IN WARM DESERTS: An Examination of Strategies and Patterns in Deserts of Argentina and the United States / *Gordon H. Orians and Otto T. Solbrig*

4 MESQUITE: Its Biology in Two Desert Scrub Ecosystems / *B. B. Simpson*

5 CONVERGENT EVOLUTION IN CHILE AND CALIFORNIA: Mediterranean Climate Ecosystems / *Harold A. Mooney*

6 CREOSOTE BUSH: Biology and Chemistry of *Larrea* in New World Deserts / *T. J. Mabry, J. H. Hunziker, and D. R. DiFeo, Jr.*

7 BIG BIOLOGY: The US/IBP / *W. Frank Blair*

8 ESKIMOS OF NORTHWESTERN ALASKA: A Biological Perspective / *Paul L. Jamison, Stephen L. Zegura, and Frederick A. Milan*

9 NITROGEN IN DESERT ECOSYSTEMS / *N. E. West and John Skujins*

10 AEROBIOLOGY: The Ecological Systems Approach / *Robert L. Edmonds*

11 WATER IN DESERT ECOSYSTEMS / *Daniel D. Evans and John L. Thames*

12 AN ARCTIC ECOSYSTEM: The Coastal Tundra at Barrow, Alaska / *Jerry Brown, Philip C. Miller, Larry L. Tieszen, and Fred L. Bunnell*

13 LIMNOLOGY OF TUNDRA PONDS: Barrow Alaska / *John E. Hobbie*

14 ANALYSIS OF CONIFEROUS FOREST ECOSYSTEMS IN THE WESTERN UNITED STATES / *Robert L. Edmonds*

15 ISLAND ECOSYSTEMS: Biological Organization in Selected Hawaiian Communities / *Dieter Mueller-Dombois, Kent W. Bridges, and Hampton L. Carson*

US/IBP SYNTHESIS SERIES | 13

LIMNOLOGY OF TUNDRA PONDS

Barrow, Alaska

Edited by

John E. Hobbie
Marine Biological Laboratory

Dowden, Hutchinson & Ross, Inc.
Stroudsburg Pennsylvania

Dedicated to three scientists who shaped and supported ecological research at Barrow, Alaska, during the 1950s and 1960s: Max Britton, Max Brewer, and Frank Pitelka.

Library of Congress Catalog Card Number: 80-26373
ISBN: 0-87933-386-3

82 81 80 1 2 3 4 5
Manufactured in the United States of America

Library of Congress Cataloging in Publication Data
Hobbie, John E.
Limnology of tundra ponds, Barrow, Alaska.
(U.S./IBP synthesis series ; v. 13)
Bibliography: p.
Includes indexes.
1. Limnology—Alaska—Barrow. 2. Pond ecology—Alaska—Barrow. 3. Tundra ecology—Alaska—Barrow. I. Hobbie, John E. II. Series.
QH105.A4H62 574.5′2632′097987 80-26373
ISBN 0-87933-386-3

Distributed world wide by Academic Press,
a subsidiary of Harcourt Brace Jovanovich,
Publishers.

Foreword

This book is one of a series of volumes reporting results of research by U.S. scientists participating in the International Biological Program (IBP). As one of the fifty-eight nations taking part in the IBP during the period of July 1967 to June 1974, the United States organized a number of large, multidisciplinary studies pertinent to the central IBP theme of "the biological basis of productivity and human welfare."

These multidisciplinary studies (Integrated Research Programs) directed toward an understanding of the structure and function of major ecological or human systems have been a distinctive feature of the U.S. participation in the IBP. Many of the detailed investigations that represent individual contributions to the overall objectives of each Integrated Research Program have been published in the journal literature. The main purpose of this series of books is to accomplish a synthesis of the many contributions for each principal program and thus answer the larger questions pertinent to the structure and function of the major systems that have been studied.

Publications Committee: U.S./IBP
Gabriel Lasker
Robert B. Platt
Frederick E. Smith
W. Frank Blair, Chairman

Preface

This book is a report of investigations of several small ponds on the arctic tundra near Barrow, Alaska. The main study, which ran from 1971 through 1973, was funded from three sources: The National Science Foundation, the State of Alaska through the University of Alaska, and individual companies and members of the petroleum industry. The NSF funding was under the joint sponsorship of the U.S. Arctic Research Program (Division of Polar Programs) and the U.S. International Biological Program (Ecosystem Analysis Program). The U.S. Tundra Biome Program was under the overall direction of Jerry Brown of the U.S. Army Cold Regions Research and Engineering Laboratory and consisted of aquatic and terrestrial sections. A companion volume to this reports the findings of the terrestrial projects (Brown et al. in press).

The principal investigators of the aquatic projects were:

Vera Alexander, University of Alaska
Robert J. Barsdate, University of Alaska
Donald A. Bierle, Sioux Falls College
James N. Cameron, University of Alaska
Raymond D. Dillon, State University of South Dakota
John E. Hobbie, North Carolina State University
C. Peter McRoy, University of Alaska
Michael C. Miller, University of Alaska
Raymond G. Stross, The University (SUNY) at Albany.

Other scientists who took part in the project were Staffan Holmgren (Uppsala University), Tom Fenchel (Aarhus University), Stanley Dodson (University of Wisconsin), John Kelley (University of Alaska), Patrick Coyne (U.S. Army, CRREL), Ralph Daley (North Carolina State University), Richard Prentki (University of Alaska), Tor Traaen (Norwegian Institute for Water Research), Donald Stanley (North Carolina State University), and Jawahar Tiwari (North Carolina State University).

Additional information on the macrobenthos came from a study in 1975, 1976, and 1977 carried out by Samuel Mozley and Malcolm Butler (University of Michigan) which was funded by the U.S. Department of Energy.

There are a number of possible approaches to the study of the ecology of tundra ponds. We concentrated first on measuring the fluxes of carbon, nitrogen, and phosphorus through the ecosystem. Next, we used a variety of manipulations, ranging from changed light conditions for plankton in a small bottle to an increase in phosphate in a whole pond, to investigate the controls of various processes. While the field work was going on, we also constructed a mathematical model of the ecosystem. This left little time for detailed studies of the ecology of individual species although several dominants, such as *Daphnia middendorffiana* and a *Chironomus* sp., were examined. Most effort was put into nutrient cycling studies and into investigations of the lower trophic levels. Some areas, such as the control of zooplankton species composition or the physiology of individual species of algae, were not well studied.

Most of this book was written during the summer of 1974, and was later edited for consistency of style and overall integration. The portion of Chapter 7 on the insect larvae was completely rewritten in 1978, and, by the author's request, the section of Chapter 6 by R. Stross was not edited.

The book is organized in a conventional fashion with the physical and chemical information first followed by the descriptions of the primary producers, secondary consumers, etc. Each chapter ends with an extensive summary; a good idea of the important parts of the limnology of the pond can be gained from these. Chapter 1 consists of a summary of the conclusions of the overall study but only those conclusions that are most interesting to an ecologist. In this way, we attempt to answer the question, "What new things did you discover?"

Excellent logistics and laboratory support were provided by the Office of Naval Research through its Naval Arctic Research Laboratory at Barrow. Two former directors of this laboratory, Max C. Brewer and John Schindler, deserve particular credit for facilitating this support. Gene E. Likens of Cornell University reviewed the first draft of this book and Colleen M. Cavanaugh and Kate Eldred of the Marine Biological Laboratory provided valuable editing services. Harold Larsen, USA CRREL, prepared the illustrations and the CRREL editorial staff under the direction of Stephen Bowen provided valuable assistance throughout the preparation of this book. Kathleen Salzberg, editor of *Arctic and Alpine Research,* provided assistance in editing the references. Typesetting was done by The Job Shop, Woods Hole, Massachusetts, and by Donna Murphy of CRREL. Special thanks go to George Llano, formerly of the National Science Foundation, for his administrative guidance and his sympathetic understanding of the difficulties of running a large scientific study in the Arctic, and to Jerry Brown for his assistance in the field aspects of the program and in publication of this volume.

John E. Hobbie

Contents

Foreword v
Preface vii
List of Contributors xii

1: Major Findings 1
J. E. Hobbie

Introduction, 1 Description of the Ponds, 2 Flux of Carbon, 5 Flux of Phosphorus, 10 Flux of Nitrogen, 13 Effects of the Arctic Environment, 13 Modeling, 15

2: Introduction and Site Description 19
J. E. Hobbie

The Tundra Biome Project, 19 Limnology of the Arctic, 21 Geography and Geomorphology, 25 Biology, 36 Energy and Nutrient Cycling, 42 Summary, 46

3: Physics 51
M. C. Miller, R. T. Prentki, and R. J. Barsdate

Geography, 51 Temperature Studies, 54 Hydrology, 62 Light, 70 Currents, 71 Summary, 72

4: Chemistry 76
R. T. Prentki, M. C. Miller, R. J. Barsdate, V. Alexander, J. Kelley, and P. Coyne

Sediments, 76 Major Ions, 79 Trace Metals, 86 Carbon Dioxide Systems, 87 Oxygen, 97 Nitrogen, 100 Phosphorus, 115 Control of Phosphorus, 136 Organic Carbon, 151 Summary, 169

5: Primary Producers 179
V. Alexander, D. W. Stanley, R. J. Daley, and C. P. McRoy

Phytoplankton, 179 Epipelic Algae, 193 Factors Controlling Algae, 201 Rooted Aquatic Plants, 224 Factors Controlling Rooted Aquatic Plants, 234 Summary, 243

6: Zooplankton 251
R. G. Stross, M. C. Miller, and R. J. Daley

Communities, Life Cycles, and Production, 251 Control of Zooplankton Production (I), 274 Control of Zooplankton Production (II), 288 Summary, 293

7: Macrobenthos 297
M. Butler, M. C. Miller, and S. Mozley

Introduction to Arctic Benthos, 297 Chironomidae Studies, 303 Tadpole Shrimp, 323 Summary, 335

8: Decomposers, Bacteria, and Microbenthos 340
J. E. Hobbie, T. Traaen, P. Rublee, J. P. Reed, M. C. Miller, and T. Fenchel

Bacteria, 340 Fungi, 356 Decomposition of Macrophytes, 357 Sediment Respiration, 363 Microbenthos, 372 Summary, 384

9: Oil Spill Effects 388
R. J. Barsdate, M. C. Miller, V. Alexander, J. R. Vestal, and J. E. Hobbie

Introduction, 388 Physical and Chemical Measurements, 391 Biological Measurements, 395 Summary, 405

10: Modeling 407
J. L. Tiwari, R. J. Daley, J. E. Hobbie, M. C. Miller, D. W. Stanley, and J. P. Reed

Modeling in the Aquatic Program of the Tundra Biome, 407 Whole Systems Models, 409 Benthic Carbon Flow Model, 430 Planktonic Carbon Flow Model, 434 Results and Discussion, 447 Summary, 456

References 457
Taxonomic Index 493
Subject Index 499

List of Contributors

Vera Alexander
Institute of Marine Science, University of Alaska, Fairbanks, Alaska 99701

Robert J. Barsdate
Institute of Marine Science, University of Alaska, Fairbanks, Alaska 99701

Malcolm Butler
Department of Ecological and Evolutionary Biology, Division of Biological Sciences, University of Michigan, Ann Arbor, Michigan 48109

Patrick Coyne
USDA/SEA/AR; Southern Plains Research Station, 2000 18th Street, Woodward, Oklahoma 73801

Ralph J. Daley
Canada Center for Inland Waters, Pacific Region, 4160 Marine Drive, West Vancouver, British Columbia, Canada V7V 1N6

Thomas Fenchel
Institute of Ecology and Genetics, Ny Munkegade, Aarhus University, Aarhus C, Denmark

John E. Hobbie
Ecosystems Center, Marine Biological Laboratory, Woods Hole, Massachusetts 02543

John Kelley
Naval Arctic Research Laboratory, Barrow, Alaska 99623

C. Peter McRoy
Institute of Marine Science, University of Alaska, Fairbanks, Alaska 99701

Michael C. Miller
Biological Science Department, University of Cincinnati, Cincinnati, Ohio 45221

Sam Mozley
Department of Zoology, North Carolina State University, Raleigh, North Carolina 27650

Richard T. Prentki
Department of Biological Sciences, University of Nevada, Las Vegas, Nevada 89154

James P. Reed
Ecosystems Center, Marine Biological Laboratory, Woods Hole, Massachusetts 02543

Parke A. Rublee
CBCES, Smithsonian Institution, P. O. Box 28, Edgewater, Maryland 21037

Donald W. Stanley
Institute of Coastal Studies, University of East Carolina, Greenville, North Carolina 27834

Raymond G. Stross
Department of Biology, State University of New York, Albany, New York 12203

Jawahar Tiwari
Department of Surgery, School of Medicine, University of California, Los Angeles, California 90024

Tor Traaen
Institute of Water Research, P. O. Box 260, Blindern, Oslo 3, Norway

J. Robie Vestal
Department of Biological Science, University of Cincinnati, Cincinnati, Ohio 45221

1

Major Findings

John E. Hobbie

INTRODUCTION

Studies in the extreme environments of mountains, tropics, and the Arctic have long been an important part of ecological research. Apart from the stimulation and enjoyment of visiting new places, ecologists have compared these extreme habitats with one another and with temperate habitats in order to test hypotheses about general principles. This approach of comparative natural history requires a large body of data collected from many habitats; both descriptions and a good understanding of processes are required. The data from extreme environments are especially valuable as they extend the range of important variables and may even allow analyses of the effect of certain factors that always vary together in temperate regions.

The IBP study of arctic ponds reported in this book is primarily a description of the habitat, the biota, and the processes by which organisms interact with other organisms and with their physical and chemical environments. In the report, the comparative aspects of the study have been deliberately de-emphasized, as constant reference to temperate and tropical lakes would have quickly doubled the size of the book. The value of the study in this comparative sense will become apparent later, when this study is referred to to find out what controlled photosynthesis, how rapidly a sedge leaf decomposed, or what the community structure was in an arctic pond.

In addition to the comparative importance of the arctic ponds, there are certain advantages to investigating aquatic processes in the Arctic. For example, low diversity of the higher plants and animals allows cohorts and age classes to be identified and followed through time; this simplifies productivity measurements. In some groups, such as most of the zooplankton, there may be only a single generation each year which also greatly simplifies growth measurements. The low diversity also permits a more complete study to be carried out with fewer scientists but does not, of course, make the study of an individual process any easier.

There are other tactical advantages to arctic research. First, the ponds freeze completely in mid-September, so they need to be studied for only 3 months a year (which fits into academic schedules quite well). During the field research months the scientists were working in crowded laboratories in an isolated location where there were few distractions from beaches, families, or television. The resulting intense interactions and scientific excitement could only be maintained for a month or so, but helped immensely to stimulate creativity and to integrate the various projects.

Description of the Ponds

Small ponds formed on old lake beds are abundant on the flat coastal plain of northern Alaska (Figure 1-1). A number of these ponds, several kilometers from the Naval Arctic Research Laboratory at Barrow, were studied for several years to improve our understanding of the controls of aquatic populations and processes that operate in this extreme environment.

The ponds are small, only about 30 x 40 m, and shallow, up to 0.5 m deep (Figure 1-2, 1-3). Each pond is surrounded by wet tundra, mostly low grasses and sedges, and is cut off from adjoining ponds by a network of

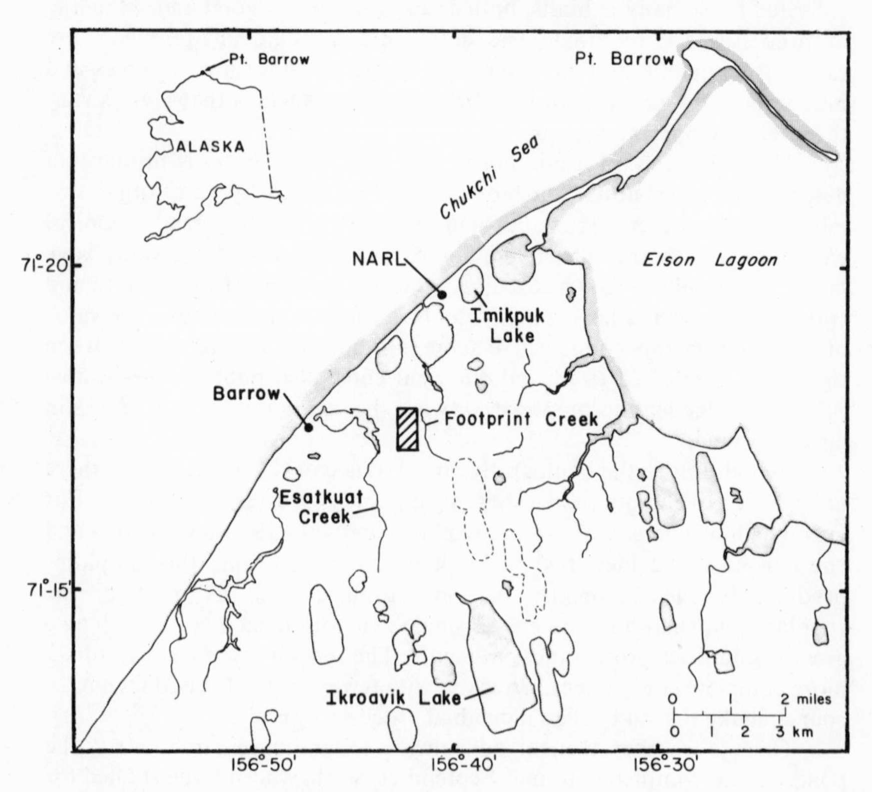

FIGURE 1-1. *Location of IBP Tundra Biome Project, showing the Naval Arctic Research Laboratory, the village of Barrow, and the research sites (cross-hatched area).*

small ridges pushed up by the growth of underlying ice wedges. The total area enclosed by the ridges is about double the pond area. Despite the minuscule drainage basins and the desert-like levels of precipitation (12 cm annual, 50% falls as rain in June through September), the surrounding tundra is often saturated and the ponds do not dry up. Water flows from one basin to another for only a few days during the spring runoff. There is no belowground water movement from basin to basin or into the sediments because of the underlying permafrost.

From late September until mid-June, the ponds and their underlying sediments are solidly frozen. Melting of the ice in the ponds occurs over a few days in the spring and water temperatures can reach as high as 16°C any time thereafter. Thawing of the sediments continues throughout the summer until 30 cm are thawed. The ponds are so shallow that the water temperature can change as much as 10°C per day in response to sunlight, air temperature, and wind. June, July, and August are cool (mean air temperature is 2.8°C), cloudy (83% cloud cover), and windy (an average of 6.1 m sec^{-1}) so the mean water temperature is low, around 6°C.

FIGURE 1-2. *The intensively studied ponds near Barrow, Alaska. A small field lab is at the upper left of the picture and experimental subponds in Pond B are at right. An aerial cable car is suspended above the subponds.*

FIGURE 1-3. *To avoid disturbing the water and sediments of Pond B, an investigator takes samples from a cable car.*

The pond waters contain small amounts of salts and have a pH around 7.3. The light, flocculant sediments are made up of 80% organic matter.

The dominant primary producers of the ponds (Table 1-1) are a sedge (*Carex aquatilis*) and a grass (*Arctophila fulva*) which live in the shallow margins of the pond and cover 30% of the surface (Figure 1-4). Benthic microalgae, mostly diatoms and blue-greens, are also important producers but their numbers are kept low by the continual mixing of the upper few centimeters by animals, which keeps most of the algae away from light. Algae in the water above are all small flagellated nannoplankton, especially greens and chrysophytes. Their total productivity is low and they are heavily grazed by zooplankton such as *Daphnia* and fairyshrimp. In turn, these herbivores are preyed upon by predaceous zooplankton (*Cyclops, Heterocope*) but there are no important vertebrate predators (although shorebirds do feed on zooplankton). The leaves, rhizomes, and

TABLE 1-1 *Annual Production of Tundra Pond Communities*

Type of community	Production $g\,C\,m^{-2}\,yr^{-1}$	Method of measurement
Phytoplankton	1.1	carbon-14
Benthic algae	8.4	carbon-14
Macrophytes	16.4	biomass changes
Zooplankton	0.20	biomass changes
Macrobenthos	1.65	biomass changes
Planktonic bacteria	0.01	biomass changes
Benthic bacteria	8.6	biomass changes
Benthic bacteria	4-16	CO_2 evolution in cores
Benthic bacteria	20	CO_2 exchange with atmosphere
Protozoa	0.3	biomass changes, lab growth rates
Microbenthos	0.2	biomass changes, lab growth rates

roots of the grasses and sedges enter the detritus food chain as there are no grazers on the live plants. Most of the detritus is mineralized by bacteria and fungi but some is consumed by chironomid larvae, the dominant animals of the sediment. These larvae eat a few percent of the bacteria and algae per day as do the microfauna of nematodes, harpacticoid copepods, and protozoans.

Flux of Carbon

The measurement of the flux of carbon is a useful way to begin an ecosystem study, as all the important elements can be identified. The techniques we used for the carbon flux and standing stock are standard ecological measurements such as ^{14}C for the primary productivity of the algae, biomass changes for rooted plants, CO_2 partial pressures by gas analysis to obtain water-air exchange, laboratory respiration studies of larger organisms, and acridine orange direct counts for the bacteria. The only component of the biota not measured was the fungi. A single measurement indicates that in sediments the mass of fungal hyphae is about equal to the mass of bacteria.

The rooted plants in the pond provide most of the input of organic carbon (Figure 1-5, Table 1-1). They release dissolved organic carbon into the water, release a large quantity of CO_2 via root respiration, and add dead leaves, stems, and roots to the detrital pool. Once it reaches the sediments, a leaf of *Carex* takes 4 years to decompose. One reason for this rather long life-after-death is the lack of shredders in the pond ecosystem. Another reason is the 9 months of cold storage each year (however, freezing and thawing does mechanically damage the leaves). When

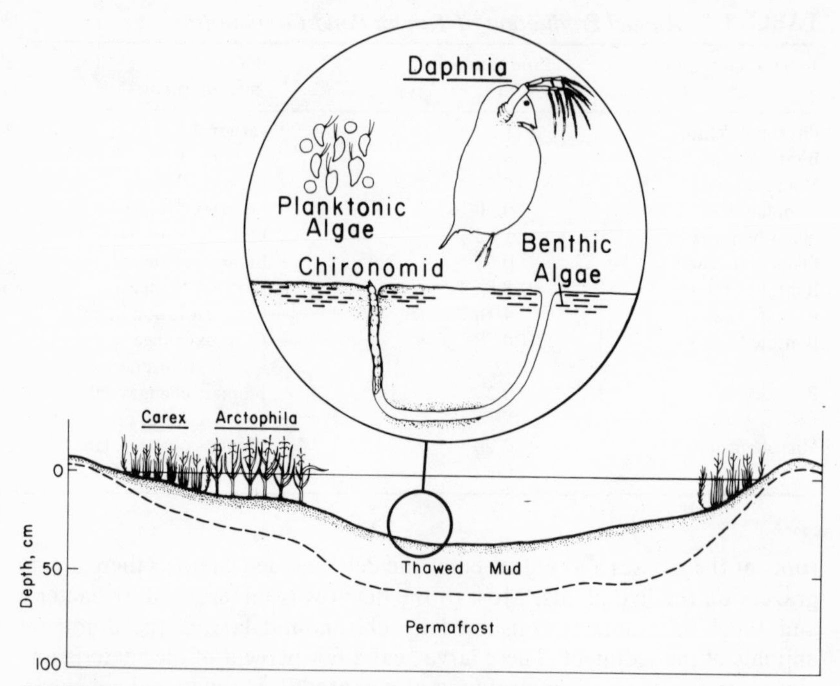

FIGURE 1-4. *A cross section of a typical pond.*

decomposition is calculated as percent per month of open water, then it appears that the rate is very similar to temperate rates.

Algal photosynthesis in the sediment surface is also an important input of organic carbon. Although photosynthesis occurs only in the top 2 mm, the algal cells were found throughout the upper 5 cm and deeper. The buildup of these benthic algae is prevented by the downward mixing of sediment and algal cells by the animals; in the absence of this mixing, an algal mat would develop, which would have a very high productivity.

Algal photosynthesis in the water column is extremely low, as low as any in the world. This is not caused by low numbers of cells, as millions per liter are always present. These are all very small cells, however, and their mass is also small. Thus, in temperate lakes the algal mass is 100 times greater than the bacterial mass; in the arctic ponds the algal mass was about equal to the bacterial mass (Note: The bacterial mass is the same in both systems). These planktonic algae also show a paradox found in other extremely oligotrophic systems, such as the Sargasso Sea. This is the rapid turnover of cells (the amount of carbon produced per day equals the carbon of the cells) in oligotrophic waters. Grazing by zooplankton, especially *Daphnia,* is likely responsible for much of the rapid turnover in the ponds.

The result of the distribution of primary production, low in the plankton but high in the macrophytes and benthic algae, is a shift of

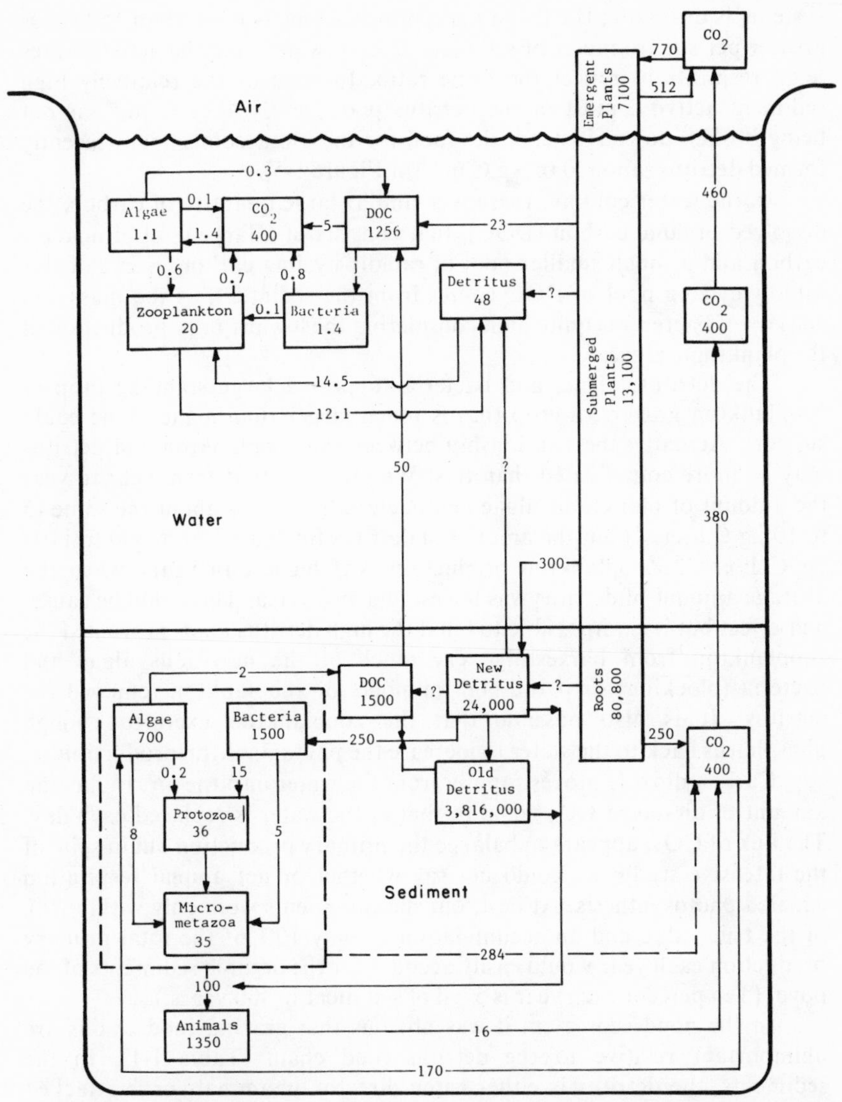

FIGURE 1-5. *Carbon flux through a typical tundra pond. Measurements were made on 12 July 1971. On this date, the average depth of the water was 10 cm and the depth of the sediment was taken as 5 cm. Units of the standing crop (in boxes) are mg C m^{-2} and transfer rates (arrows) are mg C m^{-2} day^{-1}.*

organic carbon to the sediments where it enters the detritus food web (Figure 1-5). Here, the abundant detritus is a large reservoir of food for animals while the decomposition of the organic matter provides a steady supply of nutrients for algae. The contrast between the sediment and water systems is dramatic; the living mass of organisms is more than 150 times greater per square meter of sediment than of water, and the activity rates (e.g., respiration) reflect the same ratio. In spite of the relatively high sediment activity, most of the detritus pool, nearly 4 kg C m^{-2}, is not being broken down. Instead, the food for the biota comes from recently formed detritus (about 0.02 kg C m^{-2} in Figure 1-5).

In the water column, there is a similar large quantity of carbon, the dissolved organic carbon (DOC), that consists of a large pool of inactive carbon and a much smaller pool of rapidly-cycling carbon. Some of this rapidly-cycling pool of DOC comes from the sediment, as the mass and activity of bacteria is quite high considering the low primary production of the planktonic algae.

The detritus, algae, and bacteria support a large standing crop of zooplankton grazers, a crop that is much larger than algae alone could support. Actually, the relationship between the zooplankton and detritus may be more complicated than this. We observed that from year to year the amount of planktonic algae and bacteria remained about the same (5 to 10 μg C $liter^{-1}$) but the amount of detritus fluctuated from 300 to 1400 μg C $liter^{-1}$. Zooplankton production was highest in years when the average amount of detritus was lowest and vice versa. This could be cause-and-effect but it is impossible to tell if the high detritus loads prevented the zooplankton from harvesting very much of the nutritious algae and bacteria (blocking) or if the high numbers of zooplankton removed the detritus. It is also possible that the zooplankton excreted enough phosphorus back to the water to increase the phytoplankton production.

Carbon dioxide moves rapidly from the water into the air. In fact, an amount of dissolved CO_2 equal to that in the water is replaced each day. The flux of CO_2, appears to balance the primary production but in spite of the intensive study, we could not say whether or not annual respiration equaled photosynthesis. At best, our measurements were only within 20% of the true value and an accumulation of only 10% of the total primary production each year would easily account for the organic sediments of the pond. (Ten percent each year is 5 cm of sediment in 400 years.)

In the pond ecosystem it was obvious that grazing food chains are unimportant relative to the detritus food chain (Table 1-1). In the sediments, the detritus is either eaten directly by animals or is attacked first by microbes. Our evidence for direct utilization comes mostly from studies of the energy requirements of the chironomid larvae (the "animals" in Figure 1-5). At the rate of particle ingestion that we measured, the larvae had to be digesting mostly detritus. Previous workers postulated that the animals were obtaining enough energy by stripping the microbes from the detrital particles. We actually measured the quantity of bacterial and algal biomass that is included in the detritus and found it to be only 0.06% of the total carbon (Figure 1-5). This amount of carbon is 0.3% of the organic carbon requirement of the larvae. Although these animals may select microbe-rich particles or locations for feeding, they

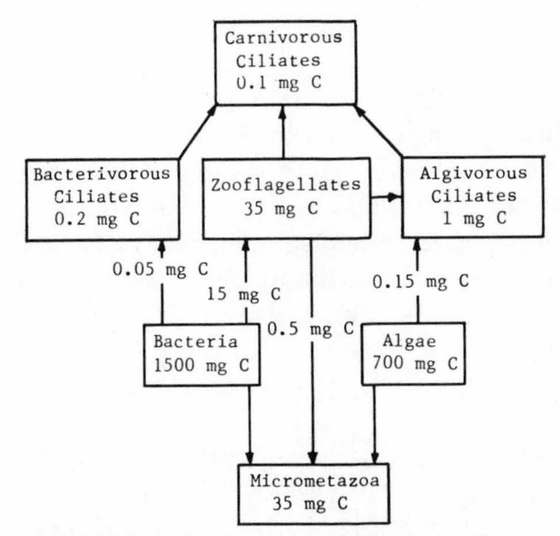

FIGURE 1-6. *Carbon flow through the protozoans and micrometazoans of the sediment of a tundra pond. Units are mg C m^{-2} day^{-1} for the fluxes and mg C m^{-2} for the standing crop.*

would have to be extraordinarily selective to meet their energy requirements from bacteria and algae alone.

The bacteria which break down detritus are (along with the benthic algae) the base of another food chain of protozoans and micrometazoans such as nematodes. The feeding rates and production of these animals have not previously been studied in the field. Here, these animals grazed only 1 to 2% of the bacteria and algae per day (Figure 1-6). This seems small yet represents 20% of the bacterial production and 5% of the algal production each day; thus, the small animals may control the bacteria to some degree.

The protozoans and bacteria interact in other ways as well. It has long been known that decomposition proceeds faster, and bacteria are more active, when grazing animals are present. One hypothesis has been that nutrients were rapidly released by the grazers and that this release allowed higher microbial activity. This hypothesis was tested in an experiment which investigated the rate of cycling of phosphorus-32 in small flasks containing *Carex*, bacteria, and one species of protozoa (Barsdate et al. 1974). When the protozoan was present, the bacterial biomass was lower, the bacteria were more active, and phosphorus was taken up faster by the bacteria (1.67 vs. 0.25 pg P $cell^{-1}$ hr^{-1}), than when the protozoan was not present. Yet, only a few percent of the phosphorus actually cycled through the protozoans. Thus, direct release of phosphorus by the protozoans did not affect the bacteria and the hypothesis was disproved. It is possible that the bacteria are kept in a phase of rapid growth by the grazing and that it is this rapid growth that is responsible for the faster decomposition.

Flux of Phosphorus

Phosphorus enters the ponds in rainfall (8 μg P liter^{-1}) and in overland flow. The quantities entering a pond are small and are equaled by the losses. We did measure an annual loss of 0.7 mg P m^{-2} but this amount is minuscule relative to the 25,000 mg P m^{-2} found in the top 10 cm of sediments. In spite of the large amounts of phosphorus present, the concentrations of inorganic phosphorus in the waters of the pond and in the interstitial water of the sediments is always extremely low, between

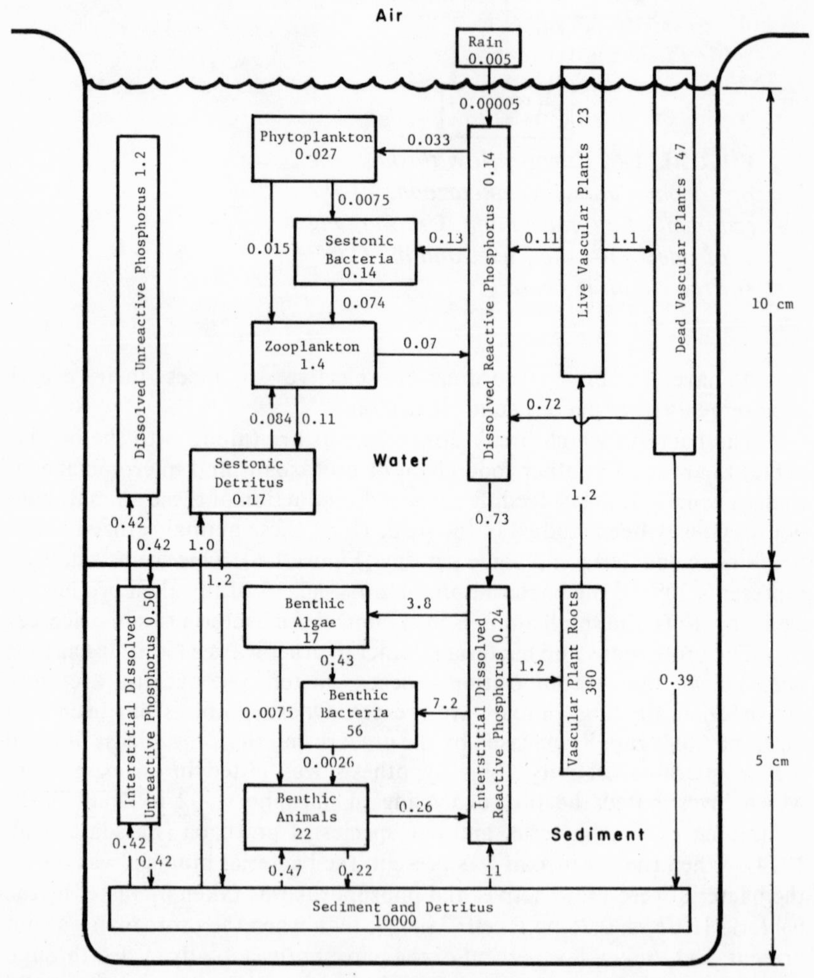

FIGURE 1-7. *Phosphorus flow diagram in a tundra pond for 12 July 1971. Units are mg P m^{-2} and mg P m^{-2} day^{-1}. Fluxes were measured whenever possible (Prentki 1976) or were based on the carbon flux data (Figure 1-5).*

0.001 and 0.002 mg P liter $^{-1}$. This is the form of phosphorus that is available to algae and higher plants; their primary production can be enhanced over several weeks by adding phosphorus to a pond. Even in the fertilized ponds, however, the concentrations of inorganic phosphorus rapidly decline to 0.001 to 0.002 mg P liter $^{-1}$. Why are these phosphorus concentrations so low and why are there such small changes over the summer?

A part of the answer is that the dissolved reactive phosphorus (DRP) cycles very rapidly in the ponds (Figure 1-7). For example, there is 0.14 mg P m $^{-2}$ in the water on the day illustrated in the figure while the bacteria and algae take up 5.8 mg P m $^{-2}$ day $^{-1}$. At the same time, there is also a transport of 0.73 mg P m $^{-2}$ into the DRP pool of the interstitial water; the DRP thus turns over 50 times per day in the ponds. During the rapid turnover of the small amount of DRP in the water, the large quantities of phosphorus in the sediment turn over very slowly and actually buffer the whole system.

The other part of the answer lies in the chemical properties of the sediment. When DRP enters the pond, it quickly moves to the sediment where much of it is sorbed onto a hydrous iron complex. The concentration of DRP and the release rate of the sorbed phosphorus are controlled by a chemical equilibrium; ponds with different amounts of iron and inorganic phosphorus in the surface sediments will have different

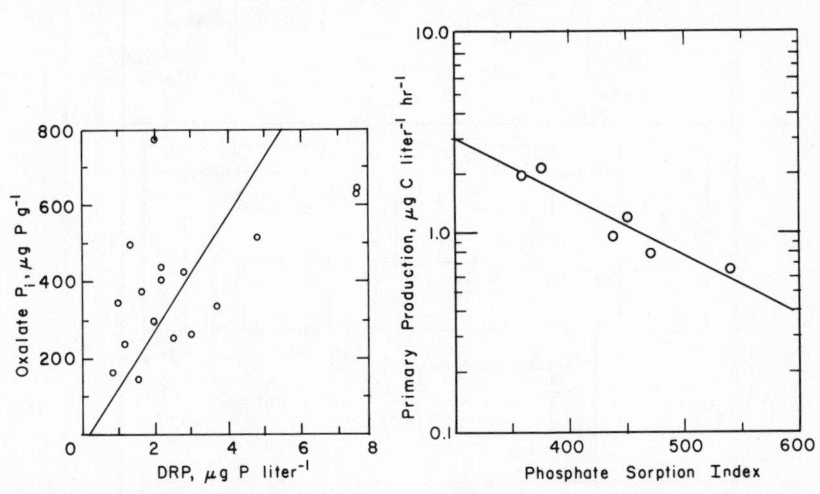

FIGURE 1-8. *Oxalate extractable phosphorus in the sediments of five tundra ponds of similar origin plotted against the dissolved reactive phosphate (DRP) of the overlying water column.*

FIGURE 1-9. *Algal photosynthesis in the water column of a series of tundra ponds plotted against the phosphate sorption index of the underlying sediments (9 August 1973). (Data are from Prentki 1976.)*

amounts of DRP in the water. In a series of intensively-studied ponds (Figure 1-8), the concentration of DRP in the water column could even be predicted from a single measurement of the sediment phosphorus that could be extracted with oxalate. In the same ponds, the oxalate-extractable phosphorus appeared to be directly related to the photosynthesis rate of planktonic algae (Figure 1-9). Thus, we conclude

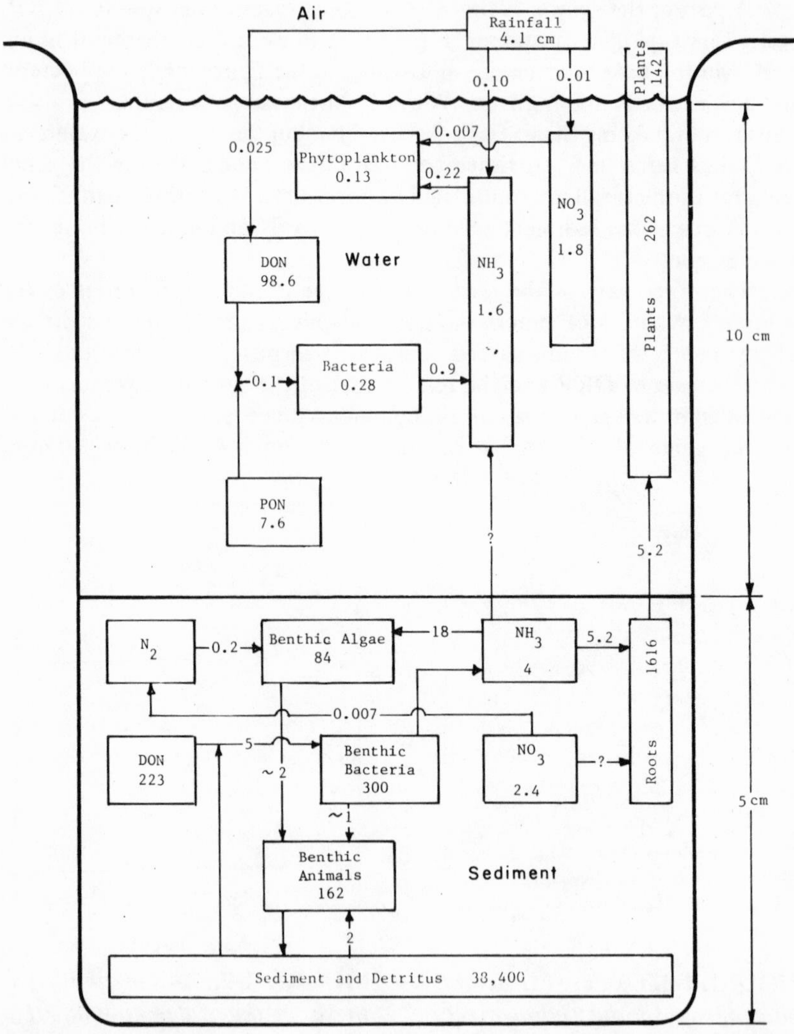

FIGURE 1-10. *Nitrogen standing crop and fluxes in the water and sediments of a tundra pond on 12 July 1971. Units are mg N m^{-2} for the concentrations and mg N m^{-2} day^{-1} for the fluxes. The amounts and flux rates were measured whenever possible; some are calculated from the data in Figure 1-5.*

that chemical reactions in the surface sediments, especially those reactions involving iron, set the concentration of DRP in the water and in this way control the productivity of the ponds.

Flux of Nitrogen

The main inputs of nitrogen to the pond came from the rain water (11.5 mg inorganic N m^{-2} yr^{-1}) and from nitrogen fixation (28 mg N m^{-2} yr^{-1}). The ponds appear to accumulate some nitrogen each year but the total, somewhat less than 80 mg N m^{-2} yr^{-1}, is very small compared to the 38,400 mg N m^{-2} stored in the top 5 cm of sediment. In the water column, ammonia was more abundant than nitrate, 20 to 40 μg NH_3 N $liter^{-1}$ vs. 2 to 13 μg NO_3-N $liter^{-1}$. Uptake in the plankton was slow (Figure 1-10) so that turnover times ranged from 30 to 100 days for the inorganic nitrogen. Measurements with $^{15}NH_3$ indicated that the rate of supply of ammonia from within the water column was high enough to replace the NH_3 in 6 to 48 hr. As expected, we found no evidence of nitrogen limitation upon the primary production of the algae in the pond. When the uptake of nitrogen was used to calculate primary productivity, by taking the ratio of C uptake to N uptake as 100 to 12, the results exactly matched the ^{14}C primary productivity measurements.

In the sediments, the interstitial water contained high amounts of ammonia except where plant roots were present. For example, there were 0.7 to 2.7 mg NH_3-N $liter^{-1}$ in the sediments in the plant-free center of the pond but only 0.01 to 0.08 mg NH_3-N $liter^{-1}$ inside a *Carex* bed at the pond edge. Based upon production calculations, the *Carex* may turn over all the ammonia each day. Despite this relatively high rate of removal, the *Carex* appears not to be limited by nitrogen concentrations.

The only evidence for a limitation by nitrogen was that nitrogen fixation by sediment algae began when the ponds were continually fertilized with phosphorus. It is likely that the algae were phosphorus-limited; when excess amounts of P were added, the uptake rate of both nutrients increased and eventually the N became limiting. At this point, the blue-green algae gained a competitive advantage by fixing nitrogen.

Effects of the Arctic Environment

The annual primary production of the ponds is low, but this is largely a result of the short ice-free season. When compared to the daily production of other ecosystems the ponds are reasonably productive. Thus, food supply is adequate in spite of the low temperatures. There is, of course, a general slow-down of metabolism because of the cold temperatures and this will affect all ecosystem processes from decomposition to predation. Yet, low temperatures (2 to 8°C) are also found in all temperate ponds; in fact, a majority of months will be cold-water months. The unique properties of arctic ponds are: (1) they never warm, (2) they are frozen for 8 1/2 or 9 months of the year, (3) there is continuous light from late April until mid-August (although the intensity does vary greatly over 24 hours).

The smallest life forms of the ponds do not seem to have any special adaptations to the Arctic. Bacteria, for example, are just as abundant in the Arctic as in the water and sediments of any temperate pond; their activity is low but it is about the same as for a temperate pond in the spring when the water is cold. Phytoplankton species are almost identical to the species found in temperate ponds in the spring. There is not even a reduction of species number as 105 species were found in the ponds. These same species, by the way, are found throughout the world and even reach the Antarctic. The physiology of the algae was also normal except that at low temperatures photosynthesis was strongly inhibited by high light levels. This is not adaptive; it may be a result of the low temperatures slowing the rate of repair of chlorophyll. It has been suggested that photosynthesis is less affected than is respiration by low temperatures and for this reason, biomass production would be very efficient in the Arctic. We were not able to measure algal respiration or biomass changes very well so we cannot say if this is true.

Protozoans were also found to have the same species and total abundance as temperate pond communities. Some forms that do not have resistant resting stages might be absent, for so little is known of protozoan life history that this could not be detected. Thus, it was surprising to find *Paramecium* in the pond, for this species has never been known to form resistant cysts.

Metazoans are affected in a number of ways by the arctic environment. The obvious way is by exclusion of some forms because of physiological limitations. Amphibia and sponges are absent while Hemiptera, Odonata and Megaloptera are rare. Ephemeroptera, Trichoptera, and Coleoptera are represented by only a few families or genera.

Another way that organisms are affected is by exclusion from certain habitats. For example, fish are found in deep lakes but not in ponds or lakes shallower than 2 m; they are excluded by the 2-m-thick ice cover in arctic fresh water. This absence of fish allows large zooplankton to exist such as the large *Daphnia middendorffiana* which reaches 3 mm in length and the fairyshrimp which can be 20 mm long. These large animals, in turn, affect the species composition of the algae. We first noticed this effect when all the zooplankton were killed in a pond by the addition of oil. The same shift in the dominant algae, a replacement of the *Rhodomonas* by *Uroglena*, also occurred when we removed the zooplankton by net.

Metazoans which are found in arctic ponds do not appear to have any particular adaptations to arctic conditions because the same forms are often found in temperate ponds as well. Rather, many of the adaptations or abilities they already possess that permit them to survive freezing or other stresses permit survival in the Arctic as well. For example, in temperate lakes, cladoceran zooplankton overwinter as diapausing eggs or embryos and the cyclopoid copepods overwinter as diapausing subadults (copepodids); the same species has a copepodid diapause stage in temperate lakes where it moves into the sediment for several months. They utilize the same methods for longer periods in the Arctic.

Many of the arctic forms of midges have adults which have reduced wings. After hatching, the adults move about on the water surface rather than make the typical laying swarms. This behavior keeps them from being blown away by the constant winds. The life histories of many higher animals are affected by the low temperatures. Zooplankton grow so slowly that there is time for but a single generation per year in most species. This puts a very specific upper limit on annual production of zooplankton because the number surviving the winter may be the single most important factor determining annual production. Chironomids, which live in the top several centimeters of sediment where the annual average temperature in the summertime is only about 3°C, are most affected by the environment. They grow very slowly; one species of *Chironomus* takes 7 years to pass through the four instar stages. In spite of these difficult conditions for growth, the number of midge species found in the ponds, around 35, is not very different from the number in a temperate pond. Many of these species have been found only in the Arctic or only in the Barrow area.

The sedge, *Carex aquatilis,* that grows in the pond is well adapted to a variety of habitats from dry meadows to the shallow water. Perhaps the most obvious adaptation to the arctic environment is its ability to begin growth when the air temperature is still close to 0°C. It also takes up phosphorus through the roots at these very low temperatures.

Modeling

A part of the original plan of the U.S. IBP was to construct predictive mathematical models of various ecosystems. This pond study offered several advantages for this approach; the ecosystem is somewhat simplified and physical factors (circulation and stratification) could be ignored; the active life of the pond organisms is completed in only 100 days each year; all the investigators worked on a common pond; controlled experiments could be carried out in replicate ponds and in the nearby laboratory. Our conclusion is that modeling is a very helpful approach but that the construction of a large, complicated predictive model is not possible with the existing gaps in our knowledge of aquatic ecology.

Even in the pond ecosystem, the number of interactions is too great and most cannot be included in any model. We decided what to include on the basis of the carbon flow results (Figure 1-5). The modeling helped by forcing all the workers to sample just one pond and by forcing us to work on some processes which were virtually unknown (e.g., the detritus food web). It quickly became obvious that we simply did not understand many of the controls of the pond ecosystem. Despite this, the deterministic model which resulted did an excellent job of mimicking or simulating the annual cycles of carbon flow through the plankton and benthic systems. For example, the submodel of benthic algal photosynthesis (Figure 1-11) was developed with physiological data from 1973 but the simulation fits the 1971 and 1972 field data very well. Unfortunately, there were several assumptions that had to be included, in spite of our attempts to measure every parameter, and the results of the calculations were very sensitive to

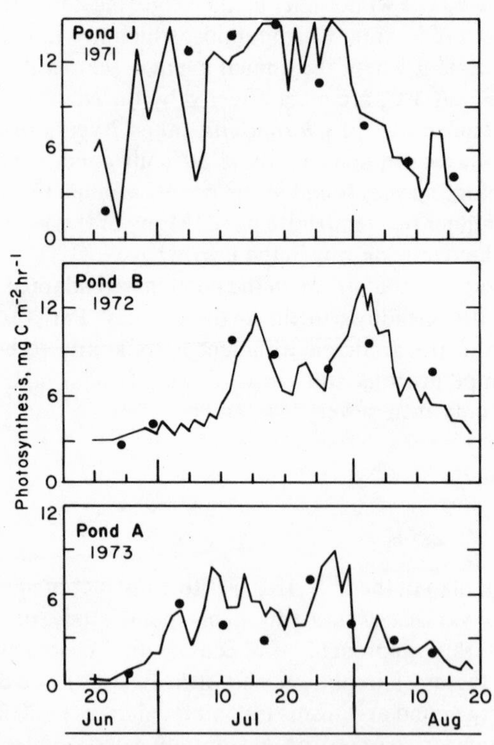

FIGURE 1-11. *Model simulation (solid line) and measured estimates (circles) of the photosynthesis of epipelic algae (mg C m^{-2} hr^{-1}). (Redrawn from Stanley 1974.)*

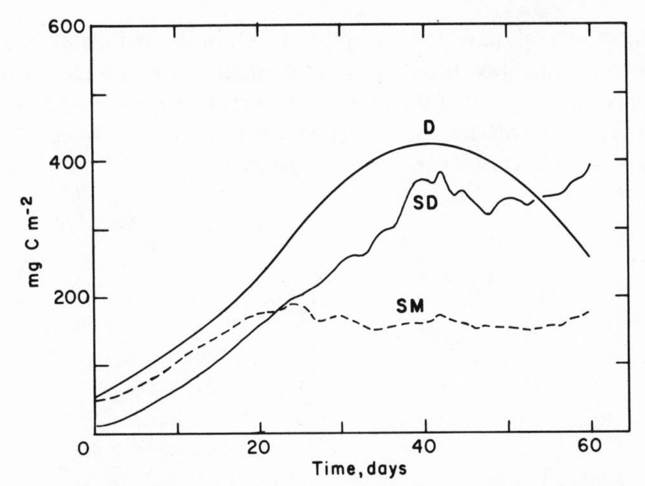

FIGURE 1-12. *Model simulation of epipelic algal biomass. The lines represent the deterministic model (D), the mean of nine runs of the stochastic version of the same model (SM), and the standard deviation of the stochastic mean (SD). (After Tiwari et al. 1978.)*

some of these. Therefore, the result shown in Figure 1-11 can be easily changed in an extremely drastic way by changing the algal respiration coefficient from 0.25 to 0.30. Even if we did have a good way of measuring algal respiration, in the field it would not distinguish between these two values. In a similar fashion, if the rate of maximum photosynthesis is changed from 0.05 to either 0.04 or 0.06, the algal biomass rapidly approaches zero. Yet we know that this maximum photosynthesis rate does change over the year in our pond.

The next step was to see what would happen to the model if all the coefficients and parameters were varied slightly. This stochastic model is probably more realistic, as variability is a property of every biological measurement and interaction. Unfortunately, we did not have the data on the mean and standard deviation of every measurement that would be necessary to implement this approach. However, when we incorporated reasonable variability into the benthic model, the mean values (Figure 1-12) were quite different from the deterministic model values.

In only a few cases could the model be used to test hypotheses. For example, we did examine the hypothesis that the mixing of the sediments was an important control of the growth of benthic algae. This mixing could not be measured directly in the field or laboratory but the modeling exercise helped to put reasonable limits on this rate.

Our conclusions from the modeling effort were that we knew too much about the pond ecosystem to be satisfied with fine-tuning a model which contained several untested assumptions. It was obvious that the interactions which controlled the results of the model were not necessarily the same ones acting to control the real ecosystem. In spite of our failure to construct a predictive model, the modeling exercise was very worthwhile and should be a part of every large-scale ecosystem project.

2

Introduction and Site Description

John E. Hobbie

THE TUNDRA BIOME PROJECT

Biome Studies

The International Biological Program (IBP) was organized with the overall goal of discovering more about the biotic resources of the world through studies of the ecology of natural communities and of man himself. These studies took many forms in many countries but were generally small-scale efforts. It was decided that a part of the U.S. effort should be a series of large-scale, tightly coordinated studies of the ecology of a unit of the earth's surface which would represent a major ecological classification (e.g., desert, grasslands). These units, called biomes, should ideally encompass a watershed or similar area where terrestrial and freshwater ecosystems, and their interaction, could be investigated. The use of mathematical models of whole systems was to be a major tool for the investigations.

Five biome studies were eventually established; one was the Tundra Biome. The site for the study, the flat coastal tundra near Barrow, Alaska, was suitable for terrestrial studies but was ideal for an aquatic study as ponds and lakes were abundant. Also, the well-equipped Naval Arctic Research Laboratory would provide logistic support and work space. The numerous ecological studies of this area over the preceding 25 years provided background information as well as a core of experienced scientists.

The goal of the Tundra Biome study was to obtain a detailed understanding of the ecology of this site. The flows of carbon, nitrogen, and phosphorus were to be quantified and a mathematical model was to be constructed which would incorporate the data of these fluxes, their controls, and the interactions with the physical and chemical environment. It was hoped that the models could then be used in a predictive manner to investigate possible changes in the environment due to man or to natural alterations in the climate.

The Tundra Aquatic Project

In 1970, a modest, pre-IBP grant from NSF allowed the pond studies to begin. The goal of this grant was to follow the effects of oil and nutrient fertilization on ponds. Two projects, that of V. Alexander on phytoplankton responses and that of R. Barsdate on nutrient and water chemistry, were started and the ponds chosen for whole-pond experimentation and for controls. One whole pond and several small subponds (plastic enclosures) were fertilized with P and N and one whole pond was treated with crude oil.

In 1971, the IBP funding began and a complete range of aquatic projects was started. The emphasis during this year was on obtaining the most complete data possible for carbon, nitrogen, and phosphorus flow for both a pond and a lake (Ikroavik). The projects of Alexander and Barsdate were continued and additional projects added that dealt with zooplankton (R. Stross), bacteria, decomposition, and benthic algae (J. Hobbie), fish (J. Cameron), dissolved carbon, particulate carbon, and benthic respiration (M. Miller), macrobenthos (D. Bierle), protozoa (R. Dillon), and macrophytes (P. McRoy). Observations on the manipulated ponds were continued throughout the entire project.

In 1972, the first modeling efforts began and a preliminary plankton model was developed. The field work at Barrow was oriented towards subpond experiments; treatments of nutrients (two concentrations), of added light, of darkness, and of higher temperatures were used. Because the outdated tracked vehicle continually broke down, travel to Ikroavik Lake became impossible and no more samples were taken.

During 1973, emphasis was shifted to modeling, and simulation models of benthic, planktonic, and zooplankton systems were developed. The field work was devoted mostly to developing the constants and rates needed for the models. Several specialists were brought in for the summer to work on areas of research that were still poorly-understood: (T. Fenchel, protozoan ecology; D. Kangas, zooplankton respiration; S. Dodson, invertebrate predation.) The summer of 1974 was spent in preparing reports.

Modus Operandi

To reach the goals of the project we first used traditional limnological techniques to identify and measure the important pathways of carbon and energy flow in the tundra ponds. Next, the modeling was begun and used both to plan future experimental research on the processes and to evaluate the importance of proposed research. The philosophy we have followed in modeling is as follows:

1. We must first understand as completely as possible the system to be modeled.
2. Only the important parts of a system, and their controls, can be modeled.
3. Wherever possible, constants, rates, and relationships included in the model must be measured and not taken from the literature.
4. The modeling exercise and the resultant simulations are regarded as tools to be used to further our understanding of how the ecosystem operates.
5. The modeling must be done by ecologists with the aid, if necessary, of a professional modeler rather than vice versa.

LIMNOLOGY OF THE ARCTIC

Circumpolar

For this report, the Arctic is defined as the region north of the tree-limit which has a mean air temperature of less than 10°C during the warmest month. Until very recently, arctic limnology was entirely organism-oriented. This was in part due to the great difficulties of carrying out anything more than expedition-type collecting activities and in part due to the particular interests of the people involved. Modern studies of the physical, chemical, and limnological processes in lakes and ponds began after 1950 or so when research stations became established (e.g., at Disko Bay, Greenland, and Barrow, Alaska). As a result, the first review of arctic limnology (Rawson 1953) mentioned only seven papers that dealt with arctic lakes.

The availability of these research stations, plus the realization that long-term studies were needed, led to a number of detailed investigations. In Scandinavia, the biology of Lapland lakes has been investigated by Ekman (1957), Holmgren (1968), and Nauwerck (1968). In Spitzbergen, water chemistry and zooplankton biology have been studied (Amren 1964a) and there are a number of reports of investigations of Greenland lakes (e.g., Hansen 1967, Holmquist 1959). Several investigators, such as McLaren (1964), Kennedy (1953), and Oliver (1964), worked in northern Canada. By far the greatest number of studies were made in northern Alaska (e.g., Livingstone et al. 1958, Hobbie 1964, Kalff 1967a, Stross and Kangas 1969, Carson and Hussey 1960) of the chemistry, biology, and physics of a wide range of lakes and ponds.

The results from all these and other arctic studies are summarized in reviews by Livingstone (1963a), Kalff (1970), and Hobbie (1973). In brief, arctic lakes and ponds have the following characteristics:

1. Arctic lakes and ponds seldom warm above 10°C and almost never stratify.
2. Arctic lakes and ponds shallower than 1.7 to 2.0 m usually freeze completely.
3. Ponds and lakes less than 2 m in depth do not contain fish. One consequence of the lack of predation is that the zooplanktonic crustaceans are almost all large species in ponds and shallow lakes.
4. The ice cover is 1 to 2 m thick and lasts for 8 or 9 months.
5. Arctic lakes and ponds usually contain low amounts of available nutrients and low total dissolved salts. However, as in the temperate regions, the total inorganic ion concentration is different for drainage basins in different types of bedrock.
6. Oxygen is usually present in saturation concentrations in open waters but becomes depleted to some extent near the end of the under-ice period. In shallow lakes the exclusion of oxygen during the freezing of the ice may result in super-saturation (200%).
7. The biota of shallow freshwater lakes and of ponds are subjected to strong physiological stresses as the ions may be concentrated 30-fold during freeze-up, while the water immediately after the spring melt may resemble distilled water.
8. Only nannoplankton are found in arctic lakes and ponds. These usually bloom beneath the springtime ice of lakes but total primary production is low and lakes and ponds are oligotrophic.
9. With a few exceptions, each species of zooplankton has a dormant phase in its life cycle.
10. Fish are very slow-growing, but large fish may live for 40 years.
11. There are no benthic animals that graze on aquatic plants or that shred large organic particles or leaves.
12. The number of animal species is small and some groups—for example, sponges, Notonectidae, Corixidae, Gyrinidae, Dytiscidae, and Amphibia—are rare or not present.
13. Decomposition rates are slow and large amounts of energy and nutrients are tied up in dead organic matter.

It is obvious that this list of characteristics was obtained from largely descriptive studies. Recently, two IBP projects were carried out that were designed to add experimental studies to the descriptive observations in order to gain an understanding of controlling factors and environmental interactions. These projects, one at Resolute Bay, Canada, and one at Barrow, Alaška, were located near airfields and studied only a single lake (Char Lake) or a small group of ponds in a single area (Barrow).

Some of the results of the Char Lake project have been published. A general description (Rigler 1972) indicates that the lake is ice-covered until early August, has a moss cover over 30% of its bottom, and has low quantities of nutrients. Phytoplankton began to increase beneath the ice cover in February and reached a peak in May (Kalff et al. 1972). Morgan and Kalff (1972) found a maximum of 2×10^5 bacteria ml^{-1} with peaks of glucose uptake in July and October. Zooplankton had low populations with *Limnocalanus macrurus* as the dominant form (Roff and Carter 1972). Most of the population hatched, grew, reproduced, and died between December and October, although a few adults were present during the entire winter. Finally, the long period of ice cover allowed the lake to be used as a sealed vessel respirometer to measure respiration of the ecosystem by changes in oxygen concentration (Welch 1974).

Northern Alaska

It was possible to carry out the IBP aquatic program only because of the experience and information provided by previous research in northern Alaska. Thus, even before the IBP aquatic project began, we knew such things as the primary productivity of the phytoplankton, the basic cycles of water chemistry, and the life cycles of many of the zooplankton species.

There are seven types of freshwater habitats in northern Alaska: deep lakes, shallow lakes, ponds, large rivers, small rivers, streams, and springs. The deep lakes, located in the mountains, were formed mostly behind end moraines that dam narrow valleys. These lakes are rather rare and may number only 20 or 30. Shallow lakes, very abundant (many thousands) on the flat coastal plain, were formed mainly by melting of the ice-rich permafrost. These are only a few meters deep and many will freeze to the bottom each winter. The area of these lakes can be large, with lengths reaching up to 10 km or so. Ponds are extremely abundant (tens or hundreds of thousands) in the coastal plain region, particularly in the old lake beds. Here, the growth of ice wedges has pushed up networks of small ridges that contain small (50 m on a side), shallow (10 to 50 cm) ponds. Most of the limnological investigations have been carried out on lakes and ponds; little is known about the flowing water systems. However, there are a number of large rivers in northern Alaska and parts of these rivers are deep enough to allow fish to survive. Small rivers and streams, in contrast, cease flowing completely each fall. Because of the flat landscape and small amount of total precipitation, the drainage is poorly developed in the Barrow area and sizable amounts of flow occur only during the melting period. The final habitat, springs, occurs only in the mountain and foothill area. Although some ten or twenty springs exist, they are a very minor part of the entire aquatic scene. The fact that they flow year-round, however, allows a rich fauna to develop and illustrates both the potential production of arctic water and, by contrast, the strong stresses on intermittent streams.

Deep mountain lakes were first investigated by Livingstone et al. (1958) who pointed out that most of the thermal, chemical, and biological events in these lakes were similar to those of oligotrophic temperate lakes. They thought that the major effect of the arctic environment was on the physiographic process affecting lake origins, sedimentation rates, and input from the drainage basins. An intensive study of two other deep lakes (Hobbie 1961, 1962, 1964) revealed that most of the yearly primary productivity of the plankton occurred beneath the ice cover in late spring and early summer. Later, both the light regime and the algal species and biomass responsible for this early season bloom were investigated in detail by Holmgren, Kalff, and Hobbie (reported in Hobbie 1973). It was found that when the snow depth was less than 10 cm, great amounts of light penetrated the ice. The light, plus the non-turbulent conditions beneath the ice, allowed large numbers of flagellates and diatoms to develop. After the ice left the lakes, conditions were poor for algal growth.

Most of the research on shallow lakes has centered on the question of the origin and development of the oriented lakes of the coastal plain. These lakes originate when permafrost melts and the soil subsides (Rex 1961) and receive their orientation from wind-driven currents which differentially erode the ends of the elongated lakes (Carson and Hussey 1960). Scattered bits of information that exist on Ikroavik Lake, near Barrow, show it to have low numbers of algae, nutrients (Prescott 1953), and benthic animals (Livingstone et al. 1958). The whitefish population has also been described (Wohlschlag 1953).

Another well-studied shallow lake is Imikpuk, which is not an oriented lake and which lies close to the Arctic Ocean. Chemistry of the lake has been reported by Howard and Prescott (1973) and Boyd (1959); the primary productivity by Howard and Prescott (1973) and Kalff (1967b); the zooplankton by Comita (1956) and Edmondson (1955); and the microbiology by Boyd and Boyd (1963).

All of the pond limnology in northern Alaska has been done near Barrow. The most extensive investigations were of the chemistry and plankton productivity (Kalff 1965, 1967a, 1971). In these ponds, phosphate, nitrate, ammonia, trace elements, and growth factors all stimulated photosynthesis at various times. Kalff concluded that nutrient deficiency did exist and that plankton productivity was extremely low (around 1 g C m^{-2} yr^{-1}). Another series of studies dealt with reproductive cycles and controls of zooplankton (Stross and Kangas 1969).

The only large rivers studied in northern Alaska have been the Colville (Kinney et al. 1972) and the Sagavanirktok (Carlson et al. 1974). These studies were mostly concerned with water chemistry but zooplankton (Reed 1962), fish (McCart and Craig 1971), and discharge (Arnborg et al. 1966) have also been looked at. The rivers contain little plankton and a scanty bottom fauna.

Smaller streams are also little known and the research has been restricted to observational limnology. The best-studied area is at Cape Thompson (68°N, 165°W) where the discharge (Likes 1966) and biology (Watson et al. 1966a) on Ogoturuk Creek were investigated. Near Barrow, Brown et al. (1968) measured the hydrology of a small watershed (1.6 km^2) over four summers and Lewellen (1972) reported on flow and chemical data from three other Barrow area watersheds.

Springs are present only in the foothills and mountains of arctic Alaska. One spring, Shublik Spring on the Canning River, has been sampled by Kalff and Hobbie (unpublished, quoted in Hobbie 1973). It flows year-round at 4.0 to 5.5°C, and contains a fantastic abundance of insects as well as a dwarf char (McCart and Craig 1973).

There are other reports that cover several aquatic habitats. Hydrology was reviewed by Dingman (1973), Kalff (1968), and Barsdate and Matson (1966).

GEOGRAPHY AND GEOMORPHOLOGY

Geographical Setting

Northern Alaska, all of which lies north of the tree limit, is cut off from the rest of the state by the east-west running Brooks Range, an extension of the Rocky Mountain System. North of the mountains, which have an area of 136,200 km^2, lie the Arctic Foothills (100,800 km^2) and between the foothills and the Arctic Ocean lies the Arctic Coastal Plain which contains 70,900 km^2 (Walker 1973). Barrow lies on the northern tip of the coastal plain, some 175 km from the foothills (Figure 2-1).

Near Barrow, the flat coastal plain is covered either by large lakes, shallow ponds, or old drained lake basins (Figure 2-2). In places, freshwater lakes and ponds cover up to 40% of the surface. Despite the abundance of water, streams are small and most flow only during the spring melt. The remainder of the area is covered by grasses, sedges, mosses, and lichens. Usually, the standing dead stems and leaves of the grasses and sedges dominate the scene and color the tundra brown.

The Naval Arctic Research Laboratory is on the coast of the Chukchi Sea, 10 km from Point Barrow, and the town of Barrow lies 5 km further southwest (Figure 2-1). Research on tundra ecology has been carried out at this laboratory since 1947, while the National Weather Service has operated a first-class station at Barrow since 1920. The pond research site (71°18′N and 156°42′W) is halfway between the laboratory and the town and 2 km inland. A shallow lake, Imikpuk, lies adjacent to the laboratory and to the ocean, and a large lake, Ikroavik, lies 7 km south.

FIGURE 2-1. *Aerial view looking north across the U.S. Tundra Biome research area. The ice-covered Arctic Ocean is in the background. The Naval Arctic Research Laboratory camp complex is in the upper right corner. The ice-covered water body is Middle Salt Lagoon. Polygonal terrain is visible in the foreground and the study ponds in the lower left corner. (Photograph by CRREL.)*

Geomorphology

The Arctic Coastal Plain consists of unconsolidated silty sand and gravel of Quaternary age (Gubik Formation) deposited in a shallow sea (Black 1964). The uppermost section at Barrow was deposited and reworked over the past 35,000 years (summarized in Brown and Sellmann 1973). Radiocarbon dates and composition analyses of peat suggest that tundra existed in the Barrow area for as long as 14,000 years (Brown 1965). Based on a number of radiocarbon dates, it is believed that most of the soils and surficial features of the present land surface are not older than 8,000 to 10,000 years and perhaps are considerably younger.

Mean annual air temperatures on the North Slope of Alaska are below freezing; thus, there is continuous permafrost (perennially frozen ground) beneath the entire area. At Barrow, the frozen layer is 400 m thick (Brown and Sellmann 1973) but there is a layer of soil 25 to 100 cm thick that does thaw each summer. The depth of thaw is influenced by the type

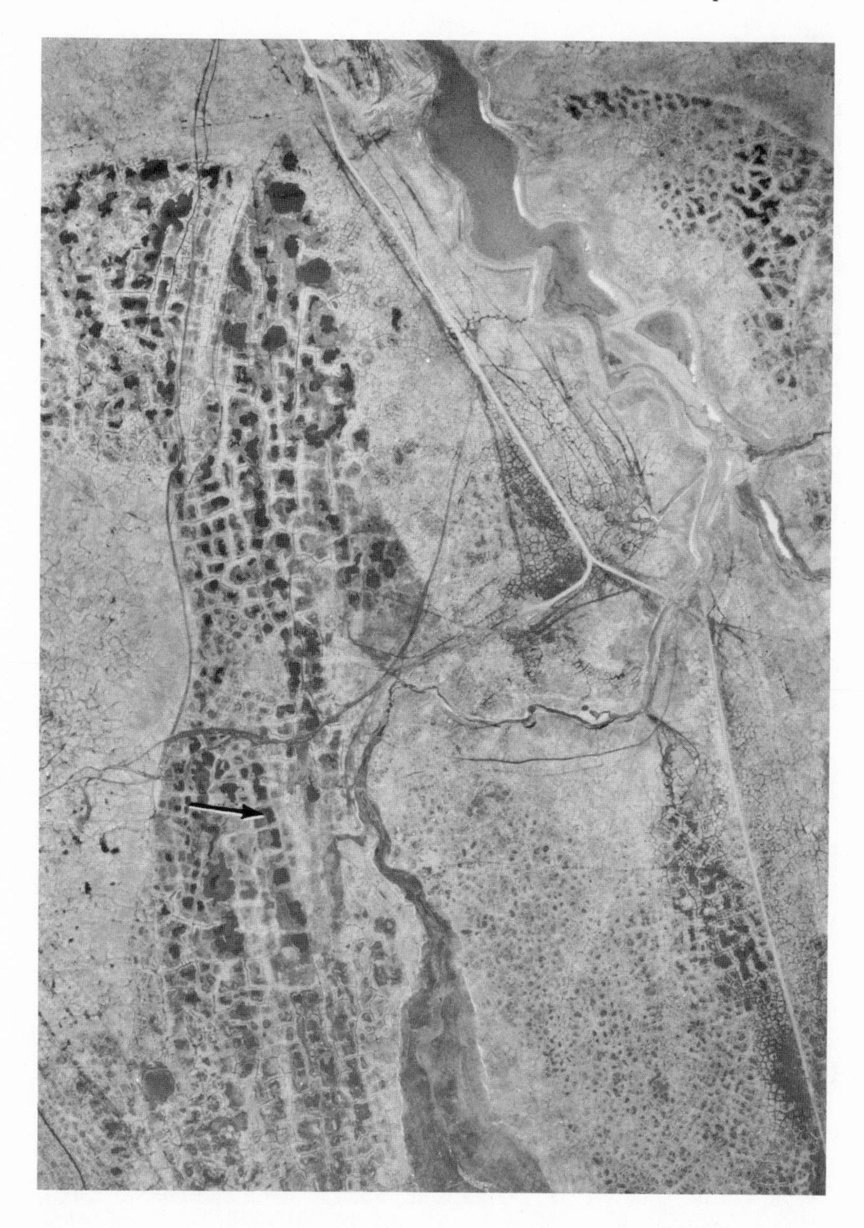

FIGURE 2-2. *Aerial photograph of Tundra Biome site. Arrow indicates Pond C.*

of vegetation, amount of insulating plant litter, and type of soil. For example, beneath a thick vegetation mat the depth of thaw may be only a few centimeters while coarse-textured, south-facing materials may thaw to a depth of 100 cm (Walker 1973). At the Barrow site, the maximum depth of thaw is around 25 cm below the grassy tundra while in the sediments of the ponds it may vary from 20 to 50 cm. Large rivers and lakes of this region may be underlain by extensive thawed areas. Brewer (1958) found 60 m of thawed material beneath Imikpuk Lake.

The presence of permafrost has important biological effects. Because the permafrost is impervious, the water cannot drain away, and the low-lying soils are saturated. Roots are restricted to the upper, thawed layer of soil which limits the total quantity of nutrients available. Finally, nutrients and energy are removed from circulation either when the permafrost level rises or when soil and sediments accumulate and become part of the permafrost.

The upper layers of permafrost on the coastal plain contain large quantities of ice. One form of this occurs as interstitially segregated ice (up to 80% of the top 3 or 4 m of permafrost) (Sellmann and Brown 1965). When this melts, due to disturbance of the plant cover or to heat transfer by flowing water, a depression is formed that may result in a pond. Another form of ice occurs when water runs into cracks formed by the winter contraction of the frozen tundra. The resulting buried ice takes the form of ice wedges that can range from a few centimeters to 8 m in width. Over many years, the wedges grow and eventually a network of ice wedges is formed. Sometimes these wedges are expressed on the surface as polygonal ground (Figure 2-3) caused by heaving or other surface processes that form troughs and ridges. Typically, these polygons may be 20 to 50 m across; polygonal ground covers almost the entire coastal plain.

In the early stages of growth of polygonal ground, the polygons are low-centered and often contain small ponds. The water changes the insulating properties of the surface and also traps heat so that the upper layers of the permafrost thaw, subsidence occurs as the ice melts, and a basin up to 0.5 m in depth is formed. These ponds frequently coalesce and may form a lake. Eventually, the lake may grow enough that a drainage divide is breached. Then the lake may drain and the polygonal ground start to form again; the whole process has been called the thaw-lake cycle (Britton 1957).

Many of the larger thaw lakes of the coastal plain display a striking elliptical shape with an elongated north-south axis. The exact reason for the orientation has been the subject of a number of studies and theories (Black and Barksdale 1949, Livingstone 1963a, Carson and Hussey 1960), but it is evident that differential erosion is still occurring today. For example, Lewellen (1972) measured a rate of elongation of 1.3 m per year in Twin Lakes. In similar lakes, Hussey and his co-workers measured currents at the ends of the lakes of up to 61 cm per second, which Livingstone (1963a) believed to be adequate to account for the elongation of the lake basins. However, Walker (1973) believes that the precise mechanism of elongation is still unexplained.

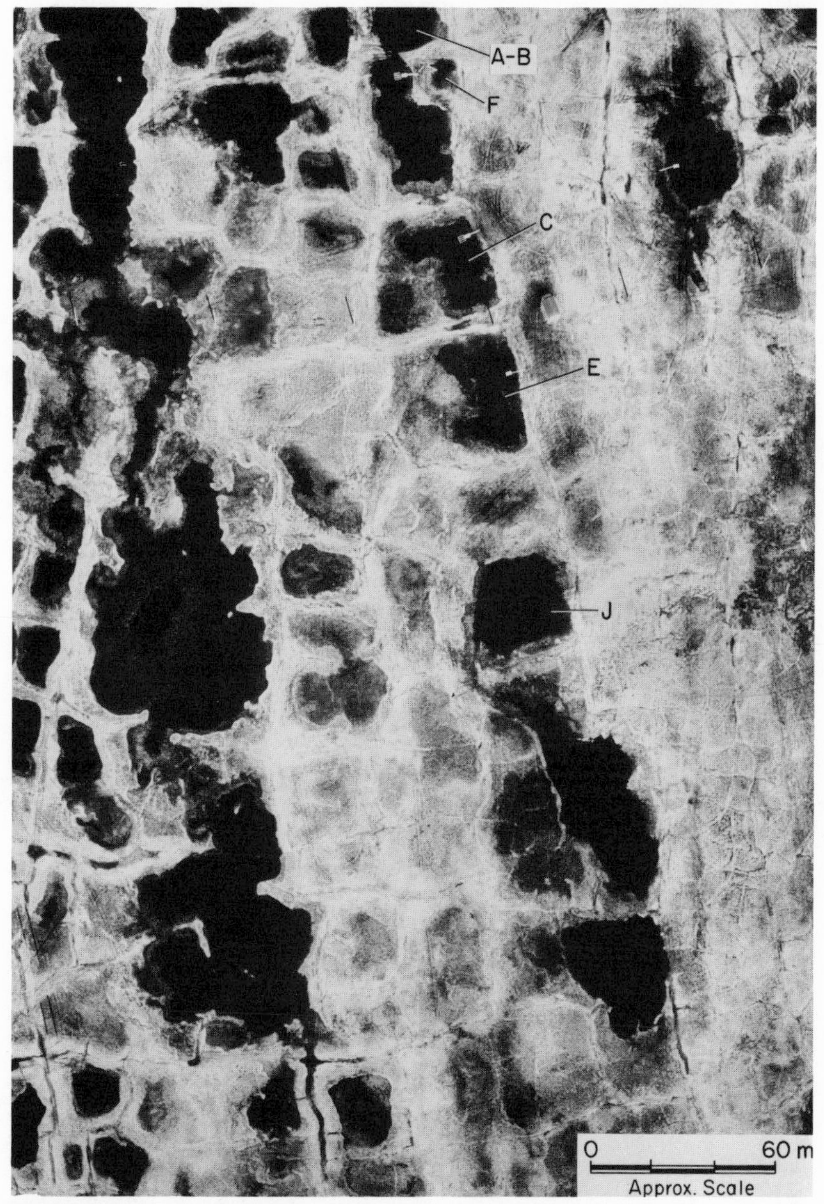

FIGURE 2-3. *Aerial photo showing polygonal ground near Barrow, Alaska. Study ponds are labeled.*

Another type of pond, called a trough pond, is formed when an ice wedge melts. Around Barrow, melting is often caused by destruction of the insulating vegetation by tracked vehicles.

TABLE 2-1 *Averages of Climatological Parameters for Barrow, Alaska**

Parameter	Jan.	Feb.	Mar.	Apr.	May	Jun.	Jul.	Aug.	Sep.	Oct.	Nov.	Dec.	Total or Average
Solar radiation (cal cm^{-2} day^{-1})	1	38	175	385	526	557	447	262	120	42	4	0	213
Temperature (°C)	-26.2	-28.3	-26.3	-18.1	-7.3	-0.9	4.1	3.3	-0.9	-8.9	-17.9	-24.1	-12.4
Precipitation (cm)	0.46	0.43	0.28	0.28	0.30	0.91	1.96	2.28	1.63	1.27	0.58	0.43	10.82
Snowfall (cm)	5.8	5.6	5.1	5.6	5.1	1.0	1.8	1.8	7.4	17.5	9.4	7.4	73.2
Relative humidity (percent sat. at 1400 hrs)	65	64	67	74	85	89	88	89	89	85	76	67	78%
Hrs fog day^{-1}	2	1	1	3	8	12	13	11	5	4	3	2	65 days
Wind ($m\ sec^{-1}$)	5.8	5.6	5.8	5.9	6.0	5.9	6.0	6.4	6.7	6.8	6.4	5.8	6.1
Direction	ESE	E	ENE	NE	ENE	E	E	E	E	E	E	E	E
Mean sky cover (% of daylight hrs cloudy)	n.l.	52	49	59	84	79	81	89	92	87	n.l.	n.l.	

*Source: National Weather Service.

TABLE 2-2 *The Snowfall, Maximum Snowpack, Mean Temperature, Mean Total Precipitation, and Rainfall from 1964 to 1973**

Parameter	1973	1972	1971	1970	1969	1968	1967	1966	1965	1964	Average
Snowfall (cm yr^{-1}), July to July	103.9	50.0	45.2	47.0	87.4	96.5	84.3	106.4	101.9	87.9	91.7
Max snowpack (cm), July to July	30.5	20.3	27.9	17.8	43.2	45.7	35.6	48.3	40.6	33.0	41.3
Mean annual temperature (°C), Jan. to Jan.	-11.4	-12.0	-13.5	-13.8	-12.8	-12.2	-11.9	-12.1	-12.7	-15.2	-12.5
Annual precipitation (cm), Jan. to Jan.	18.2	12.4	7.8	4.7	8.4	8.4	11.9	13.9	14.9	7.8	12.6
Rainfall (cm), Jun. to Sept.	13.2	6.6	4.1	1.7	3.8	3.4	5.7	8.7	5.5	2.3	6.4

*Source: Lewellen 1972 and National Weather Service.

Climate

The Barrow area has short, cool summers and long, cold winters (Table 2-1). At this latitude the sun is below the horizon from 18 November to 24 January but never sets between 10 May and 2 August. Usually, snow is on the ground for 9 months of the year.

The solar radiation is high during April and May but the albedo (reflection) of the snow cover is also high (80 to 90%) so little of the energy is available for warming or melting of the snow (Figure 2-4). After late May, the albedo gradually drops to 70% and then, during a 4- or 5-day period in mid-June, drops to 10% during the thaw (Kelley 1973). Albedos average 18% during summer but by mid-October they return to the winter levels. Solar radiation in the summer (Table 2-1) is strongly affected by the very cloudy weather and frequent fogs so that most of the annual solar radiation at Barrow occurs before the tundra ponds have melted.

Temperatures at Barrow average −12.4°C but there are only 109 days when the average temperature is higher than 0°C. Daily minimum temperatures are above 0°C for only 41 days each year so that low water temperatures and even snow can occur at any time during the summer. In winter, the low temperatures and the thin snowpack (Table 2-2) cause the ponds to freeze completely. February is generally the coldest month and July the warmest. Because of the nearness of the Arctic Ocean to Barrow and to the IBP site, the summer climate is strongly affected by the highly variable ice conditions in the ocean as well as by coastal fogs. As a result, temperatures are warmer in summers with little ice. While temperatures at the research site may be a little higher than those at the Barrow Weather Station, the main effect of the distance between the two stations seems to be lower insolation at Barrow than at the research site. Often the sun shone at the IBP site while fog covered Barrow only 2 km away.

The mean annual precipitation is 10.8 cm (as water). About 50% of this falls as snow; October and November have the highest amounts. During the summer, most of the precipitation falls as rain. Despite the low total precipitation, the relatively low evaporation and the impervious soils mean that the tundra has a great deal of water available, particularly in the early summer when the soils are saturated. The average snowpack is 40 cm in depth but there is tremendous variability due to drifting caused by the constant easterly winds. Drifts fill all depressions while ridge tops may be blown free of snow.

Winds at Barrow are almost continual and almost always from the east. The monthly averages of wind speed are remarkably uniform (Table 2-1) but some strong storms do occur in September and October. These data, however, are taken at 9 m and may not reflect the wind effect on the small ponds. Frequently we have observed a microinversion over a pond such that the pond surface was completely calm when there was wind at a 1.5 m height. Additional information on the decrease of wind close to the ground surface comes from 1971 micrometeorological data from the

IBP site 2 (Weller and Holmgren 1973). For example, on 7 July 1971 the average wind speed (m sec^{-1}) at 0.25, 1, 2, 4, 8 and 16 m above the ground was 2.5, 4.4, 4.5, 4.9, 5.2 and 5.5, respectively.

Hydrology

The vertical relief near Barrow is small; consequently the drainage is poorly developed and small ponds and lakes are common. There are no large rivers in this part of the coastal plain and the small streams are found only where polygonal ground is absent. There is some overland drainage from polygonal areas but only during the snowmelt or during rare periods of heavy precipitation when the ponds become completely filled. As a result, the two detailed studies that have been made concentrated on small streams with well-developed channels. One of these streams is located on a drained lake basin about 8 km northeast of Barrow (Brown et al. 1968). In this basin the total elevation change is 0.3 m, the area 1.57 km^2, and open water covers about 5% of the area. The other study was carried out by the U.S. Geological Survey on two creeks near Barrow Village (Dingman et al. in press).

The snowpack reaches its maximum depth in February, March, and April (Table 2-1). The actual depth of the snowpack in any one place depends upon the microrelief, the wind during the snowstorms, and the amount of time available for the snow to age and harden before the next period of high winds. Not only is snow removed by the winds, but also there is almost continual drifting so that huge drifts accumulate behind every house and small drifts behind each ridge or hummock. Thus, even the ponds, which usually have a complete snow cover, will have a variable depth of snow cover depending upon their immediate surroundings.

The snowfall and snowpack are highly variable from year to year (Table 2-2) and averaged 91.7 and 41.3 cm, respectively, over the past decade.

In April and May there is an increase in insolation (Table 2-1) but the continuous snow cover still has an albedo of 85% so little melting occurs (Figure 2-4). Finally, in late May, the rising air temperatures and the increasing solar radiation cause the first snowmelt. As the snowpack begins to decrease and becomes saturated with water, the albedo falls and more solar radiation is absorbed. Dingman et al. (in press) have summarized the heat balance for the Barrow site for six periods in 1971 (Figure 2-5). The sudden change in the amount of energy going to melt the snow is mostly caused by the sudden decrease in albedo from 85 to 48%.

The snow has a mean extinction coefficient of 0.10 cm^{-1} (Weller and Holmgren 1974). Thus, little insolation penetrates the snow, and the pond ice does not begin to melt until all the snow is gone. Most of the melting occurs at the surface but some radiation also penetrates to the bottom

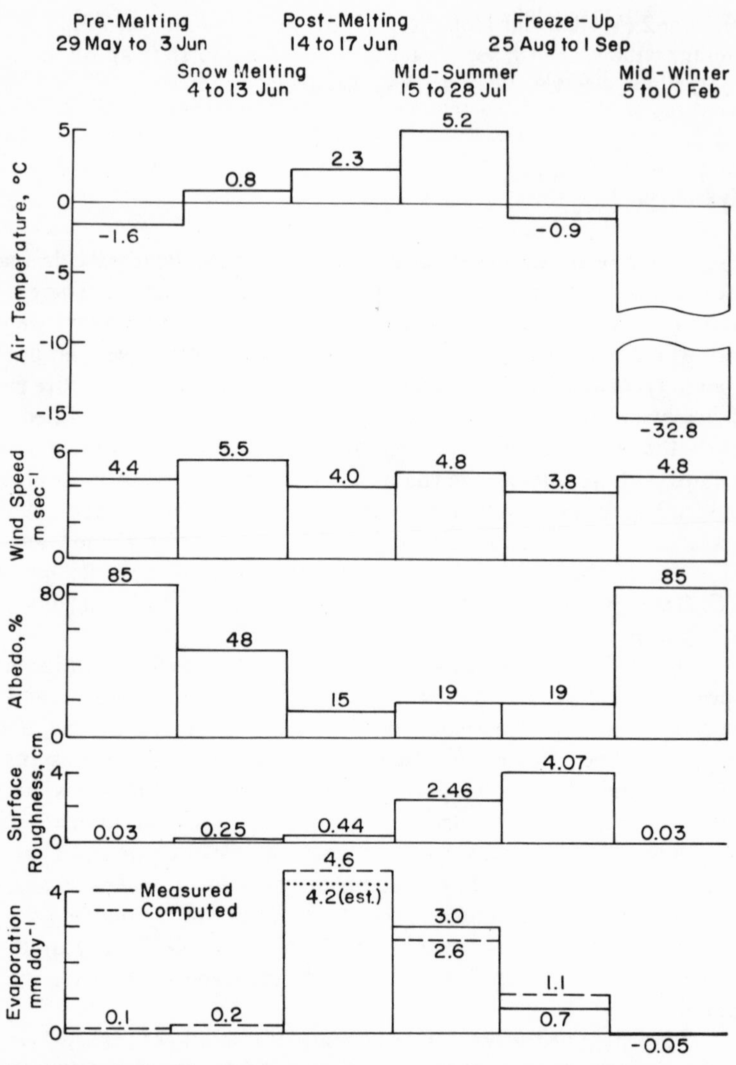

FIGURE 2-4. *The evaporation rate and factors that affect it for six characteristic periods. Evaporation rates reported are from pans. (After Weller and Holmgren 1974.)*

sediments and melts a little ice there. In general, it takes only 4 days for the pond ice to melt completely.

In lakes, the same events take place but the 2-m-thick ice sheets take much longer to melt. Consequently, the ice does not melt completely in the deeper coastal lakes until early- or mid-July, depending upon the thickness of the ice at the beginning of melt. Other complicating factors affecting the ice melt in large lakes, such as ice albedo, crystal orientation, moat formation, and wind action, have been reviewed by Hobbie (1973).

Runoff begins on the day the snowpack becomes completely saturated with water and the maximum discharge occurs within 24 hours. In the next 4 days, 40 to 60% of the total spring runoff takes place—this is still before there is any thawing of the frozen ground. Around Barrow, the runoff from polygonal ground begins in the first half of June (it was 9-11 June from 1970 to 1973) and lasts until the first week in July. Runoff in the sloughs and larger creeks is delayed and prolonged. In the Colville River, for example, the maximum runoff does not occur until three weeks after

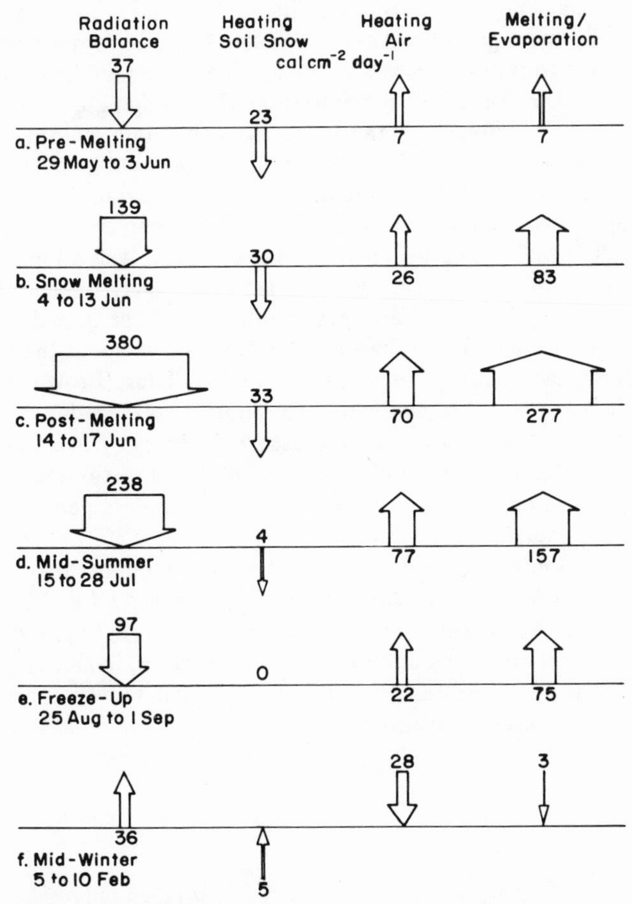

FIGURE 2-5. *Heat balance for Barrow tundra for six characteristic periods. The width and direction of the arrows and the numbers at the base of each arrow indicate energy flux directions and rates. (After Weller and Holmgren 1974).*

the onset of runoff. Following spring runoff at Barrow, the tundra is left saturated and covered with numerous flooded areas.

Precipitation during the summer season (Table 2-2) averaged 6.4 cm over the past decade (range 1.7 to 13.2 cm). Most of this fell as rain but snow can fall during any month. August and September are the wettest and cloudiest months. After the snowmelt period ended, less than 5% of the summer rainfall (3 cm) ran off from a 1.6-km^2 watershed during 4 years (Brown et al. 1968). There is virtually no water lost by runoff from ponds in polygonal ground during the summer, except in rare summers when the rainfall fills the ponds during late August. Permafrost blocks any downward water movement but some small quantities may move laterally.

One additional source of moisture during the summer is condensation of fog and dew on the vegetation. As seen in Table 2-1, the average relative humidity is high during this period and the condensation may be equal to 23 to 50% of the summer precipitation (L. Dingman, personal communication). Unfortunately, there are no good data on this, but condensation has little effect on the water balance of the ponds.

As the summer runoff is so small, the most important water loss occurs by evaporation. Brown et al. (1968) measured a pan evaporation of 16.0 cm in a "typical" precipitation year. However, it is well known that this method overestimates the true loss from land and vegetation and these authors estimated (from runoff and precipitation) that the actual losses were 6.0 cm. Mather and Thornthwaite (1958) measured about the same amount in large evapotranspirometers at Barrow. Thus, Brown et al. (1968) concluded that the summer precipitation was balanced by evapotranspiration in the watershed they studied. Rates of evaporation from the open water of ponds and lakes are close to the pan-measured rates and likely average 2 to 3 mm day^{-1}. During rainless periods of a month or so, many tundra ponds dry up.

The evaporation rate for the tundra, as well as the factors that affect it, was described by Weller and Holmgren (1974) (Figures 2-4 and 2-5). It is seen that neither wind speed nor air temperature changed appreciably from pre- to post-melting period; thus, the strong decrease in albedo when the snow melts and the resultant rise in absorbed solar radiation are the main factors in the increased evaporation.

BIOLOGY

The following information on the terrestrial biology at the IBP site is covered in greater detail in Bunnell et al. (1975) and Brown et al. (in press).

Soil

The Barrow soils (Brown and Veum 1974) have mostly formed on flat to gently sloping landscapes during conditions of low temperature and high moisture. These soils are relatively high in organic matter, of which some has accumulated in place and some has been mixed to various depths by frost churning. Additional organic matter results from the burial of pond or lake sediments in the thaw-lake cycle.

The Barrow soils have a strong thermal gradient during summer when the surface may reach 25°C while at the same time the horizons at 20 or 30 cm are below 2°C. This surface warming dries out the surface layers but the lack of drainage keeps the soil moisture contents at depths below 4 cm at greater than 85% of water-holding capacity. Consequently, vascular plants rarely lack moisture. On the other hand, the abundant moisture combined with low pore volume in mineral layers causes the deeper horizons to become anaerobic early in the summer.

Chemically, most of the soils are highly organic, strongly acid, and not very fertile (Bunnell et al. 1975). Apparently, the base nutrients are usually sufficient for plant needs; N and P, although stored in large quantities, are released only slowly from the organic matter.

Primary Producers

Plant production in the Barrow tundra has been extensively studied for many years; the data and results are reviewed in Bunnell et al. (1975), Tieszen (1978a), and Brown et. al. (in press).

In summer, the coastal tundra vegetation resembles a yellow-brown grassland, relieved only by strips of greener vegetation in troughs between polygons or at the edges of ponds. The yellow-brown color results from the large accumulation of dead plant parts of grasses and sedges. All plants are short (10 to 15 cm high) and many have broad, flat stems.

Compared with most ecosystems, the Barrow tundra is indeed uniform. Although there are 100 species of vascular plants (plus 96 bryophyte and 57 lichen species), the low relief of the coastal region produces only small environmental differences between lowlands and uplands. The result is that all species of the extensive marshes also occur on the uplands and most of the species of the driest upland sites are also found whenever hummocks appear in the wetter areas (Bunnell et al. 1975). Webber (1978) has divided this continuum of vegetation at the IBP site into eight plant assemblages. Five of these cover 91% of the area. They are: mesic *Salix rotundifolia* heath in dry, low-center polygons (7%); mesic *Carex aquatilis*/*Poa arctica* meadow on dry, flat, polygonized areas

(41%); moist *C. aquatilis/Oncophorus wahlenbergii* meadow in moist, flat sites (21%); wet *Dupontia fisheri//Eriophorum angustifolium* meadow in wet, flat sites and polygon troughs (7%); and wet *C. aquatilis/E. russeolum* meadow in low polygon centers and at pond margins (15%). The soil conditions controlling this distribution appear to be moisture, redox potential, and soluble phosphorus levels.

Only a small amount of green tissue is present at the base of the plant stems when the snow melts. Leaves are rapidly formed, however, partly by the translocation of carbohydrates stored below ground, and very rapid growth (0.2 g g^{-1} day^{-1}) begins around 15 June. This lasts for about 10 days and then declines to 0.03 g g^{-1} day^{-1}; this rate continues until about 1 August when the peak aboveground biomass of 60 to 100 g m^{-2} is reached. Since the dead remains of several years' previous growth may be still standing, the total plant material is between 150 and 300 g m^{-2}. This production is drastically reduced when there are high numbers of overwintering lemmings. Not only do they cut down the stems and leaves during the summer, but also under the snow they feed almost exclusively on the green stem-bases.

Photosynthesis continues during August but the photosynthate is mostly incorporated into belowground reserves. In addition, organic materials, minerals, and nutrients are transferred below ground. Roots appear to live 2 to 10 years; their production is around 65 g m^{-2} $year^{-1}$ (Shaver and Billings 1975). Live biomass below ground is frequently 10 times that above ground.

Vegetative reproduction is more reliable than sexual in the short growing season and unfavorable climate, so flowering and seed set are greatly reduced. In four monocotyledons only 2.5 to 10% of the shoots were flowering shoots.

Bryophytes are abundant at Barrrow but are mostly hidden by the vascular plant canopy. Yet, moss and liverwort primary production ranges up to 160 g m^{-2} yr^{-1} in wet meadows. These plants contain lower concentrations of macronutrients than do vascular plants but have higher amounts of micronutrients and calcium. This may explain the change in lemming diet to increasing amounts of bryophytes during the winter.

Lichen abundance is inversely correlated with soil moisture; they are absent in areas where there is any standing water for a week or so in the spring. On better drained meadow sites, the lichen biomass can exceed 50 g m^{-2}. In spite of their low biomass, they are important nitrogen fixers. Unlike the situation in many tundra areas, at Barrow the vertebrates consume few lichens. Lemming stomachs, for example, seldom contain more than 2% lichens.

The plants in this arctic environment show a number of specialized adaptations. In the spring, at 0°C, they are able to mobilize carbohydrates and inorganic nutrients from belowground reservoirs; this allows very rapid growth early in the growing season. Compensation levels are very low in these plants so that net photosynthesis can proceed for 24 hours a day. The leaves of the dominant graminoids are inclined at 65° from the horizontal; at the low solar altitude at Barrow (averaging 25° on 21 June), the erect leaves intercept almost all the solar radiation.

Herbivores

The brown lemming (*Lemmus sibericus*) is by far the dominant consumer at Barrow (Pitelka 1973). One leafhopper (Homoptera) and one leaf beetle do occur and root-piercing nematodes are present, but their impact on the vegetation is slight. In years when lemmings are abundant (Figure 2-6), every 2 to 5 years, their impact is great. Because they do not hibernate, they are active during winter and even reproduce as soon as there is a protective layer of snow (MacLean et al. 1974). By the end of the winter and before a summer high, the lemmings are at peak abundance and will have completely cut all standing plants. This accounts for 40% of the previous season's production; even more important, the lemmings eat the stem bases and parts of the rhizomes, which slows down the initial growth of plants in the spring. Usually, the population declines throughout the summer following a spring high and about 25% of the primary production is consumed. At low population levels, lemmings consume about 0.1% of the total aboveground production.

Lemming grazing may sustain the monocotyledon dominance at Barrow. Some adaptations of these plants, such as vegetative reproduction

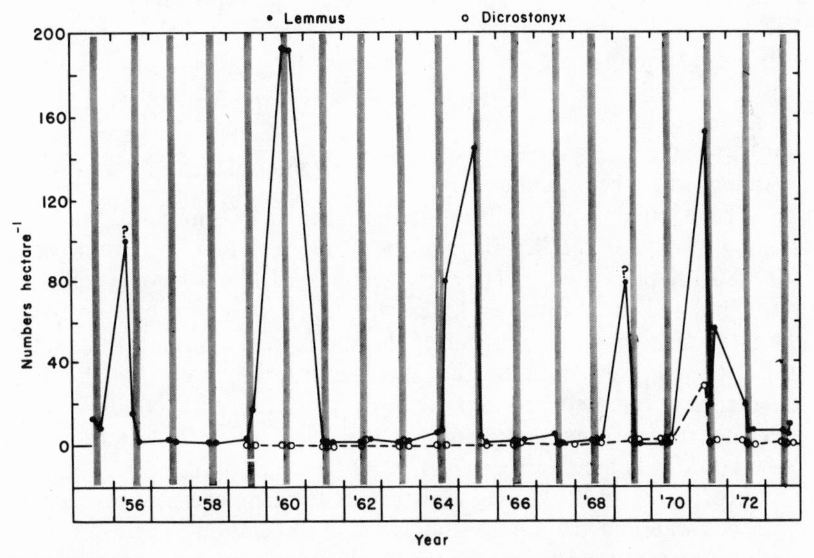

FIGURE 2-6. *Abundance of lemmings at Barrow (based on trapline data from Pitelka 1973). Bars indicate summer. (After Bunnell et al. 1975.)*

and the large belowground storage of energy and nutrients, will help resist the effects of grazing. In fact, when lemmings are excluded from experimental plots, there is a buildup of standing dead vegetation, a reduction in the depth of seasonal thaw, an increase in the thickness of the moss layer, a decrease in vascular plant productivity, and a change in species composition.

No single factor has been found to control the lemming population cycles. One contributing factor may well be the high year-to-year variation in the nutrient content of plants; at the low levels found in Barrow plants, nutrients have been found to affect lemming reproduction. Another factor is the thickness and condition of the snow cover. Temperatures as low as −25°C, frequently found at the ground surface during winter, severely stress the animals and may prevent winter reproduction. Vertebrate carnivores may kill a significant percentage of the lemming population during the winter beneath the snow (weasels) or during the summer (jaegers, owls).

Carnivores

The lemming cycles also produce a cycle in abundance of their predators (Pitelka et al. 1955). In a high year, pomarine jaegers (*Stercorarius pomarinus*), snowy owls (*Nyctea scandiaca*), short-eared owls (*Asio flammius*), least weasels (*Mustela nivalis*), ermine (*M. erminea*), and arctic fox (*Alopex lagopus*) may all be abundant. The avian predators are migratory and arrive between late April and early June. Because these large birds eat four to seven lemmings a day, they breed only when lemmings are abundant. Weasels immigrate into the area when lemmings begin to be abundant; they reproduce during the winter and their number may be high (150 km^{-2}) at snowmelt. Foxes breed inland from Barrow but appear at Barrow in late fall and will prey actively on lemmings when they are abundant.

The common smaller birds (seven species of shorebirds and two buntings) arrive in the first days of the spring melt and begin to breed at once (80-100 pairs km^{-2}). They rely at first on Diptera larvae (especially craneflies) and on fat reserves. Later, when adult insects become available, the birds take these. At the end of the summer, the birds turn again to insect larvae, this time to the chironomids (Diptera). By the time emergence is complete, the smaller birds have cropped about 30% of the adult insects from the tundra surface; they have little effect on the larval insects (taking less than 1%).

Feeding by insectivorous birds serves to bring energy and material from belowground pools into aboveground circulation. These birds are also an alternate food source for the larger avian predators. The owls and jaegers usually nest and roost on elevated sites, such as mounds and polygon ridges, which are consequently fertilized. This may be the only mechanism for nutrient movement onto the higher parts of the tundra.

Decomposition

Within 3 years of death, plant materials at Barrow lose 60% of their weight (see Brown et al. in press for details). Half of this loss occurs in the first year from *Carex aquatilis* (26.6%) and *Eriophorum angustifolium* (27.7%). Most of the first-year loss is by leaching of organic matter, but inorganic nutrients may also be rapidly lost. For instance, 70 to 80% of the phosphorus and potassium are lost during the first year. In contrast, calcium is immobilized in the cell walls and is lost very slowly. The pattern of total weight loss after the first year is the sum of two exponential decay rates; one rate is 49% per yr for rapidly metabolized compounds (ethanol soluble), the other rate is 11% per yr for recalcitrant compounds.

The factors controlling the rates of decomposition at the Barrow site include the duration of freezing, the low pH, the low oxygen concentrations in the soil, the low amounts of available nitrogen and phosphorus, and the effect of the low temperatures on microbial processes. Even though water is abundant, the standing dead plant parts are too dry for rapid decomposition. Thus, the loss of weight in standing plants is 4 to 5% per yr but this increases to 7 to 10% per yr once the material enters the litter layer.

The amount of carbon dioxide evolved from the soil was twice as high on polygon rims and in troughs as in polygon basins; evolution from meadow soils was even higher. Over a period of 85 days (26 June to 10 September) the evolution from meadow soils matched the net primary production (159 g C m^{-2}). Despite the anaerobic soils, little methane leaves the soil. Respiration, and therefore decomposition, is increased when lemmings are abundant or where vehicles have pressed down the dead vegetation.

Both bacteria and fungi are abundant in the litter and soils; in fact, it is not their biomass but their activity that limits decomposition. Bacterial numbers, 10^9 or 10^{10} cells (g dry weight)$^{-1}$ estimated as a direct count, are similar to those in temperate soils. Aerobic plate counts fall in the range 0.5 to 10×10^6 cells g^{-1}. Mycelia length (per g dry wt) is from 200 to 2700 m, but despite this amazing length the fungal biomass is only a third to a quarter that of the bacteria.

Invertebrate Biomass and Production

The major invertebrates in Barrow soils are Nematoda, annelid worms (Enchytraeidae), mites (Acarina), springtails (Collembola), and

Diptera larvae (MacLean 1974). Their total biomass lies between 1.3 and 5.1 g dry wt m^{-2}, which approximates the microbial biomass. Enchytraeidae dominate and make up 50 to 75% of the invertebrate biomass in all habitats. Most of these animals are aerobic; thus they are found mainly in the top 2.5 cm.

The relatively high densities and high biomass of invertebrates are, in part, a result of long life cycles. For example, two species of craneflies (Diptera) require at least 4 years to complete larval development. The annelid worms also have long life cycles. Despite the large biomass, the long lives of the soil invertebrates give rise to low productivity rates.

The biomass of soil macroinvertebrates (Nematoda, Enchytraeidae, Acarina, and Collembola) is positively correlated with net primary production rates but negatively correlated with accumulated organic matter. This implies that the greatest macroinvertebrate biomass occurs in habitats with the highest rates of energy and nutrient turnover. There is still an abundance of accumulated organic matter in all habitats, so the correlation of biomass with energy and nutrient turnover may reflect the better quality or greater abundance of microbes. Alternatively, the feeding of the invertebrates on the microbes may stimulate microbial activity and thus cause a greater removal of organic matter.

ENERGY AND NUTRIENT CYCLING

The information that supports the conclusions in the following sections is given in Bunnell et al. (1975) and Brown et al. (in press).

Energy

The entire aboveground biomass of the dominant tundra plants, the grasses and sedges, grows and dies each year. The belowground biomass lives longer but still turns over in 2 to 12 years depending upon the species. In addition, rootlets and root hairs last but a single season. Thus, there is a large annual input of fixed chemical energy to the system.

In view of the virtual absence of lemmings during the "lows" of the population cycle, it is amazing to discover that, on the average, the percentage of the primary production consumed by animals is higher at Barrow than in most other ecosystems. Vertebrate carnivores are, on the average, also very active relative to other ecosystems. In spite of this activity of the vertebrates, more of the energy of the system passes through the populations of soil saprovores, and especially microbivores, than through the lemmings.

Organic matter is abundant in the soil at Barrow, but it is not known whether there is a long-term accumulation or loss. In part, the difficulty is caused by the great quantity of organic matter—from 22 to 45 kg m^{-2} (to 20-cm depth). This is 50 to 400 times the net annual primary productivity. Given this large quantity, small changes are difficult to measure. In part, the difficulty in calculating a long-term energy budget is caused by the spatial variation and by the tremendous changes in such things as climate and lemming effects from year to year.

Two hypotheses have been proposed:

1. The entire system is in steady state but the terrestrial system is accumulating organic matter, while aquatic systems (lakes and ponds) are degrading organic matter. Habitats at Barrow can change from meadows to polygons to ponds to lakes and back to meadows. Thus, the long-term effect is no net gain of organic matter as a given area of land moves through the thaw-lake cycle.

2. The system is not in steady state; accumulation continues until conditions change. This accumulation is deep in the soil where the decomposition decreases sharply with depth. Since the amount accumulated at Barrow is nowhere near as great as is found in peat bogs, this hypothesis requires that either 1) the Barrow tundra system is young; thus large peat deposits have not accumulated, or 2) recurrent disturbances reverse the pattern of accumulation in any habitat.

Nutrients

Like energy, nutrients are almost all contained in the pool of soil organic matter; less than 1% of both nitrogen and phosphorus is contained in living biomass (Figure 2-7). This contrasts with the rain forest, for example, where living organisms are a significant reservoir of nutrients.

There is only a small pool of soluble soil nitrogen and phosphorus available to plants; this is taken up or turned over many times during a season. This pool is replenished from a much larger pool of exchangeable nutrients, but the non-exchangeable pools are even larger. To replenish the nutrients absorbed by plants, the pool of soluble plus exchangeable nitrogen must turn over 11 times during a growing season; in this same period the pool of phosphorus must turn over 200 times (3 times per day). It is likely that there is a close connection between the rate of supply of nutrients and plant production. Therefore, primary productivity in tundra (as in tropic ecosystems) depends on the rate of decomposition.

The Barrow ecosystem is very conservative with nutrients. For example, most vascular plants have mechanisms for retaining nutrients, particularly phosphorus, in belowground parts rather than allowing them to be lost to decomposers. In the drier habitats, plants have mycorrhizae to facilitate phosphorus uptake. Overall, nitrogen, calcium, and potassium

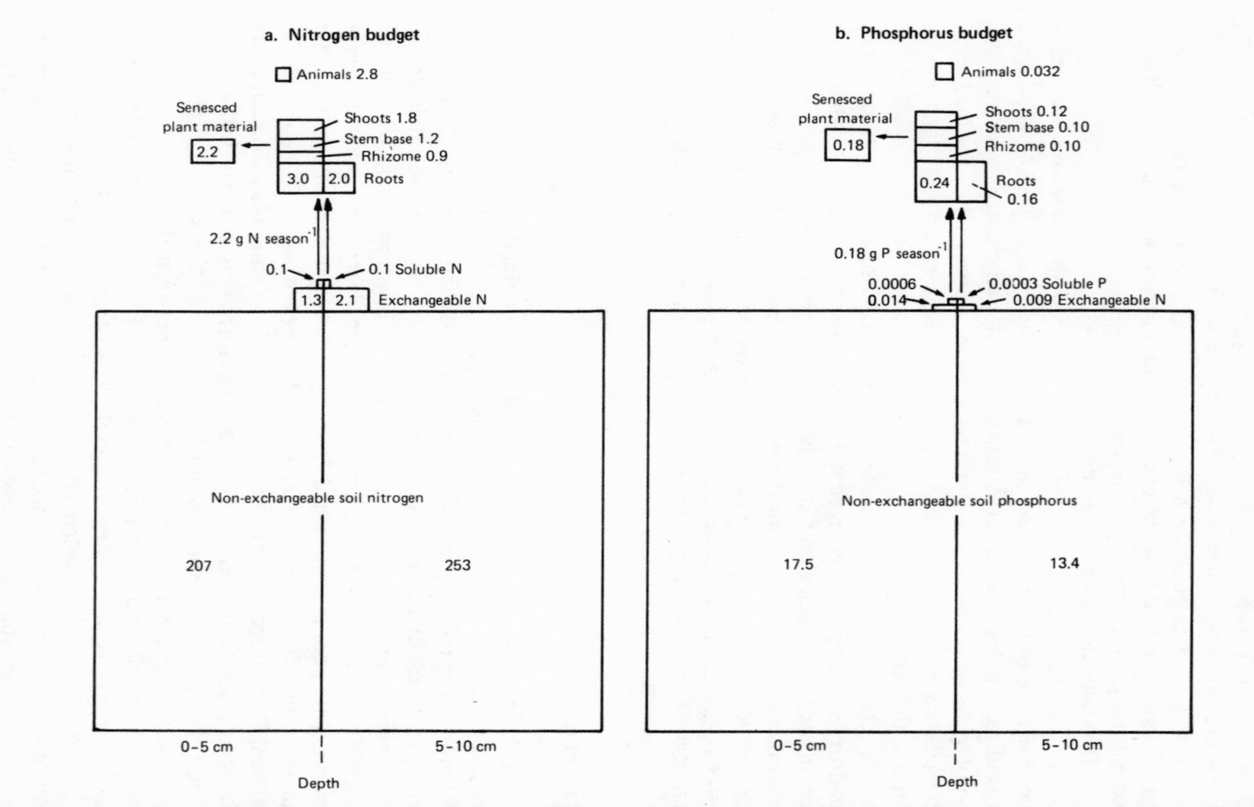

FIGURE 2-7. *Nitrogen and phosphorus budgets for the Barrow ecosystem (g m^{-2} yr^{-1}). (After Bunnell et al. 1975.)*

concentrations are similar in plants from site to site in spite of great variations in soil concentration. In contrast, phosphorus concentrations in plants are highest in the most productive sites. These data suggest that phosphorus more strongly limits production than do other nutrients. Fertilization experiments tend to confirm this hypothesis, as they result in an increase both in production and in the phosphorus concentrations in plants; other nutrients cause no change in concentration.

Despite this limitation of phosphorus on plant productivity, it does not act through altered photosynthetic capacity of each leaf. Instead, all leaves produced have the same optimal photosynthetic capacity. The limitation of nutrients appears to act by controlling the rate of production of new leaves and may also influence the rate at which nutrients are removed from older leaves.

Vascular plants may lose a significant amount of potassium and perhaps phosphorus by leakage onto the leaf surface. These nutrients, plus those leached from standing dead, are subsequently washed off the leaves and form the major nutrient input to the bryophytes. Once nutrients are incorporated into bryophyte tissue, they are released slowly as decomposition of these forms is very slow. Recycling may be speeded up by the feeding of lemmings on bryophytes in winter.

The spatial distribution of nutrients is in part controlled by lemming activities. This control occurs in winter when lemmings build nests in polygon troughs but forage in other areas as well. Most of the lemming

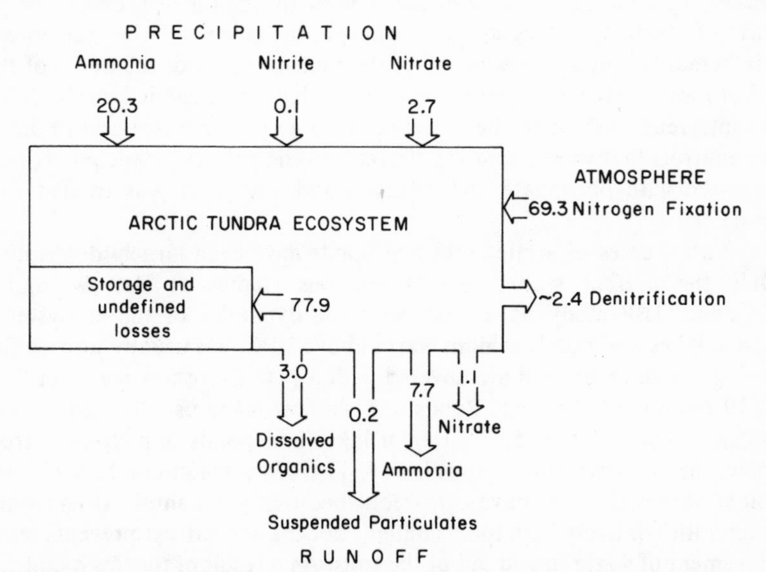

FIGURE 2-8. *Nitrogen budget for coastal tundra at Barrow. All units are mg N m^{-2} yr^{-1}. (After Barsdate and Alexander 1975.)*

feces are deposited in the troughs and in this way the nutrients from many habitats are actually concentrated in the troughs. This concentration may contribute to the higher soil nutrients (especially phosphorus) found in the troughs which in turn may contribute to the higher rates of production and decomposition found in these areas.

Some of the same hypotheses came out of a study of the entire nitrogen budget (Barsdate and Alexander 1975). Nitrogen fixation was the most important input; precipitation was only one-third as great (Figure 2-8). The outputs were very small, with denitrification and runoff of organics and ammonia the major losses. Most of the input (65%) was stored. Thus, the system conserves its nitrogen (nitrogen appears to accumulate), and microorganisms (here, nitrogen-fixing algae) are important in controlling flux rates of nutrients and may well control the entire production rate.

In conclusion, the terrestrial ecosystem of the tundra is rich in total nutrients and energy but poor in amounts actually available and circulating. The activity of decomposers regulates the system.

SUMMARY

The Tundra Biome was one of five ecosystems studied in the U.S. under the International Biological Program. Both a terrestrial and an aquatic study were carried out at Barrow, Alaska, with the goals of developing an understanding of the ecology through measurements of the flux of carbon, nitrogen, and phosphorus through the ecosystem. Mathematical modeling was one of the tools to be used; the needs of this effort meant that much experimental work had to be carried out to define the interrelationships of the processes with an environment and to define the controls that were operating. In the aquatic project, whole ponds were fertilized with phosphate and nitrogen and one pond was treated with crude oil.

Past studies of arctic lakes and ponds have been largely descriptive. Only the present study, several previous studies at Barrow, and a Canadian IBP study have dealt with the dynamics of the ecosystems. Arctic lakes and ponds seldom warm above 10°C, are usually unstratified during the summer, and are covered with a 1- to 2-m-thick ice sheet for 9 to 10 months of the year. Ponds and shallow lakes usually freeze solid. Because much of the total water in lakes and ponds is meltwater from snow, the concentration of ions is low. There is some interaction with the soil so that areas that have calcareous bedrock will contain streams and lakes with relatively high ionic content, but the permafrost prevents much movement of water in and out of the soils. As a result of the low quantities of ions, and of the relative purity of the precipitation, the nutrient concentrations are low and the lakes and ponds are oligotrophic.

The algae of the plankton are all nannoplankton, mostly cryptophytes, chrysophytes, and greens. Productivity is low, usually 1 to 30 $g\ C\ m^{-2}\ yr^{-1}$. Zooplankton are never abundant and only one calanoid copepod, one cyclopoid copepod, and one cladoceran are present while the most oligotrophic lakes have either only one species of copepod or no zooplankton at all. Fish are always present except where the body of water freezes solid. In shallow ponds, the absence of fish permits the fairyshrimp and large *Daphnia* to thrive. Chironomid larvae dominate the bottom fauna and may live for several years. Oxygen is usually close to saturation during the open water season but decreases during the winter and may disappear in the deepest part of a lake. Other primary producers (rooted plants, benthic algae) have not been studied, although mosses are abundant in deep, clear lakes.

In northern Alaska there are a few deep lakes (50 m) in mountain valleys, a few more moderately deep lakes (25 m) in the foothills, and tens of thousands of shallow lakes (2-3 m) on the coastal plains. Hundreds of thousands of shallow ponds have formed on the coastal plain in former lake beds. There are a number of rivers as well but these are virtually unstudied. The deep lakes are dominated by planktonic processes. Several receive glacial rock flour so they are quite turbid. The lack of stratification during open water causes poor conditions for algal growth; as a result, the maximum productivity occurs in the spring beneath the ice cover (light penetrates the ice sheet when the snow cover is less than 10 cm thick).

The shallow lakes research in northern Alaska has concentrated on the origin and development of the oriented lake basins (north-south). This orientation or elongation is caused by differential erosion by wind-driven currents. A non-oriented lake near Barrow (Imikpuk Lake) has been sampled extensively for microbes, algae, chemistry, and zooplankton. The few species of cladocera and copepoda enabled several life cycles to be worked out in detail. The chemistry and biology (taxonomy) of small lakes have been investigated at Cape Thompson.

Ponds near Barrow have been investigated by two projects which concentrated on nutrients, phytoplankton, and zooplankton. Primary productivity was extremely low, around 1 $g\ C\ m^{-2}\ yr^{-1}$; different nutrients stimulated primary productivity at different times of the year. The dominant *Daphnia* species, *D. middendorffiana* and *D. pulex*, have only females in the population. The resting eggs, which overwinter, need to be frozen before they will hatch in the spring.

In a small watershed near Barrow, most of the runoff occurred in the spring melt period; summer precipitation was balanced by evapotranspiration. Large rivers, which have been little studied, contain virtually no plankton, but do harbor fish which breed in smaller streams and overwinter in deep pools or in streams near springs.

Air temperatures average below freezing so that permafrost underlies the area to a depth of 400 m. Some 20 to 50 cm of soil thaws each summer but the permafrost is impervious and water cannot drain. Low-lying soils are saturated and ponds easily form. They are particularly abundant in old lake beds where ice wedges form in the soil and eventually push up the overlying soil into a ridge a few centimeters high. The ice wedges and ridges form connected polygons with "diameters" of 20 to 50 m. Each polygon is a separated basin; many contain ponds that form when the soil subsides due to destruction of the insulating vegetation.

The average temperature at Barrow is −12.4°C while the summer temperature averages are −7.3°, 0.9°, 4.1°, 3.3°, and −0.9° for May, June, July, August, and September. Approximately 50% of the 10.8 cm of annual precipitation falls in June, July, and August. Solar radiation in the summer is reduced by the cloudy and foggy weather so that most of the annual radiation occurs before the ponds melt. Winds are almost continual during the summer at 6 m sec^{-1} from the east. Microinversions frequently occur over a pond, however, so the wind speed at the water surface may be only one-third the recorded speed.

Snow normally reflects more than 85% of the solar radiation; in early June when the snow begins to melt and becomes saturated with water, the albedo drops to around 50% and the snow rapidly melts. Within a few days the snow is gone and the pond ice begins to melt. This melting is complete in 4 days. Lake ice is 2 m thick so does not disappear until mid- or late July.

The soils of the IBP site are highly organic, acid, and not very fertile. Nitrogen and phosphorus are abundant but are tied up in organic matter. Soil temperatures may reach 25°C at the surface while the horizons 20 to 30 cm below are at 2°C. There is some surface drying but soils are saturated below 4 cm.

The tundra vegetation at the site is a yellow-brown grassland. All plants are short (10-15 cm) and the large amount of standing dead vegetation hides the green plants. Although there are over 100 species of vascular plants, a few grasses and sedges dominate: *Carex aquatilis, Eriophorum angustifolium, Poa arctica,* and *Dupontia fisheri.* New leaves sprout from green tissue at the base of the plant stems as soon as the snow melts. Rapid growth occurs until about 1 August when the peak aboveground standing crop (new growth) of 60 to 100 g m^{-2} is reached. Production is reduced during lemming highs by summer grazing on stems and leaves and by winter feeding on the green stem bases. Roots live 2 to 10 years; their biomass is 10 times the aboveground weight and production is about 65 g m^{-2}. All reproduction is vegetative.

Mosses and lichens are also abundant. Moss and liverwort production may be as high as 160 g m^{-2} yr^{-1} in wet meadows; lichen productivity is low but biomass may exceed 50 g m^{-2}.

Plant adaptations to the arctic environment include the ability to translocate carbohydrates and nutrients at 0°C, a low compensation level so that net photosynthesis proceeds for 24 hours a day, and an average leaf inclination (65° from the horizontal) that allows almost complete interception of the low-angle solar radiation.

The brown lemming is the dominant consumer; their numbers increase from less than 1 ha^{-1} to nearly 200 ha^{-1} every 2 to 5 years. In "lemming high" years their impact on the vegetation is startling; they reproduce beneath the snow cover and completely cut all standing plants during the winter. Their consumption of the annual primary production varies from 40% to 0.1% but on the average they consume a higher percentage than any other grazer community on earth. No single factor has been found to control the lemming population cycles. Instead, control may occur by a combination of year-to-year variation in the nutrient content of plants, of the amount of protection by the snow, and of predation.

The lemming cycles also produce a cycle in abundance of their predators, the pomarine jaegers, snowy owls, short-eared owls, least weasels, ermines, and arctic foxes. The large birds eat four to seven lemmings per day and breed only when lemmings are abundant.

The common smaller birds (seven species of shorebirds and two buntings) arrive in the first days of the spring melt and begin to breed at once (80-100 pairs km^{-2}). They eat mostly insects such as larvae and adults of craneflies and midges. About 30% of adult insects and 1% of larvae are harvested.

On the tundra, about 60% of the weight of plant material disappears within 3 years of death. Half of this loss occurs in the first year. After the first year, the loss is the sum of two exponential decay rates, one of 49% yr^{-1} for rapidly metabolized compounds and one of 11% yr^{-1} for recalcitrant compounds. Factors controlling the decomposition rate include the duration of freezing, the low pH, the low O_2 concentrations in the soil, the low amounts of available N and P, and the low temperatures.

Soil respiration is another way to measure decomposition. Over an 85-day summer period, the evolution of CO_2-C from meadow soils (159 g C m^{-2}) matched the net primary production. Most of the respiration is by microbes. Bacterial numbers, 10^9 or 10^{10} (g dry weight)$^{-1}$ are similar to temperate soils. The amount of fungal mycelia (g dry weight)$^{-1}$ is 200 to 270 m but this is only a third to a quarter the biomass of the bacteria.

The major soil invertebrates are nematodes, annelid worms, mites, springtails, and dipteran larvae. Their total biomass is 1.3-5.1 g dry wt m^{-2} or about the same amount as the microbes. Enchytraeid worms dominate (50-75% of biomass). Most of the animals are found in the top 2.5 cm (aerobic layer). The craneflies and worms have long life cycles (up to 4 years). Thus, although the biomass is high, the productivity is low.

Despite the high average rate of lemming grazing, most of the energy passes through the soil saprovores and microbivores. There is so much organic matter in the soil (22 to 45 kg m^{-2} to a depth of 20 cm), that the total primary production for 1 year is only 0.25 to 2% of the total. Because of this large quantity of soil organic matter, the small changes each year can not be measured and it is impossible to discover whether the soil systems are gaining or losing organic matter. Two hypotheses have been proposed: the entire system is in balance but there is accumulation on land and net decomposition in ponds; the system is accumulating organic matter and this will continue until conditions change.

Nutrients are tied up in the soil organic matter; less than 1% is in the biota. To replenish the nutrients absorbed by plants, the pool of soluble plus exchangeable nutrients must turn over 11 and 200 times a year for nitrogen and phosphorus. It is likely that this supply ratio, and also the primary productivity, depend upon the rate of decomposition. Despite differences in nutrient concentrations in the soil, almost all the nutrients are present in similar amounts from site to site. The exception is phosphorus where plants from the most productive sites have the highest concentrations. This evidence, plus evidence from fertilization studies, suggests that phosphorus more strongly limits primary production than do other nutrients such as nitrogen, calcium, or potassium. Nutrients may become concentrated into troughs between polygons during the winter. Lemmings build nests in the troughs and deposit most of their feces there but forage over a larger area.

A study of the nitrogen budget revealed that the most important input was nitrogen fixation which was 3 times the precipitation input. Outputs due to denitrification and runoff were small; 65% of the input was stored in organic matter.

It was concluded that the terrestrial ecosystem of the tundra is rich in total nutrients and energy but poor in amounts of nutrients and energy actually available and circulating. Decomposition rates regulate the system.

3

Physics

M. C. Miller, R. T. Prentki and R. J. Barsdate

GEOGRAPHY

Description

The IBP study ponds lie on the west side of the Footprint Creek drainage system in an area of polygonal ground (Figures 3-1 and 2-3). Their maximum depth is a little less than 40 cm and they are approximately rectangular. The ponds occupy most of the area of the polygon but it is difficult to define the boundaries of a pond exactly as the whole of the polygon may be flooded at the time of the melt. Yet, after a dry summer, only the central depression will retain water. In a normal summer with some rainfall spread over the entire period, the flooded areas include emergent sedges (out to a depth of 10 cm). This is the area called "marsh" in Table 3-1. An average pond is about 30 m on a side.

History

The IBP pond site occupies a drained lake basin which has probably contained several lakes at different times (Brown et al. 1980). Remnants of these lakes can still be seen on aerial photographs. It is difficult to date

TABLE 3-1 *Comparison of Pond Areas*

Pond	Pond area (m^2)	Area of marsh (m^2)	Area of polygon (m^2)
A	278		
B	435		
A+B	714	1055	1333
C	332	609	995
D	500	1285	2674
E	312	494	1151
F	37		120
X	638	974	2033
J	848	1200	1600

these lakes because of reworking and movement of organic sediments, but Lewellen (1972) reported that the maximum age of a lake is probably 12,000 years (based on a ^{14}C date of a lacustrine peat overlying a marine sediment). Other recently-drained lake basins near Barrow may be 3000 to 5000 years old (Brown 1965).

The present land surface and the ponds at the IBP site are relatively younger than the basins studied by Brown. The most recent lake drained into Footprint Creek, through a small outlet on the east side of the basin (site of weir 1, Figure 3-1). There is no definitive evidence for the age of this event; it probably took place several thousand years ago.

The ponds themselves form in the center of ice wedge polygons. As the ice wedges grow in the permafrost over a period of centuries, ridges or rims are formed due to the upward displacement of soils (Lachenbruch 1962). The central depressions of these low-centered polygons form the pond basins. With time these small ponds may erode; the surrounding rims then coalesce to form larger ponds and eventually lakes, a process termed "the thaw lake cycle" (Britton 1957).

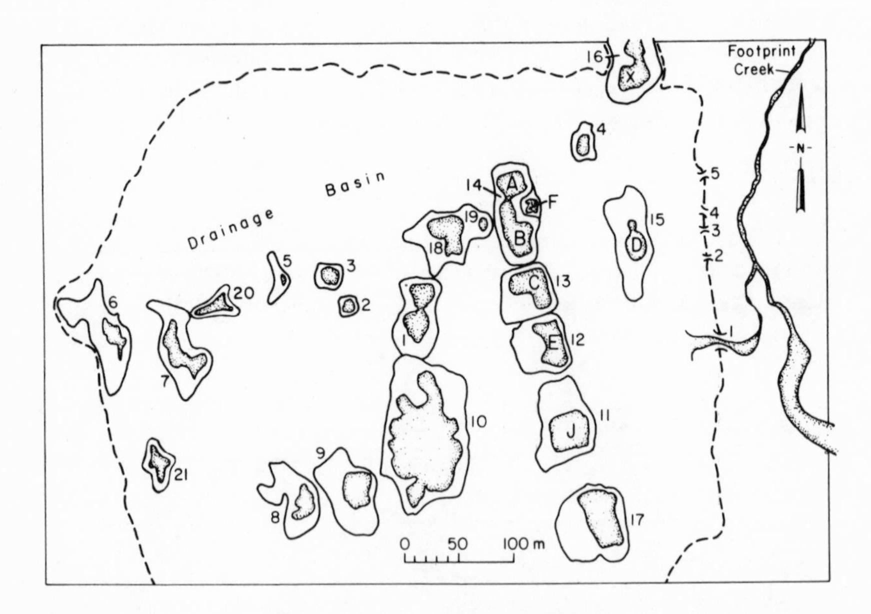

FIGURE 3-1. *Ponds and outlets of site 7 drainage basin for pond comparison, 18 August 1972. The solid lines are the borders of the polygons and the stippled area indicates open water. The dashed line encloses the drainage basin. The numbers along the right side of the drainage basin indicate weirs for water flow measurement.*

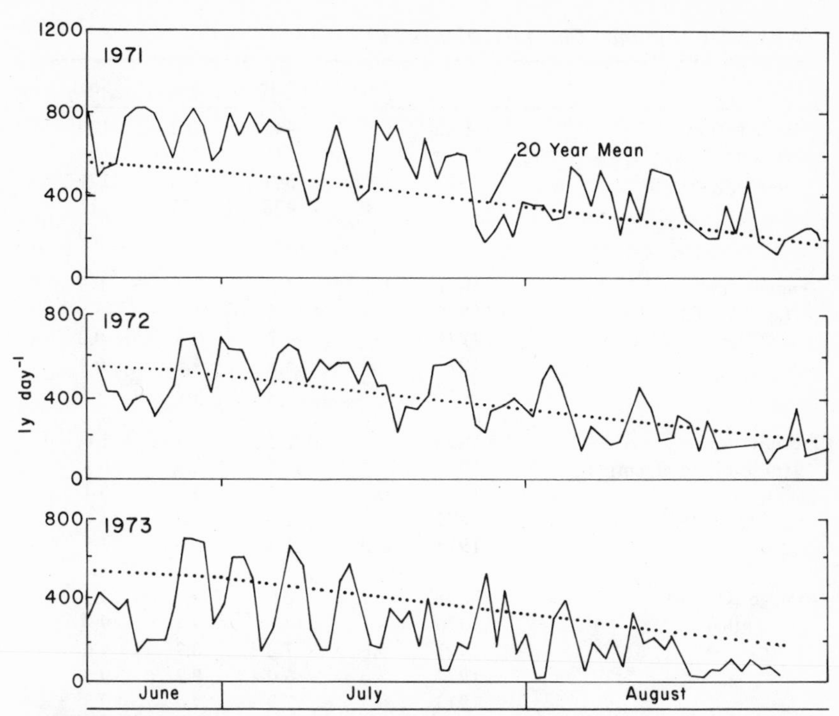

FIGURE 3-2. *Daily total solar radiation at Barrow, 1971 to 1973 (courtesy NOAA). The dotted line is the 20-year average (1950–1970).*

The IBP ponds can be separated into two types, based on accumulation of organic sediments. Beneath ponds C, D, E, and F there is an organic layer 20 to 40 cm thick. The ponds to the west have organic accumulations of only 0.5 to 11 cm. If this organic matter was in a former lake, it may be deduced that the ponds A through F were located more in the center of a shallow lake. This assumption is based on observations by Carson (1968) of present Barrow lakes which have a gentler and wider littoral zone on their west sides than on their east sides. It is likely that organic detritus deposition would be thickest in the deeper central and eastern side of the basin.

Summer Climate

The summers at Barrow during the four years of our study (1970-1973) were slightly warmer than the 30-year mean (Table 3-2). Precipitation was near normal but the average includes a very dry summer (1970) and a very wet summer (1973).

TABLE 3-2 *Summer Climate, 1970-1973**

		June	July	Aug.	Sept.
Solar radiation	Mean	557	447	262	120
(cal cm^{-2} day^{-1},	1970	-	515	229	109
monthly average)	1971	665	541	313	121
	1972	488	478	253	111
	1973	401	321	130	60
Temperature	Mean	0.9	4.1	3.3	-0.9
(average °C)	1970	0.5	3.2	1.7	-3.5
	1971	1.7	4.7	0.8	-0.2
	1972	0.3	6.1	4.8	-0.4
	1973	0.7	4.3	4.3	1.1
Precipitation	Mean	0.8	2.2	2.3	1.4
(monthly total, cm)	1970	0.1	0.4	0.9	0.3
	1971	0.3	2.5	0.9	0.4
	1972	0.1	0.3	2.8	3.4
	1973	2.0	2.7	5.6	2.9
Average sky cover	Mean	7.9	8.0	9.0	9.2
(tenths)	1970	8.4	7.1	9.4	9.7
	1971	5.8	7.2	8.7	9.6
	1972	7.6	6.0	8.9	9.5
	1973	8.3	7.9	9.6	9.3

*Source: Data and means (50-year except for solar radiation) taken from U.S. Department of Commerce, NOAA, Environmental Data Service, 1970-1971.

The cloudiness increased each summer from 1971 to 1973; as a result, the solar radiation was drastically reduced over this time (Table 3-2, Figure 3-2). This reduction is important as the water temperature is influenced more by solar radiation than by air temperature.

TEMPERATURE STUDIES

Water Temperatures

Temperatures in the water were recorded hourly in several ponds. Details of methods and data are given in Stanley (1974, 1976a).

The ponds present a rather simple thermal pattern. In the first place, temperature differences in the water column are only a few tenths of a degree as the winds continually circulate the water. On the few windless, sunny days the convective currents caused by solar heating of the sediments also keep the ponds mixed. Thus, temperature recordings at one depth are sufficient to describe the whole water column. In the second

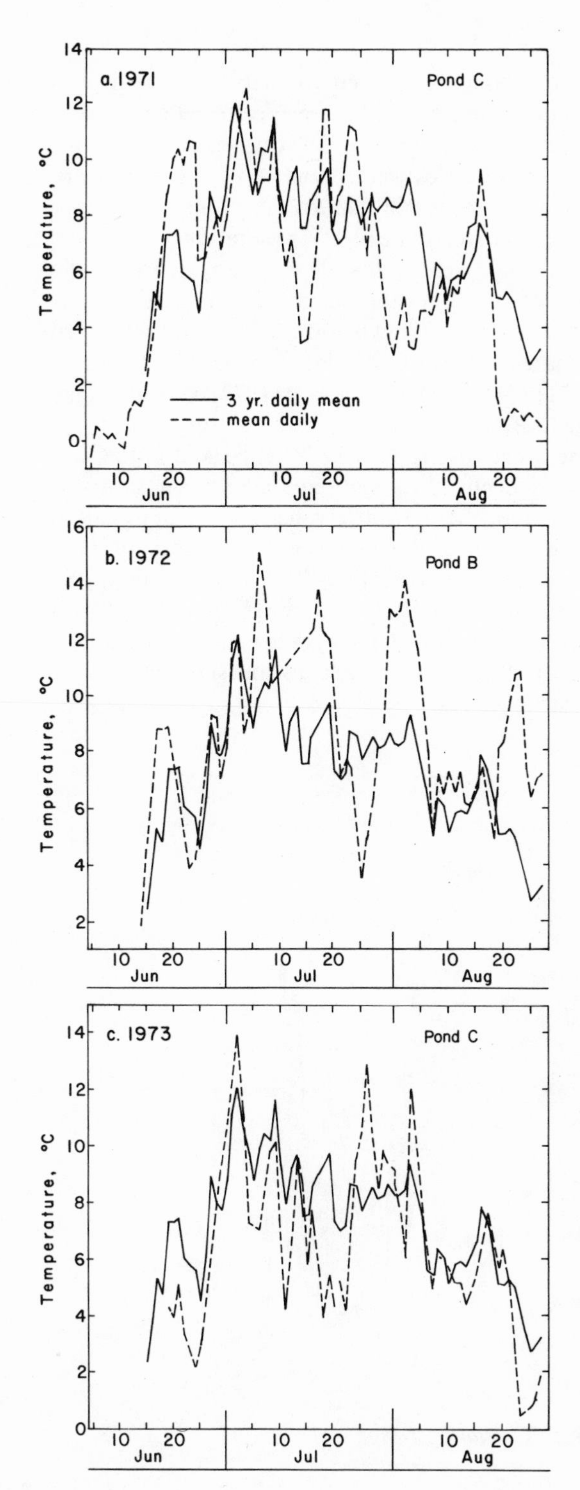

FIGURE 3-3. *Daily average water temperatures for (a) Pond C, 1971; (b) Pond B, 1972; and (c) Pond C, 1973. In all these figures the solid line is the three-year daily mean.*

place, the ponds are shallow enough that their capacity to store heat is small. Therefore, water temperatures would be expected to reflect current weather rather than to integrate weather over several weeks. When this was tested, Pond C water temperatures were found to have a significant direct correlation with air temperatures. Kalff (1965) also found that water temperatures were significantly correlated with solar radiation.

As expected, shallow ponds were a few degrees warmer than the deeper ones but ponds with about the same depth had similar temperatures. Thus, Ponds B and C differed by only 0.5°C when 68 paired measurements were compared. Because of this similarity, the water temperatures from Pond C (1971, 1973) and Pond B (1972) were averaged to give a daily mean (Figure 3-3).

The mean daily temperature rose rapidly in June, peaked in the first week in July, and then slowly declined for the remainder of the summer. Aside from this general pattern, there is little consistency from year to year. For example, the mean on a given day may vary by 10° from one year to the next. Overall, 1972 was much warmer than 1971 or 1973; this agrees with the average air temperature data but not with the solar radiation data (Table 3-2). Water temperatures are higher than air temperatures; between 5 July and 6 August 1971 the simultaneous measurements of air and water averaged 3.7° and 7.1°C, respectively.

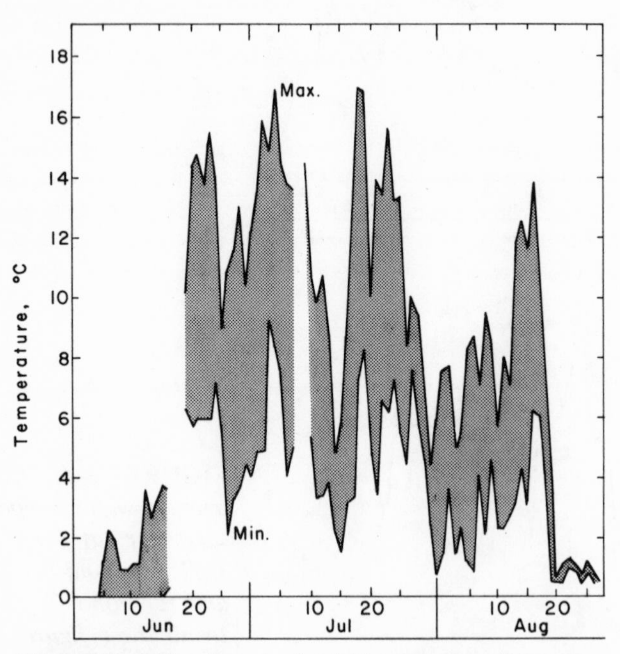

FIGURE 3-4. *Pond C daily range of water temperatures, 1971.*

The highest temperatures and greatest diel ranges occur on sunny days when changes of 8°C are common but changes of 10° are rare (Figure 3-4). In this figure, the depression of both range and magnitude in late August is due to snowdrifts in the pond following a snowstorm on 19 August. The highest temperature recorded during 4 years of study was 20°C in Pond B on 6 July 1972.

The temperature regimes in other tundra ponds appear similar to that of the IBP ponds. Kalff (1965) observed a high of 19.9°C in his Pond II on 16 July 1964. Cameron et al. (1973) recorded 22°C in flooded shallows behind the beach berm at Ikroavik Lake on 6 July 1972, the same day as the maximum in Pond B. Ponds from other arctic areas such as Cape Thompson (Watson et al. 1966a) and Bathurst Island (Danks 1971a) have similar temperature patterns.

The same pattern does not hold for lakes, however, which are similar to temperate lakes with a very small diel temperature range and a maximum temperature in August (data summarized in Hobbie 1973). In one direct comparison between a Barrow pond and a nearby shallow lake, Stross and Kangas (1969) found that pond temperatures averaged 10°C higher than lake temperatures (15 June to 15 July 1967).

Sediment Temperatures

Temperatures at various depths in the sediments were recorded hourly in Pond C in 1971 (2 cm depth using thermocouples), and in Ponds B (1972) and C (1973) at 0 and 10 cm depth using a YSI scanning telethermometer (data reported in Stanley 1974, 1976a).

The average temperatures at the surface of the sediment are very close to the average water temperature. For example, betwen 14 June and 27 August 1972 the water temperature averaged 8.4°C and the sediment temperature 8.5°C (Table 3-3). However, on sunny days the sediment temperatures can be 2° to 4°C above the water temperatures. This rise was relatively brief and as the sediments warm more slowly than the water in

TABLE 3-3 *Mean Water and Sediment Temperatures, 1971 and 1972*

	Water	Sediment		
		0 cm	2 cm	10 cm
9 July to 16 August, 1971				
Average	7.1°C		4.5	
14 June to 27 August, 1972				
Average	8.4	8.5		5.2
Average diel range	6.1	5.0		1.5

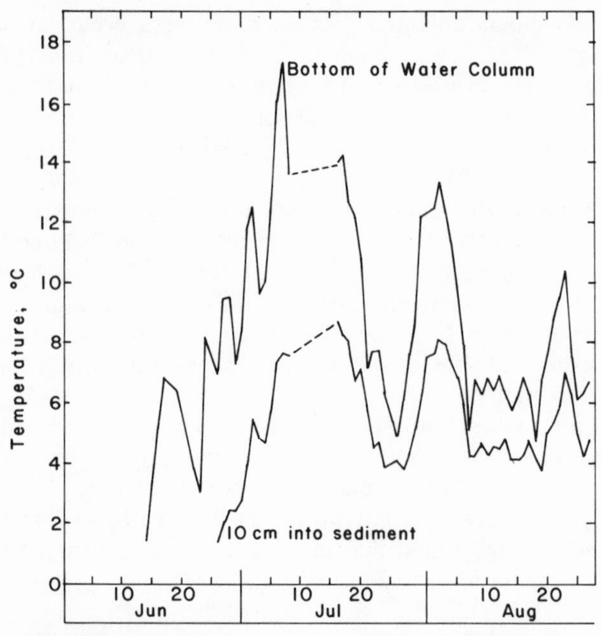

FIGURE 3-5. *Daily average sediment temperatures at the bottom of the water column and at 10-cm depth in the sediment in Pond B, 1972.*

the morning and cool more slowly than the water in the evening, these sharp rises have little effect on the overall average.

The daily temperature oscillations are also evident at the 10 cm sediment depth but these deep sediments are colder than the surface sediments and the daily temperature range is less than one-quarter of that at the surface (Figure 3-5). At 10 cm the average temperature is 2.3°C below the mean at the sediment surface (Table 3-3) and the oscillations are greatly damped. In addition, the higher temperatures move downward through the sediments in a wave so that there is a 7-hour delay between the time of the temperature maximum at the surface of the sediments and the maximum at 10 cm. As might be expected, the daily temperature range at 10 cm is only 1.5°C. Thus the wave has both a time lag and a decrease in amplitude as it moves downward.

The only comparable data are those of Brewer (1958) for Imikpuk Lake (around 2 m deep) and two other Barrow area lakes. The surface sediments were always the same temperature as the water column in these lakes and the rate of temperature decrease, approximately 2°C per 10 cm, agreed with the pond data. One difference was that the lake sediments reached their maximum temperature in mid-August, one month after the maximum in the ponds, but this is merely a reflection of the parallel situation for the water temperatures in lakes and ponds.

TABLE 3-4 *Freeze-thaw Dates for Barrow Ponds*

Year	Pond	First melt water	Last ice	Permanent ice cover	Sediments frozen	References
1962	I-III			12 Sept.		Kalff (1965)
1963	I		1st wk June	1st wk Sept.	14 Oct.	Kalff (1965)
	II, III		2nd wk June	1st wk Sept.	after 21 Oct.	
1964	I		1st wk June			Kalff (1965)
	II		11 June			
	III		2nd wk June			
1967	unnamed		14 June			Stross and Kangas (1969)
1970	B,C	20 May	16 June	16 Sept.	before 30 Sept.	
	E	1st wk June	18 June	16 Sept.	before 30 Sept.	
1971	B,C,D,E,F	5 June	19 June			
1972	B,C,D,E,F,J	26 June	13-18 June	after 15 Sept.		
1973	B,C,D,E,F,J	20-24 June	11-15 June	after 15 Sept.		

Freeze-thaw Events

The onset of the pond thaw is determined by air temperatures and also by the snow cover that effectively reflects most of the solar radiation. For example, in 1971 a laboratory hut was placed about 10 m upwind from Pond C. As a result, there were snow drifts on Pond C the following spring and thaw was delayed 1 to 2 weeks. When there is little snow, as in 1970, the ponds may become windswept and snow-free in places. Thus, in 1970 Ponds B and C had snow-free patches in May while Pond E was snow-covered. There was up to 5 cm of melt at the bottom of the ice in B and C as early as 20 May, giving these ponds a 15% longer thaw period (mid-May to mid-September) than Pond E. However, this early start did not appreciably affect the time of disappearance of the ice as Ponds B and C became ice-free only two days before Pond E. While it is difficult to determine a mean date for ice melt, it is obvious that the winters of 1970, 1971, and 1972 had much lower than normal snowfall (Table 2-2). Therefore, windswept ponds are not a usual occurrence and the first melt-water should occur in the first week in June (Table 3-4). On the average, the ponds should melt completely between 10 and 17 June.

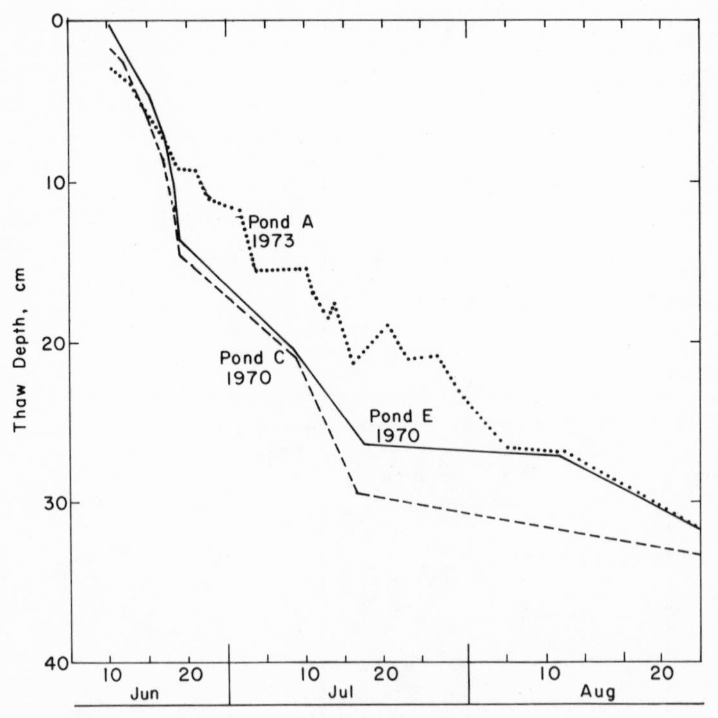

FIGURE 3-6. *Sediment thaw depth, 1970 and 1973.*

After the ponds become ice-free, the sediments thaw rapidly and half of the total thaw occurs in the first 10 to 20 days (Figure 3-6). There is little between-pond difference in a given year but there are climatic differences from year to year. Thus, the sediments of Pond C thawed more rapidly in 1970 than did Pond A in 1973 (Figure 3-6) but at the same time there was 43% more solar radiation in July of 1970 than in July of 1973. Kalff (1965) found much the same thing in a pond where the thaw depth was 41% greater and the solar radiation was 32% higher in 1964 than in 1963.

The pond sediments reach their maximum thaw depth, 25 to 36 cm, in late August (Table 3-5). After 1 September, the sediments begin to slowly refreeze from the bottom but a permanent ice cover does not form on the water until 7 to 16 September (Table 3-4). Because of the permafrost, the sediments begin to refreeze from the bottom, but this is a slow process and most of the freezing of the sediments occurs from the top down after the water is completely frozen. In 1970, it took an additional two weeks more for the sediments to become completely frozen; the two freezing fronts met at a depth of 20 to 30 cm in the sediments.

The shallow lakes in the vicinity of Barrow get their first ice cover at about the same time as the ponds but thaw about one month later. Brewer (1958) states that lakes less than 1 m deep become ice-free in late June but slightly deeper lakes, e.g., Imikpuk Lake (2.8 m), become ice-free between 17 and 28 July. Ikroavik Lake did not lose its ice until 4 August in 1970 but was ice-free on 14 July 1971. Still deeper lakes, such as the 40-m-deep Lake Peters in the Brooks Range, do not freeze until October (Hobbie 1973). Much of the difference in thaw dates is a result of the time required to melt the thick ice sheets of deep lakes. Aerial photos and ERTS

TABLE 3-5 *The Maximum Depth of Thaw Across the Same Transects in Ponds C and E, 1970 to 1972*

Year and date	Pond	Depth of thaw ± SD (cm)
1970	C	33.7 ± 8.6
28 Aug.		
	E	31.9 ± 5.1
1971	C	31.8 ± 4.4
23-25 Aug.		
	E	25.4 ± 3.9
1972	C	33.9 ± 4.7
1 Sept.		
23 Aug.	E	36.0 ± 4.5

imagery over the thaw period have been used to determine the thaw dates and in this way to distinguish between shallow and deep lakes in northern Alaska (Sellmann et al. 1975).

HYDROLOGY

Snowfall and Rain

The total recorded snowfall at the NOAA station at Barrow may be quite misleading as this is a cumulative number calculated from measurements made after every snowfall. Later, compaction plus snow loss by sublimation and drifting takes place. Thus, the total snowfall from the fall of 1959 until the spring of 1973 ranged from 45.2 to 103.9 cm snow yr^{-1} (Table 2-2) but the maximum snow pack ranged from only 17.8 to 30.5 cm. In fact, the official reading of snow on the ground on 1 June, just before the thaw began, was 5.1, 7.6, trace, and 12.7 cm for 1970, 1971, 1972, and 1973, respectively. A better measure is of the snow actually on the pond site at the beginning of melt as well as an estimate of its water content (Table 3-6). The snow depth on the pond site was 2 to 6 times greater than the depth at the Barrow station (1971 to 1973) and the water content was also higher than might have been expected. These facts

TABLE 3-6 *Snow Depth and Water Content of Snow, IBP Pond Sites*

Year	Date	Pond	Snow depth (cm)	Water equivalent (cm)	Percent water
1971	30 May	J	43.4	16.5	38
			57.0	16.5	29
		C	48.0 ± 7.9		
		E	40.0 ± 4.2		
		mean	47.1	16.5	
1972	12 June	A	12	5.2	43
		B	12.5 ± 2.1	4.9	39.5
		C	20.0 ± 1.4	8.3	41.5
		D	12.0	5.8	48.0
		J	21.5 ± 0.7	8.8	40.9
		mean	16.5 ± 4.7	6.9 ± 1.9	41.8
1973	5 June	5 transects across sites 7 and 12 n=163	28 .9 ± 9.3	11.6	40

TABLE 3-7 *Precipitation at Barrow (30 year average) and at Other Stations in Northern Alaska**

Barrow	10.6 cm
Cape Lisburne	15.7
Pt. Hope	10.2
Cape Thompson	10.8
Kotzebue	8.0

*Source: Allen and Weedfall 1966.

emphasize again the importance of the drifting and of the microrelief of the polygon ridges on the snow depth and on the volume of water available for runoff.

The precipitation during the summer is more important to the ponds than the meltwater from the snowfall because it adds nutrients and organic matter through leaching from the soils and vegetation. In addition, the amount of summer rainfall determines the water level, and therefore, the degree of biological effect of the shrinking pond volume. During the 1970 to 1973 period, the summer precipitation ranged from 1.7 to 13.2 cm of water at the NOAA weather station at Barrow (Table 2-2), where the 50-year mean is 6.7 cm. While there were some differences between the IBP site and the NOAA station, they were not significant.

About 50% of the total precipitation is received as snow at Barrow, similar to the value for other arctic sites. Thus, at two stations in northern Canada, the values were 53% (Devon Island, Cogley and McCann 1971) and 54% (Resolute Bay, Cook 1960). In the Yukon Basin to the south of Barrow, some 40% of the precipitation occurred as snow and at Fairbanks only 33.3% of the total of 31.3 cm fell as snow (Dingman 1971).

The total precipitation at Barrow over the past 30 years is comparable to that of other coastal sites in northern Alaska (Table 3-7) and also to other arctic sites.

Runoff

In 1972 and 1973 the runoff of snowmelt water was measured at five outlets from the IBP Pond Site (Figure 3-1). This was done somewhat crudely in 1972 by measuring the cross-sectional area of the outlets and timing the speed of a slug of fluorescent dye over a distance of 1 m or more. In 1973 the small outlets were equipped with "V" notch weirs which were calibrated by collecting all the flow for an interval. The large outlet (weir 1) was fitted with a Parshall Flume early in the melt period and later with a rectangular notch weir.

There were many problems associated with the flow measurements at the pond site. For example, exact size of the drainage basin is unknown because the site is so flat. Also, the 1972 flow measurements are relative only, as it was difficult to make an accurate calibration. Finally the Parshall Flume was not operative on 3 days in June (11 to 13) when the flow was highest.

For comparison, data are included from Esatkuat Creek (Figure 1-1), near Barrow. This creek drains an area of 3.8 km^2 and flows into Esatkuat Lagoon near the village of Barrow. Its flow has been monitored for several years by the U. S. Geological Survey.

In 1973 the runoff began on 10 June from the small channels on the pond site (weir 4) but began a day later in the much larger Esatkuat Creek (Figure 3-7). Flow in the large channel on the pond site (weir 1) lagged the small channel flow slightly, but overall the flow began with amazing

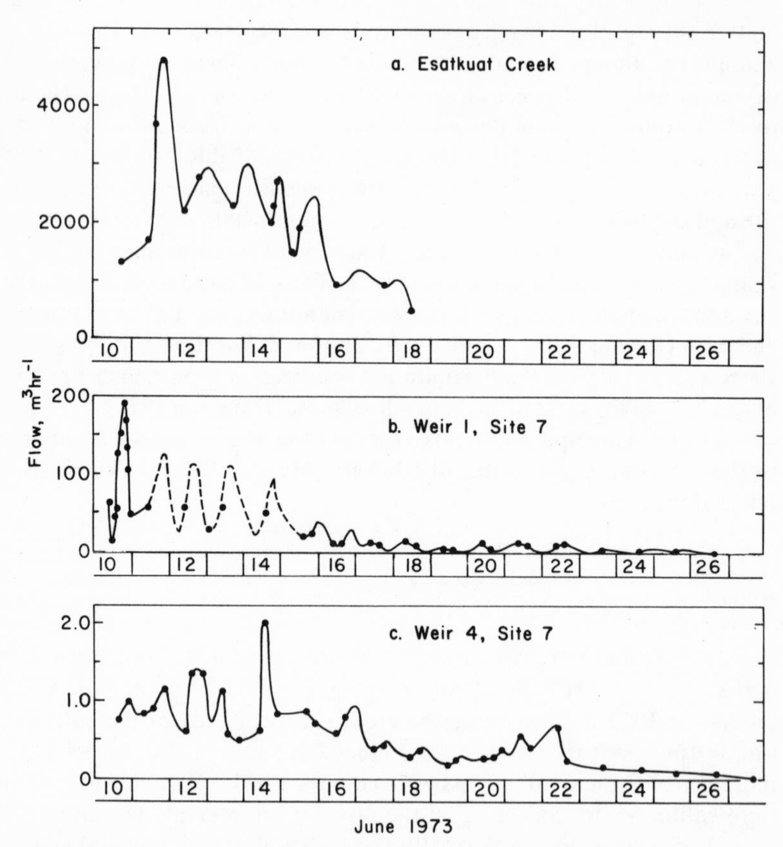

FIGURE 3-7. *Runoff from small, medium, and large streams, 1973. Esatkuat Creek data from Jones (unpublished).*

suddenness. The daily peak occurs earlier in the small channels than in Esatkuat Creek, a pattern typical of streams.

The total runoff from the pond site was 5.7 cm in 1972 and 5.9 cm in 1973 (Table 3-8). This amounted to a recovery of 83% of the expected runoff in 1972 and 51% in 1973. The expected runoff was calculated from the snowpack measurement (Table 3-6). Much of the water unaccounted for in the runoff goes to fill up the more than 100 ponds in the watershed. This amount of water is called the dead storage in the total reservoir (Dingman 1973). Usually, this must be filled before the runoff can begin. However, in these ponds, the water does not flow in defined channels so the filling and runoff appear to be going on at the same time. The total amount of dead storage in the ponds is directly related to the pond level at freezeup the previous fall. Our measurements confirm this, as the pond levels were higher in late summer 1971 than they were in late summer 1972. Thus, there was a greater dead storage in spring 1973 than in 1972, and the recovery was less (51%) in 1973 than in 1972 (83%).

The only other Barrow area studies that attempted a runoff budget were from Esatkuat Creek (Dingman et al. in press). Recoveries were 95% in 1972 and 93% in 1973. This creek, however, has a well developed drainage with a total elevation difference of 5 m. Although other hydrology studies have been made (Brown et al. 1968, Lewellen 1972), they did not begin until after the bulk of the snowmelt had run off.

After the initial peak of runoff, the streams at Barrow fall to a very low level. Brown et al. (1968) and Lewellen (1972) measured a summer runoff equivalent of 5% of the summer rainfall. There was runoff after every rainfall so presumably runoff began as soon as the soils became saturated again. This was not the case at the pond site because of the dead storage available as soon as the water levels began to fall in mid-June, and no runoff was observed during the three years from 1970 to 1972. However, in 1973, when the summer rainfall was double the average, the ponds refilled on 17 August and after that date, 86% of the rainfall during the next 7 days passed out through the five small weirs.

TABLE 3-8 *Water Budget for Pond Site During Runoff, 10 to 27 June 1972 and 1973*

Parameter	1972	1973
Snow depth of ponds (water equivalent)	6.9 cm	11.6 cm
Runoff 10-27 June (Total weirs 1-5)	12,300 m^3	12,645 m^3
Watershed area	0.215 km^2	0.215 km^2
Runoff, 10-27 June	5.7 ± 2 cm	5.9 ± 1 cm
Percent of snow recovered as runoff	83%	50.7%
Rainfall 10-27 June	0.23 cm	1.61 cm
Pan evaporation 10-27 June	2.77 mm day^{-1}	1.72 mm day^{-1}

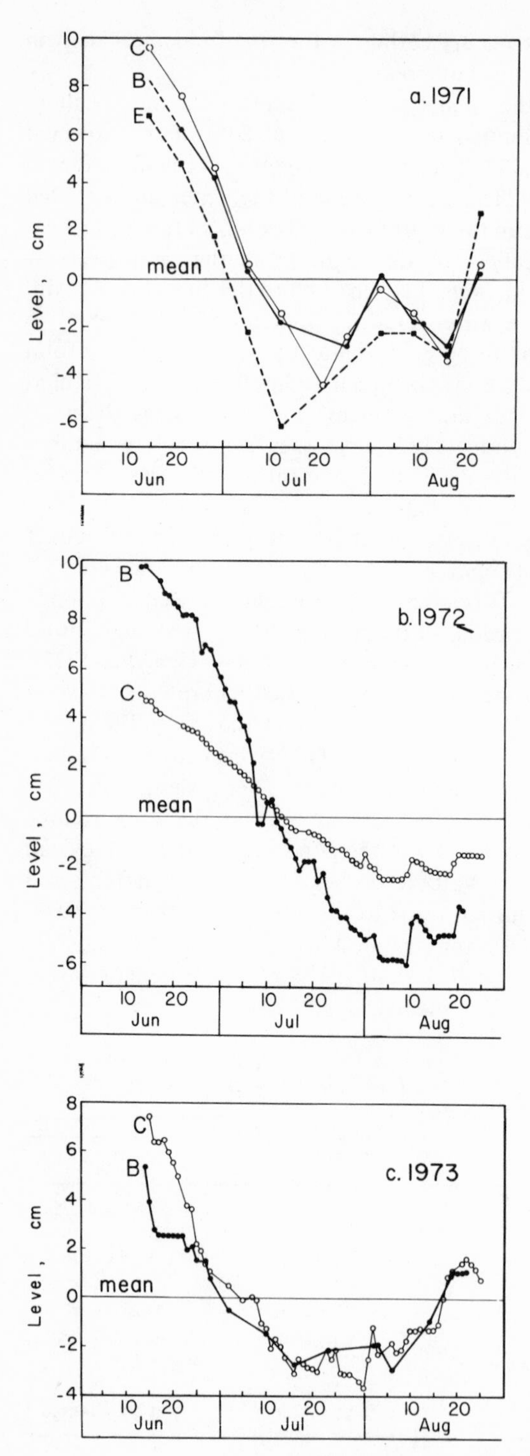

FIGURE 3-8. *Water levels in ponds, 1971 to 1973.*

Over the entire year, about 37% of the precipitation runs off the pond site drainage basin. This is much lower than other reported values for permafrost areas. For small watersheds at Barrow, Brown et al. (1968) assumed that all the winter snowfall runs off as well as 5% of the summer rainfall to give a total of 52% for the year. Dingman (1973) summarized arctic data and reported that in tundra areas in the USSR the runoff ranged from 70 to 76% of the total precipitation. In the Brooks Range where the drainage basin was steep and rocky, Hobbie (1962, 1973) found that 85% of the total precipitation ran off over the entire year, a value similar to that found by Kuzin (1960, quoted in Sater 1969) for Novaya Zemlya. For a study of nutrient budgets of a high arctic, nearly vegetationless drainage basin (Char Lake), Schindler et al. (1979) and de March (1975) assumed that all snow and rainfall ran off.

The percentage runoff from the Barrow ponds is low because of the high evaporation and the very low precipitation. Most of the surface of the watershed is standing water for some or all of the summer, so evaporation and transpiration rates are always the highest possible. This loss lowers the water level so that any summer precipitation fills up the ponds rather than running off. If the total precipitation were increased, then almost all of this increase might well run off and in this way increase the percentage. Ultimately, the low runoff is caused by the flat ground, the many small catchment basins formed by polygonal ground, and the low amounts of precipitation.

Evaporation and Pond Levels

Evaporation was measured daily in the summers of 1971, 1972, and 1973 in a standard 1.21-m-diameter pan supported above the tundra surface by a 5-cm-high wooden platform. It is well known that this method overestimates the true evaporation from a free water surface by as much as 140 to 150% in dry regions (Harding 1942) but no correction values are available for cold climates. Actually, the depth of the evaporation pan, 25 cm, was similar to the pond depths and the water temperatures were virtually the same as in the pond. Therefore, we assume that the pan does give an estimate of pond evaporation rates (Table 3-9). Other workers at Barrow have found that evaporation from a pan was much greater (250%) than the evapotranspiration from the surface of the tundra (Brown et al. 1968).

Pond water levels were read daily in 1972 and weekly in 1970, 1971, and 1973 from staff gauges anchored in the permafrost. The pattern of water level changes (Figure 3-8) is similar in all the measured ponds for a given year but there are year-to-year changes caused by the interactions of precipitation and evaporation. Thus, the ponds begin to decrease in depth in mid-June. The decrease was continuous throughout the summer in the

TABLE 3-9 *Evaporation Pan Data from Barrow, Alaska, 1971 to 1973*

Year	Location	June (cm)	July (cm)	August (cm)	Total (cm)
1971	Site 2*	0.3677 day^{-1} 11.03 est.	0.3325 day^{-1} 10.31 est.	0.1377 day^{-1} 4.27 est.	=25.61 est.
1972	Aquatic Site	0.2826 day^{-1} 8.48 est.	0.3161 day^{-1} 9.80 est	0.269 day^{-1} 8.33 est.	=26.61 est.
1973	Aquatic Site	0.2030 day^{-1} 6.09 est.	0.2245 day^{-1} 6.96 est.	0.1407 day^{-1} 4.36 est.	=17.41 est.

*Site 2 data are from Weller and Holmgren (unpublished data).

TABLE 3-10 *The Calculated (Precipitation Minus Evaporation) and the Measured Water Level Loss in Pond B for July and August 1971-1973**

	1971	1972	1973
Evaporation (Table 3-9)	-14.6	-18.1	-11.3
Precipitation (Table 3-2)	3.4	3.1	8.3
Calculated water loss	-11.2	-15.0	- 3.0
Measured water loss (Figure 3-8)	- 4.0	-11.0	0

*Data expressed in cm.

TABLE 3-11 *Water Budget for Pond B, June through August of 1972 and 1973*

	1972 cm	1973 cm
Input		
a. Snow on pond at start of melt (Table 3-8)	6.9	11.6
b. Summer precipitation (Table 3-2)	3.2	10.3
c. Extra water to fill dead storage (Figure 3-8) minus amount of snow on pond ("a")	1.8	- 5.3
d. Overfill of pond from drainage basin (Table 3-8) (equals runoff)	5.7	5.9
	17.6	22.5
Output		
e. Runoff (Table 3-8)	5.7	5.9
f. Evaporation (Table 3-9)	24.8	17.4
	30.5	23.3
Balance (Input-Output)	-12.9	-0.8

dry summer of 1972 while there was an increase in water level towards the end of the summer in 1971 and 1973. As already discussed, there was so much rain in 1973 that the ponds overflowed in late August. In 1972, there was less rain (Table 3-2), but the evaporation rate (Table 3-9) was almost 50% higher in August of that year than in 1971 or 1973.

The depth measurements indicate that the evaporation pan data approximate evaporation. This is best shown for the first month of 1971 (Figure 3-8) when some 12 cm of water was lost from the pond and 11 cm from the pan (Table 3-9). Of course, the shallow, sloping sides of the pond complicate this argument, which really should be made on the basis of volume. The calculated balance between evaporation and precipitation for July and August (Table 3-10) agrees quite well with the measured water loss for 1972 and 1973 but did not agree in 1971. It is possible that a large rain would saturate the tundra and allow runoff into the pond while a series of small rainfalls, which might add up to a larger total, would not produce runoff into the pond.

The water budget for Pond B (Table 3-11) indicates a loss of 11.7 cm of water in 1972 and a gain of 1.9 cm in 1973. The change we measured in the pond was a loss of 11.0 cm of water in 1972 and a gain of 0 cm in 1973; this is quite good agreement. From these data, we conclude that the evaporation measurements are approximately correct and that there is no additional seepage of water from the soils into the ponds, even when the water levels of the ponds are 5 or 10 cm below the surface of the tundra.

One strange finding from measurements of the water depth by a series of transects was that the bottom of the pond does not remain constant but apparently rises and falls during the thaw season. Another bit of evidence for this movement comes from litter bags that were suspended immediately above the sediment on 3 July 1972 by strings attached to poles frozen into the permafrost. By 27 August the bags were 9 to 10 cm above the sediment. One explanation for the rise and fall is that as ice crystals grow in the sediment, water moves through the sediment to feed their growth. In this way, the frozen sediment eventually contains much more ice than the original water content of the unfrozen sediment could have produced. In addition, the frost heaving or expansion of the sediments upon freezing is much greater than the 8% increase due to the freezing of water. This ice crystal growth has been shown to cause frost heaving in soils (Taber 1929, 1930).

In the ponds, the fall or compaction of the sediments changes the water level only slightly. As the sediments thaw, the excess water is excluded from the sediments but this water has 8% less volume than the original ice. Thus, the effect on the pond level is a decrease of 8% of the distance of the sediment compaction.

When the sediment movement is ignored, we calculated that the volumes of Ponds C and E were reduced by 49%, 43%, 63%, and 41% during the summers of 1970, 1971, 1972, and 1973. These data are based on mean water depths taken on several transects and they assume that the area of the ponds does not change during the summer.

LIGHT

Water

The light extinction in the ponds was measured *in situ* with an underwater selenium photocell (Schueler, Waltham, Mass.). The spectral response of this instrument resembles that of the human eye. The data are summarized as an extinction coefficient (*n*, per meter) calculated from

$$I_z = I_0 e^{-nz}$$

where I_z is the light intensity at depth z (in meters) and I_0 is the light intensity immediately beneath the water surface. Unfortunately, it is difficult to be precise with this instrument, designed for oceanic work, when measuring a water column of only 30 cm. There are also other problems, such as the rapid extinction of the infrared and ultraviolet light in the top few centimeters, that make it very difficult to obtain a good extinction coefficient for the total light over these short distances.

The extinction coefficient is high in the ponds (Table 3-12), reflecting a high concentration of colored organic compounds. In similar ponds and lakes, the organic compounds absorb strongly in the short wavelengths (UV and blue) so that, unlike clear lakes, the maximum transmission occurs in the red wavelengths. One measure of the amount of dissolved humic matter in the water is the absorption of light at 250 nm (Miller 1972) (Table 3-12). There is good agreement between the *n* and the optical density (OD) except that Pond E, the oil experiment pond, has more humic compounds than expected from the extinction coefficient.

TABLE 3-12 *Mean Extinction Coefficient and Mean Optical Density (OD) at 250 nm and a 1-cm Path Length, for 4 Ponds during 1971*

Pond	Mean extinction coefficient	Mean OD
B	3.84	0.446
C	3.73	0.444
D	4.40	0.611
E	3.04	0.482

The extinction coefficients ranged from 6.75 to 2.1 in the ponds, which is equivalent to 26 to 66% of the surface light reaching a depth of 20 cm. For Pond B, the mean *n* of 3.84 equals 46% of surface light reaching 20 cm. In contrast, the water of Ikroavik Lake has an *n* which ranges from 0.49 to 1.22 so the water is much clearer (at 2 m, this is 37 and 8% of the surface light). In arctic waters in general, the shallow bodies of water have a high extinction coefficient unless they are located in a very rocky watershed. Deeper lakes may be very clear; for example Taserssuaq Lake in Greenland (Holmquist 1959) has an *n* of 0.154 for white light and 5% of the light penetrates to 25 m. Some deep lakes that receive particulate matter from glaciers have an *n* of up to 2.5.

Most of the absorption of light in these ponds is due to the dissolved organic matter; very little absorption or scattering is due to the particulate material. Thus, there is a highly significant ($r=0.83$, $n=11$) relationship (Kalff 1965) between water color, as measured in Pt-Co units, and the extinction coefficients (extinction coefficient = (0.0096)(Pt-Co units) −30.4).

Sediments

The light penetration into the sediments was measured by placing a water-filled glass cylinder on top of the photocell and adding known amounts of sediments. From this measure, an *n* of 15.22 cm^{-1} (or 1522 m^{-1}) was measured, which implies that all of the light was absorbed in the top 3 mm. This, of course, means that all of the photosynthesis of the epipelic or benthic algae must take place in an extremely thin layer.

CURRENTS

A Pygmy Water Current meter (Gurley), placed just beneath the water surface in Pond C, was used for eight 24-hour studies in 1973. This instrument is designed for stream work so likely underestimates the pond currents. Air speed data came from a hand-held anemometer, from the

TABLE 3-13 *Average Wind Speeds at the Barrow Weather Bureau, 1970-1973**

Year	June	July	August	Sept.
1970	4.24	4.29	4.47	6.08
1971	5.54	5.28	4.78	4.83
1972	5.54	4.82	4.87	5.54
1973	5.01	4.96	4.87	4.65

*The anemometer is at the top of a 9-m pole; wind speeds expressed as m sec^{-1}.

data of P. Miller (personal communication), or from the NOAA weather station at Barrow.

The winds are remarkably constant during the summer months (Table 3-13) and vary little from month to month or from year to year. However, the NOAA data are taken at 9 m above the ground and there is a logarithmic decrease of average wind speed towards the ground. For example, the summer 1971 IBP data (Weller and Holmgren 1974) show that the average wind speed at 0.25 m above the ground was only 44% of the speed at 8 m. The emergent grasses and sedges around and in the ponds reduce the wind speed even further but we have no measurements of the speed of the wind that contacts the water surface.

There was no significant correlation between the wind speed and the current velocities during six of the eight diurnal measurements. Of course the currents did increase during high winds, but evidently our measuring devices were not sensitive enough to give the complete picture. On the two runs showing significant correlation, 18 and 19 July 1973, the currents were measured 1.5 cm below the surface while the wind speeds were from the NOAA Station at Barrow. On these dates (Figure 3-9), the current velocity was 0.07 to 0.12% of wind speed at 9 m, which is much less than the 2% of wind speed expected for currents in lakes (Hutchinson 1957). Over the 8 days of measurement, the currents ranged from 0 to 1.2 cm sec^{-1} while the winds ranged from 102 to 1028 cm sec^{-1}.

On five of the diurnal studies the surface currents of the pond dropped to zero during the night even though there was continuous wind at 9 m (Figure 3-9, 19 July). This is likely caused by an inversion or similar abrupt temperature change of the air immediately above the water.

The surface currents produced by the wind move water to the leeward side but there is also a return current along the bottom. These wind currents, along with the microturbulences set up along the shear zones and the circulation produced by sediment heating, are extremely important in resuspending detritus and bacteria from the sediments and in keeping the benthic sediments and water well oxygenated.

SUMMARY

The IBP study ponds lie in an area of polygonal ground. Each pond averages about 20 cm deep and has a maximum depth of 40 cm. At the beginning of the summer the whole polygon may be flooded; after a dry summer only the central depression will retain water.

Erosion and reworking of sediments make ^{14}C dating unreliable in the pond sediments. A peat layer beneath the oldest lake sediments is 12,000 years old. Several lakes formed and drained at this site but the ponds lie in a lake basin that likely formed 3,000 to 6,000 years ago. Thus, the ponds are likely several thousand years old.

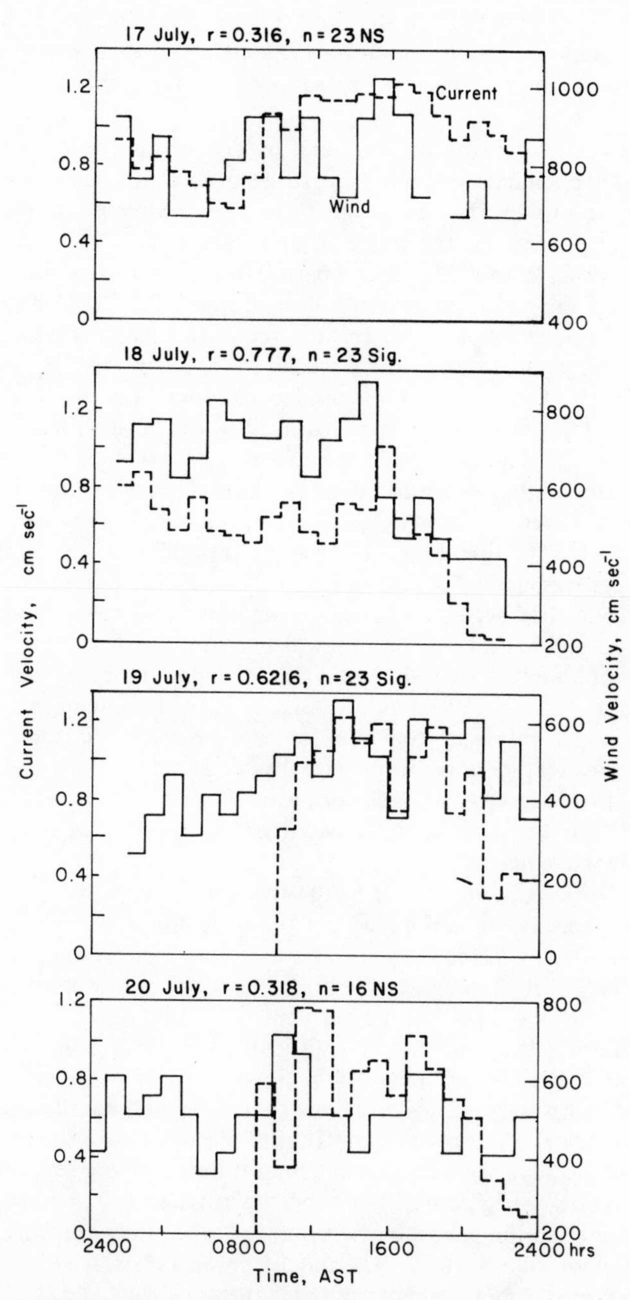

FIGURE 3-9. *Relationship between wind and current speed, Pond C, 1973.*

The summer climate at Barrow during the 4 years of our study was slightly warmer than the 30-year mean. Precipitation was near normal but the average includes both a very dry and a very wet summer. Solar radiation decreased each summer from 1971 to 1973 as a result of increasing cloud cover. This affected water temperature and photosynthesis.

The ponds are well mixed by wind and by convection. Ponds with similar depths have similar temperatures. In general, mean daily water temperatures rose rapidly in June from 2° to 12°C. After early July, temperature slowly declined; the range in July was 7° to 12°C and in August was 3° to 9°C. Within this general pattern there was a great deal of variability because the water temperature often dropped 9° to 10°C in one day as the air temperature changed. In fact, the water temperature was higher than the air temperature; over 4 weeks the average was 7.1° and 3.7°C, respectively. The highest temperature recorded was 20°C; the lowest was 0°C after a mid-August snowstorm filled the pond with snow. The average water temperature was about 6 to 8°C.

Sediment temperatures were the same as the water temperatures at their interface. However, at 10 cm below the sediment surface the average temperature was 2.3°C less than the water. In addition, the daily temperature range was also reduced (to 1.5°C).

The pond ice melted between 10 and 17 June. Pond sediments thawed rapidly thereafter and half of the total thaw occurred in the first 10 to 20 days. The thawing rate then slowed; the maximum depth of thaw, 25 to 36 cm into the sediment, was reached in late August. After 1 September, the sediments began to refreeze from the bottom. Between 7 and 16 September a permanent ice cover formed on the ponds but the ponds were not completely frozen for several weeks or more. When this occurred, the sediments froze from two directions; the two freezing fronts met at a depth of 20 to 30 cm in the sediment.

The total snowfall at Barrow is 45 to 104 cm year^{-1} but most of this is compacted, blown away, or sublimated. At the pond site, the maximum snowpack was 18 to 31 cm; the snow is 40% water. Summer precipitation, which is about half of the total, ranged from 1.7 to 13.2 cm of water from 1970 to 1973 but the long-term average is 6.7 cm.

The runoff from the pond site was difficult to measure accurately; there were no well-developed channels so the flow had to be measured at a number of sites. The watershed includes more than 100 ponds but the exact boundary is unknown. Most of the runoff takes place within 2 weeks in late June. Some of the meltwater does not run off but instead goes to fill up the ponds. As described earlier, the ponds decrease in size and depth during late summer so the ponds are far from full when they are frozen. The measured runoff of 5.7 cm in 1972 and 5.9 cm in 1973 was 83% and 51% of the expected runoff calculated from the snowpack data. This matches the water level data which showed a higher pond level in late summer of 1971 than in late summer of 1972. Usually, the pond water levels fall throughout the summer and there is no runoff after late June. However, during the wet summer of 1973 the ponds filled again in mid-August and then there were 7 days of runoff.

Most of the surface of the watershed is standing water for most of the summer. Therefore, the evaporation and evapotranspiration is maximal and the summer precipitation remains in the ponds. The result is that only 37% of the annual precipitation leaves as runoff. This is low compared with the 52% measured in a small nearby watershed with well-developed streams or compared with the 70 to 85% measured in other arctic watersheds. The evaporation measured in a 25-cm-deep, 1.21-m-diameter pan at the edge of the pond was 26.6 cm in 1972 and 17.4 cm in 1973. By our calculations, the water level increase should have been 1.9 cm but there was no actual change in the level.

The light extinction coefficients ranged from 6.75 to 2.1 in the ponds which means that 26 to 66% of the surface light reaches 20 cm. The strong absorption of light is caused mostly by the humic compounds dissolved in the water. In the sediments, all of the light was absorbed in the top 3 mm.

Water currents in the ponds increased during periods of higher than normal winds but overall the currents were low. In part, this is caused by the rapid decrease of wind speed close to the pond surface. Also, the short fetch did not allow waves to build to any size. On a number of dates, the wind, measured at a height of 9 m, continued throughout the night but there was no surface water current at all due to an inversion.

4

Chemistry

R. T. Prentki, M. C. Miller, R. J. Barsdate, V. Alexander, J. Kelley and P. Coyne

SEDIMENTS

The chemistry of the pondwater is strongly influenced by interactions with the sediments. The ponds are shallow, so there is a high ratio of sediment surface to pond volume. Also, there is a high amount of resuspension of the flocculent sediment caused by the constant wind.

Sediment Particle Size

The bottom sediments of the tundra ponds are composed mainly of dark brown, highly organic, unconsolidated material. Some organic particles can be identified as bits of leaves of sedges and grasses, roots and moss fronds. Usually, the highly organic sediments are 18 to 30 cm thick and are underlain by a layer of mixed organic matter and sand. Beneath this are layers of sand and lenses of buried peat at a depth of 40 to 60 cm.

The organic particles in the surface sediments are quite large, especially close to the edge of the pond, and are poorly sorted. The data come from a measurement in the middle of Pond A of a 0-2 cm sediment sample which was wet sieved with metal sieves (Wentworth Scale), oven-dried at 110°C, and weighed. The mean particle size was 500 μm (Figure 4-1) and there was a linear relationship between the cumulative percent dry weight and the $\log_2$ of particle size. This linear relationship and the low slope suggest that these sediments are poorly sorted by physical processes. This relationship contrasts markedly with the sigmoidal curve, steeper in slope, that characterizes highly sorted beach sand.

The ponds can be visually separated into those with mostly large organic particles in the surface sediments and those with mostly small particles. The intensively studied ponds, A, B, and C, fall into the first category and Ponds X and D fall into the second. Pond J was so disturbed by wading and destructive sampling in 1971 that it cannot be classified.

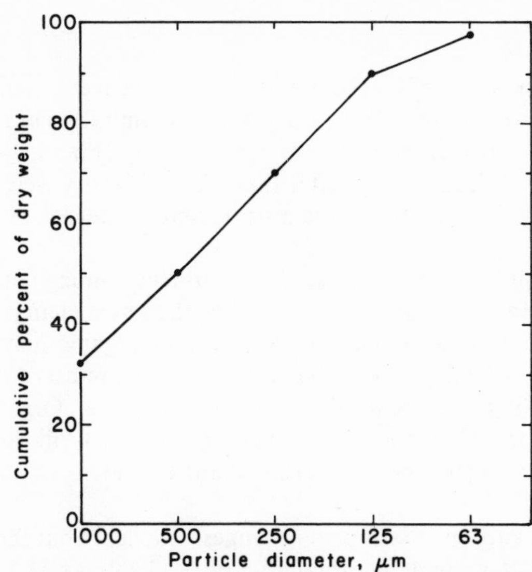

FIGURE 4-1. *Cumulative percent of dry weight of surface sediments of different sizes in Pond A, June 1973.*

Bulk Density

A measurement of the bulk density takes into consideration both the water content of the sediments and the density of the soil material; in Ponds B, C, D, and E the mean was 160 mg dry wt cc^{-1} (Table 4-1). However, these ponds were certainly on the low end of the spectrum, as 20 ponds in the old lake basin had a range of bulk density from 65 to 1096 mg dry wt cc^{-1}. The high values come from trough ponds where the sediments contain much inorganic matter; the modal class of these 20 measurements was 100 to 150 mg dry wt cc^{-1}.

TABLE 4-1 *Bulk Density, Percent Organic Matter, and Organic Carbon in the Top 1 cm of a Square Meter of Pond Sediments*

Pond	B	C	D	E	J	X
Bulk density (mg dry wt cc^{-1})	150	157	205	126	-	-
Organic content of surface sediments (% dry wt)	82.8	83.4	69.0	79.1	67.6	74.4
Organic carbon (g m^{-2})	496	523	566	400	-	-

Percent Organic Carbon

The organic matter in the surface sediments of the intensively studied ponds, measured as a percent of the ash-free dry weight (combusted at 525 to 550°C), was higher than that in other ponds in the vicinity (Figure 4-2). It is likely that these high percentages, greater than 80% of the dry weight, occur in ponds located near the center of the most recent former lake (see discussion in Chapter 3).

The changes in the bulk density of the surface sediments among ponds were caused by changes in the organic content rather than by a change in water content. Thus, the correlation between the percent organic matter and the bulk density was -0.92 ($n=20$). The percent organic matter (0 to 2 cm) equals (bulk density in mg dry wt cc^{-1})(-0.01244) + 77.4. Organic carbon was assumed to be 40% of the organic matter. Thus, a 1-cm-thick slice of the surface sediments in the intensively studied ponds contained from 300 to 566 g C m^{-2}.

The percent organic carbon of sediments changes with depth but there is an increase in the total amount of organic carbon (Table 4-2). This apparent paradox is explained by the increase in bulk density with depth. It is evident from these data that the top 4 cm are quite uniform with respect to percent organic carbon; this is a result of the stirring activity of

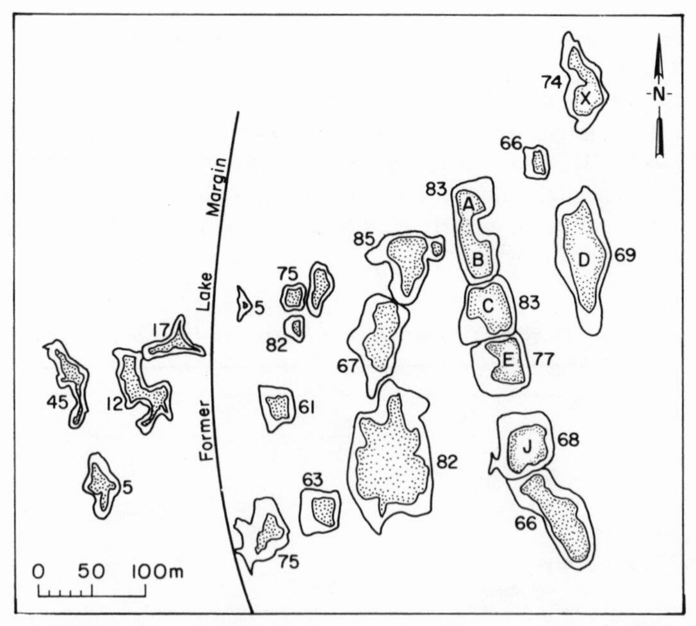

FIGURE 4-2. *Percent of organic matter in the surface sediments of Barrow ponds, 19 August 1973.*

the chironomid larvae and oligochaetes which stay mostly in the top 4 cm of the sediment (see Chapter 7).

The percent organic content continues to decrease in the top 20 cm but the pattern below this depth varies from pond to pond (Table 4-3). In Pond C, for example, the mean values for the 0-10, 10-20, 20-30, 30-40, and 40-50 cm sections were 70%, 60%, 37%, 23%, and 3%, respectively. Below 40 cm the sediment consists of sand and ice layers. In contrast, for Pond D the mean values for 10 cm sections were 67%, 42%, 66%, 42%, 33% and 58%. In this pond the sand is mixed with ice and peat layers.

MAJOR IONS

Water Chemistry

Dissolved sodium, magnesium, calcium, potassium, iron, and silica were sampled weekly in several ponds during the summers of 1970 and 1971. Cations were determined by atomic absorption and reactive silica by

TABLE 4-2 *Depth Profile of Organic Carbon, Percent Organic Content, Interstitial CO_2, and Leachable Dissolved Organic Carbon in the Sediment of Pond A, 16 August 1973*

Depth interval (cm)	Organic carbon ($g\ C\ m^{-2}$)	Percent organic content	Interstitial CO_2-C ($\mu g\ C\ ml^{-1}$)	Leachable total dissolved organic carbon ($\mu g\ C\ ml^{-1}$)
0-1 cm	297.9	81.2	3.71	13.0
1-2 cm	310.3	82.1	6.44	12.2
2-3 cm	327.6	85.1	8.29	10.2
3-4 cm	364.8	81.2	10.10	7.2
4-5 cm	426.5	74.4	11.30	15.1
5-6 cm	492.4	67.9	15.29	12.8
6-7 cm	501.5	52.7	23.30	11.4
7-8 cm	537.8	28.7	28.20	13.8
8-9 cm	490.5	18.1	34.75	16.3
9-10 cm	-	-	43.44	-
10-11 cm	-	-	48.48	-
11-12 cm	-	-	43.86	-
Total Sediment Organic carbon				
0-5 cm	1727.8 g			
0-9 cm	3749. g			

A sediment core 9 1cm diameter X 1cm was stirred once a day in 100 ml distilled water for six days at 25-28 °C. The water was filtered through Reeve Angel 984 ultra filter before analysis.

TABLE 4-3 *Chemical Composition of Sediments, in Ponds B, C and D, 15 September 1970*

Sample	Ca	Mg	Fe	Mn	Cu	Zn	Water content (% dry wt)
	mg(g dry wt)$^{-1}$						
Pond B							
0-1cm, oxidized	2.6	2.0	45	0.09	0.017	0.093	780
11-4cm, reduced	2.4	2.6	25	0.05	0.033	0.067	800
6-10cm, reduced	2.1	4.1	28	0.07	0.043	0.081	290
12-16cm, reduced	1.1	2.0	48	0.06	0.052	0.052	200
Pond C							
0-1cm, oxidized	3.6	3.2	43	0.09	0.022	0.087	1200
1-4cm, reduced	2.8	2.4	37	0.08	0.020	0.11	1030
7-10cm, reduced	1.8	5.4	31	0.10	0.046	0.11	190
12-15cm, reduced	1.8	5.3	35	0.11	0.043	0.11	230

Sample	P mg(g dry wt)$^{-1}$	Fe mg(g dry wt)$^{-1}$	% Organic
Pond C			
0-10cm, organic	680	21	70
10-20cm, organic	640	20	60
20-30cm, organic	590	17	37
30-40cm, organic > ice	540	23	22
40-50cm, sand	390	18	2
50-60cm, silt loam > ice	440	13	2
60-70cm, sand	230	6	2
70-80cm, sand	200	4	2
80-90cm, sand	190	4	10
Pond D			
0-10cm, organic	720	24	67
10-20cm, organic	390	16	42
20-30cm, organic > sand	340	12	65
30-40cm, sand > ice	430	17	42
40-50cm, sand > ice > peat	510	15	32
50-60cm, peat > ice	320	9	60

the silicomolybdate method. Conductivity was measured in 1970 on an impedance bridge at 25°C. Bicarbonate anion was calculated from alkalinity and chloride estimated by difference from the sum of strong cations minus bicarbonate.

The ponds are dilute salt solutions with chloride the major anion and sodium the major cation (Figure 4-3). Concentrations of chloride, sodium, bicarbonate, magnesium, and calcium all parallel conductivity and the concentrations of these ions are controlled by abiotic factors. Thus, during freeze-up, ions are excluded from ice and become concentrated in the remaining water. For example, Kalff (1965) found that alkalinity in

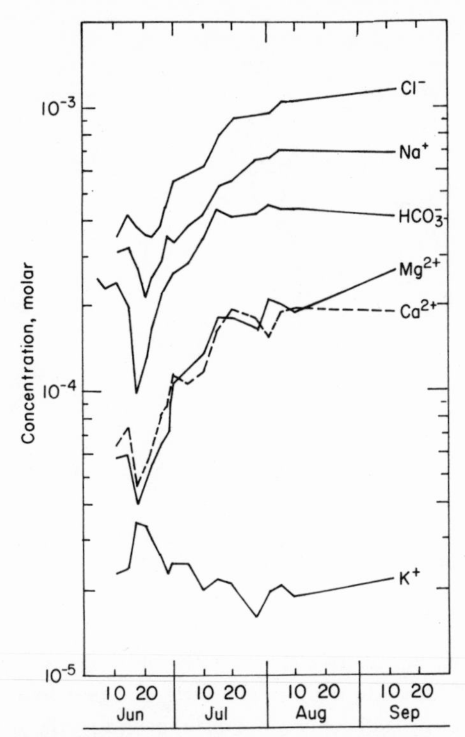

FIGURE 4-3. *Concentrations of major ions in Pond B, 1970.*

Barrow ponds increased 10-fold between the time the permanent ice cover formed and the time the ponds were frozen solid. Similar concentration increases in major ions and conductivity would be expected. The excluded ions are forced into the sediment as the water column freezes and the freezing front penetrates the sediment (see Chapter 3). The following spring, the water from the melting pond ice still has a considerable amount of ions, even though most of the electrolytes were frozen out. In fact, the conductivity of this meltwater is 5 times higher than that of water from precipitation or snow. Later, as the snowpack melts and runoff water floods the ponds, the resultant flushing produces almost uniform ion concentrations and conductance in all ponds (Figure 4-4). After the period of flooding, two processes, evaporation of the water and re-solution of the ions concentrated in the sediment the previous fall, contribute to an increase in the concentration of ions in the ponds. In the intensively studied ponds, this increase can be as much as 4-fold; in smaller and shallower ponds, such as Pond F, concentrations can increase even more. However, in very wet years, such as 1973, the ponds may be concentrated less than 2-fold during the summer (e.g., Kalff 1965).

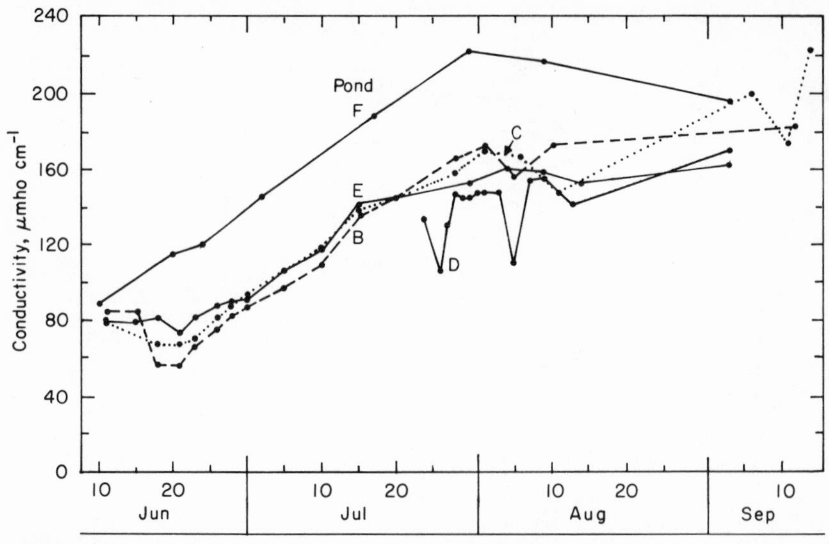

FIGURE 4-4. *Conductivity of five ponds, 1970.*

The seasonal cycles of the major ions that are not nutrients, such as sodium, magnesium, and calcium, are usually similar from year to year. For example, the concentration of sodium (Figure 4-5a) is almost the same in 1970 and 1971 for June, July, and part of August. The divergence of the two graphs in the middle and end of August is attributable to the difference in amount of rainfall. Thus, the peaks and valleys in the concentration graph during August of 1971 are the exact inverse of the pond water levels (Figure 3-8).

Potassium, iron and silica have more complicated seasonal cycles and their concentrations do not follow the conductivity. Potassium concentrations (Figure 4-3) are highest during runoff as a result of leaching from vascular plants and from lemming feces, but they decrease through the rest of the summer because of some unknown abiotic reaction (perhaps with clays?). The iron concentrations (Figure 4-5b) are low early in the season, reach a maximum in July, and then decline during August. The concentration of dissolved iron is correlated with humic color ($r=0.72$, $n=40$) and, therefore, follows the dissolved organic carbon rather than the conductivity. There is an undoubted cause-and-effect here but it can not be determined if the dissolved iron is chelated by organic material or if iron or some iron compound (ferric hydroxide?) is complexed with a colloid. There is also an excellent correlation of the small peaks in the concentration of iron with rainfall. Thus, the peaks on 10 and 28 July 1970 and on 21 July 1971 were on rainy days. Overall, the iron concentrations were 25% higher on rainy days than on non-rainy days; it is likely that the iron is being leached from the soil or from bottom sediments. The silica

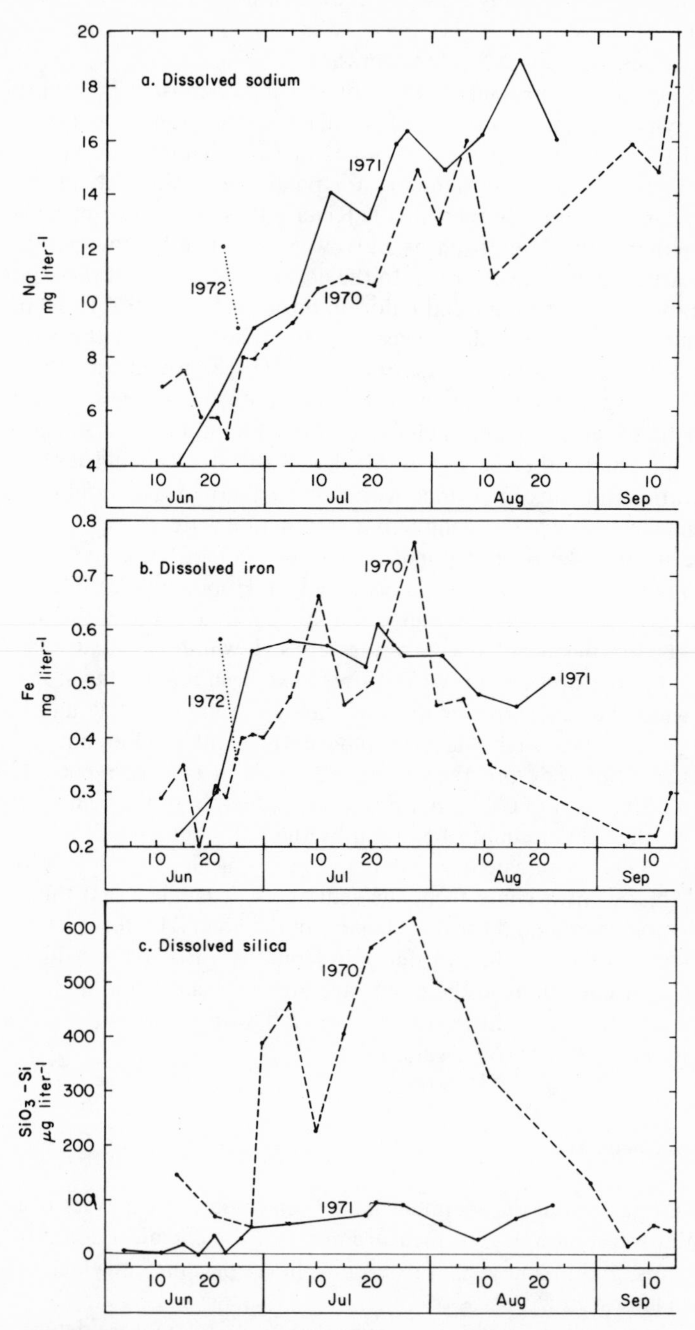

FIGURE 4-5. *Concentration of dissolved ions in Pond C, 1970–1971.*

concentrations in 1970 (Figure 4-5c) showed a large peak in all ponds in mid-July. However, in 1971 there was no peak and the overall concentrations were about 5-fold lower than in 1970.

Kalff (1965) has presented very similar water chemistry data for other Barrow ponds, Lamar (1966) and Watson et al. (1966a) made some chemical measurements in several ponds at Cape Thompson, and Reed (1962) has reported water chemistry for ponds along the Colville River. Calcium concentrations in the Colville River ponds are 3 to 5 times higher than in either Cape Thompson or Barrow ponds. The Barrow ponds are strongly affected by their nearness to the ocean and as a result these ponds have 2 times higher sodium and chloride levels and at least 3 to 10 times lower sulfate levels than do ponds at Cape Thompson to the west or Colville River area ponds to the east. Kalff (1968) found that the high chloride waters extended farthest inland south of Barrow, but that bicarbonate generally replaced chloride as the dominant ion as one moved inland from the northern coast of Alaska. Kalff also concluded that the concentrations of ions in surface water of northern Alaska and Canada were similar to those in low conductivity temperate waters.

The silica in the Barrow ponds is very low; in fact, the concentrations are the lowest of any ponds or lakes listed by Hobbie (1973) in a recent review of arctic limnology. Livingstone et al. (1958) also found the Barrow ponds to be low in silica. Concentrations are so low in these ponds that it is doubtful that diatoms can grow. Diatoms have been reported to stop their growth when the concentration of Si fell below 0.5 mg liter^{-1} (Lund 1964) and the ponds never reach this level. Indeed, the plankton does not contain any diatoms but they are the dominant forms in the sediments. If we assume a Si:C ratio of 0.4, and a primary production of the benthic algae of 10 g C, then the amount of Si used by the sediment algae is about 4 g m^{-2} yr^{-1}. This is equivalent to a demand of 13 mg Si liter^{-1} yr^{-1}. While most of this Si must come from the sediments, it is clear that there is a strong biological demand for this element in the ponds. Unfortunately, we do not have any data on the transfer of Si from the water to the sediments.

Iron concentrations in the ponds are high, at least 3-fold higher than in arctic coastal lakes; however they do fall within the range given in Hutchinson (1957) for brown water lakes.

Sediment Chemistry

Investigations in the sediment have been limited to analyses of three interstitial water samples, to measurements of the chemical composition of two cores, and to analyses of iron-phosphorus parameters which will be discussed later.

Calcium, sodium, potassium, magnesium, and iron were determined by atomic absorption spectrometry. Sediments for cation analysis were digested in perchloric acid.

TABLE 4-4 *Chemistry of Pond J Water and of Interstitial Water of Ponds E and J and Ikroavik Lake**

	Ca	Na	K	Mg	Fe
Pond J, (water column), 7 July 1971	4.2	15.8	0.45	4.9	1.5
Pond J, Central Basin, 11 July 1971, 4-12 cm	14.5	24.0	0.70	11.0	17.1
Pond E, terrestrial *Carex* stand, 20 July 1970, 7 cm					
Bore hole 1	7.8	11.8	0.75	5.72	-
Bore hole 2	6.6	14.0	0.24	4.22	-
Ikroavik Lake, interstitial water, 31 May 1971	53.4	161	3.96	19.8	13.8

*Data are expressed as mg liter^{-1}.

The Pond J interstitial water is enriched relative to either the water column above or the nearby terrestrial pore waters in all cations but potassium (Table 4-4). The high concentrations of ions in the interstitial waters from Ikroavik Lake are due to ion exclusion from the 2 m of ice formed in winter. The smaller enrichment in cations in Pond J relative to the pondwater above may also be due to cation exclusion from ice during fall freeze-up, but the values for all ions fall within the range of pore water for terrestrial Barrow area cores in general (Gersper et al. in press). The difference in cation concentration between the water column and the interstitial water is much greater than that found in lakes such as Lake Ontario (Weiler 1973) which have no mechanism for forcing ions into sediment.

The Barrow pond sediments are best described as iron-rich peats. Iron concentrations for two cores taken from Ponds B, C, and D in 1970 (Table 4-3) are 20% to 40% higher than those found in interior Alaska subarctic lakes such as Smith (Alexander and Barsdate 1971), Birch, and Harding Lakes (Barsdate and Matson 1966); but they still fall within the range reported both for Linsley Pond (Hutchinson and Wollack 1940) and for 14 soft water Wisconsin lakes (Williams et al. 1971a). Additional iron determinations discussed later indicate that surficial sediments of some ponds can contain much higher quantities, at least up to 133,000 ppm total, than reported for either the above subarctic or temperate lakes. Manganese concentrations in the Barrow ponds are low in comparison to those in the above mentioned lakes, and are remarkable in light of the high correlation usually expected between iron and manganese concentrations (Williams et al. 1971c). This may be due both to differential humic mobilization of iron vs. manganese and to a lack of secondary mineral formation in Barrow pond sediments. Both calcium and magnesium concentrations are 2- to 4-fold higher than values reported for Linsley Pond, but are well within the range normally found in noncalcareous sediments.

TRACE METALS

Total dissolved copper, lead, and zinc in Pond B in 1971 and Ponds B and C in 1972 were analyzed by anodic stripping voltammetry. Pond B in 1972 had aberrantly high trace metal concentrations, apparently caused by contamination from equipment buried close to the pond; therefore, only Pond B 1971 and Pond C 1972 data were used to construct averages.

Copper averaged 1.0 μg liter^{-1} (6 analyses), lead 0.7 μg liter^{-1} (5 analyses) and zinc 4.9 μg liter^{-1} (one analysis). The 1972 Pond B samples averaged 2- to 3-fold higher than these, with copper concentrations reaching 10 μg liter^{-1} in early August (Figure 4-6).

The trace metal concentrations in the uncontaminated ponds are very similar to those of other northern Alaskan lake waters summarized in Hobbie (1973). Kalff (1968) however has reported much higher amounts of trace metals in his Barrow Pond III, including 14 μg Cu and 55 μg Zn liter^{-1}. Brown et al. (1962) have analyzed trace metals in a large number of streams and lakes throughout northern Alaska, but the concentration procedure they used precludes comparisons outside their own data. The ponds and northern Alaskan waters in general are low in trace metals compared to most temperate fresh waters: Hutchinson (1957) quotes an average of 26 μg Pb and 62 μg Zn liter^{-1} for North American waters and

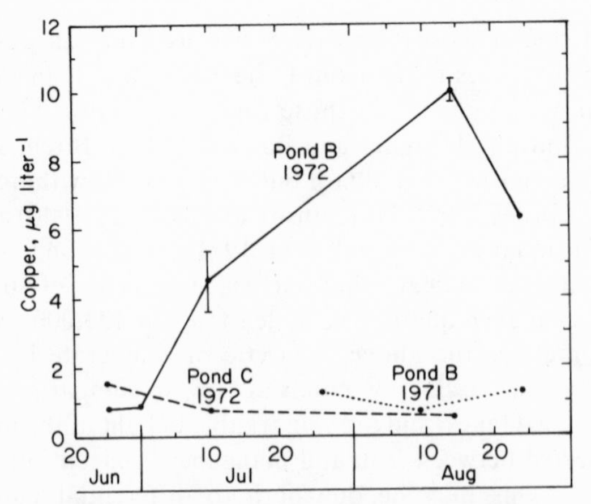

FIGURE 4-6. *Concentration of total dissolved copper in Ponds B and C, 1971–1972.*

an average of 29 μg Cu liter^{-1} (ionic plus some organic) for 136 lakes in northeastern Wisconsin. The ponds appear to be closer to the average seawater composition of 10 μg Zn, 3 μg Cu and 0.03 μg Pb liter^{-1} given by Goldberg (1965).

The copper and zinc concentrations in the pond sediments (Table 4-3) fall within the range reported for the lakes mentioned above.

CARBON DIOXIDE SYSTEM *

Alkalinity and pH

The pH and alkalinity of a number of ponds, subponds and of Ikroavik Lake were measured weekly in 1970 and 1971 in order to calculate the available carbon for photosynthesis studies.

The alkalinity of the ponds was low and generally stabilized by midsummer at 0.35 to 0.45 meq liter^{-1} (Figure 4-7a). As expected, the meltwater of the early weeks contained very low amounts of alkalinity; as the pond became more concentrated and as the equilibria with the sediments became established, the quantity rose. The next summer (Figure 4-7b) was wetter (see Chapter 3) and the alkalinity stabilized at a slightly lower level (0.32 to 0.36 meq liter^{-1}).

The pH of the ponds followed the trends of the alkalinity (Figure 4-7c) and was low early in the summer but then rose to a plateau by 1 July and remained between 7.05 and 7.45 (mean of 7.18). The next summer the pH plateaued between 7.24 and 7.62 (mean 7.37).

Because the waters are poorly buffered, even the low rates of respiration and photosynthesis in these ponds produce measurable changes of pH and alkalinity during a single day. For example, in Pond C on 23 to 24 July, 1971, the alkalinity changed by 0.02 meq liter^{-1} and the pH by 0.5. For this reason, it is important to take chemical samples at the same time each day.

Total Inorganic Carbon

The measurement of the total inorganic carbon was a rapid way to examine changes in CO_2 due to respiration and also to measure the inorganic carbon available for photosynthesis. In 1971, this total was measured weekly in four ponds and in Ikroavik Lake. It was measured twice weekly in Pond B in 1972 and weekly in three other ponds. In 1973, there were 40 measurements in Pond B and 16 in Pond C.

*J. Kelley and P. Coyne

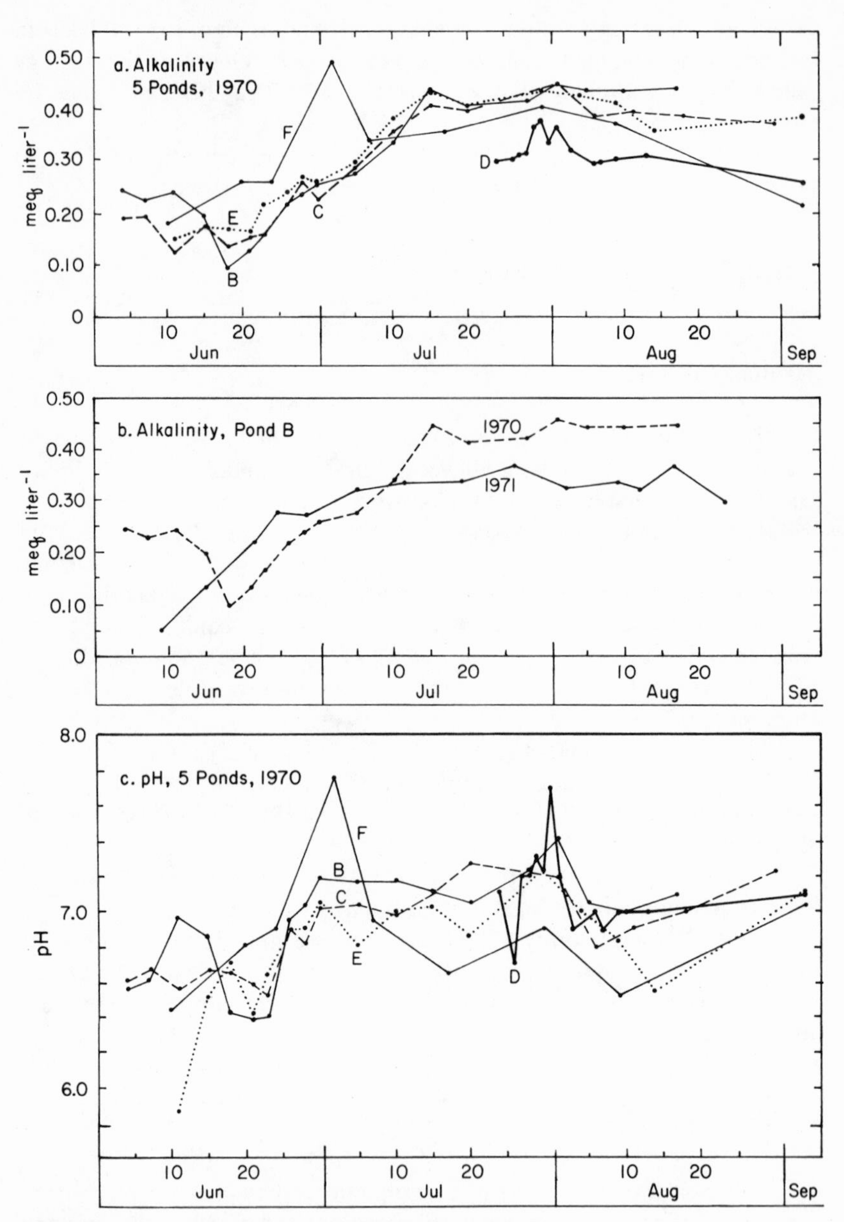

FIGURE 4-7. *Alkalinity and pH in five ponds.*

Total CO_2 was estimated in two ways: by calculation and after a conversion to gaseous CO_2. The calculation was made from pH and alkalinity data and from the equations and apparent dissociation coefficients given by Skirrow (1965). In the conversion method, a small volume of water is injected into a weak acid solution and the evolved CO_2 is measured with an infra-red gas analyzer (IRGA).

The two methods of measurement gave very similar values for the ponds (Table 4-5). Based on a group comparison "t" test, however, the values were significantly different in two out of five ponds. This was likely caused by the inclusion of organic acids in the titration value (Salonen and Kotimaa 1975).

The total inorganic carbon, as measured by the IRGA method, ranged from 1.43 to 8.08 mg C liter^{-1} in Pond B with averages of 4.10, 5.01, and 2.73 in 1971, 1972, and 1973, respectively (Figure 4-8). In general, the amounts were lowest after the spring melt, reached equilibrium values about 1 July, and then changed slightly (usually an increase) in August due to concentration of the ions as the pond volume decreased.

As mentioned above, there were daily changes in pH and alkalinity and these are reflected in similar changes in the total inorganic carbon. During the 23 to 24 July measurements, the change by the IRGA method was 0.63 mg C liter^{-1} and the change by the calculation method was 0.86. Both methods showed the lowest concentrations during the late afternoon and early evening and the highest concentrations during the night and early morning.

The concentration of inorganic carbon in these ponds is quite low. For example, the values for inorganic carbon in both north German soft water lakes (Hutchinson 1957) and for the average river water (Livingstone 1963b) are 3 times the Pond B average while the seawater has about 7 times this amount (Stumm and Morgan 1970). Lawrence Lake (Wetzel 1975), an alkaline marl lake, contains 10 times this amount or around 40 mg C liter^{-1}, while Mirror Lake in New Hampshire has around 1 mg C liter^{-1} (Jordan and Likens 1975).

TABLE 4-5 *Mean and Standard Deviation of Total Inorganic Carbon in 1971 as Determined by Two Methods**

Pond	Infrared gas analyzer	Calculated from pH and alkalinity	Group comparison 't' test $t_{observed}$ (df)	$t_{critical}$ 5% level	
B	4.10 ± 0.75	4.02 ± 0.90	$1.524_{(6)}$	2.447	NS
C	4.46 ± 0.66	4.63 ± 0.63	$1.250_{(7)}$	2.365	NS
D	2.80 ± 0.55	3.053± 0.93	$4.148_{(7)}$	2.365	Sig.
E	4.66 ± 0.74	5.06 ± 2.12	$1.742_{(7)}$	2.365	NS
F	–	5.00 ± 1.96			
J	3.01 ± 0.31	3.76 ± 0.98	n.a		
X	2.34 ± 0.60	2.99 ± 0.34	$2.278_{(7)}$	2.365	Sig.

*Data are expressed as mg C liter^{-1}

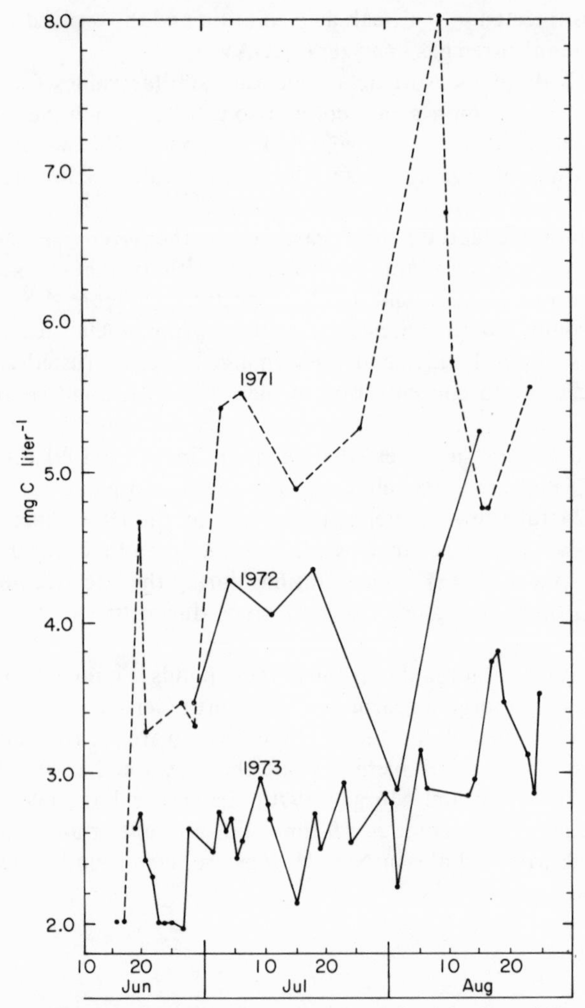

FIGURE 4-8. *Total inorganic carbon in Pond B, 1971–1973.*

Partial Pressure of CO_2

During 1971, continuous measurements of the partial pressure of CO_2 in the water (pCO_2) were carried out in Pond C from 9 July to 16 August. The next year, continuous measurements were made from 25 June until 15 September of the pCO_2 in North Meadow Lake, a shallow (70 cm) lake 2 km northeast of the ponds.

The pCO_2 was determined with an IRGA by measuring the CO_2 concentration of air in equilibrium with the water. Details are given in

Coyne and Kelley (1974). Air was circulated in a closed system from a floating cuvette to the IRGA in a hut and then eventually bubbled back to the cuvette via a fritted glass disc located in the water just beneath the cuvette. The dimensions of the cuvette were 30 × 30 × 23 cm. The CO_2 in the ambient air was also measured periodically. Both measures are reported in ppm by volume on a dry air basis. Both water and air measurements were compared with reference gases each hour.

Accuracy of CO_2 measurements in the concentration range of air (320 ppm) was approximately ±2 ppm. Pond pCO_2 values ranged from 450 to over 1,500 ppm. Accuracy of the measurements decreased in proportion to the increase in concentration range and was possibly no better than ±5% at the upper end of the range.

The pCO_2 values in the ponds are always high, yet the actual quantity of CO_2 gas dissolved in the water is quite low. This is a result of the low alkalinity and the slightly basic or neutral pH values in the ponds. For example, in early July 1971 the total CO_2-C was 4.45 mg liter^{-1} as calculated from alkalinity and pH (Figure 4-8). Most of the C was in the form of HCO_3. If the CO_2 gas were in equilibrium with the atmosphere, that is at 100% saturation, then the pCO_2 would be 320 ppm but the

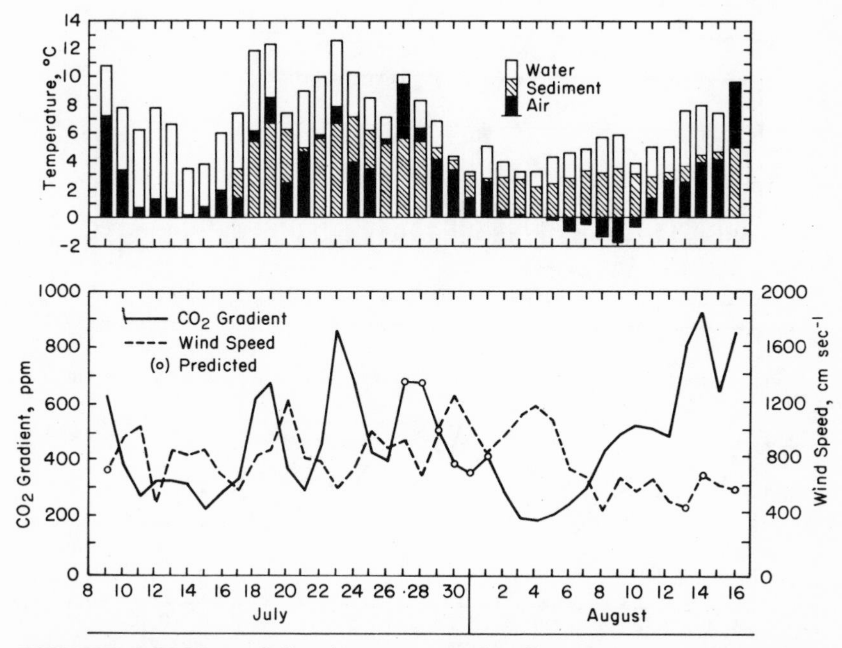

FIGURE 4-9. *Mean daily air, water, and sediment temperatures, mean wind speed, and mean gradient of pCO_2 (water minus air) in Pond C, 1971. Circles are predicted data from regression analysis. (After Coyne and Kelley 1974.)*

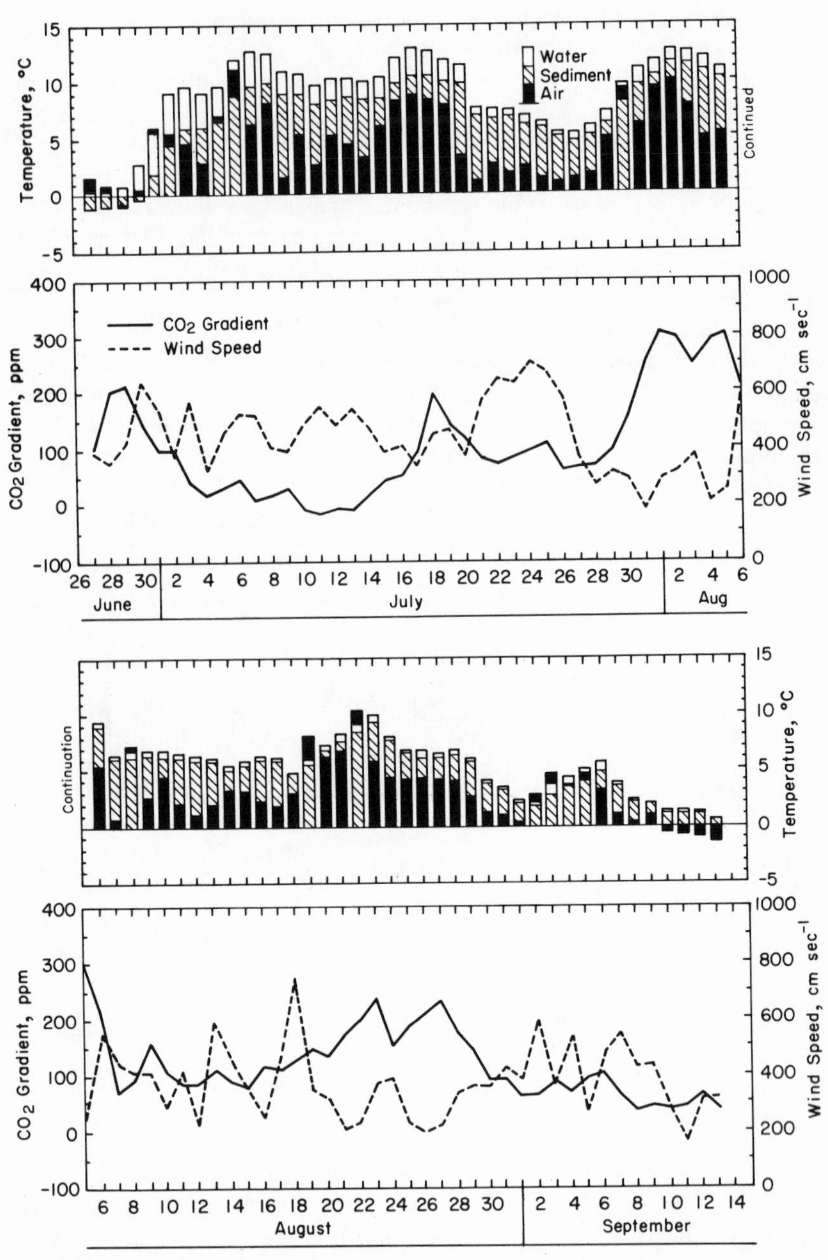

FIGURE 4-10. *Mean daily air, water, and sediment temperatures, mean wind speed, and mean gradient of pCO₂ (water minus air) in North Meadow Lake, 1972. (After Coyne and Kelley, 1974.)*

dissolved CO_2 would be only 0.37 mg C liter^{-1}. A pCO_2 of 640 ppm equals 0.74 mg CO_2-C liter^{-1} in this pond.

The data are expressed as the gradient of the partial pressures of CO_2 between the water and the air (Figures 4-9, 4-10). Because the pCO_2 in air was constant at 320 ppm, the actual pCO_2 in the water is equal to the gradient plus 320.

The water in the pond was supersaturated with CO_2 with respect to air throughout the period of sampling. This implies that there is a transfer of CO_2 from the water to the air. There must be an excess of respiration over photosynthesis in the ponds; this respiration is mostly taking place in the sediments (Chapter 8). Another implication is that despite the constant turbulence of the water in the ponds, the transfer rate of CO_2 is relatively slow so that the pond's CO_2 is not in equilibrium with the atmosphere. Over the period of the pond measurements, approximately the middle half of the open water season, the gradient averaged 397 ppm (Table 4-6).

The water of the shallow lake was also supersaturated with CO_2 (Figure 4-10) but the mean gradient was only one-third that found in the ponds (Table 4-6). The record of CO_2 for the lake spans the entire ice-free season and includes meltwater on top of the bottom-fast ice in the spring and several spot measurements of CO_2 beneath the ice in the fall. In part

TABLE 4-6 *The pCO_2 and Related Factors in Pond C (1971) and North Meadow Lake (1972)*

Parameter	Pond	Lake
Seasonal mean CO_2 gradient (water minus air)	397 ± 185 ppm	115 ± 83 ppm
Maximum mean daily CO_2 gradient	923 ± 208 ppm	306 ± 47 ppm
Maximum instantaneous CO_2 gradient	1217 ppm	404 ppm
Maximum instantaneous pCO_2	1530 ppm	722 ppm
Minimum instantaneous CO_2 gradient	136 ppm	-32 ppm
Seasonal mean temperature Water	2.5±3.4°C	4.0±3.5°C
Seasonal mean temperature Sediment	7.1 ±3.5°C	7.0±3.1°C
Seasonal mean wind speed	399 ± 139 cm sec^{-1}	409 ± 174 cm sec^{-1}

TABLE 4-7 *Third Order Partial Correlation Coefficient Matrix for Pond C and North Meadow Lake*

	Temp (°C)				
	Air	Water	Sediment	Wind speed	CO_2 gradient
POND					
Air temp.	1	0.72*	−0.16†	0.32†	0.54†
Water temp.		1	0.45†	−0.22†	−0.28†
Sediment temp.			1	0.31†	0.38†
Wind speed				1	−0.56†
CO_2 gradient					1
LAKE					
Air temp.	1	0.47†	−0.28†	−0.06*	0.29†
Water temp.		1	0.95†	0.10†	−0.29†
Sediment temp.			1	−0.09†	0.31†
Wind speed				1	−0.21†
CO_2 gradient					1

Source: Coyne and Kelley 1974
*Significant (0.05 probability level)
†Highly significant (0.01 probability level)

the lower CO_2 gradient in the lake reflects its greater depth than the ponds (70 vs. 40 cm). Mainly, however, it reflects the lower rate of respiration in the lake than in the ponds. We attribute this to the small amount of organic matter per square meter that enters the lake from the rooted aquatics along the shore. The ponds, in contrast, have a smaller area and so the rooted plants contribute a significant amount of organic matter per square meter.

There was a short period of undersaturation of CO_2 in the lake (10-13 July, Figure 4-10). There is no obvious physical cause of this undersaturation so it could have been an algal bloom in the water or sediments. Unfortunately, no additional measurements were taken.

The factors that affect the pCO_2 are respiration, water temperature, and wind speed. Respiration should increase with increasing temperatures (a Q_{10} of around 2) and thus increase the size of the CO_2 gradient. An increase in temperature when the water is already supersaturated with CO_2 would increase the tendency of the CO_2 to leave the water and therefore would appear as a high CO_2 partial pressure in the water. The rate of turbulent exchange across the water surface should increase with wind speed and thereby decrease the CO_2 gradient.

From the data on the pond and on the lake, separate multiple regression equations were calculated (Table 4-7). The third-order partial correlation coefficients show a direct relationship between sediment temperature and CO_2 gradient but indirect relationships between water temperature, wind speed, and CO_2 gradient. The regression explained only

58% of the variability in the CO_2 gradient. The direct relationship between sediment temperature and the CO_2 gradient is likely caused by the increased respiration with increased temperature. The negative correlation of water temperature with gradient might be a result of increased photosynthesis in the water when the algae are more important than the bacteria. As expected, the gradient was proportional to the wind speed. The theory of gaseous exchange between atmosphere and water has been described by Kanwisher (1963); one important control of exchange was the thickness of a thin, stagnant layer of water at the surface. Increased wind decreases the thickness of this layer and increases the rate of exchange. This inverse relationship is obvious in Figure 4-9 for the pond and also present, but less obvious, in the data for the lake (Figure 4-10).

There are only a few bits of information on pCO_2 in freshwaters. For example, Park et al. (1969) found that two rivers were supersaturated in CO_2 with gradient values up to 340 ppm. There are more data for oceans but oceans are strongly buffered against CO_2 fluctuations by high alkalinity of about 2.3 meq liter^{-1} (vs. 0.4 in the ponds). Thus, the same amount of CO_2 added to seawater produces only about 1/13 the change in pCO_2 that occurs in distilled water. Teal and Kanwisher (1966) found that the small changes in pCO_2 that result from the biological activity in the oceans persisted for long periods. Coyne and Kelley (1974) concluded that the changes in pCO_2 in freshwater are so rapid that the pCO_2 must be intensively monitored if biological activity is to be followed in this way.

The lake began to freeze on 11 September 1972 and a permanent ice cover formed on 27 September. Ice thickness was 30 cm by 26 October and the lake was completely frozen by 4 January 1973. The values of pCO_2 were 800 ppm on 28 September, 2,000 on 12 October and over 3,000 ppm on 22 October. The limit of the instrument was 3,000 ppm, so a pCO_2 value was calculated on 16 October from precise pH measurements and accurate CO_2 reference gases of 12,000 ppm (Kanamori personal communication). Thus, decomposition in this lake (and presumably in the ponds) continues until the water is frozen to the bottom.

CO_2 Evasion Rate

During 1972, more than 25 attempts were made in North Meadow Lake to assess the rate of transfer of CO_2 from the water to the air. Details of technique and calculation are given in Coyne and Kelley (1974). The technique was similar to that for measurement of pCO_2 except that the air returning to the floating cuvette (at 0.5 liter min^{-1}) entered above the water level rather than by bubbling through the surface layers. In an experiment, the instrument was set up so that air circulated through the cuvette. Next, the IRGA was set to zero, the cuvette placed in the water, and the time required for the CO_2 to come to equilibrium was measured.

Because periods of up to 6 hours were required to attain CO_2 equilibrium, some of the series of measurements could not be used as the pCO_2 in the water changed so drastically over a short period. From 17 experiments between 19 July and 6 September a mean evasion coefficient of 0.34 mg $cm^{-2} atm^{-1} min^{-1}$ was measured (standard deviation of 0.17). These coefficients were measured under a variety of temperature and wind conditions so we can make the assumption that the mean coefficient can be applied to the average gradient of 115 ppm (Table 4-6) to give an average rate of transfer from the lake of 0.56 g CO_2 m^{-2} day^{-1} (this is 0.15 g C). We also assume that the coefficient is similar for the pond. Based on an average gradient of 397 ppm, the rate of transfer from the pond was 1.95 g CO_2 m^{-2} day^{-1} or 0.53 g C.

There are no other values for CO_2 evasion rates in freshwater. Rates for seawater are widely variable (Riley and Skirrow 1975) probably because of the strong influence of turbulence which cannot be controlled from experiment to experiment. In our study, the water was generally turbulent during the rate experiments. The cuvette did prevent direct wind effect on the surface but did not attenuate the wave train to any degree. Accordingly, the measured evasion rates probably lie somewhere between those for diffusion and turbulent exchange rates. We imagine that the rates are conservative estimates, based on this information. However, as discussed in the sediment respiration section (Chapter 8), this value of 0.5 g C m^{-2} day^{-1} is about twice as high as the respiration of the sediment organisms measured in incubations of sediment cores.

Total Dissolved Inorganic Carbon

Bacterial respiration in the sediments as well as the solution of carbonate in the peat layer contribute to the total dissolved inorganic carbon (DIC) in the interstitial water. Actually, four processes are occurring. The first is the production of DIC by respiration of aerobic and anaerobic bacteria. Most of the production (77%) occurs in the top 4 cm (Chapter 8). The second process is the diffusion of the DIC through the sediment towards the water. The third is the rapid diffusion of DIC across the sediment/water interface. This is a rapid process, as the gradient is large due to the continual water circulation that renews the water film and does not allow a buildup of DIC. The fourth process is the solution of carbonates. While this occurs, we have no evidence that it is at all important. The concentrations resulting from these processes increase exponentially with depth (Table 4-2).

In the areas of the sediment where there are rooted aquatic plants the roots also will add CO_2 to the sediments. The rate of root respiration is poorly known but it may produce much of the CO_2 measured in the whole-pond transfer rates.

Some of the DIC is present as dissolved CO_2; this quantity will be frozen out of the ice and moved downwards as the sediment freezes. In one experiment, about 40% of the DIC was purged from the interstitial water within seconds, indicating that most of it was present as CO_2.

In the tundra soils, there is a large-scale release of CO_2 in the early fall and again in the spring thaw (Coyne and Kelley 1971). This is presumably caused by freezeout of CO_2 followed by movement of the gas upwards through soil cracks or along roots and stems. This would not happen in the ponds because there are few plants or cracks.

OXYGEN

Water Column

The concentration of oxygen was not measured routinely. Analyses were made of transects of Ponds C and E on 23 July 1970, one week after the oil spill on E, and on Pond B and experimental subponds on 23 August 1972. The azide modification of the Winkler method was used in 1970 (American Public Health Assoc. 1960) and a Yellow Springs Instrument probe in 1972.

Measurements in the middle of Ponds B and C gave mid-day concentrations of 10.6 and 11.4 mg O_2 liter^{-1} or 97% and 88% of saturation. In Pond C, the concentration decreased towards the shallow sides of the pond and the minimum value of 3.5 mg (31% of saturation) was reached in 2 cm of water within the *Carex* bed. This low concentration is partially a result of the reduced water circulation and partially a result of high sediment respiration in an area with only a small amount of oxygen in the water column above. Evidence for the importance of sediment respiration comes from an average ($n=2$) of 91% of oxygen saturation in two subponds with sediment bottoms and an average of 109% in two similar ponds with plastic bottoms. This is an indication that there is a net production of oxygen in the water column and a net utilization in the benthic community.

Most surface waters of the world, whether fresh or salt, are near saturation with respect to oxygen. The major exceptions appear to be those supersaturated waters with high photosynthesis-to-respiration ratios and undersaturated waters such as freshly upwelled oceanic waters and highly colored lakes (Hutchinson 1957). The oxygen concentrations in arctic tundra ponds are in agreement with the generalization that highly colored waters are oxygen-deficient. Thus, Kalff (1965) found that the oxygen concentration in his Barrow ponds ranged from 60% to 118% of saturation, but normally fell below 100%. Reed (1962) found that ponds along the Colville River area usually ranged between 60% and 70% of saturation, and Watson et al. (1966a) always found undersaturation in Cape Thompson ponds. Both Hutchinson (1957) and Kalff (1965) suggest that chemical oxidation likely maintains oxygen undersaturation in highly colored waters. Our data (see Chapter 8) suggest that high rates of benthic respiration rather than chemical oxidation is the dominant force causing undersaturation in the Barrow ponds and, by analogy, in other shallow ponds.

The summer oxygen regime of shallow arctic lakes is similar to that of oligotrophic temperate lakes, but the winter regime is strongly modified by the extreme arctic winter. In Ikroavik Lake, exclusion of oxygen during formation of 1.6 m of ice in a water column only 2.4 m deep was the major process, producing a 20% to 40% supersaturation by early December and a 74% supersaturation in early June (Barsdate et al. cited in Hobbie 1973). The opposite effect, severe deoxygenation, can also occur in shallow arctic lakes underneath ice; Lake 5 at Cape Thompson, for which zero oxygen was recorded in April 1961 by Tash and Armitage (1967), is a good example. Presumably, the degree of snow cover and its effect on photosynthesis, the water depth, and the respiratory rate of each individual lake dictate its position between these two extreme winter oxygen regimes.

Deep arctic lakes have an oxygen regime that is similar in almost every way to that of temperate oligotrophic lakes. They are saturated with oxygen throughout the open water season except when the rates of cooling are so rapid that the oxygen influx, and thus the percentage of saturation, cannot keep up (Hobbie 1962). In some cases, the long duration of the ice cover allows the decrease of oxygen caused by under-ice respiration to be quantified (Welch 1974).

Sediment pH and Eh

Eh and pH are factors of importance in sediment chemistry, particularly as they influence the iron-phosphorus relationships discussed later in this chapter. Three Eh and pH sediment profiles in Pond J on 11 July 1971 were measured *in situ* with a combination pH and reference probe modified for immersion and a platinum wire redox electrode (Fenchel 1969). Calibration and calculations for E_7, the potential that would be observed at pH 7 and 25°C, were made according to Golterman (1969). On 19 July 1973, nine additional redox profiles were measured with a multi-electrode probe similar to that of Machan and Ott (1972).

The sharp negative changes in pH seen in the 1971 profile (Figure 4-11) at the interface between oxygenated and anoxic sediments may occur as the result of oxidation of ferrous iron with the formation of α FeOOH (geothite): $2\ Fe^{++} + 1/2\ O_2 + 3\ H_2O \rightarrow 2\ FeOOH + 4\ H^+$. The oxidation potentials indicated that sediments were oxidized only to a depth of 1 to 2 cm in the deep water location but to considerably greater depths

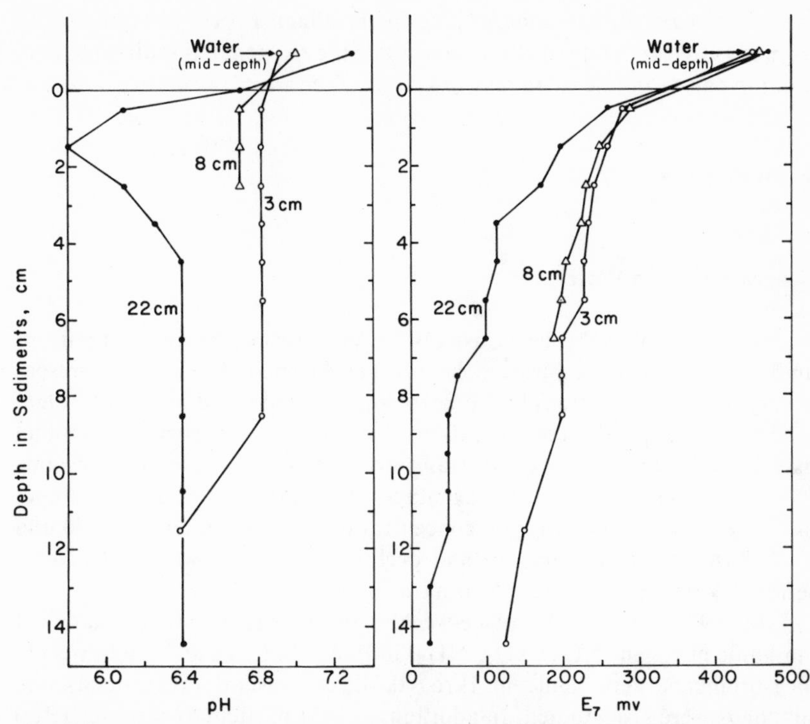

FIGURE 4-11. *Oxidation potential and pH profiles in the water and sediments of Pond J, 11 July 1971. The profiles were taken in a stand of* Carex *(3-cm- and 8-cm-deep water) and in an area devoid of macrophytes (22 cm).*

at the shallower *Carex*-covered sites. Redox potentials of the pond central basin sediments and of the deeper sediments (below the root zone) of the plant beds fall within the upper part of the Eh range of lake sediments reported in Hutchinson (1957). Measurements in 1973 substantiated these observations, and indicated that the higher oxidation potentials in the plant beds were due to the presence of *Carex,* rather than due to the water depth. Vascular plants have passages in their stems that allow oxygen to reach the roots; this facilitates the uptake and transport of phosphate and other salts (Loughman 1968, Armstrong 1964). Leakage of oxygen is evidently significant from the *Carex* in this pond, as there are relatively high oxidation potentials in sediments under stands of pond vascular vegetation in ᒻ ⁴' Pond J (1971) and Pond C (1973). In other *Carex* species the leakage of oxygen stops the transport of ferrous iron into roots. Without this mechanism, the ferric iron would precipitate within the roots and immobilize the plant's phosphorus (Jones 1975).

Relatively high oxygen potentials have also been reported in beds of *Lobelia* in Denmark (Wium Andersen and Andersen 1972) and in beds of

Isoetes lacustris in a mountain lake in the Black Forest (Tessenow and Baynes 1975). In some of these situations, the oxygenated sediments may result in part from the extremely low rates of sediment respiration.

NITROGEN *

Concentrations in Water

Although phosphorus is usually thought to be the most important limiting nutrient in freshwater, the nitrogen compounds were also present in very low concentrations in the Barrow ponds and could be limiting algal production. Also, the concentration of a particular nitrogen compound may interact with the specific transport systems of algae to allow one species to have a competitive advantage over another. For this reason, an intensive study of the nitrogen concentrations was made during 1970 and 1971. In addition, the uptake and cycling of nitrogen was studied in a series of experiments using ^{15}N as a tracer.

In 1970 and 1971, twice-weekly measurements were made of inorganic nitrogen (NO_2, NO_3, NH_3) in Ponds B, C, D, and E and weekly measurements were made in Ikroavik Lake. Several other ponds and subponds were also studied. In addition, in 1971 particulate nitrogen (PN) and dissolved organic nitrogen (DON) were measured six times in B, C, and D.

The nitrate and nitrite analyses were carried out in 1970 and 1972 with the copper-cadmium column technique (Strickland and Parsons 1965). In 1971, a sodium phenate reduction modified for high levels of dissolved organic substances was used (Barsdate and Alexander 1971). Ammonia analyses followed Solórzano (1969) with a sodium hypochlorite and phenol treatment and measurement of the blue indophenol. Particulate nitrogen was measured on a Coleman Automated Nitrogen Analyzer and dissolved organic nitrogen after oxidation with a strong UV lamp.

Ammonia is generally the predominant inorganic nitrogen form in the ponds; concentrations reached 85 μg N liter^{-1} on occasion and never fell below 9.5 μg N. Monthly averages for 1970 and 1971 are given in Table 4-8 while the actual data for Pond B, 1971, are given in Figure 4-12. Nitrate levels were often below the limit of detection, although during the summer of 1971 the concentrations remained higher than in 1970. Seasonally, nitrate levels tend to be maximal early in the season and to decline as the season progresses; ammonia shows two maxima, one early in the spring as soon as the ice melts (preceding the nitrate maximum) and

*V. Alexander

TABLE 4-8 *Average Concentrations of Various Forms of Nitrogen in the Water of Barrow Ponds**

Pond	1970			1971		
	June	July	August	June	July	August
Nitrate						
B	7.5	0.9	1.3	25.0	33.0	1.0
C	10.5	3.7	0.7	33.0	30.0	9.0
J				40.0	88.0	19.0
Nitrite						
B	0.9	1.1	0.6	0.3	0.2	0.2
C	0.8	0.9	1.0	0.4	0.2	0.2
J				0.2	0.4	0.1
Ammonia						
B	48.0	31.0	22.0	27.0	25.0	23.0
C	43.0	31.0	20.0	22.0	28.0	30.0
J				22.0	39.0	77.0
Dissolved organic nitrogen						
B					920	870
C				920	1040	570
J					760	
Particulate nitrogen						
B				108	36	39
C				59	60	59
J				55	61	64

*Data expressed as μg N liter^{-1}

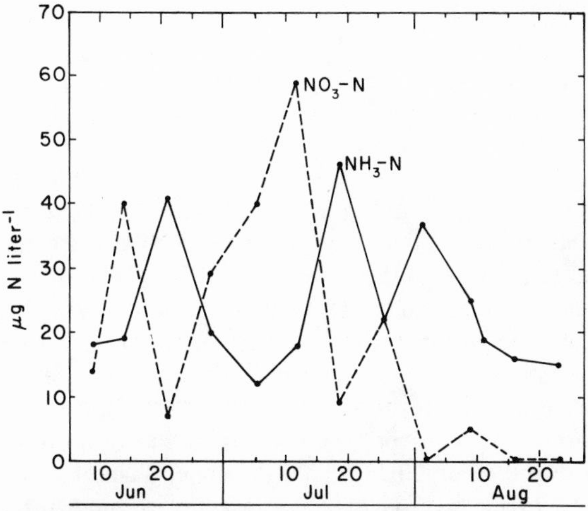

FIGURE 4-12. *Concentrations of nitrate and ammonia in Pond B, 1971.*

a second in late July and early August. There was an additional maximum of ammonia in the melt water from the snow pack. In 1971, the NH_3-N was above 100 μg liter^{-1} from 2 to 5 June and reached 204 μg liter^{-1} on 5 June. It is likely that the NH_3 in the snow is easily mobilized by the first water that begins to move through the melting snow (Barsdate and Alexander 1971). The nitrite concentrations are insignificant at all times.

Most of the nitrogen in Ponds B and C was present as organic nitrogen. In fact, taking mean values for 1970 and 1971, the dissolved organic N (DON) was 89% of the total and the particulate organic N (PON) was 6.8%, about 2 times the concentrations of the dissolved inorganic N. The range of concentration of DON was 800 to 1,400 μg liter^{-1} while the PON levels ranged from 30.7 to 159.9 μg liter^{-1}.

There have been only a few measurements of nitrogen compounds in other arctic waters and these are difficult to compare with the ponds because of the usual build-up of nitrate and ammonia beneath the ice of lakes that do not freeze to the bottom. For example, nearby Ikroavik Lake contained 1,625 μg NH_3-N liter^{-1} and 8.3 μg NO_3-N liter^{-1} on 14 June (1970) when the ice still covered the lake. By August, when the lake was ice-free, the surface water contained only 14 μg NH_3-N and 0.0 μg NO_3-N liter^{-1} which is similar to the ponds at that time of year. Char Lake, in the Canadian Arctic, contained less than 1 to 30 μg NO_3-N and less than 2 μg NH_3-N liter^{-1} (Schindler et al. 1974). Particulate nitrogen was higher (mean of 100 μg PON liter^{-1} samples) in Ikroavik Lake than in the ponds and Char Lake had 7 to 27 μg PON liter^{-1}. Dissolved organic nitrogen in Ikroavik Lake was similar in concentration (996 μg N liter^{-1} average for 14 samples) to the ponds while values in Char Lake were much lower. Schindler et al. (1974) found an average total dissolved N of 75 μg liter^{-1} so DON is likely around 60 μg.

Sediments

Pond sediments typically contain high amounts of dissolved organic nitrogen (DON) and ammonia but low concentrations of nitrate. For example, in the 4 to 12 cm level of the sediments of Pond J (11 July 1971), the interstitial water contained 5.6 mg DON liter^{-1}. On 19 June 1971 the 1 to 5 cm level of the sediments contained 1.3 mg DON liter^{-1}. During the winter months, interstitial ammonia concentrations as high as 7 mg NH_3-N liter^{-1} were found in the surface 6 cm, but these high values are greatly reduced during the summer in the areas of the beds of rooted aquatic plants. For example, on 20 July 1973 the upper 8 cm of sediments contained only 14 to 75 μg NH_3-N liter^{-1} in the *Carex* and *Arctophila* beds but from 700 to 2730 μg N between the beds and in the pond center (Table 4-9). The quantities of NO_3-N liter^{-1} in the interstitial water

TABLE 4-9 *Concentration of Nitrogen Compounds in the Sediment of Pond J, 1973, for a Transect from the Edge to the Center of the Pond*

Date	Description	Water depth (cm)	Sediment depth (cm)	NH_3 (μg N liter^{-1})	$NO_3 + NO_2$ (μg N liter^{-1})
30 July	*Carex* bed	1	0-8	75	6
			8-16	53	6
20 July	*Carex* bed	5	0-8	63	1
			8-16	56	1
30 July	Between *Carex* and *Arctophila* beds	18	0-8	2660	5
			8-16	3290	4
20 July	*Arctophila* bed	19	0-16	14	4
30 July	Pond center	25	0-8	2730	11
			8-16	3080	5
21 July	N.E. central basin near sampling platform	21	0-8	700	4
			8-16	1880	3

followed a similar pattern except that the amounts were much lower (1 to 6 μg N in the plant beds and 4 to 11 μg N in the pond center). From this evidence, it appears that sediment nitrogen is being cycled by macrophyte uptake.

The exchangeable nitrogen is another potential source of nitrogen in the sediment. This pool is measured by extraction with 0.5 N HCl for 1 hr at 25°C and includes the relatively small amount of nitrogen in the pore water. In Pond C, the exchangeable nitrogen in the top 16 cm was more than 99% ammonia and measured 26 μg N (g dry wt)$^{-1}$ in the *Carex* bed and 45 μg N (g dry wt)$^{-1}$ in the center of the pond.

Nitrogen Fixation

One process by which nitrogen could be increased in the ponds is nitrogen fixation. This process was studied with the acetylene reduction method, an indirect technique which yields results closely correlated with nitrogen fixation (Stewart et al. 1967). The details of the method used here are reported in Schell and Alexander (1970); briefly, samples were taken and placed in a small tube, acetylene was added and the samples incubated in the light for 6 to 24 hours. After this, gas samples were taken and the amount of ethylene formed was determined in a gas chromatograph equipped with a flame-ionization detector. Some experiments carried out

TABLE 4-10 *Nitrogen Fixation Rates in the Barrow Tundra Pond Sediments, 1972**

	7/31	8/4	8/14	8/17	8/21	8/24
Pond B	-	7.1	14.3	-	25.4	-
Pond C	0.0	-	-	0.0	-	0.0
Pond D	0.0	-	-	0.0	-	0.0
Pond E	-	9.2	0.0	-	0.0	-

*Data are expressed as μg N m^{-2} hr^{-1}.

with ^{15}N proved that the method was indeed measuring fixation and that the correct conversion factors were being used.

Of the four ponds sampled for nitrogenase activity (Table 4-10), two of them, C and D, contained sediments that showed no activity at all during July and August, 1972. Pond E sediments had activity (0.99 μmole ethylene produced m^{-2} hr^{-1}) at one sampling but not at two others. Pond B, however, had activities of 0.76, 1.53, and 2.72 μmoles m^{-2} hr^{-1} on 4, 14, and 21 August, respectively. These values are low compared with terrestrial fixation rates of up to 50, but may add up to respectable amounts of nitrogen over the ice-free season. Thus, if we assume that 3 moles of acetylene are reduced for each mole of N_2 fixed and that daily rates can be calculated by multiplying the observed hourly rates by 20 (based on diurnal studies carried out in the summer of 1972), then the average fixation of 0.31 mg N m^{-2} day^{-1} is equal to 28 mg N m^{-2} yr^{-1}. Barsdate and Alexander (1975) calculated an average value for terrestrial tundra at Barrow of 48 mg N m^{-2} yr^{-1} so the pond value is reasonable (however, three other ponds showed virtually no fixation at all). However, it appears that this fixation in the pond is low in relation to the amount of nitrogen already present in the sediments. One measurement of the interstitial waters gave 2730 μg NH_3-N $liter^{-1}$ and 11 μg NO_3-N in the upper 8 cm of a pond sediment (Table 4-9). This is 168 mg N m^{-2} for the top 8 cm alone (concentrations were even higher in the 8 to 13 cm layer) so it appears that the inorganic nitrogen is abundant and the fixation relatively unimportant in providing N for algal and plant growth.

Denitrification

One possible reason that the sediments contained low amounts of nitrate is that denitrification occurs. This process was measured in Pond J by taking a core and thoroughly inoculating it with $K^{15}NO_3$. After incubation for 16 days in the original core-hole, the part of the core that had been injected was extruded into a core squeezer and the interstitial water was collected. Gas was stripped from the water (Goering and Dugdale 1966) and analyzed for ^{15}N on an AE1 MS-20 mass spectrometer. The two measurements gave 0.17 and 0.19 μg N liter^{-1} day^{-1} or an average of 32 μg N m^{-2} day^{-1}. Because of these low rates of denitrification, the mean mass ratio of 28:29 changed very little; there was a larger change in the mass ratios of 30:28 and so this ratio was used as the indicator of denitrification.

Hauck et al. (1958) have established that molecular nitrogen produced through denitrification has mass 28, 29, 30 distribution determined by the nitrogen source and that isotopic equilibration with the preexistent N_2 pool does not occur. With this, and the further assumption that the nitrogen gas produced came only from the interstitial water nitrate pool, the amount of nitrogen gas produced per liter of water was calculated: μg N_2 liter^{-1} day^{-1} = excess $^{15}N_2$ (at %)$\times N_2$ (μg liter^{-1})$\times$(100)(at % $^{15}NO_3$)$^{-1}$. The N_2 concentration in the interstitial water was not measured but was assumed to be 2.032×10^4 μg liter^{-1} (saturation value at 5°C, Weiss 1970). The initial atoms percent $^{15}NO_3$ used in the equation above was calculated from the amount of tracer originally added to the core, the nitrate concentrations, and the water content.

We did not include N_2O in our measurements. Under certain circumstances such oxides of nitrogen may be produced, and for most denitrifying organisms N_2O is a precursor of N_2 (Alexander 1971). However, Cady and Bartholomew (1960) found complete reduction of N_2O to nitrogen even in their experiment in acid soil, although previously N_2O reduction had been found strongly inhibited below pH 7 (Wijler and Delwiche 1954). Any error in our results due to production of N_2O would be in the direction of an underestimate, but for the reasons discussed above, we feel that such error is likely to be small.

Another possible source of error is the effect of the added nitrate on the process. Hart et al. (1965) and Clasby (personal communication) concluded that low levels of nitrate do not limit the rate of denitrification as long as measurable nitrate is present so we believe that the added $^{15}NO_3$ had no effect (levels were kept low, however). In fact, rather than the level of nitrate limiting the rate, it is more likely that the denitrification process is limited by other nutrients. For example, when tundra soils were tested with added glucose and phosphorus, the denitrification rate rose 4-fold.

Because the water above the sediment always contains abundant oxygen, denitrification never occurs except in the sediments. This is not true of shallow subarctic lakes where anoxic conditions occur beneath the ice in late winter. Goering and Dugdale (1967) measured rates of denitrification as high as 15 μg N liter^{-1} day^{-1} in the water column and even higher rates in the presence of lake sediment.

TABLE 4-11 *Concentration of Inorganic Nitrogen Compounds in Summer Precipitation at Barrow, Alaska*

Date		NH_3	NO_3	NO_2
Rain (Dugdale and Toetz 1961)				
5-6 Aug 60		553	22.2	-
11-12 Aug 60		69	7.2	-
	Mean	311	14.7	
Mixed rain-snow (Kalff 1965)				
1 Aug 64		-	40	-
Rain				
9 Jul 71		269	36	-
14 Jul 71		235	16	0.8
30 Jul 71		147	29	1.1
	Mean	217	27	1.0
	Overall Mean	255	25	1.0

*Data are expressed as μg N liter^{-1}.

Budget

The additions and losses of nitrogen to a pond are the sum of nitrogen fixation, nitrogen added in summer rainfall, nitrogen added in spring runoff, and nitrogen lost in denitrification.

The inorganic nitrogen in the summer precipitation at Barrow has been measured by three different projects (Table 4-11). We have also made several measurements of DON that average 60 μg N liter^{-1}. During the summer of 1971, the precipitation was 4.1 cm. Taking the overall means from Table 4-11 and the DON value, the total input in rain of dissolved inorganic nitrogen (or DIN) was 11.5 mg N m^{-2} yr^{-1}, while the input of DON was 2.46 mg N m^{-2} yr^{-1}. The ammonia concentrations, which make up 91% of the total DIN input in rain, are similar to those reported by Junge (1958) for various locations within the U.S. In contrast, precipitation at the Hubbard Brook Experimental Forest, New Hampshire, contained nitrate as the most abundant DIN form (76-87% of the total DIN)(Fisher et al. 1968). The nitrate in rain was 50 times higher at Hubbard Brook than at Barrow and the rainfall was 10 times higher. The net result was that the input at Hubbard Brook was 880 mg N in 1964-65 and 2090 mg N m^{-2} in 1965-66.

Another possible source of nitrogen is from seepage of water from the pond drainage basin. This is potentially important as there is a high concentration of DON in the tundra soil. However, as discussed in Chapter 3, the drop in water level matches the evaporative water loss fairly well and other experiments indicate that there is very little percolation or movement of water through the soil into the ponds. Accordingly, we have decided that this input is not important.

The final input to be considered is from the meltwater that enters the pond in mid-June. As discussed in Chapter 3, the ponds retain some of the meltwater, as they normally are below capacity at the beginning of the melt season. This retained water amounted to an average of 5.8 cm (the difference between the fall low water level at the end of 1970 and the spring peak in 1971). The measured nitrogen quantities were as follows: ice in the ponds contained 33 μg DON liter^{-1}; snowmelt water contained 37 μg DIN liter^{-1}; and runoff water, after contact with the soil, contained 16 μg DIN and 339 μg DON liter^{-1} on 10 to 15 June 1973. Since the water flushed out of the ponds had about the same concentrations of nitrogen as the water entering the ponds, the DIN and DON added from the meltwater was only 1.7 mg and 39 mg m^{-2} yr^{-1} respectively.

The loss of nitrogen in the runoff (10 to 27 June 1972) was 1.53 mg DIN and 37.9 mg DON m^{-2}. The average flow-weighted concentration of DIN was 14.4 μg and of DON was 357 μg liter^{-1}, which agree with the initial pond concentrations in Figure 4-12.

A budget of measured and estimated influxes and losses from Pond B shows that most of the DIN comes in from rainfall and most of the DON from spring runoff (Table 4-12). From these data, it would appear that the pond accumulated nitrogen each year but the 61 mg total N m^{-2} estimate is very small when compared with the 219,000 mg N m^{-2} in the top 10 cm of sediment. Char Lake, the only arctic lake studied in detail, accumulated 87 mg N m^{-2} (de March 1978) while temperate lakes may accumulate 400 to 1300 mg N m^{-2} (Likens and Loucks 1978.)

Uptake of Inorganic Nitrogen

Ammonia and nitrate uptake rates were measured with the methods of Dugdale and Dugdale (1965). In this technique, ^{15}N labeled nitrate and ammonia were added to samples, the particulate material recovered on a

TABLE 4-12 *Nitrogen Budget for Pond B*

	DIN mg N m^{-2}yr^{-1}	DON mg N m^{-2}yr^{-1}
Nitrogen fixation in sediments		+ 28.0
Summer rainfall	+ 11.5	+ 2.5
Runoff, spring (amount retained)	+ 0.9	+ 20.6
Denitrification	- 2.8	-
Net input	+ 9.6	+ 51.1

filter after incubation *in situ*, and the nitrogen converted to a gas for mass spectrometer isotope ratio determination.

Ammonia was the preferred nitrogen source for the plankton (Table 4-13), with a mean 1971 uptake rate of 0.157 μg N liter^{-1} hr^{-1} and a maximum of 0.41 μg N liter^{-1}. At these rates, the ammonia available in the water seems adequate for the phytoplankton production; assuming steady state conditions and a mean ammonia concentration of 21 μg N liter^{-1}, the turnover time is 150 hours. The mean nitrate uptake rate was 0.01 μg liter^{-1} hr^{-1} which implies that the turnover time was thousands of hours (the mean concentration was 13 μg N liter^{-1}).

The uptake of these nutrients in Ikroavik Lake was 2 to 3 times higher than in the ponds (a mean of 0.48 μg NH_3-N and of 0.01 μg NO_3-N liter^{-1} hr^{-1}). Compared with other systems, the uptake rates in the ponds are quite low. For example, the northwest Atlantic data of Dugdale and Goering (1967) had ranges of NO_3 uptake of 0.002 to 5.6 μg N liter^{-1} hr^{-1} while the NH_3-N uptake rates ranged from 0.14 to 2.3. In Sanctuary Lake, Pennsylvania, peak ammonia uptake rates as high as 4.1 μg N liter^{-1} hr^{-1} were measured while peak nitrate rates were 0.40 μg N liter^{-1} hr^{-1} (Dugdale and Dugdale 1965).

Inorganic nitrogen is also taken up by algae on the surface of the sediment and by the roots of the sedge and grass growing in the pond. The primary productivity of the sediment algae and rooted plants is almost 2 orders of magnitude above that of the plankton algae and their uptake of nitrogen is proportionately higher.

The rate of uptake of nitrogen may be calculated from the primary productivity because the ratio of C:N is approximately constant. In the phytoplankton of Pond B, for example, the annual primary production is about 1 g C m^{-2}. This is equal to 120 mg N m^{-2} taken up or 1.2 mg m^{-2} day^{-1} for a 100-day growing season. If the average depth of the pond is 30 cm, then this is 300 liters m^{-2} of pond or an uptake of 4 μg N liter^{-1} day^{-1} or 0.17 μg N liter^{-1} hr^{-1}. This is very close to the average uptake measured with ^{15}N of 0.156 (Table 4-13). Thus, there is agreement between the ^{15}N uptake data and the N uptake calculated from the ^{14}C productivity.

If we assume that benthic algae behave the same way as plankton algae with respect to nitrogen uptake, then the primary productivity of about 10 g C m^{-2} yr^{-1} is equal to an uptake of 1.2 g N m^{-2} yr^{-1} or 12 mg N m^{-2} day^{-1}. The large plants (*Carex*) have a total production of about 450 g dry wt m^{-2} yr^{-1} (roots and shoots) in the plant beds or about 180 g C m^{-2} yr^{-1}. Only 2% of that dry weight is nitrogen and the plants translocate and conserve about half of the nitrogen from senescent tissue (Chapin et al. 1975). Therefore, the total net uptake is 1.8 g N yr^{-1} or 18 mg m^{-2} day^{-1}. This compares with 12 mg N m^{-2} day^{-1} for the sediment algae and 1.2 mg N m^{-2} day^{-1} for the plankton.

There is abundant inorganic nitrogen in the interstitial water except in the plant beds (Table 4-9). As a first approximation, we can assume that

TABLE 4-13 *Nitrogen Concentrations and Uptake Rates in Tundra Ponds in 1970 and 1971**

	NO_3-N		NH_3-N	
	Concentration	Uptake rate	Concentration	Uptake rate
Pond C - 1970				
7 June	4.5	0.003	49	0.661
18 June	1.5	0.007	45	0.317
22 July	16.4	0.001	47	0.329
28 July	2.9	0.032	42.6	0.201
1 August	1.8	0.067	24.1	0.915
18 August	0.7	0.006	17.5	0.176
$\bar{x}$	4.6	0.019	37.5	0.433
Pond B - 1970				
7 June	0.3	0.024	9.4	0.019
18 June	0.7	0.010	32.5	0.244
23 July	9.4	0.019	25.3	0.196
29 July	0	0.002	21.4	0.125
1 August	1.12	0.010	29.2	0.196
17 August	0	0.003	-	-
$\bar{x}$	1.92	0.011	23.6	0.156
Pond B - 1971				
21 June	18.5	0.012	34	0.409
28 June	13.3	0.001	16	0.156
7 July	20.1	0.004	16	0.070
26 July	11.7	0.002		
2 August	7.9	0.002	30	0.106
11 August			16	0.130
16 August	6.8	0.001		
23 August			12	0.069
$\bar{x}$	12.9	0.004	21	0.157

*Concentrations are expressed as μg N liter^{-1} and uptake rates as μg N liter^{-1} hr^{-1}

the sediments are 50% water and contain 60 μg NH_3-N liter^{-1} in the *Carex* beds and 2,500 μg NH_3-N liter^{-1} in the rest of the pond. This is 18 mg N m^{-2} in the top 20 cm of sediment in the *Carex* beds and 500 mg outside the beds. Additional quantities of exchangeable inorganic nitrogen, 810 mg N m^{-2} in *Carex* beds and 1400 mg N m^{-2} outside the beds, are available as well. Obviously the sediment algae will have adequate NH_3 in the pore water but the *Carex* roots will not (they can remove all the interstitial NH_3 once a day). Unfortunately, our measurements of uptake of ^{15}N by isolated roots gave an extremely high value of 7.4 g N m^{-2} day^{-1} which is hundreds of times too high to agree with the productivity values. Similar results were obtained by Morris (1978) for *Spartina* roots. It is likely that lowered oxygen concentrations *in situ* cause a low rate of N uptake.

Ammonification and Nitrification

It is obvious from the uptake rates and concentrations that ammonia was being rapidly regenerated. The regeneration rates were measured by the isotope dilution method in which labeled ammonia is added to a sample of water. A part of the water is immediately removed and the isotope ratio of the ammonia fraction is determined. Following incubation, the isotope ratio of the ammonia is again determined. The results give a rate of dilution of the labeled ammonia by unlabeled ammonia from other nitrogen fractions within the water.

In Pond B in 1971 the ammonia supply rate averaged 1.9 μg liter^{-1} hr^{-1} (3 samples) while the overall range of eight pond measurements was 0.4 to 3.26. In Ikroavik Lake the mean of four measurements was 0.74 and the range was 0.28 to 1.05 μg liter^{-1} hr^{-1}. These values are higher than the uptake rates but of the same order of magnitude. Given these rapid rates of resupply of ammonia in the ponds, it is very doubtful that nitrogen would ever be limiting to phytoplankton growth.

Unfortunately, the evidence for nitrification is indirect for the ponds. One piece of evidence that nitrification is occurring is that high nitrate levels are found in the ponds and in Ikroavik Lake early in the year, suggesting that formation has outstripped uptake. Another bit of evidence is that nitrifying bacteria have been isolated from tundra soils at Barrow (Norrell personal communication). The last evidence is that Kinney et al. (1972) have found significant nitrification in fresh and saline waters some hundreds of kilometers east of Barrow. They were able to follow conversion of added ammonia to nitrite and nitrate, as well as to follow changes in water without added ammonia. The maximum potential nitrification rate found was 3.0 μg liter^{-1} day^{-1} under ice in the winter. While this is a slow rate, it is proof that nitrification occurs in arctic waters. All available evidence suggests that nitrification is an active, although slow, process in tundra ponds and lakes.

Fertilization Experiments

Based upon the data on concentrations of nitrogen in the pond water (Figure 4-12, Table 4-8) and in the sediment (Table 4-9), we found that most of the inorganic nitrogen was in the sediment. Yet the photosynthesis rates in the water were so low that photosynthesis could be supported by the DIN in the water with only a little DIN added from ammonification. To gain more insight into the process of nutrient cycling, we asked, "What would happen if phosphorus were added to a pond?"

In the whole pond fertilization experiment carried out on Pond D in 1970, phosphorus addition of 0.3 mg P liter^{-1} was followed by an almost immediate increase in particulate nitrogen in the water. This preceded an increase in photosynthesis or phytoplankton biomass by several days (Figure 4-13). Inorganic nitrogen uptake, measured with ^{15}N techniques, was depressed for 24 hr after the initial fertilization and then increased rapidly. The implication is that the additional nitrogen required for this uptake and the accumulation of nitrogen in the particulate phase probably came from the sediments. No nitrogen fixation was detected in the water or in the sediments during this period, nor was there any great change in inorganic nitrogen concentration in the pond water.

The response of nitrogen uptake to a steady infusion of phosphorus into a pond was tested in subponds (250 liters) of Ponds B and C in 1970. Phosphate was supplied daily at 2 and 10 μg P per liter of pond, and the effects on the phytoplankton nitrogen regime were compared with control subponds. Here again, there was a positive response of nitrogen uptake to phosphorus addition; the maximum response occurred at the lower phosphorus infusion rate (Figure 4-14). This increase is, of course, independent of biomass as the results are expressed as N uptake per microgram of N in the total particulate matter in the water. Thus, nitrogen uptake responded to increased phosphorus availability and allowed an increase in phytoplankton productivity and biomass.

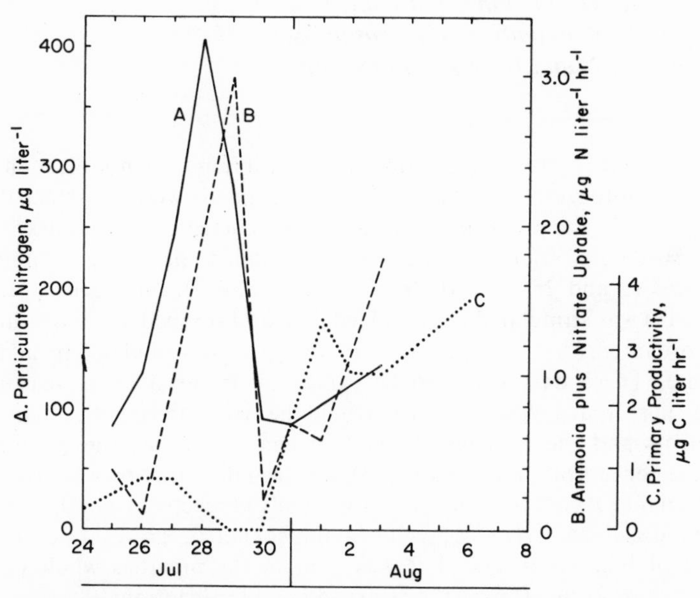

FIGURE 4-13. *Particulate nitrogen, rate of nitrogen uptake, and primary productivity in the plankton of Pond D, 1970. On 25 July, 0.31 mg P per liter of pond volume were added and on 28 July an additional 1.5 mg P per liter of pond volume were added.*

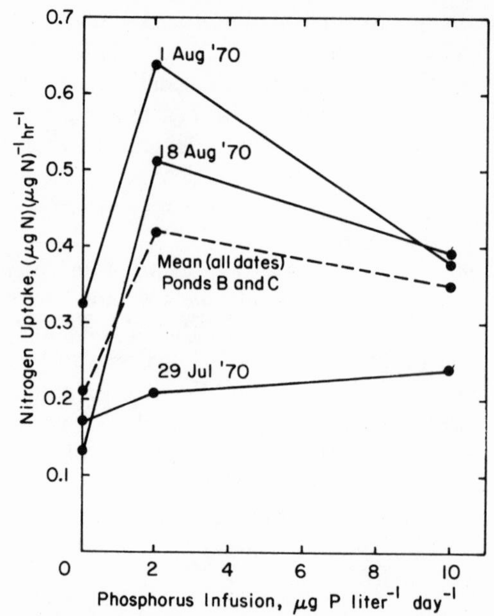

FIGURE 4-14. *Rate of nitrogen uptake [μg N (μg N)$^{-1}$ hr^{-1}] in Pond B and C subponds, 1970, at daily phosphorus infusion rates of 0, 2, and 10 μg P per liter of pond volume.*

To try and resolve the question of a possible nitrogen limitation to planktonic photosynthesis, larger (675 liter) subponds were constructed in Pond B in 1972. The experimental treatments included complete darkness, partial shading (50%), heating (+4C°), added lights, and added phosphorus (5 and 25 μg P (liter of pond)$^{-1}$ day^{-1}). Nitrogen fixation (Table 4-14) was stimulated by the added light and stopped by the shading. Even more interesting, nitrogen fixation was strongly stimulated by added phosphate. These data can also be expressed as the average seasonal fixation per hour. When results from the two phosphate addition experiments and the controls are plotted against the average seasonal phosphate concentrations (Figure 4-15), it is seen that fixation was directly proportional to P concentration both for dissolved reactive P ($r = 0.98$) and for total dissolved P ($r = 0.95$). It should be noted, however, that no response of fixation to added P was seen in the previous whole-pond fertilization experiments or in the 1970 sub-pond fertilizations.

The stimulation of N fixation by high amounts of P agrees with the data reported in the review by Schindler (1977). He views the occurrence of stimulation in lakes as one stage in the evolution of some lakes as they receive larger and larger quantities of P from their drainage basins. In the Barrow ponds, it is likely that the algae in the pond are usually phosphate-

TABLE 4-14 *Nitrogenase Activity in Sediments of Ponds and Subponds, 1972**

Ponds	16 July	22 July	27 July	4 Aug	14 Aug	21 Aug
B-1 (heated)				0.13	0.07	0.25
				1.21	0.65	2.33
B-7 (shaded)	0.00	0.08	1.10	0.00	0.00	0.00
	0.00	0.75	10.26	0.00	0.00	0.00
B-10 (control)	0.21		3.15	0.00	0.00	0.00
	1.95		29.39	0.00	0.00	0.00
B-11 (opaque)	0.00	0.00	0.00	0.00	0.00	0.00
	0.00	0.00	0.00	0.00	0.00	0.00
B-12 (lighted)	0.01		6.28	5.12	0.00	4.01
	0.00		58.60	47.78	0.00	37.4
B				0.76	1.53	2.72
				7.09	14.28	25.3
E				0.99	0.00	0.00
				9.24	0.09	0.00

Ponds	22 July	31 July	7 Aug	17 Aug	24 Aug
B-4 (high PO_4)		3.23	16.92	2.84	0.02
		30.14	157.91	26.50	0.19
B-5 (high PO_4)	9.85	9.68	12.99	8.40	3.21
	91.93	90.34	121.23	78.39	29.96
B-6 (control)		3.17	2.07	0.00	0.00
		29.58	19.32	0.00	0.00
B-8 (low PO_4)		0.39	0.97	0.00	0.00
		3.64	9.05	0.00	0.00
B-9 (low PO_4)	0.08	0.23	3.07	0.00	0.00
	0.75	2.15	28.65	0.00	0.00
C		0.00		0.00	0.00
		0.00		0.00	0.00
D		0.00		0.00	0.00
		0.00		0.00	0.00

*Upper values are μmole ethylene produced $m^{-2}\ hr^{-1}$ and lower values are μg N fixed $m^{-2}\ hr^{-1}$.

limited. When P is added, the nitrogen eventually becomes limiting; at this point the blue-green algae, which can fix nitrogen, gain a competitive advantage and begin to grow.

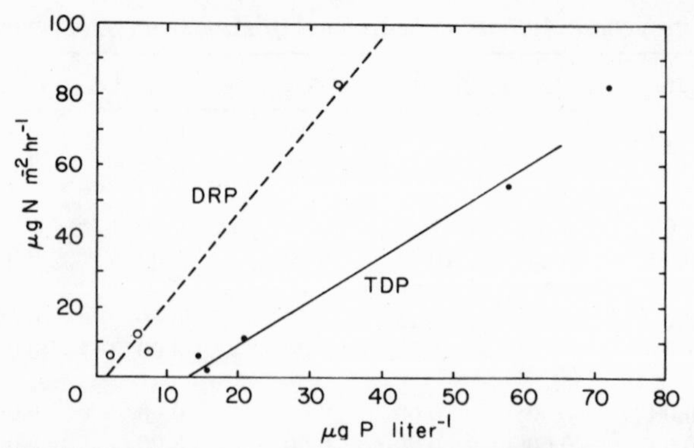

FIGURE 4-15. *Nitrogen fixation rates in Pond B subponds with various concentrations of phosphorus (DRP and TDP).*

TABLE 4-15 *The Primary Production, Chlorophyll Concentration, and NH_3–N and PO_4–P Amounts in Two Swimming Pools in 1971**

	Date	Net Production (μg C liter^{-1}hr^{-1})	Chlorophyll (μg liter^{-1})	NH_3–N (μg liter^{-1})	PO_4–P (μg liter^{-1})
Pond 7 (control)	26 July	0.30	0.4	28	2.2
	9 August	0.27	0.3	39	1.2
	9 August	0.32	0.2	22	1.1
	16 August	1.12	0.5	20	0.6
	23 August	6.88	0.4	14	1.0
Pond 8 (fertilized)	26 July	1.05	1.1	69	60.2
	2 August	0.41	0.6	25	64.0
	9 August	1.97	0.8	374	63.0
	16 August	18.83	2.6	73	15.8
	23 August	19.51	4.3	16	3.6

*Pond 8 was fertilized with PO_4 on 17 July and with NH_3 on August 7.

The importance of the sediments in supplying nitrogen is illustrated in an experiment carried out in two plastic swimming pools (each 6.4 m^3) filled with lake water but containing no sediments (Table 4-15). On 17 July 1971, 109 μg PO_4-P (liter of pond)$^{-1}$ were added to Pond 8 while Pond 7 was maintained as a control. Although a slight increase in productivity and chlorophyll content occurred in Pond 8, this was very minor until 500 μg NH_3-N (liter of pond)$^{-1}$ was added on 7 August. Following this, there was a striking increase in primary productivity compared with the control, although there was also a slight increase in the control. The ratio of the removal rates of N:P based on chemical data was 72:1 in Pond 7 when the phosphorus was limiting photosynthesis. After fertilization with ammonia in Pond 8 the ratio fell to 14:1 and then to 10:1 after 1 week. This low ratio may indicate that nitrogen was becoming limiting. We conclude that phosphorus alone does not produce a great fertilizing effect unless either sediments are present or nitrogen is supplied.

It appears that the sediments are controlling the nitrogen concentrations in the pond waters. This control acts through decomposition of the abundant organic matter in the sediment. Some of the ammonia released during decomposition is taken up by grasses and sedges (Table 4-9), some is taken up by epipelic algae, and some diffuses into the water column. The regulation of these processes is unknown.

PHOSPHORUS *

Concentrations in Water

The tundra has low phosphate concentrations in both aquatic and terrestrial environments. Thus, we thought that phosphorus could be limiting to plants in the ponds and that phosphorus cycling should be intensively studied. The first 2 years of the study concentrated on descriptions of phosphorus compartments and seasonal cycles in the water column; the later 2 years emphasized sediment chemistry and phosphorus pathways.

Concentrations of dissolved reactive phosphorus (DRP), dissolved total phosphorus (DTP), and particulate phosphorus (PP), were measured. Dissolved unreactive phosphorus (DUP) was calculated by subtraction of DRP from DTP.

Dissolved reactive phosphorus in samples was analyzed by the single solution phosphomolybdate technique (Strickland and Parsons 1965) followed by extraction into isoamyl alcohol. Reagent contact time prior to color extraction was strictly limited to 5-10 minutes to minimize

*R. T. Prentki

hydrolysis of dissolved organic phosphorus. Samples collected prior to 23 June 1970, samples with high DRP concentrations, and those otherwise so noted were not extracted prior to spectrophotometric determination. Both dissolved unreactive phosphorus and particulate phosphorus retained on membrane filters were oxidized with persulfate (Menzel and Corwin 1965), after which analyses were made by the single solution method. Zooplankton were refluxed with perchloric acid and P was determined as phosphomolybdic acid (Strickland and Parsons 1965).

Dissolved reactive phosphorus determinations usually overestimate actual phosphate concentrations, often by orders of magnitude. Hydrolysis of dissolved organic phosphorus compounds or arsenic interference are the most commonly accepted causes (Rigler 1966, 1968, Downes and Paerl 1978, Chamberlain and Shapiro 1969). In the tundra ponds, however, extensive tests showed that the DRP measurement included only dissolved phosphate. First, arsenic should not interfere in extraction phosphomolybdate analyses of unpolluted waters such as the Barrow tundra ponds. Next, Prentki (1976) has demonstrated that extracted DRP in tundra ponds does not include colloidal P or XP. In addition, application of Rigler's (1966) radiobioassay test to pond samples did not result in apparently lower phosphate uptake velocities at higher added phosphate concentrations. Finally, other ^{32}P kinetic experiments that assumed the DRP equal to phosphate gave internally consistent mass balances.

Dissolved reactive phosphorus concentrations are always low in the ponds, usually between 1 and 2 μg P liter^{-1}; monthly averages over the 2 years of intensive sampling show only modest seasonal or inter-pond differences (Table 4-16). In addition, diel variations in dissolved DRP concentrations are of the same magnitude as seasonal variations, thereby masking any seasonal trends. The only distinguishable seasonal features occur as a result of the thaw in June (Figure 4-16a). At this time, DRP concentrations are generally high; they are maintained by phosphorus entering in runoff and leaching from standing vascular vegetation. Next, the rate of phosphorus supply decreases in late June as both runoff and leaching taper off. Thus, as the phosphorus demand increases with the onset of the phytoplankton bloom, the DRP concentrations are depressed to below 1 μg P liter^{-1}.

Dissolved unreactive phosphorus (DUP) is initially re-introduced into the ponds each spring through litter decomposition, sediment leaching, and runoff. Concentrations early in the thaw season in 1970 to 1972 rank in the same order as snow pack depth for those years, with 1971 having the greatest snow depth and highest average June DUP concentration and 1970 having the least and lowest. This is very evident in Figure 4-16b, where DUP concentrations in Pond B in 1971 are almost twice those in 1970. In part, this DUP may be resorbed by the sediment and soil surface during runoff, and this sorption would be favored by greater sediment-

TABLE 4-16 *Phosphorus Concentrations in Barrow Ponds, Monthly Averages for 1970 and 1971**

	1970			1971		
Pond	June	July	August	June	July	August
Dissolved reactive phosphorus						
B	1.9	2.3	2.3	2.3	1.8	2.0
C	1.9	2.2	2.9	1.4	1.5	1.1
J				0.6	1.4	1.6
Dissolved unreactive phosphorus						
B	6.6	7.0	10.4	12.2	10.5	9.2
C	5.5	10.4	13.2	15.5	13.9	12.8
J				14.4	11.2	11.5
Particulate phosphorus						
B	8.0	5.8	5.2	10.6	5.6	5.2
C	9.3	7.4	10.1	10.9	8.4	8.5
J				6. 8	8.3	3.9

*Data are expressed as μg P liter^{-1}

water contact time associated with a shallow snow pack and resulting slower flow rates. Such a flow dependence was observed in 1972 for runoff at weir 1 (see Figure 3-2 for location) where DUP concentrations increased from 8.3 μg P liter^{-1} at no flow to 24.0 μg P liter^{-1} at 80 m^{-3} hr^{-1}

DUP is the predominant phosphorus form in the ponds, averaging 12.2 μg P liter^{-1} or 56% of the water column phosphorus (excluding zooplankton phosphorus) in the three control ponds listed in Table 4-16. Diel oscillations in concentrations range between 4 and 5 μg P liter^{-1} and appear to be related inversely to temperature. Approximately 70% of DUP is probably refractory; it is resistant to both naturally occurring phosphatases (Prentki 1976) and photo-oxidation by sunlight (Barsdate and Prentki personal communication). The other 30% of this phosphorus is in the form of XP and colloidal P; these forms can be rapidly autolyzed and enzymatically hydrolyzed to phosphate (Prentki 1976).

Particulate phosphorus (PP), another important fraction of the total phosphorus, showed no discernible pattern of monthly average or seasonal concentration (Figure 4-16, Table 4-16). The grand average of 10.1 μg PP liter^{-1} for the three ponds is assumed to be the sum of phytoplankton, sestonic bacteria, and detrital phosphorus. However, it may overestimate

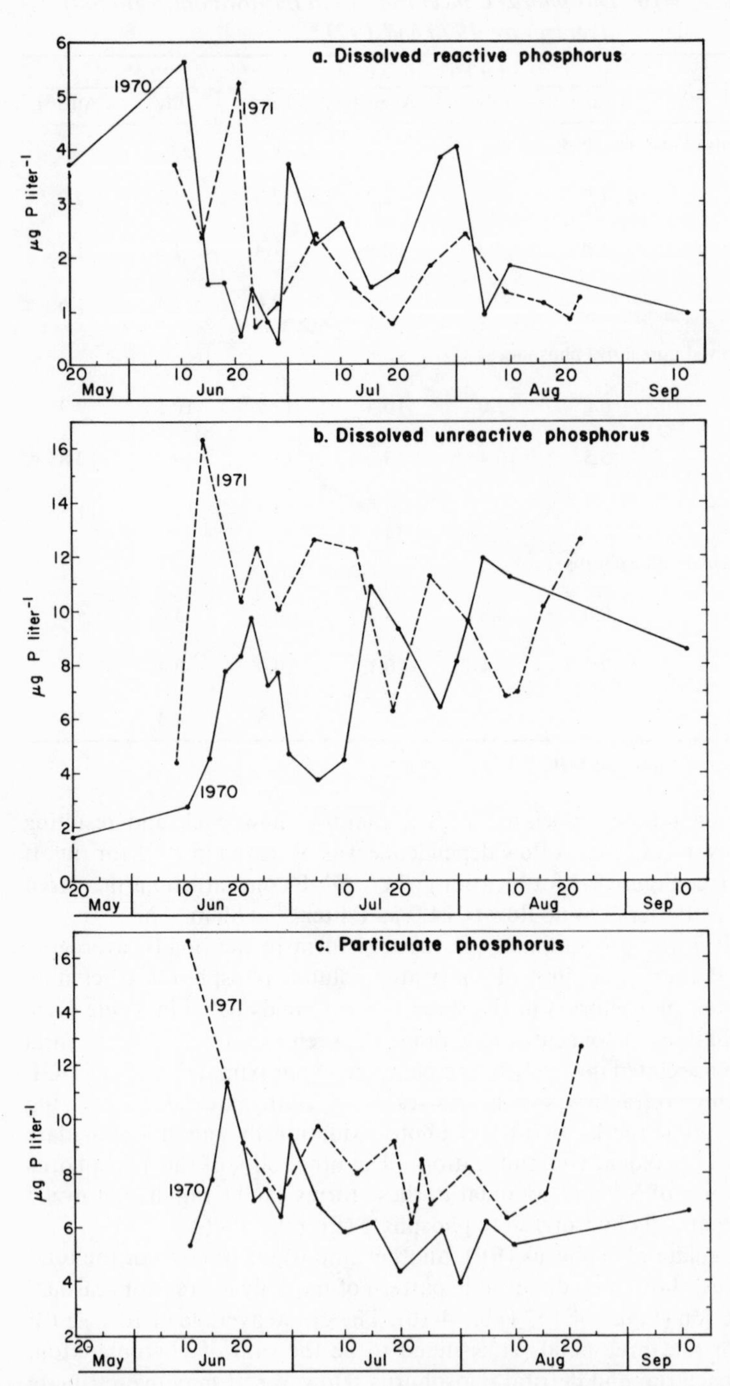

FIGURE 4-16. *Concentration of phosphorus in Pond B, 1970–1971.*

actual PP concentrations by 1 or 2 μg liter^{-1} since the membrane filters used for this (and only this) analysis retain some dissolved organic phosphorus (Prentki 1976). The measurement specifically excludes zooplankton large enough to be picked off filters.

In comparison with the intensively studied ponds discussed above, phosphorus concentrations are somewhat higher in other ponds we studied. These differences ultimately are related to sediment chemistry and will be discussed in more detail below.

Phosphate concentrations in an 18 August 1972 transect (Figure 3-1) ranged from 0.9 to 3.0 μg DRP liter^{-1} in 15 low-centered polygon ponds and from 3.7 to 7.6 μg P liter^{-1} in three trough ponds. In one low-centered polygon pond with trough inputs the concentration was 4.8 μg P liter^{-1}. The greater concentration in polygon troughs than in low-centered polygon ponds parallels similar observations for many other phosphorus parameters in both Barrow terrestrial and aquatic environments. Kalff (1965) reports DRP concentrations of 6 and 13 μg liter^{-1} for two trough ponds in this watershed. Our investigations indicate that the non-extractive phosphate techniques used by Kalff overestimated pond phosphate concentrations by approximately 45% (see Prentki 1976) due to hydrolysis of organic phosphorus. When adjusted by this factor, Kalff's phosphate concentrations would be 3 and 7 μg P liter^{-1}. The latter value agrees well with the 7.2 and 7.5 μg P liter^{-1} we have measured upon two occasions in the same pond.

The dissolved unreactive phosphorus in all IBP ponds ranged between 5 and 10 times their DRP concentration. The DUP concentrations in the 18 August 1972 transect were also highest in trough ponds or in ponds with trough inputs: in the trough ponds it was 18.6 to 60.6 μg P liter^{-1} while in the polygon ponds it was 9.8 to 24.1 μg P liter^{-1}.

Three analyses of PP on two trough ponds had a range of 20.3 to 31.5 μg P liter^{-1}, 2 to 3 times that found in low-centered polygon ponds.

Phosphorus concentrations in arctic lakes are similar to those of low-centered polygon ponds. In Ikroavik Lake near Barrow DRP concentrations ranged between 0.5 and 4.4 μg P liter^{-1} for 39 samples taken on 17 dates, while 15 analyses from four Prudhoe Bay lakes ranged from 0.2 to 1.9 μg P liter^{-1}. These concentrations are similar to those of other arctic surface waters reviewed in Hobbie (1973), but Char Lake had undetectable amounts of less than 0.7 μg liter^{-1} (Schindler et al. 1974). In arctic Alaska, lakes in the Colville River area had 0 to 2.8 μg DRP liter^{-1} in the summer but had 4.6 to 12.1 μg DRP liter^{-1} in the winter (Kinney et al. 1972). However, DUP concentrations in arctic lakes are lower than those of tundra ponds. Our analyses of Ikroavik water ranged as low as 0.1 μg P liter^{-1} in melted lake ice but normally were 4 to 10 μg P liter^{-1}. Although the major ions are concentrated more than 10-fold due to the exclusion from the winter ice cover, the concentration of DRP and DUP remain relatively constant, suggesting that sediment-water interactions buffer their concentration in the lake. The DUP concentration for the four

Prudhoe Bay lakes had a summer average of 4.4 μg P liter^{-1}, and Schindler et al. (1974) report an average of 2 μg P liter^{-1} for total dissolved P (mostly DUP) in Char Lake. The PP concentrations in arctic lakes appear to vary much more than the other phosphorus parameters, being most obviously influenced by mechanical and thermokarst erosion and suspension of particulates. Excluding samples containing organic soil or sediment taken near the shore during storms, Ikroavik concentrations ranged from 1.9 μg P liter^{-1} to 57.5 μg P liter^{-1} in the summers of 1970 and of 1971, with a mean of 17.3 μg P liter^{-1} (34 analyses). The Prudhoe Bay lakes, on the other hand, averaged only 7 μg P liter^{-1} and ranged from 1.6 to 15 μg P liter^{-1} (15 analyses), while Char Lake (Schindler et al. 1974) ranged from 1 to 5 μg P liter^{-1}.

Alaskan coastal plain lakes and ponds differ in their predominant forms of phosphorus. Thus, PP predominates in lakes and DUP in ponds. Hutchinson (1957) states that most uncontaminated lake districts have surface waters containing 10 to 30 μg P liter^{-1}, with 10% of this being phosphate (DRP). Of the arctic waters discussed above, only trough ponds and Char Lake fall outside these limits.

Sediment

Sediment studies concentrated on interstitial water analyses, chemical characterization of sediment phosphorus, and total phosphorus determinations.

Interstitial water was squeezed by N_2 gas pressure from sediment cores within 1 hour of collection. Dissolved reactive phosphorus was determined as in the water column; however, turbidity blanks were necessary due to precipitation of organics from solution upon addition of the reagent.

Sediment total phosphorus was determined by the single solution phosphomolybdate technique of John (1970) on diluted perchloric or perchloric-nitric acid digests of sediments (Sommers and Nelson 1972, Jackson 1958).

Fractionation of inorganic phosphorus in the sediments was accomplished using the Williams et al. (1971a) modification of the Chang and Jackson (1957) extraction scheme with some additional modifications described in Prentki (1976). Phosphate concentrations in the Chang and Jackson extractions were determined colorimetrically using John's reagents. Internal standards and turbidity blanks were run on all samples and boric acid was added to reduce fluoride interference in the NH_4F extracts. Citrate dithionate (reductant soluble) extracts required digestion with persulfate to eliminate interference. Careful adjustment of sample concentration and pH were also required in order to prevent precipitation of the large amounts of iron present. Corrections for phosphorus activated but reabsorbed during the NH_4F extraction were attempted using the procedure outlined by Williams et al. (1971a); however, Barrow sediments are high in iron and the phosphorus from both NH_4F and first NaOH fractions was apparently tied up by excess iron before the extractions were completed and then released in the following reductant soluble extraction. The correction equations were therefore expanded to include this additional resorption.

The Chang and Jackson soil phosphorus fractions are usually interpreted as follows: the NH_4F-P fraction is considered to be composed of Al phosphates such as variscite; the first NaOH-P fraction is composed of Fe phosphates such as strengite, or of loosely sorbed phosphorus on hydrated iron oxides; reductant soluble-P and the second NaOH-P fraction is made up of phosphate occluded within matrices of more highly crystalline iron minerals; and acid extractable-P is made up of calcium phosphates such as apatite. However, Fife (1959) and Williams et al. (1971b) have suggested that some phosphorus sorbed on hydrated iron oxides appears in the NH_4F extraction, and Syers et al. (1973) have speculated that the distribution of phosphorus among NH_4F, NaOH, and reductant soluble fractions in some cases may be a reflection of the phosphate-binding strength of a single retaining complex against the individual extractions.

Total inorganic phosphorus in the sediment was calculated from the sum of the above Chang and Jackson inorganic phosphorus fractions. Organic phosphorus was calculated as the difference between this summed total and the total derived from the perchloric-nitric acid digest.

Within one pond basin, interstitial DRP, DTP, and DUP (Table 4-17) do not appear to differ greatly. Neither presence nor absence of active roots nor oxidized or reduced sediments could be correlated significantly with phosphorus concentrations. However, between ponds the limited data suggest that there was a real difference in interstitial DRP; Pond J concentrations were significantly higher than the single values measured for Ponds A and C. Sediment phosphorus, iron, and phosphate sorption isotherm data discussed later support this observation. Interstitial DRP concentrations are 3- to 5-fold higher than DRP in the Pond J water column; however, the situation is reversed in Ponds C and A where water column concentrations are almost twice interstitial concentrations. The DRP concentrations (Table 4-17) also vary between ponds and the DUP concentrations are 2- to 10-fold higher than those in the water column above.

Results of the chemical fractionation of the sediments indicate that retention of the inorganic P is by sorption of phosphate on hydrous iron oxide rather than by formation of distinct iron, aluminum (clay), or calcium phases. Evidence for this comes from experiments where $^{32}PO_4$ was added to Barrow sediment samples prior to fractionation or anion-exchange resin equilibration. After correction for resorption in fractionation experiments, 98.5% of the added ^{32}P was found in the NH_4F and first-NaOH pools. The specific activities of these two pools were

TABLE 4-17 *The Dissolved Reactive (DRP) and Dissolved Total Phosphorus (DTP) in the Interstitial Water of Several Ponds* [1]

Pond	Location	Core depth	DRP	DTP
J	*Carex* marsh	0-8 cm (oxidized)	4.3 ± 1.0 (4)	-
		9-16 cm (reduced)	3.7 ± 0.9 (4)	-
	Arctophila bed	0-16 cm (o + r)	4.7 (1)	-
	Central basin	0-8 cm (o + r)	5.7 ± 0.8 (3)	138 (1)
		9-16 cm (reduced)	1.9 ± 0.7 (3)	-
		4-12 cm (reduced)	1.5 (1)	63 (1)
		8-13 cm (reduced	-	89 (1)
C[2]	*Carex* marsh	0.10 cm (oxidized)	0.9 (1)	24 (1)
		10-18 cm (reduced)	2.4 (1)	46 (1)
A-B	Central basin	0-5 cm (o + r)	1.2 (1)	-
		5-16 cm (reduced)	10.9 (1)	-
E[3]	Terrestrial *Carex* stand	7 cm (o + r)	3.0 ± 1.6 (2)	26.4 ± 7.6 (2)
		22 cm (reduced)	4.2 (1)	38.3 (1)

[1] Values are the mean, the standard deviation and the number of samples.
[2] From Barel and Barsdate (personal communication).
[3] Samples collected from bore holes open at 7 cm and 22 cm below tundra surface.

identical; this indicates that there is but a single binding complex rather than two chemically distinct pools. The phosphorus in the combined NH_4F and first-NaOH pools appears to be identical to the resin-exchangeable inorganic phosphorus. The only sorbent for the phosphorus in the intensively studied ponds is extractable iron (as Fe_2O_3) which accounts for 18 to 61% of the ash weight in surficial sediments. Clay concentrations in the sediments are almost unmeasureable and calcium concentrations (as CaO) account for only about 1% of ash weight. The iron-sorbed phosphorus is in dynamic equilibrium with phosphorus in solution and is potentially mobile and responsive to various limnological conditions.

These results are similar to the expected values for an unpolluted temperate latitude lake which receives little detrital input (Williams et al. 1971a, Frink 1969, Harter 1968), although the reductant soluble phosphorus and organic P are high in respect to NH_4-P and NaOH-P, and the Barrow phosphorus levels are in the lower portion of the range of reported values. Interstitial DRP concentrations, however, are exceptionally low compared to sediment values of temperate lakes (Holdren et al. 1977, Glass and Poldoski 1975, Barko and Smart 1980, in press). Barrow terrestrial soil samples are quite similar to the pond sediments, perhaps reflecting both a common origin as ancient lake sediments and the present-day similarity of environments. Distinct

TABLE 4-18 *Phosphorus Fractions in Selected Soils and Sediments from Barrow**

Site**	Description	NH_4F-P	1st NaOH−P	Reductant soluble−P	2nd NaOH−P	NHCl−P	Total inorganic P	Organic P	Total P
Pond B	*Carex* marsh, 0-0.5 cm surface floc	635	(−64)	454	19	29	1074	117	1192
Pond C	Central Basin, 0-1 cm, peat	136	70	59	13	4	283	379	662
Pond C	Central Basin, 0-16 cm, peat	56	19	93	4	24	196	467	663
Pond C	*Carex* marsh, 0-16 cm, peat	120	37	59	4	5	225	743	968
Pond 6	Central Basin, 0-16 cm, peat	121	143	210	6	48	530	374	904
Soil 22	0-8 cm, peat	17	40	42	4	6	109	503	612
Soil 23	0-15 cm, peat	143	29	170	3	6	351	1124	1475

*Data are expressed as μg P $(g \text{ dry wt sediment})^{-1}$

**Pond 6 is a trough pond and soils S22 and S23 are low-centered polygon and trough soils, respectively.

differences exist between interpolygonal troughs (Pond G and Soil 23 in Table 4-18) and low-center polygon areas, (Ponds B and C, Soil 22) and these differences transcend the aquatic-terrestrial divisions. The high inorganic phosphorus in the sediment of troughs parallels the generally high dissolved reactive phosphorus concentrations in the water of troughs as compared to low-centered polygon ponds. There are broad patterns, both in solution and sediments, which correlate with iron and other parameters; these will be discussed later.

Sediment surface layers in samples from the central basins of the intensively studied ponds usually contain about 1100 μg P (g dry wt)$^{-1}$ with a range from 662 to 1530. The underlying reduced sediments contain less phosphorus and only Pond B sediment averages more than 750 μg P (g dry wt)$^{-1}$ over the first 10 cm. Organic phosphorus constitutes 10 to 87% (117 to 805 μg P (g dry wt sediment)$^{-1}$) of the total phosphorus present; the lowest values occur in a surface floc which covers up to 25% of the sediment surface. If no floc is present, then amounts increase with depth and toward plant stands. Total inorganic phosphorus has been measured at 1074 μg P (g dry wt sediment)$^{-1}$ in the surface floc covering portions of Pond B, but ranged between 200 and 400 in the other Pond B and C samples. This floc appears to precipitate out of the water column and accumulate along the downwind shore of ponds as summer progresses. In Pond C, the total P in the sediments was 25,000 mg m^{-2} to a depth of 10 cm, the total P in interstitial water was 2 mg P m^{-2}, and the total P in the water column (20 cm) was 5 mg P m^{-2}.

Pond Phosphorus Budget

Over the long term the phosphorus status of the ponds is dependent on the net balance of phosphorus entering and leaving the pond basins, although the retention or loss of specific forms of phosphorus is influenced strongly by pond processes. The events and processes involved in phosphorus movement through the ponds are rather well understood in a qualitative fashion; however, a synthesis of our insight into how the system operates, together with the actual field measurements of concentrations and rates, results in a budget which is distinctly speculative. This is in part due to the difficulties in the measurement of water movement and in part due to the paucity of phosphorus data, particularly in such things as surface runoff during the summer and suspended particulates during runoff. Hydrologic factors related to precipitation and water balance vary greatly both within season and between years, and these factors strongly influence the phosphorus budget. Warm, dry summers concentrate phosphorus in the ponds at the expense of surrounding tundra while heavy winter snowpack results in high loss of phosphorus from the tundra and in an even greater loss from the ponds. Insofar as possible, this budget is

TABLE 4-19 *Concentration of Dissolved Reactive Phosphorus (DRP), Dissolved Unreactive Phosphorus (DUP), and Particulate Phosphorus (PP) in Accumulated Snow Cover and Precipitation at Barrow, Alaska**

Date	DRP	DUP	PP
Snow cover			
30 May 71	1.1	0.4	1.1
30 May 71	1.4	1.0	1.5
30 May 71	2.1	1.1	1.7
30 May 71	1.5	2.0	1.4
Overall mean, winter	1.5	1.1	1.4
Mixed rain-snow			
1 August 64 **	9.0	-	-
14 July 71	5.3	0.0	-
Rain			
10 July 71	12.3	0.1	-
30 July 71	11.3	-	-
10 August 72	3.0	2.0	-
Snow			
15 July 71	5.5	-	-
19 August 71	4.1	-	-
Overall mean, summer	7.2	0.7	-

*Data are expressed as μg P liter^{-1}.
**Kalff (1965).

calculated for a "mean" year with some reference to the effects of extremes.

As described in Chapter 3, a fraction of the water from winter snow accumulation will be retained during the spring melt to fill the pond to its holding capacity. Over 4 years the winter pond levels have been observed to be 0 to 11 cm below capacity, (a mean of 5.8 cm). The water content of the snow ranged from 6.7 to 16.5 cm in 1971 to 1973 (Table 3-6). With the rare exception of periods after heavy summer rains (e.g., 1973), breakup is the only time in which water actually flows out of the ponds. During breakup, meltwater floods most of the tundra surface and a relatively complicated situation exists in which the concentration of both DRP and DUP is somewhat higher in the ponds than in the water flowing into or away from them, and all of these concentrations are higher than those found in winter precipitation (Table 4-19). At this time of year, dissolved phosphorus is leached rapidly from the vegetation and litter of both terrestrial sites and pond margins. As a result the phosphorus concentration in the meltwater is increased. However, because of the shallow water depth and rapid water movement over the tundra surface, the residence time of water is much less than in the ponds (3 hours vs. 36 hours) and the contact with the soil surface is great. As a consequence, phosphorus sorption is more rapid over the tundra surface than in the ponds (Prentki 1976) and the concentrations in the meltwater are lower than those in the ponds by about 0.13 μg of DRP and about 1.8 μg of DUP liter^{-1}. As meltwater moves across the watershed, phosphorus is thus removed from the ponds and resorbed by the terrestrial soils; the extent of removal depends on both the difference in phosphorus concentration between ponds and tundra and on the volume of water flowing through the ponds. With an area of watershed above Pond B of 14,000 m^2 and a 5.8 cm potential runoff (Table 3-6) the estimated volume flowing through the pond during breakup is 800 m^3, and the loss of DRP and DUP is 0.1 and 1.4 mg P m^{-2}, respectively.

Summer precipitation at Barrow (Table 4-19), averages 7.2 μg DRP liter^{-1} and 0.7 μg DUP liter^{-1}. With a mean of 6.5 cm of precipitation in summer, the input into the ponds from this source amounts to 0.47 mg m^{-2} DRP and 0.05 mg m^{-2} DUP (Table 4-20). Unlike the winter snowpack data, these figures do not include dry fallout between periods of precipitation, and so the possibility exists that an additional amount of phosphorus enters the tundra system as dust. However, the very low winter bulk P concentrations suggest that dry fall must be very low.

Another possible input exists. As discussed earlier (Chapter 3), soil-water could enter the ponds during dry periods when the water level falls drastically. This was not considered important for the nitrogen budget, and is likely not important for the phosphorus budget either. It is an area of some uncertainty, however, so calculations are presented below of the quantities involved. There are four measurements of interstitial water in the upper layers of soil adjacent to the ponds; the averages are 4.4 μg DRP liter^{-1} and 26.3 μg DUP liter^{-1}. If, in a moderately dry summer such as 1971, 6 cm of water enters the pond, this would be added inputs of 0.26 mg DRP m^{-2} and 1.58 mg DUP m^{-2}. The size of this is about equal to the annual budget (Table 4-20) but is still very low when the large amounts of P in the sediments are considered.

Annually, the ponds gain 0.5 mg DRP while losing 1.3 mg DUP m^{-2} (Table 4-20), suggesting that phosphate retention mechanisms of the ponds are extremely effective but that dissolved organic phosphorus cannot be as efficiently retained. The annual loss of phosphorus calculated here, 0.7 mg, is only 0.003% of the 25 g P m^{-2} present in the top 10 cm of Pond B sediments, suggesting that the pond system is very close to a steady state condition in respect to phosphorus. The loss is especially small when compared to the P cycling each year in the biota. In the primary producers alone, at least 500 mg P m^{-2} is fixed each year (40 g C m^{-2}). A similar conservation of P was found in Hubbard Brook watershed where less than 1 mg P m^{-2} also was lost each year despite the hundreds of milligrams of P circulating (Hobbie and Likens 1973). In Char Lake, 18

TABLE 4-20 *Annual Phosphorus Budget for Pond B*

Parameter	Phosphorus form	Concentration g P liter^{-1}	Volume liters m^{-2}	Amount mg P m^{-2}yr^{-1}	
Winter snow	DRP	1.5	58	0.087	
	DUP	1.1	58	0.064	
	PP	1.4	58	0.081	
					0.232
Spring runoff (entering pond)	DRP	2.4	760	1.82	
	DUP	13.6	760	10.34	
					12.16
Spring runoff (leaving pond)	DRP	2.5	760	−1.90	
	DUP	15.4	760	−11.70	
					−13.60
Summer precipitation	DRP	7.2	65	0.468	
	DUP	0.7	65	0.046	
					0.514
Net balance					−0.694

mg P m^{-2} yr^{-1} (40% of input) is retained in lake sediments and individual streams entering the lake annually supply 1 to 7 mg P m^{-2} of watershed (deMarch 1975, 1978). Thus pool retention of phosphorus may be characteristic of arctic water bodies.

Phosphorus Cycling

In this section we report data on the uptake and release of phosphorus by planktonic and benthic animals and plants. The next section (Control of Phosphorus), deals with the movement of phosphorus between the sediment and the water.

Details of all the methods are given in Prentki (1976); isotope techniques for the plankton are also reported in Barsdate et al. (1974) and for the plants in McRoy and Barsdate (1970).

The seasonal picture of DRP concentrations is a chaotic one (Figure 4-16a). The reason for this is that the DRP cycles very rapidly. Some indications of this come from a series of DRP measurements in a pond over 32 hours (Figure 4-17). The changes were greater than 2-fold and concentrations appeared to increase when photosynthesis decreased (Figure 4-18). Some of the water was also incubated for 3.0 hours in plastic bottles in the ponds (Figure 4-17). The changes in these bottles were almost exactly the same as those in the pond; this indicates that processes in the water itself, rather than some advective or benthic process, caused the changes.

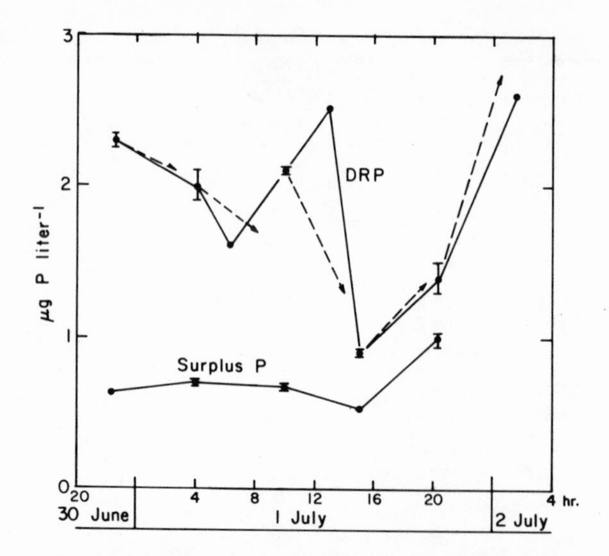

a) Concentrations of phosphate (DRP) and surplus phosphate. The dashed lines ending in arrows denote the direction of the change in concentration and the final phosphate concentrations in 3-hr bottle incubations. The lines at each point are the range of duplicate analyses.

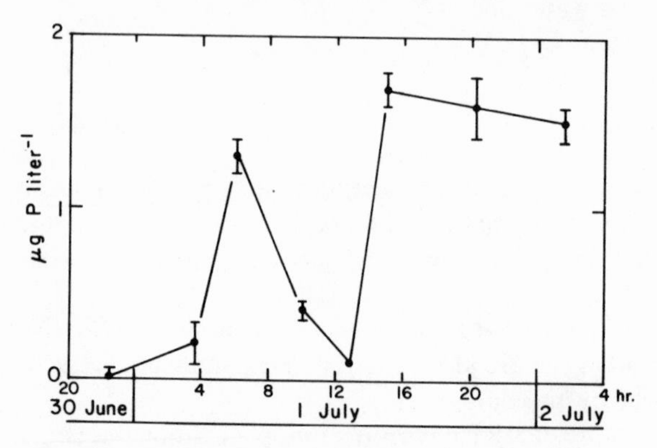

b) Concentrations of colloidal P in the same samples. The lines at each point are the standard error of duplicate difference calculations.

FIGURE 4-17. *Concentrations of DRP, surplus phosphate and colloidal P in Pond B, 30 June through 2 July 1972.*

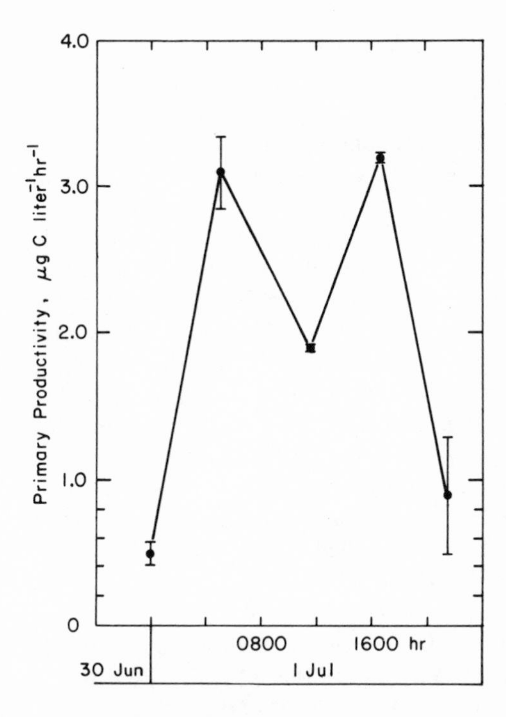

FIGURE 4-18. *Primary productivity in 3-hr bottle incubations, 30 June through 1 July 1972. The water was taken from Pond B, 3.1 μg phosphate-P liter^{-1} was added, and the bottles were incubated in the pond. The lines at each point indicate the range of duplicate samples.*

If the changes of DRP in the water are linked to photosynthesis, then the ratios of uptake of C to P are extremely low. For example, between the hours of 1200 and 1600, the photosynthesis was about 8 μg C liter^{-1} while the change in DRP was 1.5 μg P liter^{-1} (ratio is 5.3:1). If the phosphorus is being incorporated into all structures, then a (weight) ratio of 40:1 is expected. Therefore, much of the P is either being excreted or is being stored (luxury uptake).

Luxury consumption of phosphorus, however, did not account for any of the phosphorus uptake. Luxury storage was monitored throughout the experiment by following cellular surplus phosphorus (SP) concentrations (method of Barsdate et al. 1974). The SP procedure extracts the polyphosphates which constitute the luxury storage pools for P in algae (Fitzgerald and Nelson 1966, Rhee 1972). Throughout the duration of the experiment, SP averaged 0.7 μg P liter^{-1} and increased from 0.63 μg P liter^{-1} at the beginning of the experiment to 0.98 μg P liter^{-1} at the end. For the period from 1200 to 1600 on 1 July, SP decreased by 0.15 μg P liter^{-1} so the luxury storage can not account for the loss of 1.5 μg DRP liter^{-1}.

The radioisotope ^{32}P was used to investigate further the planktonic rates of cycling. The uptake rates were calculated from the initial exponential change in the concentration of $^{32}PO_4^{--}$ (Barsdate et al. 1974). For 17 measurements in the ponds we observed uptake rates of 13 to 320 μg P liter^{-1} day^{-1}, corresponding to turnover times of 0.23 to 4.6 hours. These are about 200-fold greater than necessary for growth of the planktonic algae. These turnover times were very similar to the 0.7 to 1.7 hours found for Char Lake (Rigler 1972) but were greater than the 1 to 8 minutes generally found in temperate lakes (Rigler 1964, 1973, Lean 1973a).

While some of the inorganic P is returned to the DRP pool from the particulate phase, most of it is transformed into dissolved organic phosphorus. These fractions in the ponds were first followed by a rapid and simple technique based on that of Kuenzler (1970). Water samples containing ^{32}P were acidified and extracted with isoamyl alcohol. Next, the orthophosphate reagents and more alcohol were added to the water and the phases separated (Barsdate et al. 1974). The fractions were also studied with a ^{32}P technique developed by Lean (1973b) which uses Sephadex columns to isolate high and low molecular weight pools. Both of these pools cycle rapidly; the high molecular weight material is called colloidal P and the low weight pool is called XP. Comparison of the extraction technique with Lean's method indicated that the extractable organic P pool was colloidal P and the non-extractable organic P pool was XP (see Prentki 1976 for details). In DRP analyses, pondwater colloidal P was hydrolyzed by the non-extraction but not by the extraction procedure.

The rate of cycling of the DRP to organic P and back to DRP is extremely rapid (Figure 4-19). The data are given in units of mg P m^{-2} for concentrations and mg P m^{-2} day^{-1} for rates. This assumes that each m^2 is overlain by 100 liters of water; these units are used so that later comparisons may be made with sediments. Some of the exchange between DRP and water particulates may be due to exchange with living cells rather than with detritus as indicated (Figure 4-19). However, in either case, the cycling through the organic P phase is not only rapid but is quantitatively the most important pathway (Figure 4-19); this is the first time in freshwater that a return through the organic phase to DRP has been found to be greater than the direct return from plankton.

Watt and Hayes (1963) found that dissolved organic phosphorus production from plankton was approximately 3-fold higher than phosphate production, but in this marine system there was no direct hydrolysis of dissolved organic phosphorus to phosphate. Lean (1973b) found phosphorus compartments and pathways in eutrophic Heart Lake similar to those present in the Barrow ponds. However, the direct return as phosphate from the particulates following uptake was 70-fold greater than dissolved organic phosphorus production.

The XP and colloidal P both lyse to phosphate. The XP is excreted directly by photoplankton (the dissolved organic phosphorus of Kuenzler (1970)) and bacteria (Barsdate et al. 1974). Lean (1976) hypothesized that colloidal P is composed of filaments broken off algae during filtration. However, in pondwater the concentration of colloidal P is much greater than total algal and bacterial phosphorus (Figure 4-19). Thus an earlier

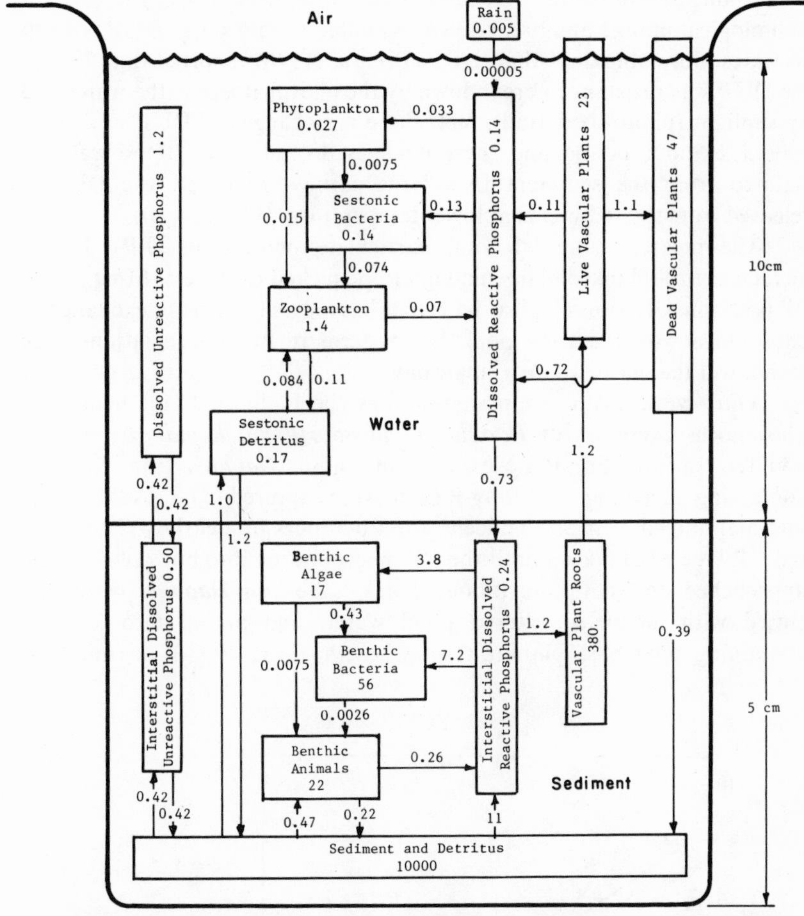

FIGURE 4-19. *Phosphorus flow diagram for Pond B, 12 July 1971. Rates are in mg P m^{-2} day^{-1} and concentrations are in mg P m^{-2}; a 10-cm-deep water column is assumed.*

suggestion of Lean (1973a) that colloidal P is formed by combination of XP and colloids appears more likely in the Barrow ponds.

The XP and colloidal P occur in significant quantities throughout the season; their sum was never observed to be less than that of DRP (Table 4-16, Figure 4-19). Since phosphorus in these two labile organic pools rapidly cycles through DRP, this means that phosphorus available for plankton growth is usually 2 to 4 times the observed DRP concentration. This alone could account for the lack of correlation between DRP and plankton productivity.

Other information on labile organic P comes from the experiment described by Figure 4-17 and 4-18. The highest amounts of colloidal P (and also XP) are formed during the periods of highest photosynthesis.

About 70% of the dissolved unreactive phosphorus (DUP) is resistant to biological uptake and breakdown. As noted (Table 4-16), DUP is up to 10 times more abundant than DRP. Prentki (1976) reported that 70% of the DUP was resistant to breakdown by the phosphatases in the water and by sunlight (photo-oxidation). Yet, there are changes in DUP in a pond over a 24-hour period and these changes do not occur if the water is isolated from the sediment in a large carboy. Evidently the DUP is released from the sediments at lower temperatures (Figure 4-20).

The measurements of total particulate phosphorus (PP) do not include the zooplankton large enough to be picked off filters. Most of this PP is sestonic detritus (Figure 4-19) that has only a low rate of exchange of phosphorus with the DRP pool. No patterns in PP concentrations were found over the season or on a single day.

Our work with zooplankton has been limited to measuring phosphorus excretion of *Daphnia middendorffiana, Lepidurus arcticus,* and fairyshrimp. For these experiments, adult *Daphnia* from Pond B, containing an average of 1.9 μg P each, were captured in late August 1971 and maintained at *in situ* concentration outdoors in a carboy. They were fed ^{32}P labeled plankton until the ^{32}P incorporation into body phosphorus approached an asymptote (about 2 days). Several *Daphnia* were then rinsed with unlabeled filtered pond water, and placed into a bottle containing unlabeled plankton spiked with 1 mg PO_4-P to minimize

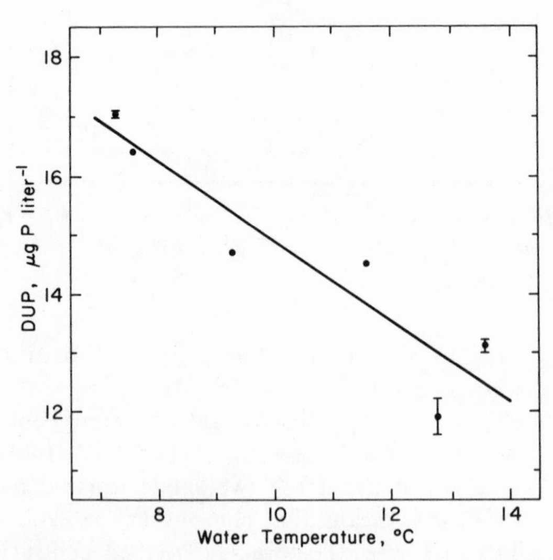

FIGURE 4-20. *Dissolved unreactive phosphorus concentrations and water temperatures during a diel cycle, Pond D, 23–24 July 1971. The lines at three points are the range of duplicate analyses.*

sestonic uptake of released phosphate. The added phosphate had no short-term effect on the *Daphnia* excretion rate. After 10 minutes, the *Daphnia* were removed from the bottle, the water was filtered and DRP-^{32}P was extracted from the filtrate. After correcting for sestonic uptake of ^{32}P, 60% of the excreted ^{32}P was found to be in the form of PP and 40% in the form of DRP. No excretion of DOP was found; this agrees with data from other species of *Daphnia* (Rigler 1961, Peters and Lean 1973).

In 1971, *Daphnia middendorffiana* made up 80% of the zooplankton biomass in Pond B in mid-July while fairyshrimp and copepods constituted only 10% each. Like *Daphnia,* fairyshrimp appear to excrete dissolved phosphorus predominantly as DRP while *Lepidurus* excretes DOP. However, on a quantitative basis, the movement of phosphorus through zooplankton other than *Daphnia* is minor and has been calculated from the same ratio of excretion to biomass as for *Daphnia.*

The overall turnover time of phosphorus in pond zooplankton is 29 days (Figure 4-19), a very low excretion rate. Direct dissolved phosphorus excretion measurements on *Lepidurus* in 1-hour experiments show the same thing as the rates were only 14-20% of that predicted by the Johannes (1964) regression model for zooplankton. Similar "too low" P excretion rates were reported for animals in coral reefs, another low phosphorus ecosystem, by Pomeroy and Kuenzler (1969). The low amounts of phosphorus in the reefs had their greatest effect on herbivore excretion; herbivorous fishes received just enough phosphorus in their diet to meet growth and reproduction requirements.

In Figure 4-19, it appears that phosphate excretion from zooplankton is sufficient to meet 100% of algal and bacterial phosphorus requirements for growth (0.13 μg P liter^{-1} day^{-1}). Barlow and Bishop (1965) also found that zooplankton in late summer in Cayuga Lake could regenerate sufficient phosphorus to meet growth requirements of phytoplankton during that period. Martin (1968), however, found that zooplankton excretion could exceed phytoplankton demand for phosphorus during periods of simultaneous low productivity and low phosphate, but not during periods of high phytoplankton abundance in Narragansett Bay. The situation varies from lake to lake; studies reviewed by Larow and McNaught (1978) found that zooplankton excretion provided 19-200% of the phosphorus needed for summer algal growth. In our study, the requirement for growth is negligible compared to the total quantity of phosphorus cycling through algae and bacteria.

Benthic bacteria were studied via decomposer microcosms (details in Barsdate et al. 1974). Phosphorus kinetics were investigated with ^{32}P in systems with only bacteria, with bacteria plus the ciliate *Tetrahymena pyriformis,* and with bacteria plus mixed microfauna populations. The systems were sampled at intervals and the radioactivity fractionated [see Figure 2 of Barsdate et al. (1974)]. Phosphate-^{32}P first appeared in cells in an organic form that had alcohol and phosphate-reagent extraction characteristics similar to XP. The XP-like organic phosphorus plus the inorganic phosphorus that quickly appears constitute a metabolic pool in the bacteria which equilibrates with extracellular phosphate many times faster than the whole organism does. Phosphorus in the XP-like phosphorus pool is then either excreted as XP or is incorporated into a bound phosphorus pool.

The results of the experiment with protozoan grazers and bacteria (Barsdate et al. 1974) showed that the grazers induced a higher rate of P turnover in the bacteria. For example, in the experiments, the uptake rate of bacteria cultures alone was 7.3 μg P liter^{-1} hr^{-1} (this is 2.5×10^{-7} μg P cell^{-1} hr^{-1}). In cultures with bacteria plus a ciliate, the uptake was 9.3 μg P liter^{-1} hr^{-1} (but the number of bacteria was smaller so the uptake was equal to 16.6×10^{-7} μg P cell^{-1} hr^{-1}). In the grazed system, the ciliates could ingest (and presumably excrete) 0.403 μg P liter^{-1} hr^{-1}. Thus P excretion was not responsible for the increased bacterial activity. This activity was very high indeed, and is greater than the 0.6 to 1.0×10^{-7} μg P cell^{-1} hr^{-1} reported for bacteria by Fuhs et al. (1972). The difference is presumably a result of physiological differences between rapidly dividing bacteria in grazed systems and relatively static bacteria of ungrazed systems.

In aquatic systems with macrophytes, the rooted plants may move otherwise unavailable nutrients from sediments into water either indirectly, by decomposition of litter (Pomeroy et al. 1972), or directly by live plant excretion into the water column (McRoy and Barsdate 1970; McRoy et al. 1972).

A *Carex* enclosure in Pond C, isolated from runoff by walls and from sediment by bottom ice, peaked at 122 μg DRP liter^{-1} during snowmelt. This concentration was 50-fold higher than that of pondwater or *Carex* exclosures, and corresponded to at least a 3.1 mg P m^{-2} release from *Carex* litter during snowmelt. Since this release of litter P takes place in part during spring runoff, an appreciable portion of the phosphorus mineralized is flushed out of the ponds and lost from the system, as described in the P budget. Phosphorus release from green-harvested *Carex aquatilis* litter in a pond was also rapid in summer (34 μg P (g dry wt)$^{-1}$ day^{-1}), and more than 60% of the initial plant phosphorus was released during the first month of immersion. Release apparently is a leaching process, rather than microbial decomposition, as controls in sealed, mercury-poisoned containers in the pond lost phosphorus at a similar rate. Both the magnitude and time span of the phosphorus release observed here are in good agreement with field measurements made on aquatic vegetation in the Kiev Reservoir (Korelyakova 1968).

Excretion of DRP into pond water by the leaves of live *Carex* plants was demonstrated in a series of laboratory and *in situ* experiments; the average rate (0.1 μg P (g dry wt)$^{-1}$ day^{-1}) was much less than the loss rate from dead *Carex* (see above). A somewhat greater amount (up to 1 μg P (g plant)$^{-1}$ day^{-1}) appears on the subaerial portions of the leaves. Some of this P release may be through guttation (Chapin and Bloom 1976). We have found 1 mg P liter^{-1} in drops on the leaf tips of tundra plants. We have assumed that DRP on the plant surface is washed into the water by rain or dew. The production of phosphorus from this source increases with increasing vascular plant standing stock, reaching a maximum daily production rate of 0.14 mg P m^{-2} of plant stand in early August. This input is important in this low phosphate environment and *Carex* is a phosphorus pump comparable to the freshwater *Nuphar luteum* at 0.22 mg P m^{-2} day^{-1} (Twilley et al. 1977). However, both are very poor phosphorus pumps relative to marine *Spartina alterniflora* at 600 mg P m^{-2} day^{-1} (Reimold 1972) and *Zostera marina* at 60 mg P m^{-2} day^{-1} (McRoy et al. 1972) in marine systems.

Phosphorus leaching from standing dead vascular vegetation is minimal at the mid-summer date of 12 July (Figure 4-19), but is much higher immediately after thaw and again in late summer as leaf senescence increases. Over a 100-day thaw season an average input of 0.72 mg P m^{-2} day^{-1} has been computed for this pathway. Therefore, when rates are compared over the entire summer, litter decomposition, which is several times greater than runoff and precipitation, is the most important source of phosphorus entering the water column.

The release of P from the *Carex* margin of the ponds can be compared to release by the *Myriophyllum spicatum*-dominated littoral of Lake Wingra, Wisconsin. Over the ice-free period and on a whole lake basis, weedbeds in Lake Wingra release 760 kg DRP + DOP or 2.5 mg P m^{-2} day^{-1} to the central lake basin (Prentki et al. 1979). Thus the contribution of phosphorus by *Carex* in the ponds is only about 3-fold less than that of macrophytes in a weedy, temperate, eutrophic lake.

The turnover time for DRP as a result of sediment sorption is 0.8 day, calculated from the disappearance of phosphate in whole-pond fertilizations. The pond water concentration divided by turnover time (0.14 mg DRP m^{-2} $\div$ 0.8 day = 0.18 mg P m^{-2} day^{-1}) is an estimate of the net uptake and release of DRP by sediment. An independent measure of the flux of DRP between water and sediment comes from a mass balance of other DRP pathways in Figure 4-19. This calculation indicates that there is a net flux of 0.17 mg P m^{-2} day^{-1} from water to sediment.

The turnover time of DRP is very short in the ponds. Hayes et al. (1952) reported a range of 2.4 to 40 days for 40 lakes, with the shortest turnover time occurring in the shallowest nonstratified lake. The 0.8 day turnover time in the ponds is due to their shallow depth and rapid convective and wind mixing.

Unlike the water column, the sediment is dominated by its abiotic components (Figure 4-19). Most of the standing stock of phosphorus in the sediment is sorbed inorganic P, reductant-soluble inorganic P, or dead organic P. In anion exchange resin experiments with shaken sediment, DR^{32}P and sorbed P equilibrated within 4 hr. Thus, the rate of exchange in units of Figure 4-19 was greater than 12,600 mg P m^{-2} day^{-1}. The *in situ* rate of exchange in undisturbed sediment would obviously be much slower but still more rapid than any other rate. Benthic bacteria in the ponds cycle large quantities of DRP through DOP (mostly XP) which lyses back to DRP. However, neither uptake nor release of P by the sediment algae plus bacteria compartment have been measured *in situ*. Uptake of DRP by vascular plant roots occurs throughout the sediment active layer and the 0.39 mg P m^{-2} day^{-1} represented in the flow diagram is for the entire thaw zone rather than just for the 0-5 cm layer.

CONTROL OF PHOSPHORUS

Introduction and Methods

Phosphorus availability in aquatic ecosystems is normally controlled by both extrinsic and intrinsic factors. In tundra ponds the extrinsic factor, phosphorus loading through runoff and precipitation, is unimportant because imports of allochthonous phosphorus are so small and are equalled or exceeded by pond exports (Table 4-19 and Figure 4-19). The intrinsic factors, those controlling the equilibrium and cycling rates, are most important.

The dominant source of recycled phosphorus in the water of tundra ponds is the vascular plants. These take up phosphorus from the sediment into their roots and then secrete a large amount of the phosphorus into the water while the plant is alive. The phosphorus is also leached from the plant material within a few days after the plant dies. The phosphorus secreted by the plants into the water is enough to drive the entire phytoplankton production for the year.

Thus, the supply of phosphorus to the water is adequate; the question is, however, what happens to the phosphorus after it reaches the water? Why are the phosphorus concentrations so low and why are there only small changes in the concentrations throughout the summer?

In this section, we will show that the phosphorus concentrations and supply rates are controlled by the buffering of the sediments. Almost all the phosphorus that enters the water column quickly moves to the sediment. There it is either strongly sorbed to the sediment or is retained in reductant-soluble (probably occluded within hydrous iron hydroxides) or organic form. In these three cases, the phosphorus is unavailable to planktonic organisms. The sorbed phosphorus will exchange with the overlying water but its availability is limited by the low concentration of DRP in equilibrium with the sediment and by the rate of sediment-pondwater exchange. The cold climate affects phosphorus cycling in the ponds most directly by slowing mineralization of organic phosphorus. Mineralization of this phosphorus to inorganic form would significantly increase the sorbed phosphorus pool and consequently DRP in the water column. A detailed discussion of this whole problem is given in Prentki (1976).

The factors controlling this partition of phosphorus between water and sediment have been investigated by correlation, isotherm, and mapping techniques. Sediments used in these investigations were sliced from cores or scraped from the sediment surface. Sediment sorption isotherms and a one-point derivative phosphate sorption index (PSI), were measured on wet sediments by procedures of Bache and Williams (1971). The inorganic phosphorus extracted by oxalate was determined in additional oven-dried (105°) subsamples of the same sediments utilized in PSI measurements by use of the procedure of Williams et al. (1971c), modified for smaller samples. This fraction is approximately equivalent to the sum of NH_4F-P through reductant-soluble-P in pond sediment (these are the Chang and Jackson phosphorus fractions discussed earlier). Organic phosphorus in sediment was usually measured as the difference between the phosphate concentration in the oxalate extract and that in an oxalate extraction of an ashed subsample (550°C). A few organic phosphorus determinations were also made by the difference between total phosphorus and inorganic phosphorus in the Chang and Jackson procedure described earlier and in Prentki (1976). The sediment inorganic phosphorus that is resin-exchangeable was determined by equilibrating sediments with an anion exchange resin and radiophosphate.

Phosphate in water and in digests was determined by the methods given earlier. Iron was measured by atomic absorption in many of the extracts and digests.

Phosphorus-Iron Correlations

Phosphorus and iron in tundra sediments and soils show strong correlation. Almost all the variation in total phosphorus content of the sediment within or between ponds could be related to parallel changes in sediment iron concentration. Thus, a regression (Figure 4-21) of all measurements of total sediment phosphorus against sediment iron concentrations, including those of samples collected in permafrost underlying the ponds, resulted in a correlation coefficient of 0.81 and the best fit line: $P(\mu g\ g^{-1}) = (18)\ Fe\ (mg\ g^{-1}) + 297$. The relationship between iron and phosphorus was much stronger than that between either of these and sediment organic content (as % organic weight) ($r = 0.71$ for phosphorus and $r = 0.47$ for iron). Thus the correlation could not be attributed to any coincidental association of both iron and phosphorus with the organic or inorganic sediment fractions. Exclusion of the sediment data from 0 to 10 cm from the iron-phosphorus regression improved the correlation and fit dramatically: $P = (22)\ Fe + 147$, $r = 0.91$.

Shallower sediments are enriched relative to deeper sediments in both iron and phosphorus (Figure 4-21). The weaker correlation and decrease in slope of the resulting correlation when surface sediment data are included indicate that the scatter in the upper sediments is caused by anomalous enrichment in iron. In Pond B, 0 to 10 cm sediments averaged 820 μg P g^{-1} (S.E.=77, n=7) and 26 mg Fe g^{-1} (S.E.=4.7, n=10) and 0 to 1 cm sediments 1080 μg P g^{-1} (S.E.=23, n=8) and 50 mg Fe g^{-1} (S.E.=2.5, n=12). These surface enrichments demonstrate a 1.9 concentration factor for iron but only a 1.3 factor for phosphorus. Thus iron appears to be differentially mobilized. The positive correlation within the water column of dissolved iron with dissolved organic carbon, with humic color, and with rainfall also suggests differential mobilization of iron, possibly through release from vascular vegetation of either soluble iron or compounds that complex iron. Such a mechanism would result in a net movement of iron from the sediments of the plant zone into the vegetation-free central basins. Only two sediment samples plotted in Figure 4-21 were collected from vascular plant zones. These two were depleted in iron relative to their phosphorus content (820 μg P, 15 mg Fe g^{-1} and 968 μg P, 16.2 mg Fe g^{-1}); therefore, the data are consistent with the hypothesis that the plant zone is the source of the iron that enriches the central basins.

Very few of the sediment samples discussed above, and none of the samples of deeper sediments for which the iron-phosphorus correlation was even stronger, were further fractionated into organic and inorganic phosphorus. Thus, there are no data on the mechanism behind the relationship.

Inorganic phosphorus and iron measurements indicate that sediment inorganic phosphorus is fixed onto a single phosphate-sorbing complex by

TABLE 4-21 *Correlation Matrix for the Results of a Fractionation of Phosphorus (Chang and Jackson) of 0 to 1 cm Pond C Sediments*

	1st NaOH–P	Reductant-soluble-P	NH_4F & 1st NaOH–P	Inorganic P	Organic P	Total Fe
NH_4F–P	−.67	−.62	.90	.80	.55	.82
1st NaOH	-	-	−.28	−.09	−.95	−.17
Reductant-soluble-P		-	−.22	−.02	−.91	−.08
NH_4F and 1st NaOH–P			-	.85	.16	.96*
Inorganic P				-	.00	.97*
Organic P					-	.15

*Significant at the 95% level of probability

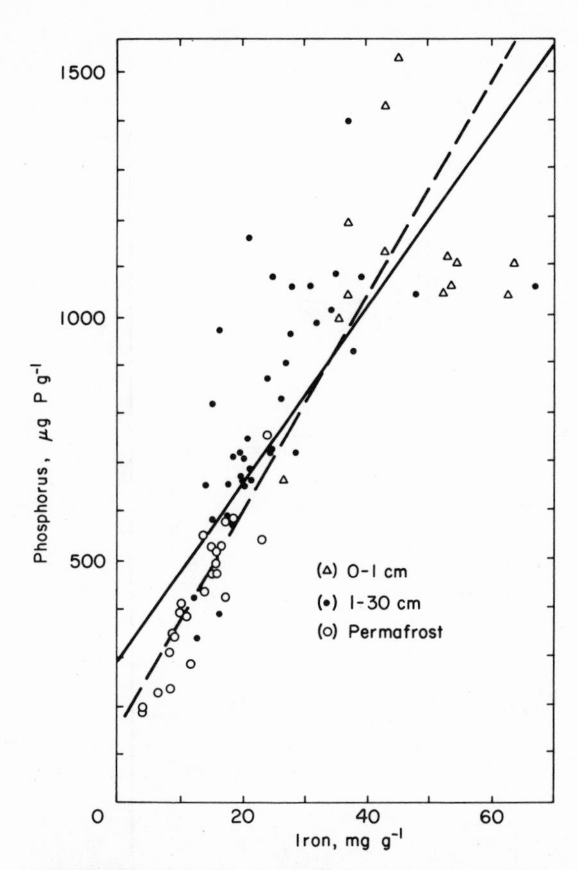

FIGURE 4-21. *Total phosphorus per gram of sediment of Barrow ponds vs. total iron per gram of sediment. The solid line denotes the least squares fit of all the data points while the dashed line denotes the least squares fit of data from sediments deeper than 10 cm.*

sorption, rather than by precipitation as a discrete mineral. This conclusion is supported by a correlation matrix of phosphorus fractionations (Chang and Jackson) and iron concentrations for the 0 to 1 cm Pond C sediment sample given in Table 4-18, and for three other 0 to 1 cm Pond C sediment samples described in Prentki (1976). The resulting correlation matrix for the four sediments (Table 4-21) shows a significant correlation with total iron for sediment inorganic phosphorus ($r=0.97^*$) and the sum of NH_4F-P+first NaOH-P ($r=0.96^*$), the two pools that contain most of the total inorganic phosphorus as well as the kinetically active portion of the inorganic phosphorus. These positive within-pond correlations with total iron support the hypothesis that sediment phosphate is bound by iron.

TABLE 4-22 *Correlation Matrix for the Fractionation of Phosphorus (Chang and Jackson) of Barrow Pond Sediments and Soils*

	NH_4F–P	1st NaOH–P	Reductant-soluble-P	NH_4F & 1st NaOH–P	Inorganic P	Organic P	Total P	Extractable Fe	Total Fe
Resin-exchangeable–P	.87**	.44	.50	.82**	.47	.72**	.84**	.57	.01
NH_4F–P	-	.46	.44	.92**	.52	.65*	.80**	.60*	.12
1st NaOH–P		-	.36	.77**	.51	–.01	.22	.28	.13
Reductant-soluble-P			-	.48	.94**	.16	.57	.78**	.64*
NH_4F and 1st NaOH–P				-	.94**	.46	.67*	.56	.14
Inorganic P					-	.17	.47	.68*	.62*
Organic P						-	.89**	.51	–.14
Total P							-	.76**	.16
Extractable Fe								-	.68*

*Significant at the 95% probability level
**Significant at the 99% probability level

A second correlation matrix (Table 4-22) constructed for the first five pond sediments and the soils (Table 4-18) and seven additional soils described in Brown et al. (in press) suggests similar relationships over the entire Barrow IBP site. For this matrix two additional parameters were available for regression: the resin-exchangeable phosphate, which is the phosphorus exchangeable with the radiophosphate sorbed on an anion exchange resin; and the extractable iron, which is the sum of iron solubilized by ammonium fluoride, plus the iron in the first sodium hydroxide extraction plus the iron in the reductant-soluble extractions (Chang and Jackson). In this matrix there were again strong correlations between phosphorus and iron parameters, especially the extractable iron. Extractable iron was better correlated with total phosphorus (r=0.76**) and reductant soluble phosphorus (r=0.78**) than either inorganic phosphorus (r=0.68*) or organic phosphorus (r=0.56); however, the relationship between total phosphorus and total iron found earlier for all sediment analyses (Figure 4-21) was not evident in these soils and sediments. The earlier relationship may have reflected homogeneity within a single drained lake basin.

The various correlations with the resin-exchangeable phosphorus are strong evidence that phosphate is bound to sediments by sorption and is not precipitated as discrete aluminum, iron, or calcium phosphate minerals. Resin-exchangeable phosphate is a measure of that portion of inorganic phosphorus which is sorbed on soil or sediment particles but which is still capable of desorption and interaction with the water phase. The NH_4F-P and first NaOH-P fractions are capable of exchange with phosphate dissolved in the water or of interaction with the water phase only in soils or sediments in which they make up sorbed phosphate pools. In contrast, in soils or sediments in which they make up discrete mineral phases there would be (1) no expected relationship between resin-exchangeable phosphate and any Chang and Jackson phosphorus parameter, and (2) no appreciable pool of resin-exchangeable phosphate. In these ponds the resin-exchangeable phosphorus pool accounts for 70% of the phosphorus in the NH_4F-P plus the first NaOH-P fractions. In addition there are highly significant correlations between the resin-exchangeable-P and the NH_4F-P fractions (r=0.87**) and between the resin-exchangeable-P and the NH_4F-P plus the first NaOH-P fractions (r=0.82**). All of these indicate that there is no discrete mineral formation. The actual regression between NH_4F-P plus the first NaOH-P fractions and the resin-exchangeable phosphorus indicates almost a 1 to 1 correspondence such that NH_4F-P plus the first NaOH-P=(1.1) resin-exchangeable-P+24.

The importance of sorption-binding in pond sediments, as suggested by these correlations, agrees with the theoretical stability field calculations for aluminum, iron, and calcium phosphate minerals given in Brown et al. (in press) for the pH, Eh, and ranges of cation concentrations found in the pore waters of Barrow sediments and soils. These calculations indicated that the phosphate concentrations existing in solution were too low to have been set by solubility criteria and, therefore, must have been set by sorption phenomena.

Other investigators have generally found that sorption rather than discrete mineral formation is the primary means of phosphate fixation within sediments, at least in lakes without intense calcium carbonate precipitation (Syers et al. 1973). Li et al. (1972) found that phosphate that could be exchanged with $^{32}PO_4^{--}$ constituted 14 to 43% of inorganic phosphorus in the sediments of four lakes, which is about the same range found for Barrow sediments and soils (see Table 4-18). Williams et al. (1971a, b, c, 1976) concluded that sorption to hydrated iron oxides was primarily responsible for phosphate fixation in aerobic and anaerobic sediments of both calcareous and noncalcareous lakes. Two earlier but similar studies (Frink 1969 and Harter 1968) considered aluminum to be more important than iron in phosphate fixation, but this conclusion depended upon sequential NH_4F and NaOH extractions releasing only aluminum-bound and iron-bound phosphorus, respectively. We and Williams et al. have found this assumption to be incorrect.

Phosphate Sorption to Sediments

In a sorption system, the relationship between water concentration, sorbent, and sorbed pool can usually be described by one of three mathematical functions (Freundlich, Langmuir, or Temkin isotherm) which are derived from different assumptions about the energy with which individual ions or molecules are held (Trapnell 1955). The applicability of these three isotherms to phosphate sorption has been discussed by Bache and Williams (1971) and need not be repeated here.

For the pond sediments, Prentki (1976) concluded that the Temkin isotherm best modeled the relationship between DRP and sorbed phosphate over the environmental range of concentrations. The Temkin isotherm may be represented by the equation (Bache and Williams 1971):

$$XX_m^{-1} = (RTb^{-1})(\ln g\, C),$$

where

X = sorbed phosphate per unit sorbent
X_m = sorption maximum
C = equilibrium DRP concentration
R = gas constant
T = temperature in K, and
b,g = constants.

This isotherm implies that energy of adsorption decreases linearly with increasing surface coverage; therefore, a plot of X against ln C or log C should be linear. Data from one experiment are plotted in this format in Figure 4-22. The other two, more frequently used isotherms assume either a constant binding energy (Langmuir) or an exponentially decreasing energy of adsorption with increasing surface saturation (Freundlich).

The Temkin equation gives a good fit over 3 orders of magnitude of phosphate concentration (Figure 4-22). Thus for this sediment, the equilibrium phosphate concentration is proportional to the antilog of the quantity of sorbed inorganic phosphorus; the less phosphate present, the more strongly it is held.

The meager literature on phosphate Temkin isotherms allows only a very limited comparison with the Pond B isotherm. The critical parameter for this comparison is the isotherm slope $dX(d \log C)^{-1} = 2.3\ RT\ X_m\ b^{-1}$, because with three unknowns in the isotherm equation the two that give the slope, X_m and b, cannot be separated. Thus, the slope is the only term available for comparison against the Temkin isotherms of other sediments or soils. The slope expressed either as the differential or the ratio of the constants is essentially a measure of the phosphate buffering intensity of the sediment. That is, it is a measure of the tendency of a solution in contact and in equilibrium with a sediment to resist a log C change resulting from addition or withdrawal of sorbed phosphate. Bache and Williams (1971) presented either Temkin isotherm slopes or graphs from which the slope could be calculated for four soils, at least three of which were cultivated (presumably fertilized) soils. These slopes ranged from 65 to 310 (μg P g^{-1})/(log μg P liter^{-1}), much less than the 690 calculated for

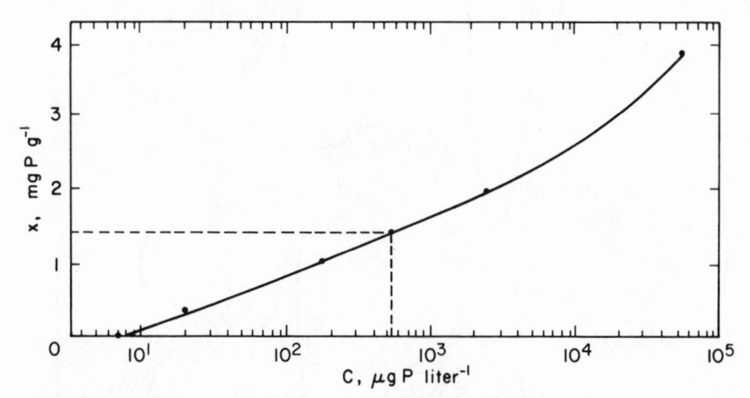

FIGURE 4-22. *Results of a sediment–phosphate sorption experiment in Pond B (0 to 10 cm). Data are plotted as a Temkin isotherm. The intercept gives the position of a one-point PSI determination: 1490 (μg P g^{-1}) ÷ (log 537)(C in μg P liter^{-1}) = 545.*

Pond B sediment. This indicates that all of these soils had a much smaller phosphate buffering capacity than the Barrow sediment.

The Temkin isotherm slope $dX(d\log C)^{-1}$ can be approximated by a one point estimator, the phosphate sorption index (PSI), equal to the ratio of X to log C obtained by adding phosphate to a sediment at a ratio of 1500 μg P to each gram of soil. Bache and Williams (1971) found that this estimator was highly correlated with isotherm slope (r=0.97) in 42 soils. A PSI of 545, calculated for the Pond B sediment in Figure 4-22, underestimated the real isotherm slope by 21%. This underestimate is not severe and the relative error involved in comparing two PSI values would normally be less, especially if similar phosphate concentrations were in equilibrium with the two sediments before addition of phosphate. The PSI determinations on 0 to 3 cm sediments from ponds in the IBP watershed (Figure 4-29) ranged from 318 to 728 (average of 532). The PSI of the intensively studied ponds A-B, C, D, E, J, and X ranged only from 359 to 540. In comparison, the highest PSI obtained by Bache and Williams (1971) from 42 soil samples was only 382 (our units). Meyer (1979) in a study of Bear Brook in the Hubbard Brook Experimental Forest found that sediment PSI averaging 10.3 for silt and 2.1 for sand were sufficient to buffer dissolved phosphorus concentrations at 2 μg P liter^{-1}. The phosphate buffering intensity of pond sediments is 50 to 250 times higher.

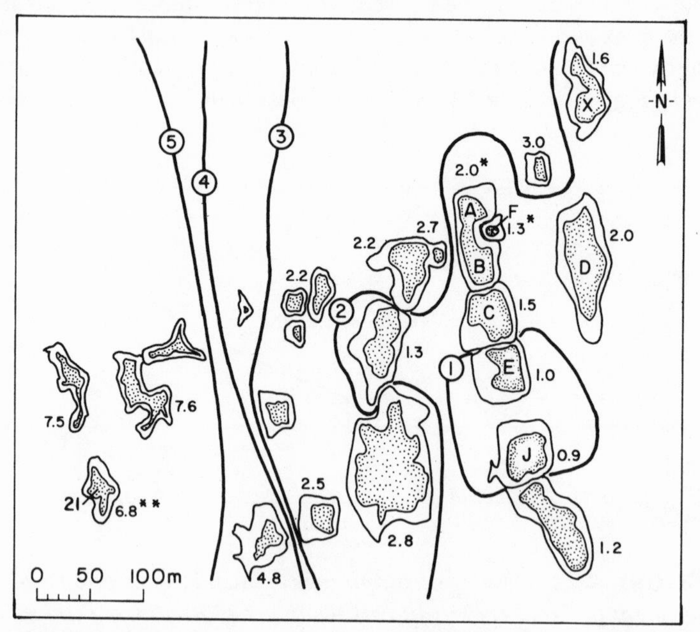

FIGURE 4-23. *Concentrations of DRP (μg P liter⁻¹) in the water of ponds. Concentrations in Pond B and in Pond F (*) are averages from August 1970 and 1971. The concentration in Pond 21 (**) is from 28 August 1970.*

DRP Variations in IBP Ponds

Based on the relationships identified in the previous sections, it is possible to explain the variations in the dissolved reactive phosphorus (DRP) among the IBP watershed ponds.

Pondwater DRP concentrations across the watershed are not randomly distributed, but decrease downslope (Figure 4-23). Although this gradient is in part related to geomorphology (high-centered polygons and high-phosphorus trough ponds occur upslope and low-centered polygons and low-phosphorus ponds occur downslope), the gradient should have an underlying geochemical basis. The six ponds, A-B, C, E, F, J, and 17 (see Figure 3-1), all lie within the youngest drained lake basin and therefore are the watershed's youngest ponds; in addition, all fall within the lowest two DRP contours. The immediate factor controlling DRP concentrations across the watershed should be the phosphate sorption by sediments which in turn is controlled by the concentration of exchangeable inorganic phosphorus present in the sediments. Thus, a plot of DRP against the amount of oxalate-extractable inorganic phosphorus in the sediment (oxalate Pi) as a Temkin isotherm (Figure 4-24) results in significant linear correlations for both: oxalate Pi (μg P g^{-1})=(1151) log DRP (μg P liter^{-1}) -22 with $r=0.56$*. Several important factors may account for the 69% of the variance not explained by the regression. One factor is that the oxalate reagents used extract the reductant soluble phosphorus (after Chang and Jackson) in addition to phosphate in the two Chang and Jackson extracts (NH_4F-P plus NaOH-P) thought to constitute the exchangeable inorganic pool. This overestimate is highly variable as the Chang and Jackson extraction data for the four surficial Pond C sediments used to construct Table 4-21 would indicate a potential overestimate of 0 to 43%. Additional scatter in Figure 4-24 would be expected due to the combination of sediment heterogeneity and the extremely small samples used in obtaining the oxalate Pi numbers (single 0 to 3 cm section of 2.5 cm-diameter cores). Scatter should also result from changes in the DRP of the water

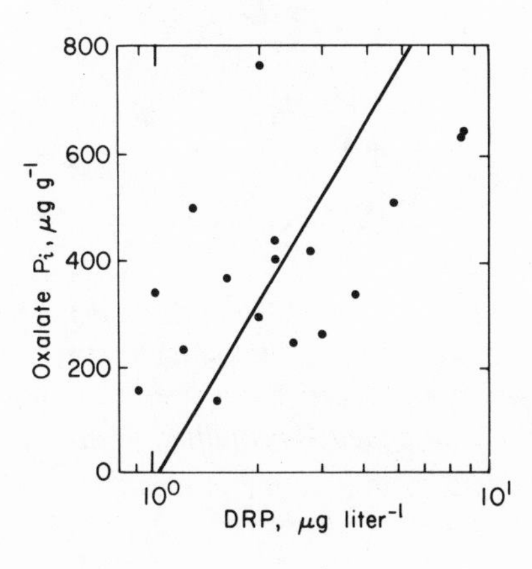

FIGURE 4-24. *Measurements of DRP and oxalate Pi in ponds of the IBP watershed plotted as a Temkin isotherm.*

pool caused by the other phosphate cycling pathways denoted in Figure 4-19 (plankton uptake, organic phosphorus hydrolysis, vascular plant secretion, litter leaching, etc.). In spite of the scatter caused by these pathways a significant correlation still exists betwen oxalate Pi and log DRP. This illustrates the effectiveness of the buffering and control of pondwater DRP concentrations by pond sediments.

Sediment Inorganic Phosphorus in IBP Ponds

Secondary control of phosphate concentrations is also exerted by whatever factors influence concentrations of inorganic phosphorus in the sediments across the watershed. Oxalate Pi concentrations are lowest within the most recently formed ponds and then increase both up- and downslope (Figure 4-25). Overall, the contours are similar to those for DRP. Sediment organic phosphorus (OP) concentrations are highest in a broad band across the watershed center that contains all but two of the low-centered polygon ponds and none of the trough ponds sampled (Figure 4-26). As a result, there is a negative correlation between oxalate Pi and OP:

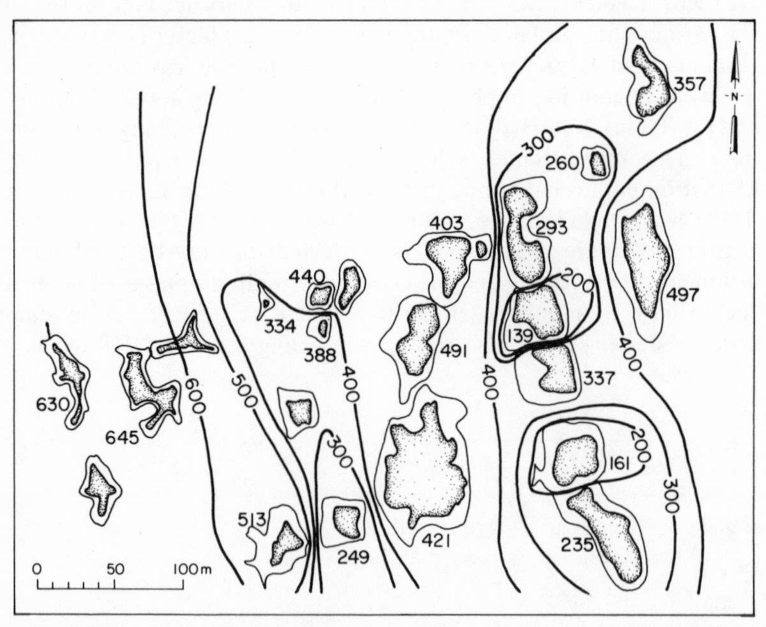

FIGURE 4-25. *Concentrations (μg P g⁻¹) of oxalate Pi in the top 3 cm of sediments of ponds in the IBP watershed. The phosphorus added in fertilization experiments in Ponds D and X has been subtracted under the assumption that all the added P remained in inorganic form in the top 3 cm.*

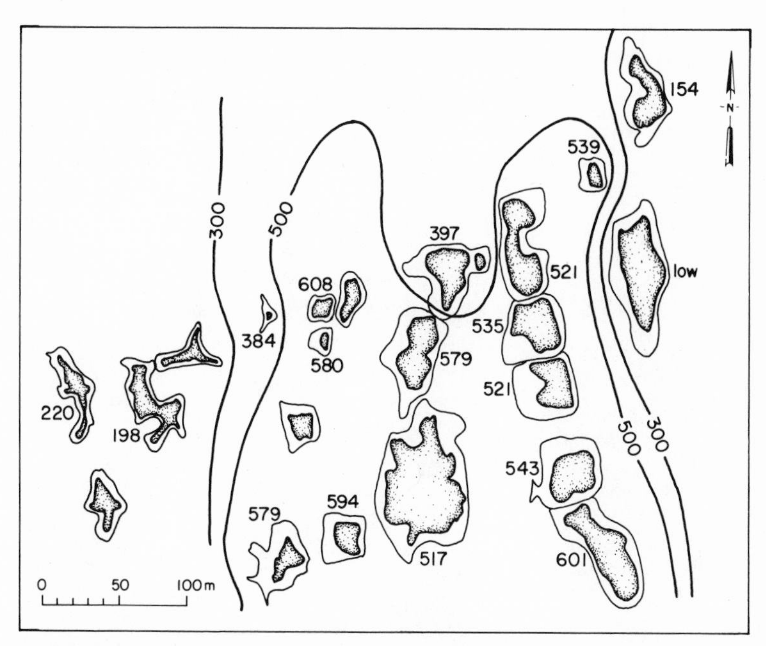

FIGURE 4-26. *Concentrations (μg P g⁻¹) of organic phosphorus in the sediments of the IBP watershed ponds.*

oxalate Pi (μg P g^{-1})=(-0.41) OP (μg P g^{-1})+564, (r= -0.50*), suggesting a low rate of *in situ* mineralization. If Pond B is assumed to be 3,000 years old (Chapter 3), then the annual net mineralization rate necessary to produce the organic and inorganic phosphorus concentrations would be less than 0.1 μg P (g sediment)$^{-1}$. This mineralization hypothesis would require low-centered polygons, such as those that existed before the ponds, to retain most of their phosphorus in organic form. The single analysis of a low-centered polygon soil indicates that 82% of the soil phosphorus is in organic form; this is consistent with this hypothesis. Additionally, the five "new" pond sediments from within the former lake bed have five of the seven highest OP to total P ratios (0.61 to 0.79) in the 18 samples of the watershed sediments.

An alternative hypothesis arises from the Pond B phosphorus budget (see Table 4-20). The average net gain of inorganic phosphorus of 0.5 mg P m^{-2} yr^{-1} and the net loss of organic phosphorus of 1.2 mg P m^{-2} yr^{-1} suggest a slow leaching of organic phosphorus and an even slower capture and binding of transient phosphate anions. The gain of 0.4 unit of inorganic phosphorus per unit loss of organic phosphorus is identical to the regression slope of -0.4 for oxalate Pi against OP for ponds in the watershed. This hypothesis implies that these tundra ponds are not in steady state, but are extremely slowly, over thousands of years, increasing their inorganic phosphorus concentrations. This does not necessarily mean

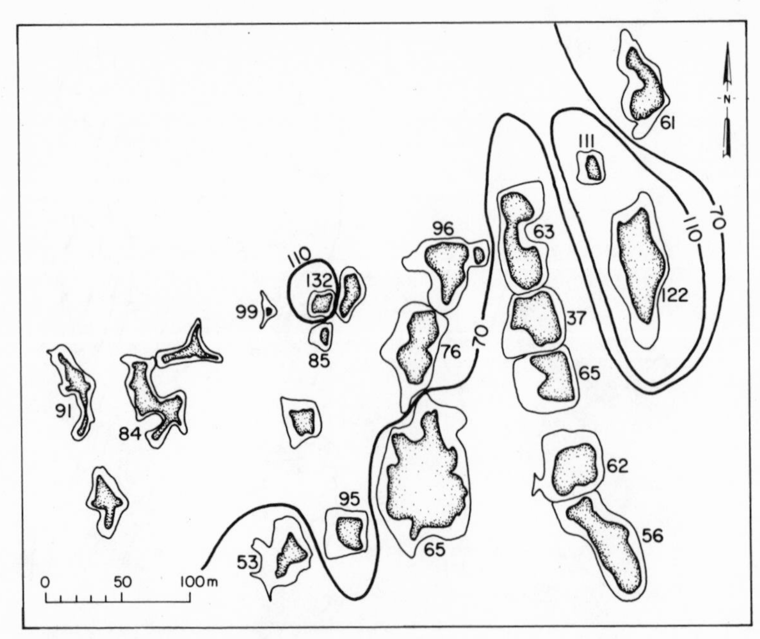

FIGURE 4-27. *Concentrations (μg Fe g⁻¹) of oxalate Fe in the top 3 cm of sediment of ponds in the IBP watershed.*

that they are becoming less phosphate-limited, for pond aging may also be a factor in the accumulation of iron in the surface sediment. The young ponds within the former lake basin are relatively low in surficial iron as they average only 57 mg oxalate-extractable iron g^{-1} (oxalate Fe), some 30% below the mean of watershed ponds (Figure 4-27). This oxalate-extractable iron is measured by atomic absorption on oxalate Pi extracts.

In these surficial pond sediments, unlike the sediments previously presented in Tables 4-21 and 4-22, inorganic phosphorus and iron parameters were not significantly correlated (r=0.34). If, as suggested above, neither sediment inorganic phosphorus nor iron concentrations are at steady state but are instead slowly increasing, then ratios of phosphorus to iron would be partially dependent upon pond age and any dependence of the distribution of inorganic phosphorus on iron would be masked by the aging effect. Sediment with a lower ratio of inorganic phosphorus to iron-sorbent would also be expected to remove more phosphate from solution and would bind it more strongly. Therefore, in ponds of similar origin and age those ponds with higher iron content should also accumulate more inorganic phosphorus.

This hypothesis was tested by graphing concentrations of oxalate Pi against oxalate Fe concentrations of those ponds in the watershed that make up a single age class and that have a common origin (B, C, E, J, and 17, Figure 4-28). The resulting correlation was not significant (r=0.70)

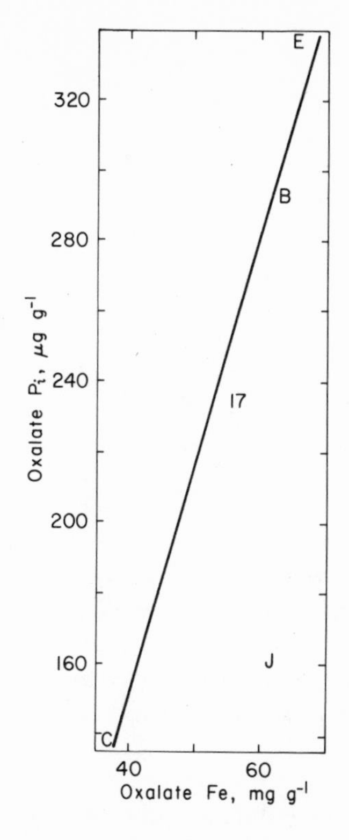

FIGURE 4-28. *Concentration of oxalate P and oxalate Fe in the sediment of some ponds of the IBP watershed. Oxalate Pi = (6.5) oxalate Fe - 110.*

unless Pond J, with a "low" oxalate Pi to Fe ratio, was dropped from the regression. When this was done, the data from the remaining four ponds all fell close to the regression line: oxalate Pi ($\mu g\ g^{-1}$)=(6.5) oxalate Fe (mg g^{-1}) -110, (r=0.98**). Pond J was destructively sampled during the IBP program and the low ratio of oxalate Pi to Fe could be due to destruction of a surface layer that was built up over 3,000 years.

Sediment sorption control of pondwater DRP concentrations across the watershed can also be illustrated by an indirect bioassay approach utilizing primary productivity rates (provided by M. C. Miller). In Figure 4-29 the log of midsummer phytoplankton primary production has been plotted against phosphate sorption indices of the individual pond sediments. All of the intensively studied ponds fall on or near a line denoting an inverse, semi-logarithmic relationship between index and phytoplankton production. This in turn suggests that there is a linear correlation between the low, midsummer primary production and the natural equilibrium phosphate concentration maintained by the sediment.

Sediment sorption equilibria can set only a lower limit on phytoplankton productivity. The rate of equilibration between water and sediment is identical to the 0.8 day turnover time of pondwater DRP. In spring, when runoff is high, and in late summer as both zooplankton

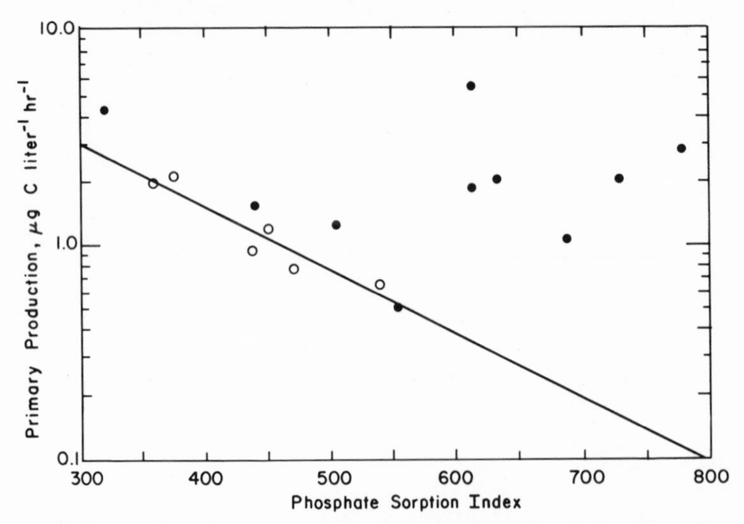

FIGURE 4-29. *Pondwater primary productivity (9 August 1973) and phosphate sorption indices for the underlying sediment. The intensively studied IBP ponds (A–B, C, D, E, J and X) are denoted by hollow circles and the other watershed ponds by solid circles. The regression line shown is for the intensively studied ponds only.*

grazing and macrophyte senescence peak, phytoplankton can greatly exceed this lower limit for productivity.

If additional watershed ponds, especially those with phosphate sorption indices over 625, are included in the regression, the correlation disappears. Those ponds with open centers and small macrophyte populations tend to fall closest to the original regression line. No pond falls much below the line, as is to be expected if the regression line defines a lower limit for phytoplankton productivity. Few details are known of these additional ponds; either higher internal loading of phosphorus or 1-3 week earlier fall blooms in the disparate ponds than in the intensively studied ponds would account for their apparently elevated productivity.

Conclusions

Only small changes are observed in seasonal pond phosphorus concentrations. This is not so much a result of low rates of phosphorus supply and utilization as it is of strong buffering action by sediment. Partly because of this buffering, phosphorus-sensitive biological indicators such as primary productivity more closely follow seasonal trends in phosphorus supply rates than seasonal trends in concentration.

Phosphate concentrations in the water column can, to some degree, be related back to inorganic phosphorus levels of the surface sediment.

Low pond water phosphate concentrations are caused more by retention of most sediment phosphorus in unavailable organic or reductant-soluble inorganic form and by strong sorption of the rest by sediments than by low sediment total phosphorus levels. Either mineralization of sediment organic phosphorus or transformation of reductant-soluble phosphorus to an exchangeable form would result in a several-fold larger pool of exchangeable inorganic phosphorus and in an increase in phosphate concentration in the water column. Thus, the low productivity of the ponds is related to the low temperatures which keep much of the phosphorus tied up in the sediments. Higher temperatures would cause higher decomposition rates and a consequent higher rate of supply of phosphorus. This would not necessarily remove the phosphorus limitation on growth of algae but would allow much higher rates of primary production.

ORGANIC CARBON

Dissolved Organic Carbon

Most of the carbon in the pondwater is present as dissolved organic carbon (DOC). The DOC is composed of a tremendous variety of compounds but can be roughly divided into two parts; the larger of these is high molecular weight compounds such as polymerized humic acids, and the smaller is low molecular weight compounds such as simple carbohydrates, amino acids, and small organic acids.

One reason for the high concentrations of high molecular weight compounds is that they are resistant to biological breakdown. Many of them originate from the terrestrial system and enter the ponds during runoff. Thus, they are the terminal products of decomposition in soils. Of course, these resistant compounds are eventually broken down, perhaps by an enzymatic hydrolysis or a photo-decomposition by UV, and the smaller products are used by bacteria.

In contrast, the low molecular weight compounds are readily used by bacteria; this holds the concentrations at a very low level. These labile compounds are derived mainly from algae and macrophytes by leakage and decay although a portion also arises from breakdown of the long-chain compounds.

Total DOC

The organic material passing through a Reeve Angel 984 H glass fiber filter, which has a mean pass of 0.8 μm (Sheldon 1972), is called dissolved. Organic material retained by the filter is called particulate although it is likely that some organic colloids are also retained by the filter. Total DOC was measured as CO_2 released from an acidified filtrate by a wet combustion with acid-persulfate at 105°C in nitrogen-purged ampules (in 1971) or by a combustion (650-750°C) in a tube furnace (in 1972, 1973). The amount of CO_2 was measured with an infrared gas analyzer (Menzel and Vaccaro 1964). In 1971, triplicate 1-ml ampules were prepared weekly for each of several ponds. For each sample in 1972 and 1973, a minimum of four replicate injections of either 25, 50 or 100 μl from a Hamilton syringe were averaged.

The seasonal cycle of DOC (Figure 4-30a) begins with low concentrations during the melt period in June and ends with high concentrations at freezeup in September. Snowmelt water contains little

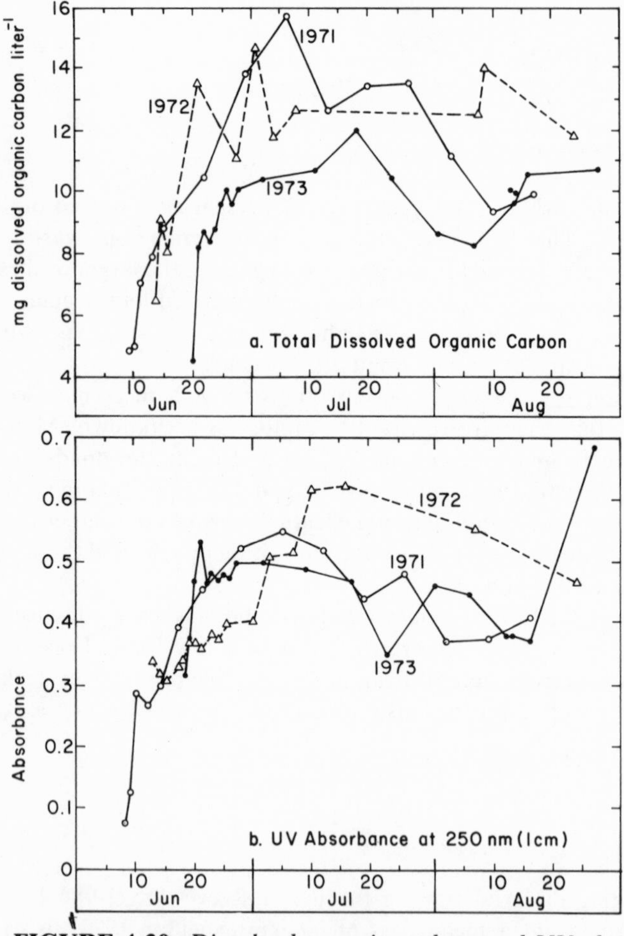

FIGURE 4-30. *Dissolved organic carbon and UV absorbance at 250 nm (1 cm) in the water of Ponds A and B, 1971–1973.*

DOC (measurements of 0.4 and 3.3 mg C liter^{-1} in Pond B, 1971) but once the sediments begin to thaw the DOC concentrations increase rapidly. In Pond B, the maximum was reached on 5 July 1971 (15.6 mg C liter^{-1}), 30 June 1972 (14.6 mg C liter^{-1}) and 16 July 1973 (11.9 mg C liter^{-1}). Through much of the summer the concentration was fairly constant at 10 to 14 mg C liter^{-1}. However, in late August of 1971 and 1972, the concentration of DOC was changed by rainfall and its runoff (Figure 4-30a). We have no data for concentrations during freezing, but Kalff (1965) has shown that in the autumn under the ice cover, the color (measured in Pt units, a relative measure of DOC) increased to seasonal maxima just before the ponds froze solid.

The sources of organic carbon are: (1) leachates from the sediments including re-solution of the previous year's DOC, (2) leachates from dead emergent sedges in the flooded peripheral areas around the ponds, (3) leachates of soil, sedge and lemming feces washed into the ponds by overland snowmelt and rainfall runoff, (4) leakage of photosynthate from aquatic sedges and grasses (*Carex aquatilis* and *Arctophila fulva*), (5) leakage by planktonic and epipelic algae during photosynthesis, and (6) excretions and autolysis of living and dead chironomids, oligochaetes, flatworms, zooplankton, tadpole shrimp, and other fauna and flora.

Much of the increase in DOC early in the thaw season comes from the re-solution of DOC which was frozen out in the previous autumn. Evidence for this comes from a series of 1.4-m^2 subponds, set up in 1970 in ponds B and C. These excluded runoff water but included sediments. Hence all the total DOC in these subponds in the early summer of 1971 had to be redissolved from the sediments as there could be no new DOC from new grass detritus and runoff water. By 21 June 1971, nine days after the ponds thawed to the bottom, the subponds contained only 50% and 64% of the DOC in the open-pond waters. By 7 July 1972, they contained 55% and 38% of the open-pond DOC. In another type of measurement, surface sediments (Pond J) were incubated in distilled water for six days (25 to 28°C) and the DOC determined. This leachable DOC was 1% of the total sediment carbon in June but was only 0.1% in August. Thus, the leachable DOC in the sediments decreased by an order of magnitude over the summer.

A second major source of DOC appears to be the leachates from dead emergent grasses and sedges. Each fall, the leaves of emergent aquatic plants die but most remain as standing dead. The next spring, there are three types of leaves present: green (grown during the previous summer but frozen in ice over the winter), yellow (grown during the previous summer but overwintered beneath snow), and brown (grown during the summer before the previous summer). When the green leaves of *Carex aquatilis* were placed in 20-μm mesh litterbags and put into the pond, the leaves lost 47% of their initial weight in 33 days. By means of tests described in Chapter 8, it was determined that 48% of the loss was caused by bacterial activity and 52% by leaching.

When older leaves were placed in plastic bags and allowed to decompose, it was found that 15 to 20% of the plant carbon remained as DOC; we believe this DOC is resistant to bacterial breakdown (refractory). This was shown in experiments with yellow *Carex* leaves placed into plastic bags (Whirlpak) along with distilled water and incubated next to the litterbags. The loss of dry weight from the *Carex* in the plastic bags was very close to losses from the litterbags. This loss equalled the DOC that accumulated in the plastic bags (controls of plastic bags plus distilled water had no increase in DOC). After one summer of incubation in plastic bags, the yellow *Carex* lost 25% of its dry weight and 19% of that dry weight was found as refractory DOC. Over the same period, brown *Carex* lost only 16% of its initial dry weight while 15% remained in the plastic bags as DOC.

There is also evidence that most of the loss from the overwintering leaves occurs in the first 2 weeks after the thaw. It is likely that in the older leaves this rapid loss is caused by mechanical rupture during the freeze-thaw process. In the green leaves, the initial loss is likely from the easily soluble material.

From the plastic bag and litterbag experiments and from production estimates for the emergent plants, we estimate that there is a total input to the pond of 3.9 g C m^{-2} of refractory DOC from sediments and macrophytes and 5.6 g C m^{-2} of labile DOC from the grasses and sedges (Table 4-23). This was calculated from the *Carex* annual production [80 g C m^{-2} (see Chapter 5)], the size of the *Carex* bed (40% of total pond), and the loss rate from the bag experiments. Thus, the refractory DOC added to the ponds from yellow *Carex* is 1.8 g C m^{-2} and that from brown *Carex* is

TABLE 4-23 *Summary of the Sources of Dissolved Organic Carbon in Pond B in 1971**

Source	Refractory humic DOC (g C m^{-2} yr^{-1})	Usable or labile DOC (g C m^{-2} yr^{-1})
Sediment re-solution	1.4	0
Macrophyte leaching	2.5	5.6
Phytoplankton	-	0.1
Epipelic algae	-	0.9
Emergent plants	-	0.9
Zooplankton	-	1.0
Subtotals	3.9	8.5
Grand total	3.9	+ 8.5 = 12.4 g C m^{-2} yr^{-1}

*Depth of the pond was taken as 20 cm.

0.6 g C m^{-2}. During the third year, some 54% of the dry weight of the plant matter is lost as small particulate material (less than 20 μm) whose leaching rate is probably much like that of the surface sediments (Table 4-23). The leaching of old leaves and the re-solution from the sediments add humic compounds which are resistant to further decomposition and which accumulate in the pond. From all these calculations, we estimate that 2.4 g C m^{-2} of refractory DOC enters the ponds from grasses and sedges [this is 87% of the 2.8 g DOC m^{-2} found in Pond B in early July (20 cm average depth)], and another 1.4 g C comes from re-solution from the sediments. In view of the relatively small amount of labile DOC that is estimated to be present at any one time (maybe 10%), it is surprising that leaf leaching contributed 5.6 g C m^{-2} of labile DOC each year. Most of this comes from the early season leaching of the grasses and sedges and almost all of it is used by planktonic and benthic bacteria (see Chapter 8). Macrophytes were also an important source of DOC in Lake Wingra where they cover 31% of the lake. Prentki et al. (1979) measured a movement of 15 g C m^{-2} of labile DOC from the submersed weedbeds to the open water.

In August of all 3 years, there was sufficient rain to raise the pond water level; yet, contrary to our expectation, the DOC concentration increased. Sometimes the rainfall appears to dilute the DOC, and at other times it appears to increase DOC by leaching vegetation. One explanation may be that the August rains frequently occur after early frosts have damaged terrestrial vegetation so that cell contents are easily leached. Another possibility is that overland runoff, which quickly reaches equilibrium with the soil solution, is entering the ponds. This runoff was observed on 19 to 26 August 1971, for example. The rain can not add appreciable DOC, for even if it contained 1 mg DOC liter^{-1}, 5 cm of rain during the summer would add only an insignificant 0.05 g DOC m^{-2}. (Table 4-23).

In 25 ponds contained in the same drained lake basin as the IBP ponds, total DOC concentrations ranged from 10.2 to 34.5 mg C liter^{-1} in 1972 and 1973. Some of the highest concentrations were found in the ponds formed in polygon troughs.

Humic Compounds

The ponds contain very brown water, equivalent to the color of water in bogs. The color can reach 370 ppm on the Pt-Co scale (Kalff 1965). Although we can only speculate, it is likely that the origin of the color is similar in these ponds and in bogs—the leachate from dead emergent vegetation and organic deposits in the littoral and benthic zones.

These colored materials are soluble humic compounds and tannins, mainly of terrestrial origin, which are resistant to bacterial breakdown. They have been classed generally as aromatic polyhydroxy methoxy

TABLE 4-24 *Total Dissolved Organic Carbon Concentration and Humic Absorbance in 1973 Pond Transect, 9 August According to Pond Origin and Dominant Vegetation*

Pond origin:	Trough ponds			Low-centered polygon ponds		
Dominant vegetation:	Moss ponds		*Arctophila* ponds		*Carex* ponds	
	$OD_{250\ nm}$	mg C $liter^{-1}$	OD_{250nm}	mg C $liter^{-1}$	OD_{250nm}	mg C $liter^{-1}$
Average	0.922	20.4	0.560	17.5	0.561	16.4
n	5	5	10	10	10	10
SD	0.200	7.2	0.148	4.6	0.204	6.7

carboxylic acids (Black and Christman 1963, Ghassemi and Christman 1968). They frequently polymerize reversibly and form colloids as a function of pH, temperature, and divalent cation concentration. Most of the humic compounds have a quinone-like nucleus, that is, an unsaturated double bond coordinated with an aromatic ring, which absorbs ultraviolet light strongly and fluoresces.

For relative quantification of humic compounds dissolved in pond waters, we used their absorbance of ultraviolet light at 250 nm in filtered water (984 H glass fiber filters). Samples were read at 25°C at the usual pond pH's.

Ultraviolet absorption of pond water over the season tended to follow the DOC (Figure 4-30a and 30b), ranging from an absorbance of less than 0.070 at 250 nm (1 cm) in unthawed Pond B to 0.545 by 5 July 1971. Like the total DOC concentration, the UV absorption increased rapidly in the post-thaw period. The relationship between DOC and absorbance at 250 nm was significant in 1971 ($r = 0.81$ to 0.94) for Ponds B, C, D, E. In 1972 and 1973, this correlation was not significant.

In general, trough ponds all have higher amounts of color than low-centered polygon ponds and frequently have some of the highest concentrations of DOC. Trough ponds drain the ice-wedge troughs which act as a repository for leachable lemming-cut grass, winter nests of lemmings, and lemming feces. In addition, the input of runoff and

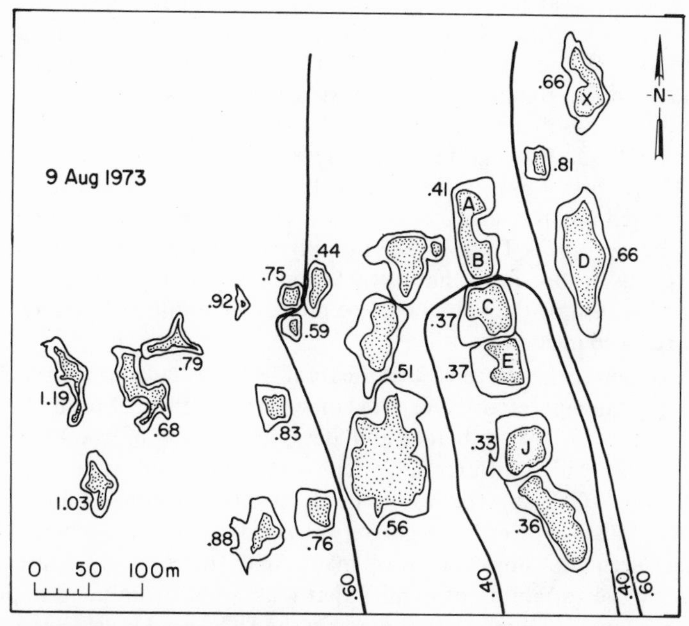

FIGURE 4-31. *UV absorbance at 250 nm in the water of some ponds near the IBP aquatic site, 9 August 1973.*

drainage water per unit of pond volume is much higher for trough ponds than for low centered polygon ponds.

If the ponds sampled in the transect in 1973 (Figure 4-31) are classed by dominant vegetation in or around the pond, then the moss ponds are significantly more colored than the *Carex* and *Arctophila* ponds ($t = 4.05$, $t_{.01(23)} = 2.807$) (Table 4-24). The conclusion could be drawn that the moss vegetation forms more materials that absorb UV light. However, an interpretation based on dominant vegetation is confounded by the origin of the two pond types, as all moss ponds are trough ponds and all *Carex* and *Arctophila* ponds are low-centered polygon ponds.

Furthermore, these pond types and vegetative types are geographically separated on the shelf of an old lake basin. The ponds with the lowest UV absorbance (color correlates with absorbance at 250 nm) in 1972 and 1973 lie in the most recent lake basin formed on the shelf of the much older lake (Ponds A, B, C, E, J, 24 and 25) (Figure 4-31, Figure 3-1). The ponds on either side of the old lake shelf (Ponds X, 4 and D on the eastern side of the shelf and Ponds 2, 23, 1, 10, 26, 9, 8, 22 and 3 on the western or in-shore side) had a higher UV absorbance. Interestingly enough, these intermediately colored ponds (absorbance at 250 nm was 0.4 to 0.8) around the most recently drained lake basin are most often *Arctophila* ponds. The ponds with the most color were primarily trough ponds above the western shore of the old basin located between high mounded polygons which are much older (14,000 yr).

The small year-to-year variations in the humics appear to be a function of differences in the water budget, especially in the amount of overland runoff later in the summer. Thus, 1973 was a wet year with overland runoff beginning by 19 August, while 1972 was a dry year with little mid- or end-of-summer rainfall. On the average, the color in 1973 (Figure 4-31) was not significantly different from the color in the drier year, 1972 (Figure 4-32). From 1972 to 1973, the intensity of color in the lowest colored ponds decreased, while it increased for the highest colored ponds (Figure 4-32). Thus, when rainfall runoff was the greatest, the central ponds whose immediate vegetation-covered drainage basin was smallest (because of the abundance of ponds) were diluted the most and became reduced in color.

We calculate that 50 to 70% of the total DOC is made up of these UV-absorbing humic materials. The data come from the ratio of total DOC to the absorbance at 250 nm before and after the addition of small amounts of anion resin which differentially removed the colored materials. This resin (Biorad AG50-8x) irreversibly binds the polycarboxylic humic acids and perhaps also some of the negatively charged non-UV-absorbing DOC. The measurement, however, was made only twice in replicate and considering the variability of humics that must occur from pond to pond, these percentages of humic compounds in the DOC pool must be taken as an indication of a general range rather than as an absolute value.

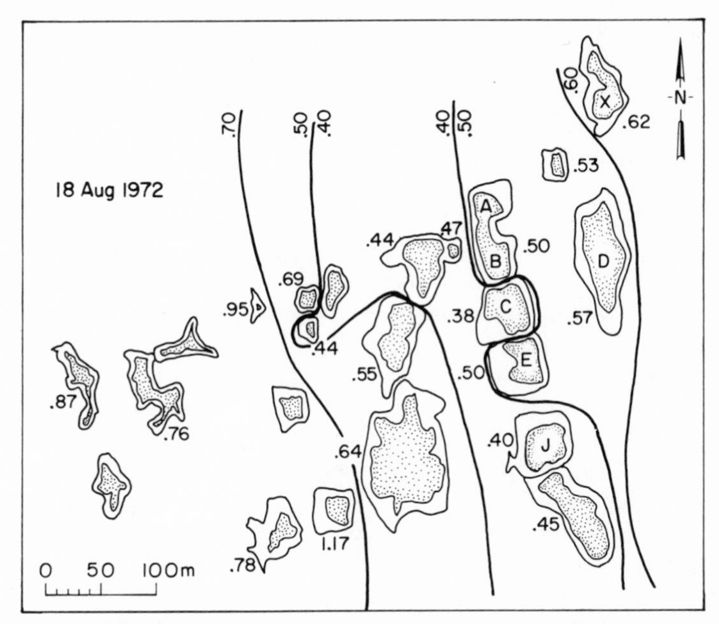

FIGURE 4-32. *UV absorbance at 250 nm in the water of some ponds near the IBP aquatic site, 18 August 1972.*

Labile Compounds

Simple organic molecules, such as sugars, amino acids and simple acids, make up a very much smaller proportion of the total DOC than the humic compounds. These simple compounds are biologically very mobile in contrast to the refractory humics and, as a result, bacterial uptake maintains them near the lower concentration threshold for active uptake by bacterial transport mechanisms. Thus, the concentration of individual compounds may only be a few μg C liter^{-1}. These labile molecules are produced by many processes, primarily by leakage from phytoplanktonic algae, from epipelic algae, and from emergent plants, although the egestion, excretion, and autolysis of animals can contribute significantly.

Only a few direct analyses of these simple compounds have been carried out anywhere, but these all confirm that there are only low amounts present in natural waters. Using an enhanced enzymatic assay, Hicks and Carey (1968) found that glucose in natural waters ranged from 1.3 to 3.8 μg C liter^{-1}. Vaccaro et al. (1968) reported average values ranged from 2 to 9 μg C liter^{-1} in Atlantic ocean waters. Crawford et al. (1974) measured the concentrations of 15 amino acids in the Pamlico Estuary for 1 year using an ion exchange chromatographic technique. Total dissolved free amino acids (DFAA) ranged from 10 to 30 μg C liter^{-1}, of which 50% was ornithine, glycine and serine. The DFAA made up less than 0.2% of the total DOC. Generally, the concentration of DFAA over the year varied with primary production in the estuary, suggesting that the amino acids originated from algal cell leakage or algal cell decay. Similarly, Gardner and Lee (1975) found an average of 10 μg C liter^{-1} in the DFAA in Lake Mendota, a eutrophic system.

To obtain some idea of the turnover of the labile DOC compounds by the natural microflora, we used a kinetic analysis of the dark uptake of ^{14}C-glucose, acetate, proline, and aspartate at four or five concentrations (Wright and Hobbie 1966, Hobbie 1967). The equation used for the plot of data was $tf^{-1} = (K_t + S_n)(V_{max})^{-1} + A\ V_{max}^{-1}$ where A is added substrate concentration, f is fractional uptake of the added radioactive substrate by bacteria and t is time of incubation in hours. The uptake was corrected for the $^{14}CO_2$ respired during incubation periods at the *in situ* temperature (Hobbie and Crawford 1969). If the plot of tf^{-1} vs. A was linear and the line had a significant correlation, then the data fit a Michaelis-Menten function for first order uptake. The slope of the regression was V_{max}^{-1}, the Y intercept was the turnover time of the substrate at its natural concentration ($S_n v^{-1}$), and the X intercept was an estimate of the sum of the half-saturation constant (K_t) and the natural substrate concentration (S_n) in μg C liter^{-1}. Unfortunately, the ($K_t + S_n$) cannot be taken as the concentration of S_n as K_t may be equally large or even larger than the S_n.

The turnover times for these small molecular weight compounds in Barrow ponds and the estimate of $K_t + S_n$ were no different from those for temperate lakes despite the lower pond temperature. The maximal estimates of $K_t + S_n$ ranged from 5 to 28 μg C liter^{-1} and the turnover times from 20 to 229 hours, the latter times being determined for acetate *in situ* without adding in the $^{14}CO_2$ respired (hence may be an overestimation) (Table 4-25). Hobbie (1967) found turnover times of 10 to 1000 hours over the year in Lake Erken and estimated from ($K_t + S_n$) that substrate concentrations were less than 2.4 μg C liter^{-1} for glucose and 4.0 μg C liter^{-1} for acetate. In an oligotrophic marl lake, Miller (1972) determined an average $K_t + S_n$ to be 2.7 μg glucose-C liter^{-1} and 3.9 μg acetate C liter^{-1}. In a dystrophic kettle lake, Duck Lake, also in Central

TABLE 4-25 *Average Turnover Times* (T_t) *and Estimates of Maximum Natural Substrate Concentrations* ($K_t + S_n$) *in Pond B, 1971*

	Glucose		Acetate[1]		Proline		Aspartate	
	T_t (hr)	$K_t + S_n$ (μg C liter^{-1})	T_t (hr)	$K_t + S_n$ (μg C liter^{-1})	T_t (hr)	$K_t + S_n$ (μg C liter^{-1})	T_t (hr)	$K_t + S_n$ (μg C liter^{-1})
Mean	55.7	15.9	229.4	27.9	70.3	5.3	20.0	15.1
SD	40.2	14.2	213.4	17.6	35.3	3.2	11.7	7.5
CV	72%		93%		50%		58%	

[1]Not corrected for $^{14}CO_2$ respired during the incubation.

Michigan, the K_t+S_n was 2.7 and 43.0 μg C liter^{-1} for glucose and acetate, respectively. The average turnover times for the year ranged from 16 to 35 hours in the two Michigan lakes.

Crawford et al. (1974) measured concentrations directly in an estuary and found that the concentration of 15 amino acids was much lower than the estimated K_t+S_n calculated from the kinetic analysis of the uptake of radioactive substrates. For proline, the average of S_n $(K_t+S_n)^{-1}$ was 5.9%; for aspartate, 7.5%. If we correct our mean estimate of K_t+S_n by the same percentages as for proline and aspartate, then proline=0.31 μg C liter^{-1} and aspartate=0.08 μg C liter^{-1}. If the measured DFAA in the estuary ranged from 10 to 30 μg C liter^{-1} when proline was 0.56 μg C liter^{-1} and aspartate 1.02 μg C liter^{-1}, then, in proportion, the DFAA in Pond B would be 9 to 27 μg C liter^{-1}. It should be mentioned that amino acids seem to differ from the general rule that the concentration of S_n is close to that of K_t.

All the data show that compounds readily metabolized by bacteria are maintained at very similar, very low concentrations in all natural waters. At most, the pool of amino acids, simple sugars, and two or three carbon acids (acetic, lactic, glycollic) found in the ponds could not be more than 120 μg C liter^{-1}. This pool is produced rapidly, but is used just as rapidly by the 10^6 bacteria per ml (Hobbie and Rublee 1975). Hence, the turnover of this pool is measured in tens or hundreds of hours, and its total concentration is at a near steady state between production and utilization.

The sources of this small pool of readily-used DOC include a proportion of the leachate of dead vascular plants (5.6 g C m^{-2} yr^{-1}), leakage by the photosynthesizing plants, and excretion by grazing animals (Table 4-23). Phytoplankton algal secretion or leakage ranged from 16 to 36% of the photosynthate (Table 4-26). In Pond B this is equivalent to 0.14 g C m^{-2} yr^{-1} when the planktonic production is 0.7 g C m^{-2} yr^{-1} (Stanley 1976a). The annual contribution from epipelic algal production, 8.9 g C m^{-2}, is much more significant and an average secretion rate found in 1971 was 10%. Hence, 0.89 g C m^{-2} yr^{-1} of usable DOC was released from this source (Table 4-26).

Higher aquatic plants also leak a portion of their recent photosynthate; this rate of loss could vary from 1 to 100% of photosynthesis rate as a function of divalent cation concentrations (Wetzel 1969a, Wetzel and Manny 1972). Glucose was a major compound produced by *Najas flexilis* in the laboratory (Wetzel 1969b). On the average, Wetzel et al. (1972) found an average of 4% leakage in the field. At Barrow, in a single set of experiments in the laboratory at 16°C with 450 foot-candles of light, completely submerged *Carex* plants released 3.75% of their photosynthate as DOC during a six-hour incubation with $^{14}CO_2$ and a subsequent six-hour incubation in non-labeled filtered pond water. If the production of *Carex* ranged from 600 to 1000 mg C m^{-2} day^{-1} in Pond J in 1971, then the leakage would be 22.5 to 37.5 mg C m^{-2}

TABLE 4-26 *Leakage of Dissolved Organic Carbon From Phytoplankton and Epipelic Algae During Photosynthesis*

Pond	Plankton[1] production ($g C m^{-2} yr^{-1}$)	Mean annual[2] leakage (%)	Plankton production of DOC ($g C m^{-2} yr^{-1}$)	Epipelic algal production ($g C m^{-2} yr^{-1}$)	Leakage[3] (%)	Epipelic algal production of DOC ($g C m^{-2} yr^{-1}$)
					range 1-25%	
B 1971		19%			mean 10%	
1972	0.7		(0.14)	8.9	(10%)	0.89
C 1971		16%				
1972	0.5		(0.08)	8.1	(10%)	0.81
D 1971	0.	17%				
1972	0.6		(0.10)	8.7	(10%)	0.87
E 1971		20%				
1972	0.9		(0.18)	9.3	(10%)	0.93
X 1971		36%				
Ikroavik 1971	2.2	7%	0.15	2.3	(10%)	0.23
J 1971	0.6	(18)	(0.11)	10.1	(10%)	1.01

( ) = means estimated using the previous years' algal secretion rate.
[1] 1971 data from Miller and Reed (unpublished) 1972 data from Stanley (1976a).
[2] 1972 data estimated from Miller and Reed, measurements taken in 1971.
[3] Measurements made on pond J, 1971, and applied to all ponds.

day^{-1}. Since the plant beds cover 40% of the pond area, this would amount to 9 to 15 mg C m^{-2}day^{-1} over the pond area or about 900 mg C m^{-2} yr^{-1} for a 75-day growing season (Table 4-23). This 45 to 75 μg DOC liter^{-1} day^{-1} (assuming a 20-cm-deep pond) is a large input to the DOC pool.

The excretion from the zooplankton may be the only significant animal input into the open water (Table 4-23). We have no direct measure of this, but if the annual net production of zooplankton is 1 mg C liter^{-1}, if the zooplankton ingest 20 times their net production, and if release of DOC from its metabolism or from mechanical destruction of cells passing through the gut is 40% of ingestion, then the zooplankton may add 1.0 g DOC m^{-2} yr^{-1}.

We were not able to measure the total input of DOC from the sediment. Obviously the maximum microbial activity occurs here but we could measure only the production of refractory DOC. A large quantity of labile DOC could have been produced and rapidly used. This amount, not entered in Table 4-23, is likely 2 or 3 mg C liter^{-1}.

The input of total DOC is 12.4 g m^{-2} yr^{-1} or 4 times the maximum standing crop of total DOC (Table 4-23). This input is made up of an estimated 3.8 g C of refractory DOC, primarily humic materials from sediment re-solution and from sedge and grass leaching, and 8.5 g m^{-2} yr^{-1} of labile DOC, primarily from the leaching of plant matter in the littoral weed beds and from leakage of autotrophs. Obviously bacterial utilization of the labile DOC must be very large (see Chapter 8). Other sources of loss include washout during snowmelt in the first 2 weeks after the thaw and photo-decomposition by sunlight. Based on June measurements on the concentration of humic material in Pyrex bottles left in the sunlight, we calculate that all of the dissolved humic material is decomposed every 28 days.

In conclusion, because of the importance of the refractory humic pool in the total DOC budget, the observed concentrations of DOC appear to be largely controlled by physical and hydrologic processes. In general, the concentration of DOC in shallow lakes with large drainage basins is determined more by allochthonous inputs than by autochthonous ones (Birge and Juday 1934). The tundra ponds do not follow this rule as they are very shallow, low-volume aquatic systems whose DOC pool is largely dominated by vegetation decay within the pond itself and by sediment solution of surficial bottom deposits. This pool, however, is mainly the refractory DOC remaining after the bacteria have removed the labile material.

Particulate Organic Carbon

Most of the material in suspension in the water is particulate organic carbon (POC). This is separated from the dissolved material by filtration.

TABLE 4-27 *Particulate Organic Carbon during Runoff, 1973**

Date	#of samples	Weir #1		Weir #6	
		mean	SD	mean	SD
10 to 14 June	14	651	129	1187	870
15 to 18 June	7	571	68	436	231
19 to 27 June	11	346	146	-	-
19 August	1	415			
27 August	1	222			
Total discharge through weir 10-27 June		11,875 m^3		319 m^3	

*Data are expressed as μg C liter^{-1}.

Thus, the particulate material is composed of particles of all sizes and the border between the DOC and POC is an arbitrary one that is defined by the retention characteristics of the filter. The 984 H filter retains all particles above 0.8 μm in diameter. For analysis, the filtered material was oxidized in a chromic acid-sulfuric acid solution and the resulting color change calibrated against glucose in a spectrophotometer (Strickland and Parsons 1968). The POC was measured in two or three ponds twice weekly from 1971 to 1973.

Both the snowmelt and the runoff water were high in POC, especially at the start of the thaw. Snow contained 1124 μg C liter^{-1} at the beginning of the melt. The runoff water (Table 4-27) from Weir no. 6 (see Figure 3-1 for location) had an average of 1187 μg C liter^{-1} in the first 5 days of the melt season but this dropped to 436 during the next 4 days. A similar pattern, though with lower concentrations, was measured for Weir no. 1 (the weir that passed large quantities of water). The same high concentrations early in the year were also measured in Pond B during 1971 (Table 4-28). The yearly means, 769, 266, and 391 μg C liter^{-1} for 1971, 1972, and 1973 (Table 4-29), reflect the great differences from year to year; these differences are unexplained.

The POC is composed mostly of non-living material. Assuming that algae are 5% carbon (of wet weight) and that bacteria contain 1.2×10^{-8} μg C cell^{-1}, then only 5.3% of the POC was made up of algae and bacteria in 1971, 13.4% in 1972, and 9.9% in 1973 (see Chapter 5 for biomass data). We have observed that the biomass of the organisms was about equal during these 3 years and so the concentration of non-living organic matter changed.

The environmental or biological factors that control the concentration of POC are not obvious in these ponds. As already noted, there was no correlation between the POC and algal or bacterial biomass. In addition, when a series of ponds was studied in 1971 (Table 4-29), there was no correlation between the productivity and the POC. Attempts to

TABLE 4-28 *Particulate Organic Carbon in Pond B (1971, 1972) and in Pond A (1973)**

Pond B				Pond A	
1971		1972		1973	
8 June	1741	10 June	127	19 June	608
9 June	339	13 June	96	20 June	445
10 June	2148	14 June	42	21 June	897
12 June	1135	15 June	105	22 June	150
14 June	659	17 June	102	23 June	527
17 June	401	18 June	62	24 June	320
21 June	594	19 June	182	25 June	458
28 June	499	20 June	613	26 June	232
		21 June	252	27 June	301
5 July	504	23 June	594		
12 July	480	24 June	248	2 July	351
19 July	518	26 June	275	9 July	339
26 July	559	29 June	307	17 July	376
		30 June		23 July	88
2 August	712	Run 1	307	31 July	320
9 August	946	Run 2	537		
16 August	301	Run 3	281	12 August	
		Run 4	622	Run 1	364
		Run 5	669	Run 2	376
				Run 3	489
		7 July	243	13 August	
				Run 1	348
				Run 2	319
				Run 3	110
				16 August	719
				27 August	464

*Data are expressed as μg C liter^{-1}.

TABLE 4-29 *Mean Particulate Organic Carbon in Ponds, 1971 to 1973*

Year	Pond A-B	Pond C	Pond D	Pond E	Pond K
1971	769	531	671	763	1450
SD	530	170	306	303	371
1972	266	-	206	-	-
SD	172		156		
1973	391	484	-	-	-
SD	187	141			

*Data are expressed as μg C liter^{-1}.

find a correlation between POC concentration and the average daily or maximum wind speed were equally unsuccessful. Certainly there are daily changes (e.g., 13 August 1973, Table 4-28) in the POC and similar changes have been noted for the bacteria throughout a single day (see Chapter 8). The causes of these changes—wind, water currents, and microturbulence—are difficult to measure.

In spite of the extreme shallowness of the ponds and the accompanying close contact with the sediments, the POC concentrations are very similar to those found in the epilimnion of moderately productive lakes. For example, Lake Erken in Sweden (Hobbie 1971) had a concentration that was usually between 500 and 800 μg C liter^{-1} (a peak of 1500 during an algal bloom) and Lawrence Lake in Michigan had concentrations of 200 to 800 μg C liter^{-1} in the circulating water (Wetzel et al. 1972). In Lake Tahoe, an extremely oligotrophic lake, POC ranged from 19 to 38 μg C liter^{-1} while coastal ocean waters had 30 to 300, open ocean waters 5 to 50, and Arctic Ocean water 1 to 7 μg C liter^{-1} (Holm-Hansen 1972). Algal biomass was estimated to be 10% of the total POC (Hobbie 1971) in Lake Erken while Miller (1972) calculated that 7% of the POC was made up by living organisms in Lawrence Lake.

Sedimentation of POC

The POC in the ponds is produced by the growth of algae and bacteria, by their death, by defecation and molting of the zooplankton, by

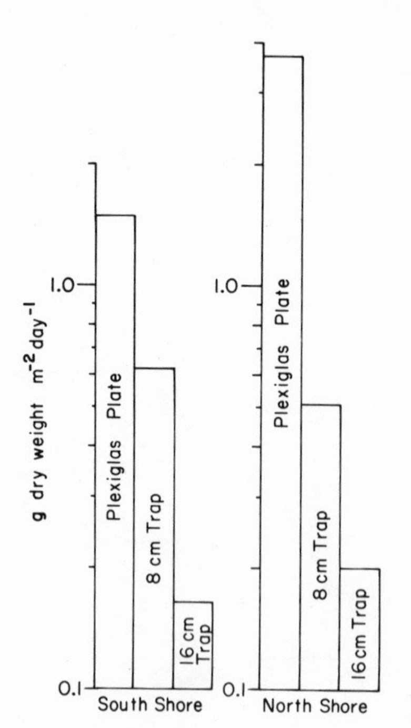

FIGURE 4-33. *Rate of POC sedimentation onto a flat plate and into sediment traps 8 cm and 16 cm high; Pond C, 3 to 5 August 1973.*

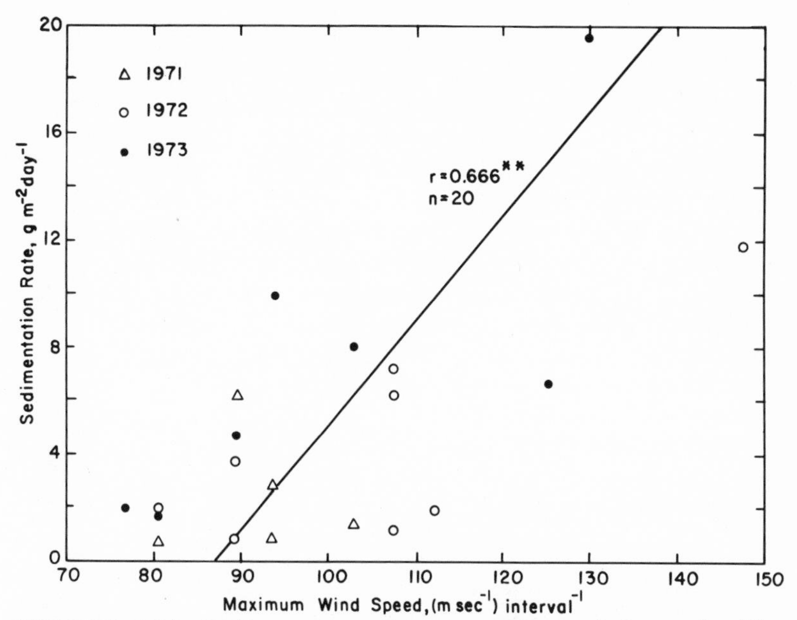

FIGURE 4-34. *Sedimentation rates at various wind speeds. The speeds are the maximum recorded speed over the sampling interval.*

precipitation of an iron-humic floc, and by resuspension of the sediment. POC is lost from suspension by settling out when water currents are minimal and by ingestion by zooplankton.

One way to examine the resuspension is to measure the POC settling out in the ponds. Because the concentration of the POC in suspension is relatively constant, the amount settling to the bottom must equal the amount resuspended. The problem becomes a methodological one, however, as any sediment traps put into the pond inevitably affect the rate of sedimentation. For example, small diameter traps suspended in lakes are supercollectors (Miller 1972, Rich 1970) and collect much more POC than actually sinks. This is especially true during circulation periods when the traps collect a portion of the POC passing across the traps, not merely the amount sinking down.

One other problem is that in the ponds the suspended material remains quite close to the bottom and can not be sampled by tall traps. For example, white Plexiglas plates placed on the bottom collected much more POC over 3 days than did 8-cm or 16-cm tall traps nearby (Figure 4-33).

Our standard trap was a 4.7-cm-diameter cylinder 12 cm or 16 cm tall, placed on the bottom. These collected from 0.46 to 28 g dry wt m^{-2} day^{-1}. In Pond D, the averages were 3.3 in 1971, 6.8 in 1972, and 4.4 g in 1973. As noted in Figure 4-33, this is an underestimate of the total resuspension but is still a very large quantity of material. In fact, it is so much greater than the total POC sampled in the water, that we suspect that most of this material is resuspended during brief periods of high wind and faster currents. A correlation study of the wind speed and the

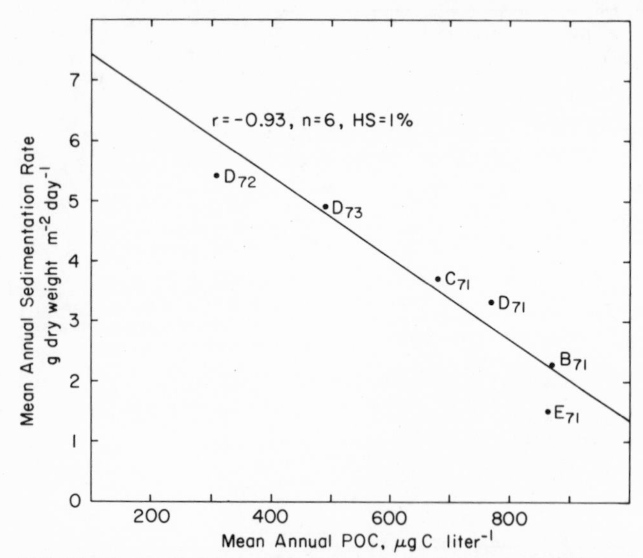

FIGURE 4-35. *Mean annual concentration of POC vs. the mean annual sedimentation rate for six ponds.*

sedimentation showed a direct relationship (r=0.66, HS) with the maximum wind speed (Figure 4-34) and no correlation with the average wind speed.

Ponds with the highest sedimentation rate had the lowest average POC in the water and vice versa (Figure 4-35). The mean annual sedimentation rate is inversely related to the mean annual POC concentration with r = −0.93 (n = 6, HS). One explanation is that the large particles that fall to the bottom (average diameter of 200 μm) actually clear the water as they fall. This could be an adhesion of the small particles to the large ones or a co-precipitation. This co-precipitation was also found in an arctic lake (Hobbie 1973) but inorganic silt, rather than organic matter, was involved.

Zooplankton Grazing

The grazing of the zooplankton might be another control on the POC in these ponds. As will be discussed (Chapter 6), the zooplankton are capable of filtering all the water in a pond every few days. If these animals are affecting the POC concentration, then the highest zooplankton populations should be found in ponds with the lowest POC concentrations. This is actually the case (see Chapter 6) and there is a highly significant inverse correlation between POC concentration and maximum zooplankton biomass (r = −0.95, n = 8). This same relationship can also be used to argue that the zooplankton had a higher production when lower quantities of non-living detritus were present. We really do not know which is cause and which effect here. Is the zooplankton grazing reducing the POC or is the low POC allowing high zooplankton production?

SUMMARY

Sediments

Because of the high ratio of sediment surface area to pond volume, the pond chemistry is controlled by the events in the sediments. These sediments are highly organic at the surface, are up to 80% organic in the top 4 cm, and are as low as 18% organic matter at 8 to 9 cm. The surface layers are dark brown, unconsolidated material containing some remains of grasses, sedges, mosses, etc. Mean particle size is 500 μm and bulk density is 160 mg cc^{-1}. If organic carbon is 40% of organic matter, then a square meter slice 1 cm thick contains 300-500 g C. There is some mixing of the top 4 cm as a result of animal activity.

Inorganic Ions

The pond water is a dilute salt solution with Na and Cl the major ions. The concentrations of Na, Mg, Ca, Cl, and HCO_3 are controlled by abiotic processes, mainly by dilution during rainfall and runoff, and by concentration during evaporation and freezing. During freeze-up, ions are excluded from the ice and are eventually forced into the sediment when the water all freezes. Dilute meltwater fills the ponds in the spring but evaporation of the water and re-solution of ions from the sediments increase the concentrations of ions as much as 4-fold by late summer. The chemistry is a little unusual, as the chloride is higher in concentration than bicarbonate, but this is typical of ponds near the sea; most ponds in the Arctic have more bicarbonate than chloride. In any event, this dominance of the chloride has no practical importance because pH and buffering capacity are still a function of the bicarbonate at all times.

Other ions have more complicated seasonal cycles than the above. Potassium is highest during runoff, presumably due to leaching from vascular plants and lemming feces. Iron is low early in the summer, reaches a maximum in July, then declines. The concentrations are correlated with DOC, so some complexing or chelating process is at work. Iron concentrations increased after rain too, perhaps as a result of soil leaching. Silica concentrations are very low; the level is always below the concentration supposedly needed for diatom growth and, in fact, there are no planktonic diatoms. There are diatoms in the sediment, however, and their production would reduce the concentration of silica in the water.

The pond sediments are iron-rich peats. Concentrations of all ions except potassium in the interstitial water are higher than in the water column but are within the range of pore water from the Barrow terrestrial site. Particulate iron concentrations are exceptionally high, up to 133,000 ppm, while the low manganese suggests both a differential mobilization of iron by humic compounds and a lack of secondary mineral formation.

Trace metal analyses revealed that the pond water was similar to sea water; there was 1.0 μg Cu liter^{-1}, 0.7 μg Pb liter^{-1}, and 4.9 μg Zn liter^{-1}.

Inorganic Carbon

The alkalinity was low, between 0.32 and 0.45 meq liter^{-1} after the effect of the meltwater disappeared. The pH was low early in the summer but stabilized in the range 7.05 to 7.62. Because of the low buffering capacity, photosynthesis changed the alkalinity by 0.02 meq liter^{-1} and the pH by 0.5 during a single day. Total inorganic carbon varied from 1.43 to 8.08 mg C liter^{-1} and averaged 4.10, 5.01, and 2.73 over the 3 years. These are low values (one-seventh of seawater).

Continuous measurements of the partial pressure of CO_2 in the water (pCO_2) were carried out for 6 weeks in 1971. The pCO_2 ranged from 450 to more than 1500 ppm so the pond was always supersaturated compared with 320 ppm in water in equilibrium with air. This does not imply a large amount of CO_2. For example, in this pond when the total CO_2-C is 4.45 mg liter^{-1} (from alkalinity and pH), most of the carbon is found as HCO_3. When the pCO_2 is 320 ppm, then the dissolved CO_2 is 0.37 mg C liter^{-1}. A pCO_2 of 640 ppm equals 0.74 mg CO_2-C liter^{-1}. In the pond water in 1971, the difference between the saturation value (320 ppm) and the measured pCO_2 averaged 397 ppm. This implies that there was a continual transfer of CO_2 from the water to the air as a result of the excess of respiration (mostly in the sediments) over photosynthesis. Also, the transfer rate is relatively slow as the pond's CO_2 is not in equilibrium with the air. The pCO_2 was directly related to the sediment temperature and indirectly related to wind speed. There was higher microbial and root respiration at higher temperatures.

An evasion coefficient of 0.34 ± 0.17 mg CO_2-C cm^{-2} atm^{-1} min^{-1} was measured under a variety of wind and temperature conditions. This gave an average rate of transfer of 0.53 g CO_2-C m^{-2} day^{-1}. The average is likely conservative even though it is about twice the estimated value based on studies of the respiration of separate groups of organisms. Much of the CO_2 transferred to the air may arise from the respiration of roots but this rate is poorly known.

Oxygen

The oxygen in the pond water was near saturation in the deep areas of the pond but decreased to one-third of saturation in the shallow *Carex* bed. This was the result of high sediment respiration rates and poor circulation of the shallow water. Overall, the water was undersaturated due to sediment respiration.

Eh and pH

The *Carex* beds also influenced the sediment pH and Eh. In the center of the ponds, only the top 1 to 2 cm is oxidized. But beneath the plant beds, the oxidized zone extends to 8 or 10 cm. This results from the downward transfer of O_2 within the roots. The E_7 at 14 cm was about 50 mV in the pond center and 125 mV beneath the *Carex* roots. Thus, the sediments are reducing but not strongly reducing. The pH fell to a low of 5.8 at the top of the anaerobic layers of the sediment but then rose slightly to 6.4 in all the deeper layers.

Nitrogen

The various compounds of nitrogen are present in the pond water in low concentrations. Nitrite was always insignificant as levels were below 0.5 μg N liter^{-1}. Nitrate concentrations were high early in the year (up to 60 μg N liter^{-1} with an average of 7 to 40 μg N) but always fell late in the summer (1 to 9 μg N liter^{-1}). Ammonia has two peaks; one occurs early in the spring and precedes the nitrate maximum and the other occurs in late July and early August. Concentrations were always at moderate levels (average monthly values from 20 to 50 μg N liter^{-1}). There was also an ammonia peak in the meltwater (100 to 200 μg n liter^{-1}). However, most of the nitrogen is present in organic form. The dissolved organic nitrogen (DON) was most abundant (600 to 1000 μg N liter^{-1}) while the particulate organic nitrogen (PON) was about the same concentration as the total inorganic forms (40 to 100 μg N liter^{-1}). These concentrations of nitrogen are much higher than those in arctic lakes; in part, the regeneration of ammonia from the sediments keeps the concentrations at these moderate levels.

Concentrations were high in the sediments from the center of the pond (5580 μg DON liter^{-1}, 3000 μg NH_3-N liter^{-1}, 11 μg NO_3-N liter^{-1}). The rooted sedges and grasses take up ammonia so the amount in the sediment near the plants was low (50-70 μg NH_3-N liter^{-1}).

There is nitrogen fixation in the sediments but the rate is very low. The average value of 0.31 mg N m^{-2} day^{-1} is equal to 28 mg N m^{-2}. yr^{-1}

and compares well with the terrestrial soil value of 48 mg N m^{-2} yr^{-1}. However, exchangeable inorganic N in the top 20 cm of sediments was high: 810 mg N m^{-2} in the weedbed and 1400 mg N m^{-2} in the central basin. Another input of nitrogen to the system is rainfall. During the summer, the concentrations were 255 μg NH_3-N $liter^{-1}$ and 25 μg NO_3-N $liter^{-1}$; in contrast, rainfall at Hubbard Brook, N.H. contains more NO_3 than NH_3. The total input of dissolved inorganic nitrogen (DIN) at Barrow was 11.5 mg N m^{-2} yr^{-1} in rainfall. Seepage from the drainage basin was not important but an important amount of DON was retained in the ponds during the runoff of the meltwater (39 mg m^{-2}). The DIN retained was only 1.7 mg m^{-2}. Denitrification also occurred in the sediments but the daily rate, 0.18 μg NO_3 $liter^{-1}$, would not have affected the NO_3 pool (11 μg NO_3-N $liter^{-1}$) very much.

The total budget for the input and output of nitrogen from the ponds was calculated from the fixation, denitrification, rainfall, and runoff values. There was a net input of 10.5 mg DIN m^{-2} and of 69.6 mg DON m^{-2}. The largest inputs were summer rainfall, nitrogen fixation, and spring runoff, while runoff was the biggest export. The amounts retained (80 mg N m^{-2}) were small compared with the total amounts in the pond (219 g N m^{-2}) and its sediments but the total amount added each year, about 80 mg N m^{-2}, was enough to supply 66% of the DIN needed for plankton photosynthesis in the pond.

Ammonia is the preferred nitrogen source for the plankton. The mean uptake rate was 0.16 μg NH_3-N $liter^{-1}$ hr^{-1} which replaces all the NH_3 in 150 hours. Nitrate was taken up much more slowly (0.004 μg NO_3-N $liter^{-1}$ hr^{-1}, turnover time more than 3000 hours). The measured nitrogen uptake rate agrees well with the calculated rate based on the photosynthesis. If the C:N ratio is 8:1, then the plankton productivity of 1 g C m^{-2} yr^{-1} gave an averge uptake rate of 0.17 μg N $liter^{-1}$ hr^{-1}, very close to the measured value.

Similar calculations for the benthic algae give an average uptake rate of 500 μg N m^{-2} hr^{-1} and for the rooted plants gives 750 μg N m^{-2} hr^{-1}. There is adequate NH_3 for the benthic algae (1400 mg NH_3-N m^{-2} in the top 20 cm) but barely enough NH_3 for the rooted plants (only 18 mg NH_3-N m^{-2} in the top 20 cm of the plant beds). Obviously, ammonia is being rapidly regenerated. This regeneration or supply rate for the water column was measured with isotope dilution experiments (^{15}N). The average supply rate was 1.9 μg NH_3-N $liter^{-1}$ hr^{-1} which is more than adequate to supply the plankton. There are no data for the sediments.

Nitrification likely occurs in the ponds but the evidence is indirect. Nitrifying bacteria are present in the Barrow soils and high nitrate levels occur early in the year. This process is certainly a slow one compared to ammonification.

When phosphorus was added to a natural pond, there was an immediate increase in the uptake of nitrogen by the plankton; later, there was an increase in the photosynthesis rate. In experimental ponds with added lights, the nitrogen fixation was increased, which implies that there is a link between photosynthesis and fixation. When high amounts of phosphorus were added each day, the fixation rate increased in direct proportion to the amount added. We interpret this as indicating that the algae are usually phosphorus limited. When excess P is added, the rate of supply of N becomes limiting and algae, such as blue-green algae, that can fix N gain a competitive advantage.

When P was added to an artificial pond containing no sediment, productivity did not increase until extra N was added. This illustrates the importance of the sediments in supplying nitrogen to the water.

Phosphorus

The concentration of dissolved reactive phosphorus (DRP) in the pond water was always low, generally between 1 and 2 μg P liter^{-1}. However, there was a brief period in mid- or late June when the concentrations reached 4 or 5 μg P liter^{-1}. The P came from runoff and from leaching of the standing dead vegetation; when the algal photosynthesis increased and the rate of supply decreased in late June the concentration of DRP dropped below 1 μg P liter^{-1}.

Dissolved unreactive phosphorus (DUP) is re-introduced into the ponds each spring through litter decomposition, sediment leaching, and runoff. When the snowpack is deep, the runoff is large and the DUP remaining in the ponds is low. The concentration in the ponds averaged 12 μg DUP liter^{-1}; this was 56% of the total phosphorus. About 70% of the DUP was refractory while the rest could be easily broken down and made available to algae.

Another important phosphorus fraction is the particulate (PP). This was fairly constant in concentration and averaged 10 μg PP liter^{-1}. Large zooplankton were not included so this includes only algal, bacterial, and detrital phosphorus.

Trough ponds are strikingly different from polygon ponds in that all forms of phosphorus are more abundant. Concentrations were 3.7 to 7.6 μg DRP liter^{-1}, 18.6 to 60.6 μg DUP liter^{-1}, and 20.3 to 31.5 μg PP liter^{-1}; these are 2 to 7 times the concentrations in the low-centered polygon ponds.

The concentrations of phosphorus in the interstitial water are much higher than those in the water column above but most of the difference is due to high DUP. Thus, the DRP was mostly 1 to 5 μg liter^{-1}, while the DUP was 24 to 138 μg P liter^{-1}. There were no observed differences due to the presence of roots or reduced sediments.

The sediment phosphorus was investigated by chemical fractionation and ^{32}P. An iron-rich sediment holds nearly all of the inorganic P in the ponds. It is held by sorption of phosphate on or occlusion within hydrous iron oxide rather than by the formation of iron, aluminum, or calcium minerals. The sorbed phosphorus is in dynamic equilibrium with P in solution and can be mobilized. The amount of sorbed P in the sediments was large, about 3,600 mg P m^{-2} in the top 10 cm of Pond C while there was 25,000 mg total P m^{-2}. In the same pond, the interstitial water contained 2 mg P m^{-2} and the water column contained 5 mg P m^{-2} (20 cm of water).

The phosphorus enters the pond through winter snow, summer rain, and spring runoff. It leaves the pond in spring runoff. The winter snow contains about 4 μg P $liter^{-1}$ (equal amounts of DRP, DUP, and PP) while the summer rain contains 7.2 μg DRP $liter^{-1}$ and 0.7 μg DUP $liter^{-1}$. As the meltwater from the winter snow moves in sheets across the tundra, phosphorus is leached rapidly from vegetation, fecal pellets, and litter on the tundra. The water entering the pond has 2.4 μg DRP $liter^{-1}$ and 11.3 μg DUP $liter^{-1}$. In the pond, concentrations are somewhat higher so the water leaving the ponds contains 2.5 μg DRP $liter^{-1}$ and 13.1 μg DUP $liter^{-1}$. The budget, in units of mg P m^{-2} yr^{-1} is: winter snow, +0.23; spring runoff entering, +10.41; spring runoff leaving, −11.86; summer precipitation, +0.51. The net balance is −0.70 or a slight loss from the ponds. The ponds are in equilibrium, really, as all the terms in the budget will change slightly from year to year. When the large amount (25,000 mg P m^{-2}) of P in the sediment is considered, it appears that much of the phosphorus in the sediments was there before the pond was formed. The amount of P lost, less than 1 mg P m^{-2} yr^{-1}, is extremely small compared to the total amount of P circulating in the ponds. For example, if the primary productivity is around 40 g C m^{-2} yr^{-1}, then this represents approximately 500 mg P m^{-2} yr^{-1} circulating in the algae and rooted plants.

The phosphorus cycled very rapidly through the plankton; in fact, the rate of uptake was about 200 times greater than the amount needed for algal growth. The isotope studies showed that DRP is taken up by bacteria and algae and most is immediately released again. The released P is mostly dissolved organic phosphorus. The organic P released by the phytoplankton has a low molecular weight (XP) and can either break down again to DRP or combine with colloids to form colloidal P (high molecular weight). The XP and colloidal P pool is as large or larger than the DRP pool. Phosphorus cycles through these organic fractions as much as 50 times a day. Uptake rates of DRP into plankton were 13 to 320 μg P $liter^{-1}$ day^{-1} where the DRP was only around 2 μg P $liter^{-1}$. Obviously the algae have actually 2 to 4 times the amount of P in the DRP available for growth. There were also changes in the 70% of the DUP which is not in the XP or colloidal P form but this pool is resistant to biological breakdown.

Zooplankton excrete phosphorus mainly as DRP. In the pond, the zooplankton excreted about 0.13 μg P $liter^{-1}$ which is enough to account for all of the P needed for algal productivity.

An experiment in a microcosm demonstrated that bacteria took up DRP and excreted an XP-like organic compound. In a mixed culture of bacteria alone, the rate of uptake was 2.5×10^{-7} μg P cell^{-1} hr^{-1} but when a ciliate was introduced the rate rose to 16.6×10^{-7} μg P cell^{-1} hr^{-1}. However, the ciliates could excrete only 0.4 μg P liter^{-1} while bacterial uptake was 9.3 μg P liter^{-1}. Thus, release of P by the ciliates was not responsible for the increased activity of the grazed bacteria. Instead, the difference is probably due to the physiological differences between the rapidly dividing bacteria in a grazed system and the relatively static bacteria of ungrazed systems.

The rooted aquatic plants cycle phosphorus from the sediments to the water through leaching from dead plants and by excretion from live plants. *Carex* leaves harvested when green lost 60% of their phosphorus during the first month of immersion (34 μg P (g dry wt)$^{-1}$ day^{-1}). Live *Carex* lost 1.1 μg P (g dry wt)$^{-1}$ day^{-1} or a maximum daily rate of 0.14 mg P m^{-2} of plant stand. This is a large amount relative to the phytoplankton needs as the quantity approximates the amount of DRP present. Leaching from the standing dead plant leaves averaged about 0.72 μg P m^{-2} day^{-1} so this pathway is even more important than the excretion pathway. Phosphorus transfer rates from the sediments to the water are rapid enough so that all the DRP can be replaced in about 0.8 day.

Control of Phosphorus

To understand the interactions of phosphorus in the ponds, it is not enough to know the quantities that are added to the pond water; we also have to know what happens to this added phosphorus. Why is the DRP concentration so low? To answer these questions requires a detailed study of the sediment because almost all the phosphorus that enters the pond moves rapidly to the sediments. There it is either retained in an organic or occluded inorganic form or is strongly sorbed to the sediments. The sorbed phosphorus is available for interaction with the water phase; most lakes appear to have sorption rather than precipitation as the primary means of phosphate fixation in the sediments.

Chemical adsorption phenomena can usually be described by an isotherm equation derived from certain assumptions about the energy with which individual sorbed ions or molecules are held. The best fit to our data was given by the Temkin isotherm equation:

$$X X_m^{-1} = RTb^{-1} (\ln g C)$$

where X is the sorbed phosphorus per unit of sorbent, X_m is the sorption maximum, R is the gas constant, T is the temperature in °K, b and g are constants, and C is the equilibrium phosphate concentration. This equation implies that the energy of adsorption decreases linearly as the amount of the sediment particle covered by P increases and that a plot of X against log C should be linear. Therefore, the less phosphorus there is the more strongly it is bound. Comparison of the slope of the Temkin isotherm for Barrow pond sediments with the slope for fertilized soils revealed that the soils had a much smaller phosphate buffering intensity than the sediments. In this sense, buffering intensity is a measure of the tendency of a solution in contact and in equilibrium with a sediment to resist a log C change resulting from addition or withdrawal of sorbed phosphorus.

The buffering intensity, which is the slope of the Temkin isotherm, can be approximated by a single number, the phosphate sorption index (PSI). This is the ratio of X to log C and is obtained by adding 1500 μg P to a gram of sediment and measuring the amount taken up. The higher the PSI, the more strongly the sediment binds phosphate. In the intensively studied ponds the PSI was 359 to 540 while in the other ponds investigated the PSI was 318 to 728. These PSI's are 50 to 250 times higher than those found to control phosphorus concentrations in a stream at Hubbard Brook, N.H. Thus, the sediments of the Barrow ponds are strongly buffered.

It is likely, then, that DRP concentrations in the pond are controlled by the phosphate sorption by sediments, and that this, in turn, is controlled by the concentration of exchangeable inorganic P in the sediment. A survey of DRP and oxalate extractable P (which approximates total inorganic P) showed a correlation coefficient of 0.56 over a series of ponds so there is certainly some relationship. A negative relationship was found between organic phosphorus and oxalate extractable phosphorus in the sediment so it is possible that it is the rate of mineralization of the organic P that controls the amount of inorganic P. Another process that may be operating is the trapping of DRP during the spring runoff. Perhaps the best demonstration of the importance of the sorption equilibrium was the excellent correlation between the log of the plankton primary production and the PSI of sediments from the intensively studied ponds.

Organic Carbon

The dissolved organic carbon (DOC) was defined as the material passing through a glass fiber filter; the effective cutoff was 0.8 μm. Meltwater contains little DOC so concentrations in the pond were low, 4 to 5 mg DOC liter^{-1}, immediately after the ice melted. As soon as the sediments thawed, the DOC quantity increased; a peak of 15.6 mg DOC liter^{-1} was reached on 5 July. The concentration then became relatively constant at 10-14 mg DOC liter^{-1} until late August when rain sometimes diluted the water. Concentrations became very high as the ice sheet thickened in September.

The amount of DOC in the pond thus changed from 0.8 to 3.1 g C m^{-2} (for a 20-cm-deep pond). However, DOC has two fractions. One fraction (90%) is refractory DOC made up of compounds, such as humic acids and tannin, that are resistant to microbial breakdown. The other fraction (10%), easily broken down and removed by microbes, is called the labile fraction and is composed of peptides, sugars, amino acids, and short-chain organic acids. The DOC that is measured in the pond is mostly refractory material, as the microbes of the sediment remove the labile compounds nearly as quickly as they are produced.

Re-solution from the sediments was determined from changes in subponds open to the sediment. Tests were run with leaves and stems of *Carex* in litterbags (mesh) and plastic bags. From these and from the production values it was calculated that 3.8 g refractory DOC m^{-2} enter the pond from macrophytes (two-thirds) and sediments (one-third) and 5.6 g labile DOC m^{-2} come from macrophytes. A survey of ponds revealed that those with the most color in the water, which correlates with humic material, are trough ponds. Ponds on either side of the old lake shelf also had high amounts of color while the intensively studied ponds had low amounts of color.

Labile DOC is made up of a large number of compounds, each present at concentrations of a few μg $liter^{-1}$. Studies of the kinetics of uptake of these compounds indicated that the sum of the concentration of substrate plus a half-saturation constant for uptake was: 16 μg glucose-C $liter^{-1}$, 28 μg acetate-C $liter^{-1}$, 5.3 μg proline-C $liter^{-1}$, and 15 μg aspartate-C $liter^{-1}$. This means that the concentration of glucose was below 16 μg glucose-C $liter^{-1}$. The same experiments showed that these compounds were completely removed from the water every 56 hr (glucose), 229 hr (acetate), 70 hr (proline), and 15 hr (aspartate). Thus, the microbes keep the pool of labile DOC at a low level even though large amounts of DOC may move through this pool.

Labile DOC leaches from *Carex* and *Arctophila* (5.6 g C m^{-2} yr^{-1}), leaks from the photosynthesizing plants (1.9 g C m^{-2} yr^{-1}), and is excreted from grazing animals (1.0 g C m^{-2} yr^{-1}). Another source, but one that could not be measured, was from the decomposition of detritus (non-plant) in the sediments.

The total input of DOC to the ponds was 12.4 g m^{-2} yr^{-1} or 4 times the maximum amount of DOC. Most of the loss was due to microbial decomposition in the sediment but there was also some photo-decomposition. The limited data indicate complete photo-decomposition of the humics every 28 days. This loss could be as much as 3-4 g DOC m^{-2} yr^{-1}.

The changes in the DOC can be accounted for by the leaching and decomposition within the pond. Thus, the DOC is autochthonous.

Most of the material in suspension in the water is particulate organic carbon (POC). It is defined as the amount retained on a glass fiber filter (0.8 μm effective pore size). Amounts were high, 1100 to 1200 μg POC liter^{-1} in snow and runoff water, but during the second half of the runoff season the concentration dropped to 436 μg. The yearly means ranged from 266 to 769 μg POC liter^{-1} in the pond water. Most of the POC is nonliving; algae and bacteria made up only 5 to 13%. POC is produced by the growth and death of algae, of bacteria, and of zooplankton, from zooplankton molts and fecal pellets, by resuspension of sediments, and by precipitation of iron-humus flocs. The concentrations in the pond are similar to those in other moderately rich freshwaters.

Because the concentration of POC was relatively stable throughout the summer, the resuspension rate should equal the sedimentation rate. This rate was measured with 12-cm-high cylindrical traps, 4.7 cm in diameter. These underestimated the sedimentation rate compared with measurements made on white Plexiglas sheets. The amounts of POC in the water column averaged 90 to 200 mg POC m^{-2} day^{-1} but the sedimentation rate averaged 1200 to 2400 mg C m^{-2} day^{-1}. Thus, much of the material that appeared in the traps must have been resuspended each day. Sedimentation rate was significantly correlated with the maximum wind speed, so the resuspension may only have occurred during brief periods of each day.

POC was lowest in ponds with the highest sedimentation rates. It is possible that large particles may capture the smaller particles as they fall. Another possible control of POC is grazing by zooplankton. These animals are abundant and active; they filter all the pond water every 2 days. However, their effect on the POC is unknown.

5

Primary Producers

V. Alexander, D. W. Stanley, R. J. Daley, and C. P. McRoy

PHYTOPLANKTON*

Populations

Seasonal variations in algal species and biomass were measured weekly from 1970 to 1973 in four or five IBP study ponds. In addition, species distribution in a larger number of ponds was investigated on two occasions. Dr. Staffan Holmgren, then at the Institute of Limnology, Uppsala University, provided taxonomic help with the nannoplankton during 1971.

A settling chamber technique, based on the technique of Utermöhl (1958), was used throughout the study to count phytoplankton. In 1970, the technique was used with an inverted microscope. Later, the technique was modified (Coulon and Alexander 1972) so that a permanent slide was produced that could be examined with a conventional phase-contrast microscope. With this technique, even the most delicate flagellates such as *Rhodomonas minuta* and various species of *Chromulina* showed no change in appearance. Cell biomass estimates were obtained from Nauwerck (1963) and from calculations involving the measured average lengths and widths of each species and the volumes of geometric figures. These volume estimates (in μm^3) were changed to wet weight values by assuming a density of 1.0 (thus, $10^9 \mu m^3 = 1$ mg).

Chlorophyll measurements provided another estimate of phytoplankton biomass. The technique was basically that described by Strickland and Parsons (1965). Glass fiber filters (Gelman A) were used for concentrating the algae and the chlorophyll was extracted in 90% acetone at 5°C for 24 hours. After centrifugation, the absorption of light was measured in a scanning spectrophotometer.

The chlorophyll content of the ponds invariably showed a rapid rise after the spring melt (Figure 5-1). Next, the concentrations fell to very low values in mid-July followed by a rise in August. In some years, a year-end peak was measured. Unfortunately, there are only a few samples from September so the constancy of the late summer peaks is not known. Overall, the very low levels indicate extremely small quantities of algae

*V. Alexander

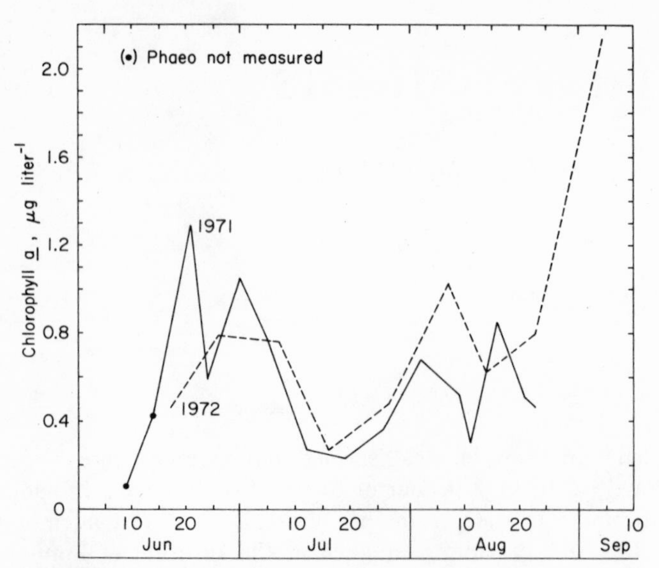

FIGURE 5-1. *Chlorophyll* a *in the water of Pond B, 1971 and 1972.*

and an ultra-oligotrophic situation. The concentrations are comparable with those found in the open ocean, in lakes at Prudhoe Bay, Alaska (0.13 to 1.46 μg Chl *a* liter^{-1}), and in Ikroavik Lake (0.03 to 2.37). In lakes Peters and Schrader, deep lakes in the Brooks Range, Hobbie (1959) found up to 1.6 μg Chl *a* liter^{-1} during an under-ice bloom but concentrations dropped to 0.8 μg during the open water period. Char Lake contained 0.4 to 0.6 μg Chl *a* liter^{-1} during the summer and 0.1 to 0.2 μg in the winter (Kalff and Welch 1974).

The microscopic examination confirmed the earlier work of Kalff (1967a) who found that almost all of the phytoplankton were small nannoplankton forms. An even earlier study of the freshwater algae of the Barrow area (Prescott 1953) sampled only with a net and also used formalin for a preservative. Not only do the nannoplankton pass through most nets, but the flagellates are destroyed by formalin. For this reason, the dominant algae we found in the ponds were not the same ones reported by Prescott. In addition, the settling chamber technique reveals millions of algae per liter while the net collects only a few hundred or thousand per liter. This dominance of the nannoplankton over the net plankton is characteristic of arctic lakes and ponds (Hobbie 1973).

Although some variability in the species did exist from pond to pond, there was a strikingly consistent pattern of early season dominance by the Chrysophyta and a later shift to dominance by Cryptophyta (Figure 5-2). The important Chrysophytes included *Chromulina* sp. and *Ochromonas* sp., with *Pseudokephyrion* sp. and *Mallomonas* sp. also present. The Cryptophytes included a number of species of *Cryptomonas* and *Chroomonas* but the most abundant form was *Rhodomonas minuta*. All

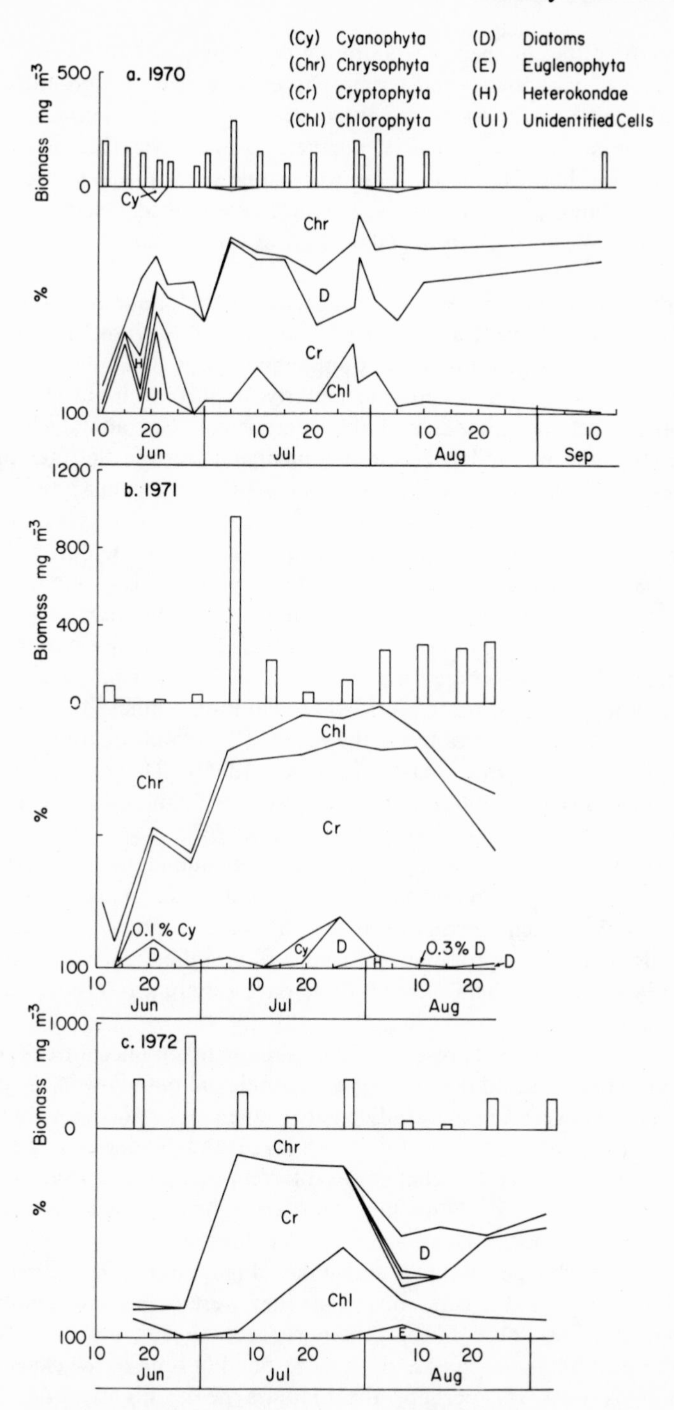

FIGURE 5-2. *Biomass (wet weight) and proportional distribution of algal groups in the phytoplankton of Pond B.*

told, some 105 different species were found in the ponds (Table 5-1). The dominance of Chrysophytes and Cryptophytes is typical of high-latitude or high-altitude bodies of water and the species found in these ponds were almost all the same ones found in northern Scandinavia (Skuja 1963, Holmgren 1968). Unfortunately, there were some small forms that could not be given names, possibly because they are undescribed species. These were lumped as 4-μm or 6-μm flagellates, for example, and often were the most abundant forms.

In other studies in northern Alaska, Coulon and Holmgren (quoted in Hobbie 1973) looked at a transect of 10 lakes from the Brooks Range to the coast. They also found a dominance of Chrysophytes and Cryptophytes but diatoms were particularly important in larger and deeper lakes such as Chandler. Lakes Peters and Schrader, Alaska, characteristically have under-ice blooms of algae. Despite the fact that these lakes are similar in water chemistry and depth and are separated by a narrow channel only 1.5 km long, the bloom alga in Peters was a diatom (*Synedra*) and the bloom organisms in Schrader were Chlorophytes (*Chlamydomonas, Pyramidomonas* and *Ankistrodesmus*). Comparisons of these and other arctic lakes (Table 5-2) indicate that the Barrow ponds are not atypical of arctic conditions and that their phytoplankton biomass is even higher than the average for all arctic water bodies. Compared with those of temperate lakes, these arctic phytoplankton biomass figures are very low. For example, the range in the ponds is 25 to 400 μg liter^{-1} (but two samples during spring peaks did reach 1,000). The range in the eutrophic Lake Esrom is 400 to 3,000 (Jonasson 1972) and in mesotrophic Lake Erken is 300 to 6,000 μg liter^{-1} (Nauwerck 1963).

The small phytoplankton algae are extremely abundant despite their low total biomass. Thus, on one date in 1970 Pond C had an algal biomass of 600 μg liter^{-1}, which corresponded to 4 million cells liter^{-1}. The range over the summer was approximately from 1 to 6 million cells. As will be described later, the numbers of planktonic bacteria are a thousand times higher than this.

Phytoplankton data represent a single sample taken once a week; the tremendous work involved in counting one sample did not allow us to take daily samples. However, in one study where tremendous effort was made a dramatic day-to-day change was found in a large lake (Rodhe et al. 1958). These workers attributed the changes to different water masses that were transported by the wind. Hopefully, this areal heterogeneity is not a problem in a small pond. There are, however, other complications as we found important changes within a single day (Figure 5-3). These changes are regular enough that it is doubtful that they were caused by sampling error. One possible cause is that rapid reproduction is occurring and indeed, the peak in algal biomass does correspond in time to the expected photosynthesis peak (or perhaps the biomass peaks slightly after the photosynthesis). Another possibility is that the algae are being heavily

TABLE 5-1 *Phytoplankton Species Found in the Barrow Area in July 1971*
P=Pond Site I=Ikroavik Lake M=North Meadow Lake

	P	I	M
Bacteriophyta			
Achromatium oxaliferum SCHEW	P		
Macromonas mobilis (LAUT.) UTERMOEHL	P		
Cyanophyta			
Chroococcus prescotti DROUET et DAILY	P		
Chroococcus turgidus var. *maximus* NYGAARD	P		
Aphanotheca cf. castagnei (BREB.) RBH	P		
Aphanotheca clathrata W et G. S. WEST	P		
Merismopedia glauca (EHR) NAEG	P		
Gomphosphaeria lacustris CHOD	P		
Gomphosphaeria robusta SKUJA	P		
Coelosphaerium kuetzingianum NAEG			M
Anabaena cf. lapponica BORGE	P		
Pseudanabaena spp.	P		
Oscillatoria cf. borneti ZUKAL	P		
Chlorophyta			
Pedinomonas minutissima SKUJA	P		
Nephroselmis discoides SKUJA	P		
Scourfieldia cordiformis TAKEDA	P		
Chlamydomonas sagittula SKUJA	P		
Chlamydomonas frigida SKUJA	P		
Chlamydomonas carolae n. sp.	P		
Chlamydomonas liloeae n. sp.	P		
Chlamydomonas spp.	P		
Sphaerellopsis cf. fluviatilis (STEIN) PASCHER	P		
Chlorogonium spp.	P		
Pandorina morun (MUELL) BORY	P		
Volvox aureus EHR	P		
Gonium pectorale MULLER	P		
Paulschulzia pseudovolvox SKUJA	P		
Korshikoviella gracilipes (LAMBERT) SILVA	P		
Gloeocystits planctonica (W et G. S. WEST) LEMM		I	
Gloeococcus schroeteri (CHOD) LEMM		I	M
Pediastrum duplex MEYEN			M
Pediastrum boryanum (TURP) MENEGH			M
Pediastrum kawraiskyi SCHMIDLE			M
Pediastrum integrum NAEGELI			M
Oocystis submarina var. *variabllis* SKUJA	P		
Oocystis arctica PRESCOTT			M
Oocystis lacustris CHODAT	P		
Oocystis gigas ARCH	P		
Oocystis borgei SNOW	P		
Tetraedron minimum (A BR) HANSG			M
Scenedesmus acuminatus (LAGERH) HOD	P		
Scenedesmus armatus (CHOD) G. M. SMITH			M
Dictyosphaerium pulchellum WOOD			M
Dictyosphaerium simplex SKUJA	P		

(continued)

TABLE 5-1 (Continued)

Crucigenia tetrapedia (KIRCHN) W et G. S. WEST	P		
Ankistrodesmus falcatus (CORDA) RALFS	P		
Ankistrodesmus acicularis A BR	P		
Ankistrodesmus spiralis (TURNER) LEMM	P	I	
Ankistrodesmus nannoselene SKUJA	P		
Planctonema lauterborni SCHMIDLE	P		
Gonatozygon monotaenium DE BY	P		
Closterium kuetzingii BREB	P		
Closterium spp.	P		
Xanthidium antilopaeum (BREB) KUETZ	P		
Staurastrum pachirhynchum NORDST	P		
Staurastrum polymorphym BREB	P		
Staurastrum spp.	P		
Hyalotheca dissiliens (SM) BREB	P		
Euglenophyta			
Euglena cf. viridis EGR	P		
Euglena pisciformis KLEBS	P		
Euglena gracilis KLEBS		I	
Euglena cf. oxyuris SCHMARDA	P		
Euglena cf. intermedia (KLEBS) SCHMITZ			M
Lepocinclis cf. ovum (EHR) LEMM	P		
Phacus pyrum (HER) STEIN	P		
Phacus sp.	P		
Trachelomonas volvocina EHR	P		
Trachelomonas hispida (PERTY) STEIN	P		
Trachelomonas spp.	P		
Astasia spp.	P		
Menoideum spp.	P		
Chrysophyta			
Chromulina diachloros SKUJA	P		
Chromulina spp.	P		
Chrysococcus rufescens KLEBS	P		
Chrysococcus cf. cordiformes NAUM	P		
Chrysococcus cystophorus SKUJA	P		
Kephyrion rubri-claustri CONRAD	P		
Kephyrion spp.	P		
Stenokalyx monilifera GERL SCHMID	P		
Mallamonas akrokomos RUTTNER	P		
Mallomonas pumilio var. *canadensis* n. var.	P		
Erkenia subaequiciliata SKUJA (=*Chrysochromulina parva* LACKEY?)			M
Ochromonas elsae n. sp.	P		
Ochromonas spp.	P		
Synura sphagnicola KORSCH	P		
Synura lapponica SKUJA	P		
Uroglena americana CALKINS	P		
Chrysomoron epherum SKUJA	P		
Pseudokephyrion enzii SKUJA (= *Pseudokephyrion hyalinum* HILLIARD)	P		
Pseudokephyrion parvum HILLIARD	P		

TABLE 5-1 (Continued)

Dinobryon sociale var. *americanum* (BRUNNTH) BACHM	P		
Dinobryon sertularia EHR	P		
Dinobryon sertularia var. *protuberans* (LEMM) KRIEGER	P		
Cercobodo spp.	P		
Parabodo attenuatus SKUJA	P		
Heterokontae			
Botryococcus braunii KUETZ	P		
Tribonema spp.		I	M
Chlorobotrys regularis BOHLIN			M
Chloromonadophyceae			
Monomastix ophiostigma SCHERFF	P		
Crytophyceae			
Chroomonas coerula (GEITLER) SKUJA	P		
Chroomonas nordstedtii HANSG	P		
Rhodomonas minuta SKUJA	P		
Rhodomonas minuta var. *nannoplanctonica* SKUJA	P		
Rhodomonas sp.	P		
Cryptomonas ovata EHR	P		
Cryptomonas obovata SKUJA	P		
Cryptomonas borealis SKUJA	P		
Cryptomonas Marssonii SKUJA	P		
Cryptomonas sp.	P		
Sennia parvula SKUJA	P		
Dinophyceae			
Amphidinium cf. lacustre STEIN	P		
Amphidinium spp.	P		
Gymnodinium palustre SCHILLING	P		
Gymnodinium cf. lacustre SCHILLER	P		
Gymnodinium triceratum SKUJA	P		
Gymnodinium uberrium (ALLM) KOFOID et SWEZY	P		
Gymnodinium cf. veris LINDEMAN		I	
Gymnodinium spp.	P		
Hemidinium nasatum STEIN	P		
Glenodinium uliginosum SCHILLING	P		
Peridinium cinctum (O. F. M.) EHR	P		
Peridinium willei HUITF -- KAAS	P		
Peridinium palustre (LINDEM) LEFEVRE	P		
Peridinium inconspicuum LEMM	P		

TABLE 5-2 *Phytoplankton Biomass[1] and Dominant Groups in Arctic Lakes and Ponds[2]*

Lake and reference	Spring (under ice)		Early summer		Late summer		Late fall and winter (under ice)	
	Group	Biomass	Group	Biomass	Group	Biomass	Group	Biomass
Nedre Laksjön (Holmgren 1968)	1. Chlorophytes	100	1. Chrysophytes	550	Diatoms	400	Diatoms	124
	2. Cryptophytes	400	2. Cryptophytes	550	-	-	-	-
Char Lake (Kalff and Holmgren 1971)	1. Dinoflagellates	150	Diatoms	150	-	-	Diatoms	50
	2. Chrysophytes	150	-	-	-	-	Dinoflagellates	100
Lake Peters (Holmgren, Kalff, Hobbie, unpubl.)	Diatoms	400	Diatoms	250	-	-	-	-
Wolf Lake (Holmgren, Kalff, Hobbie, unpubl.)	Chrysophytes	50	Chrysophytes	150	Cryptophytes	150	-	-
Lake Schrader (Holmgren, Kalff, Hobbie, unpubl.)	Chlorophytes	50	Chrysophytes	200	Chrysophytes	150	-	-
Barrow ponds (This study)	-	-	Chrysophytes	100	Cryptophytes	200	-	-

[1]Biomass data are expressed as mg wet weight m^{-3}.
[2]Groups are listed in chronological order of dominance.

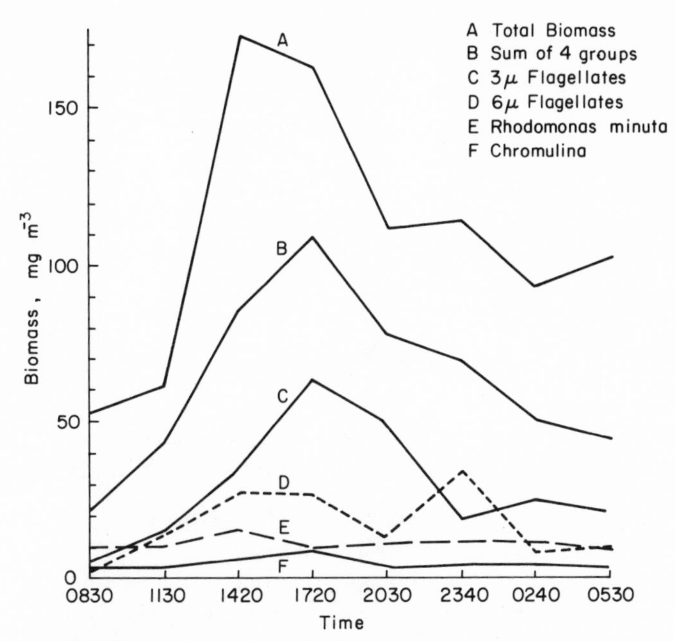

FIGURE 5-3. *Phytoplankton biomass of a pond measured at different times on 24-25 July 1972.*

grazed by zooplankton. Finally, there is the possibility that the very small forms are migrating within the water column or moving back and forth from the sediment. Among the dominant forms given in Figure 5-3, the *Rhodomonas* and the *Chromulina* are well known as planktonic forms and their biomass shows little diurnal change. In contrast, the small flagellates (3 and 6 μm) are rarely important in lake plankton and do show great diurnal changes. We think it most likely that the small forms move back and forth from the algal-rich sediment. As will be presented later, the algae of the sediments are about 60 times more abundant (per milliliter) than the planktonic algae so there is certainly a large reservoir of algae. These small flagellates could move by themselves or could be mixed into the water column by wind currents. This mixing by wind currents has been shown to be very important for bacteria in these same ponds. Hobbie and Rublee (1975) found a 3-fold increase in bacterial numbers over several hours, so it is reasonable that small flagellates would follow the same pattern.

The seasonal pattern of early Chrysophyte and later Cryptophyte dominance is typical of these ponds in general but different patterns were seen in a few ponds. Several transects of 16 ponds each were sampled during the study; one pond with diatom dominance and one with a large biomass of the blue-green alga *Microcystis* were found. We do not know why these differences exist. However, in Pond E a change in dominance occurred (Figure 5-4) after crude oil was added in midsummer 1970. Here

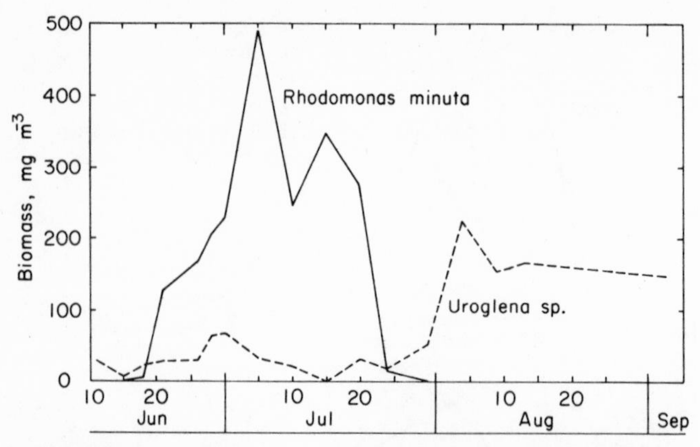

FIGURE 5-4. *Biomass of* Rhodomonas *and* Uroglena *in Pond E, 1970. Oil was added to the pond on 16 July.*

the Chrysophyte *Uroglena* replaced *Rhodomonas*. In other ponds, *Uroglena* was usually present but always rare.

This experimental oil addition was repeated in the summer of 1975 with the same results: *Rhodomonas* was replaced by *Uroglena*. However, the main effect of the oil was always on the zooplankton which were all killed immediately (see Chapter 9). For this reason, the zooplankton were removed from an experimental pond (Miller et al. 1978a, 1978b); the *Rhodomonas* again was replaced by *Uroglena*. There are two possible explanations of this relationship: either the grazing of the zooplankton indirectly controls the *Rhodomonas*, perhaps by keeping the *Uroglena* at a low level, or there is a control through nutrient regeneration by the zooplankton. Because the sediments seem to control the nutrient supply rate here, we believe that the grazing of zooplankton controls the *Rhodomonas*, either directly or indirectly.

Primary Productivity

Phytoplankton photosynthesis was measured with $^{14}C\text{-}HCO_3$. All measurements were made in the morning with a 3- to 6-hour incubation in 125 ml bottles placed on their sides on the bottom of the pond. During 1970 and 1971, subsamples were filtered through 0.45 μm HA Millipore filters, the filters rinsed with 0.005 N HCl, and the activity taken up determined with a gas-flow planchette counter. In 1972 and 1973, the filtration step was omitted (Schindler et al. 1972); instead, the incubated sample was acidified with 12 N HCl to a pH of 1 and then flushed with air to remove the inorganic ^{14}C. A portion of this subsample was then mixed with Aquasol and counted on a liquid scintillation counter. This procedure measured not only the ^{14}C fixed and retained within the algal cells but also the ^{14}C lost from the cells as dissolved organic carbon. The uptake data obtained from these experiments were converted into hourly photosynthesis rates from measurements of pH and alkalinity and appropriate calculations (e.g., Vollenweider 1968). Duplicates of each light bottle were run and the uptake values for a single dark bottle were subtracted from the average of the light bottles.

The seasonal course of the primary productivity closely followed the chlorophyll and biomass curves (Figure 5-5). In some cases, for example the increase in phytoplankton seen at the very end of August, the increase in primary productivity appeared to precede the increase in chlorophyll and in algal biomass. In general, the pattern was similar in all the ponds.

At least three different patterns of photosynthesis rates over 24 hours were seen in the ponds. These will be discussed in detail later when photosynthesis-light relationships are taken up. Two of these patterns, one with early morning and late evening peaks and one with a noon-time peak, are clearly related to optimum and inhibiting light levels. One other pattern (Figure 5-6), however, may also be related to the biomass of the algae. Here, the low photosynthesis in the morning (1130) may have been caused by the low biomass at the same time (Figure 5-3).

The algae from the detailed diurnal study (Figures 5-3 and 5-6) were studied with track autoradiography. First, the algae were preserved with acid Lugol's solution, settled, and a permanent slide made (Coulon and Alexander 1972). Next, the slides were coated with photographic emulsion (Rogers 1969); exposure lasted from 3.5 to 13 hours at 5°C. Development

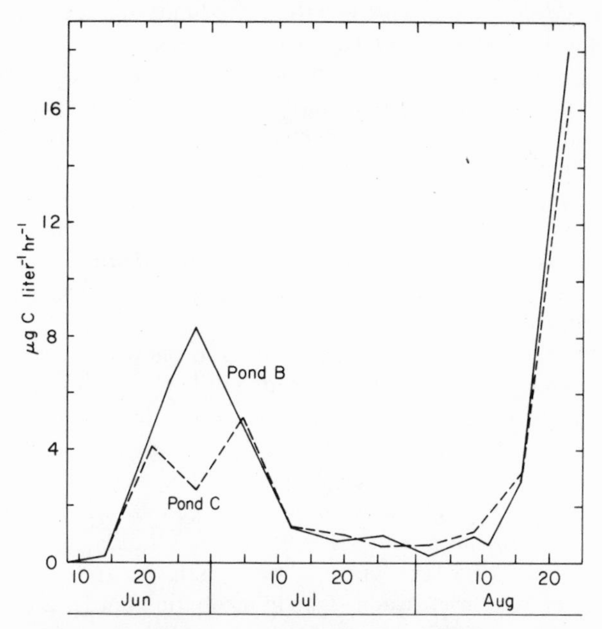

FIGURE 5-5. *Rates of planktonic primary productivity in Ponds B and C, 1971.*

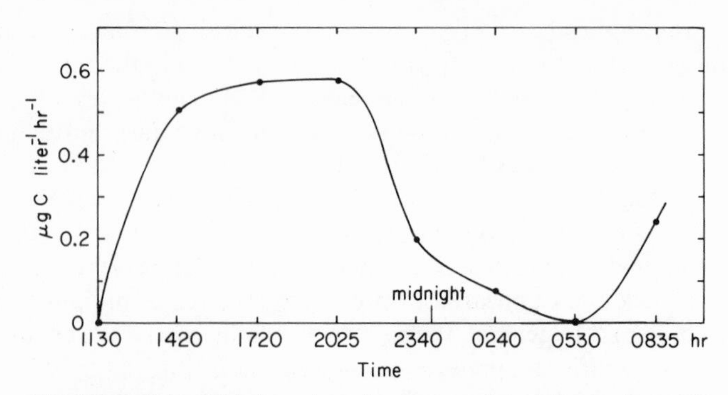

FIGURE 5-6. *Planktonic primary productivity in Pond B on 24 July 1972.*

also followed procedures of Rogers (1969) except that the slides were soaked only 5 min in 20% glycerol. Finally, the preparations were mounted in Farrant's medium (Meynell and Meynell 1965) and dried for several days before viewing. Only tracks with at least four exposed grains were counted. Tests with three species of flagellates in culture showed that no significant amounts of ^{14}C were lost by the preservation with Lugol's. Also, there was a linear increase in tracks per cell as the exposure time of the emulsion increased. The maximum exposure for the planktonic algae, 13 hours, gave about three tracks per *Rhodomonas* cell and about one track per cell of *Mallomonas*. This is within the limits for the method suggested by Knoechel and Kalff (1976a, 1976b).

The track autoradiography (Figure 5-7) shows the strong dominance of the 3 μm flagellate group during the afternoon. Their biomass (Figure 5-3) was also high at this time but the productivity of this group is obviously higher per individual than that of the other groups. When the data are expressed as production per unit of biomass (Figure 5-8), the *Rhodomonas* and 6 μm flagellates have a photosynthesis peak in the early afternoon, the 3 μm organisms peak in late afternoon, and the *Chromulina* peaks in the early evening. This is one of the ways that the algae share the resources, here the light, in the pond.

Given the changing relationships between light, photosynthesis, and algal biomass, it is difficult to translate from an hourly photosynthesis rate measured in the morning to a value for an entire day. If indeed a significant fraction of the algae are moving back and forth from the benthic to the planktonic systems, then an individual alga cell *in situ* may not be exposed to inhibiting light levels for as long a period as a cell held in a sample bottle for 3 or more hours. The situation is similar to that of algae in the epilimnion of a lake where *in situ* organisms are circulated through a variety of light levels while algae in incubation bottles are held at a constant (and often inhibiting) level of light. In both these situations, it is difficult to estimate the true daily productivity either from a 3-, 12- or 24-

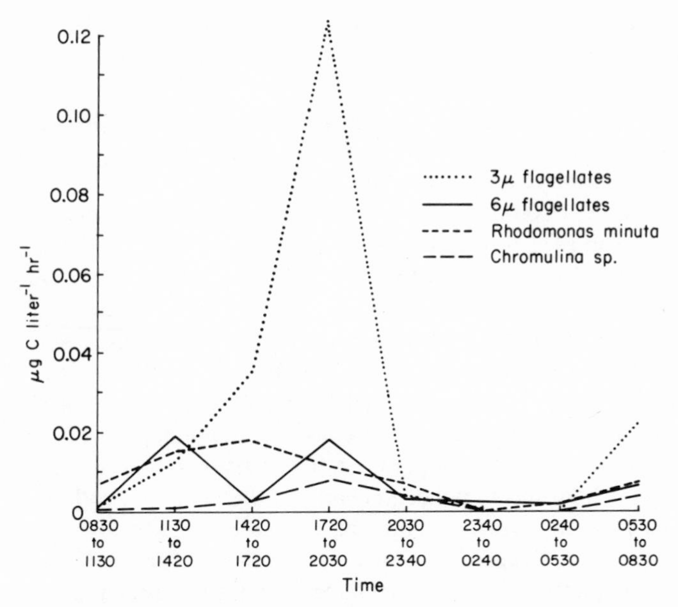

FIGURE 5-7. *Primary productivity of four groups of phytoplankton studied by track autoradiography. This is the same experiment as the one shown in Figure 5-3.*

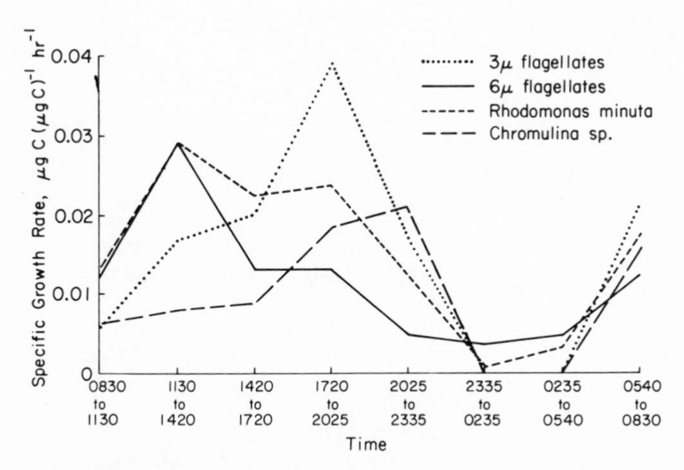

FIGURE 5-8. *Primary productivity per unit of biomass (specific growth rate) for four groups measured by track autoradiography (see Figures 5-3 and 5-7).*

hour incubation. The procedure chosen here was to set up a proportionate relationship such that (insolation during incubation)÷(total daily insolation) equals (primary productivity during incubation)÷(total daily primary production). The errors are likely due to underestimations but the total primary production due to phytoplankton was so small that it was not worth attempting to obtain a more accurate measurement.

If we assume a 100-day growing season and a mean depth of 20 cm for the ponds, the 1971 production estimate was 1216 mg C m^{-2} for Pond B and 1024 mg C for Pond C. In 1972, Pond B had a production of 900 mg C m^{-2} while in 1973 Pond C's production was 466 mg C. These estimates agree with those of Kalff (1967a) for a nearby but different series of Barrow ponds over a 2-year study (380 to 850 mg C m^{-2}). Overall, these are among the lowest values measured for phytoplankton production in any body of water including other arctic lakes and ponds (Table 5-3). It is obvious, however, that the maximum rates of photosynthesis per cubic meter are quite high in the ponds. Therefore, the shallow depth and the short growing season are the cause of the exceptionally low seasonal primary production. Deeper lakes have lower rates per unit volume but a much greater euphotic zone (up to 20 m or so in Char Lake) and a longer

TABLE 5-3 *Phytoplankton Primary Production of Some Arctic Lakes and Ponds*

Lake	Maximum mg C m^{-3} day^{-1}	Year	Annual g C m^{-2}	Reference
Char Lake	–	1970	4.3	Kalff and
	–	1971	4.7	Welch
	–	1972	3.6	(1974)
Lake Peters	5.6	1961	0.9	Hobbie (1964)
Lake Schrader	14.0	1959	7.5	Hobbie
	5.4	1961	6.5	(1964)
Imikpuk Lake	60	1961	8.5	Kalff (1967b)
Tundra Ponds II	130	1963	0.7	Kalff
	260	1964	0.8	(1967a)
III	60	1963	0.4	
	65	1964	0.8	
Tundra Pond B	200	1971	0.6	This study
(Barrow) C	–	1971	0.5	
D	–	1971	0.7	
E	–	1971	0.3	
Meretta Lake	–	1970	11.9	Kalff and
	–	1971	12.5	Welch
	–	1972	9.4	(1974)

growing season. Char Lake (Kalff and Welch 1974) and lakes Peters and Schrader (Hobbie 1964) have 8- or 9- month growing seasons and most of their primary production actually occurs beneath the ice cover. Meretta Lake, which receives sewage, indicates that the potential production of arctic lakes can be quite high.

EPIPELIC ALGAE*

Biomass of Algae

Epipelic microalgae, mostly greens and blue-greens, are abundant in the pond surface sediments. Some forms are motile, but the majority are attached to sediment and detritus particles. For this reason, direct cell counts were made in 1971 by dilution of a mixed sediment sample (to 5 cm), filtration onto a 0.45-μm black Millipore filter, and counting with an epi-illuminated fluorescence microscope. This was rather a crude method as only the red fluorescence of the chloroplasts could be seen and separate species could not be identified. On the other hand, the fluorescence made it easy to count all of the algae on the surface of the sediment particles. The numbers were converted to biomass from estimations of the volumes of the dominant algae (a specific gravity of 1.0 was assumed). This technique was used on weekly samples from Pond J collected at the same time and place as the productivity samples (Stanley 1976a). A slightly different technique, a direct count without the filtration step, was used by Fenchel (1975) in Pond A in 1973 (7 weeks). The results agreed quite well. The algal species were identified and their relative abundance determined by C. Coulon on several occasions during 1972 by using diluted samples and a conventional light microscope.

Chlorophyll *a* concentrations, another indicator of algal biomass, were measured in 1972 in Pond B. The upper 5 cm of a sediment core was sectioned at 1 cm intervals and extracted, with frequent shaking, for 12 hours with 90% acetone. After centrifugation, the optical density of the sample was read at 665 nm. Turbidity, phaeophytin corrections, and calculations were according to Lorenzen (1967).

The Pond B epipelic assemblage was dominated by Chlorophyta and Cyanophyta. These two groups made up 50% and 40%, respectively, of the total biomass. Some of the most common blue-green forms are *Microcystis, Gomphonema,* and *Aphanozomenon,* while the common green algae are *Chlamydomonas, Closterium,* and *Ankistrodesmus* (Table 5-4). However, as Figure 5-9 illustrates, the composition can vary considerably from one pond to another. Obviously the dominant benthic

*D. Stanley

TABLE 5-4 *List of Common Epipelic Algae Species Found in Pond B in 1972* *

Cyanophyta	Bacillariophyceae
Anabaena lapponica	*Cymbella* sp.
Aphanocapsa sp.	*Eunotia lunaris*
Aphanotheca clathrata	*Fragilaria virescens*
Aphanotheca sp.	*Fragilaria* sp.
Aphanozomenon flos-aquae	*Navicula* sp.
Chroococcus turgidis	*Nitzschia linearis*
Coelospherium kuetzingianum	*Pinnularia mesolepta*
Gleococcus schroeteri	*Pinnularia* sp.
Gomphospheria naegeliana	*Stauroneis* sp.
Microcystis flos-aquae	*Tabellaria fenestrata*
Gomphospheria sp.	
Oscillatoria agardii	Chrysophyta
Oscillatoria sp.	*Pseudokephyrion undulatissimum*
Synechocystis notatus	*Synura uvella*
Chlorophyta	Pyrrophyta
Ankistrodesmus falcatus	*Amphidinium* sp.
Ankistrodesmus spirale	*Peridinium inconspicuum*
Chlamydomonas frigida	
Chlamydomonas lapponica	Cryptophyta
Chlamydomonas sessila	*Cryptomonas* sp.
Closterium aciculare	
Closterium moniliferum	Euglenophyta
Cosmarium botrytis	*Lepocinclis* sp.
Cosmarium granatum	
Cosmarium ornatum	
Elakatothrix lacustris	
Euastrum binale	
Euastrum elegans	
Oocystis lacustris	
Staurastrum gracile	

*Identifications by C. Coulon.

algae are very different from the dominant planktonic algae (Chrysophyta, Cryptophyta).

Most of these cells are very small ($<10 \mu m$) so that in 1971, for example, even though their numbers ranged from about 2×10^{10} to 4×10^{10} m^{-2} the equivalent carbon biomass was calculated to be only 0.5 to 1.0 g m^{-2} (Figure 5-10). The pattern of gradual seasonal increase in the biomass was repeated in Pond A in 1973, where Fenchel (1975) found a change from 2.5 to 8×10^{11} cells m^{-2}. Although the numbers Fenchel obtained in 1973 are higher than the numbers in Figure 5-10, his estimate of algal biomass (700 mg C m^{-2}) agreed very well with the 1971 estimates.

There is considerable variation in the abundance of algae, measured by the chlorophyll *a* concentration, from pond to pond. For example, Pond B sediments contained almost twice as much chlorophyll as those from Pond D in 1972 (Table 5-5).

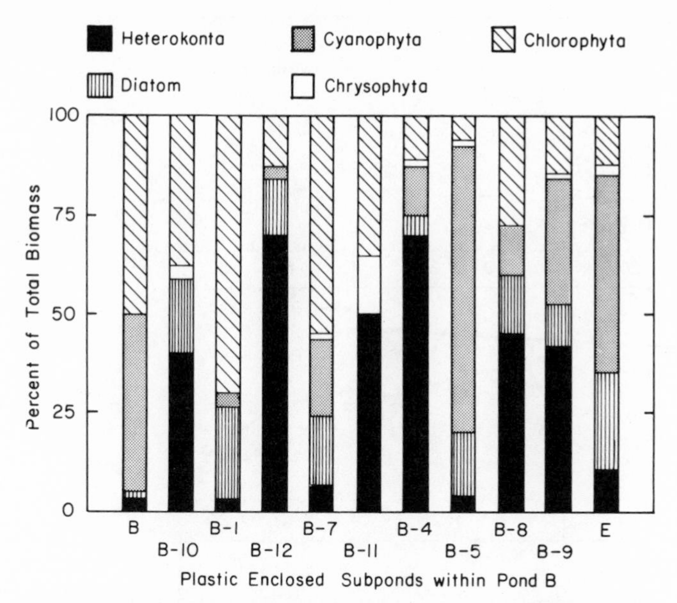

FIGURE 5-9. *Proportional distribution of epipelic algal groups in tundra subponds within Pond B, July 1972.*

There is neither a uniform distribution of epipelic algae in the top 5 cm of sediment nor an algal mat at the surface. Instead, the cell density is highest near the surface, with a steady downward decline (Table 5-5). In 1971, about half the biomass was in the upper 1 cm, and nearly 80% in the top 2 cm. This was determined from both chlorophyll *a* and photosynthesis measurements on 1-cm-thick sediment core sections (Figure 5-11). In 1973, Fenchel (1975) determined that 64% of the total number of algae was in the top 2 cm and the remainder was in the 2 to 4 cm layer. Others (e.g., Round 1964, Gruendling 1971, Fenchel and Straarup 1971) have found similar vertical distributions of algae in temperate lake and estuarine sediments. For these photosynthetic organisms, this means that only a small fraction of the total can be producing at any one time since the great majority are buried below the 1% level of light extinction, estimated to be only about 2 mm in the tundra pond sediments. Yet these cells are certainly alive, as evidenced by their appearance under the microscope and their ability to photosynthesize as soon as they are placed in sunlight (Figure 5-11). Algae can survive for months in the dark (Bunt and Lee 1972) and here it appears that cells produced at the surface are continuously being mixed into the deeper sediments by a number of mechanisms. Hence, it is not surprising that the seasonal patterns of total biomass and productivity are very dissimilar (Figure 5-10).

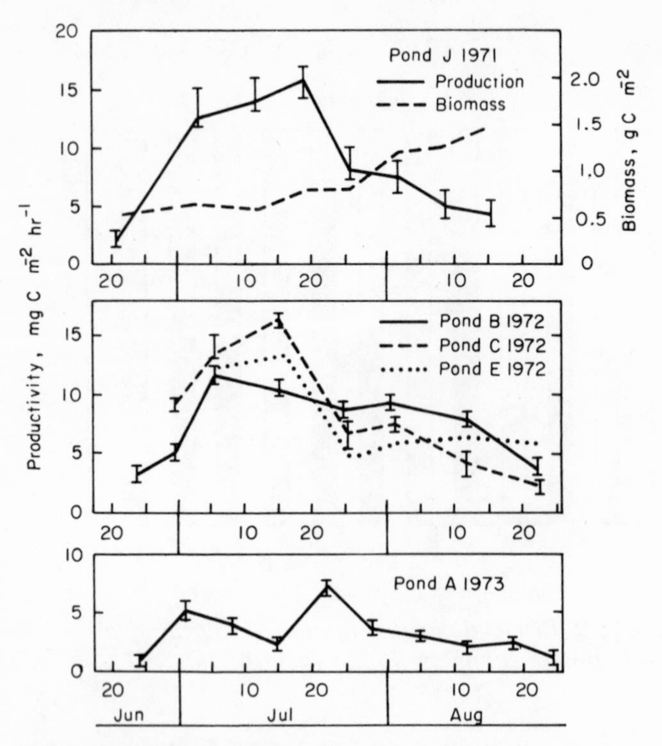

FIGURE 5-10. *Productivity and biomass of the epipelic algae in Pond J, 1971, productivity in ponds B, C and E, 1972, and Pond A, 1973.*

Productivity

Several summer-long epipelic productivity studies were carried out in the ponds between 1971 and 1973 with weekly sampling beginning as soon as the ice disappeared in late June and continuing into late August. A modified carbon-14 technique was used for these measurements. Details are given in Stanley (1976a). Four randomly chosen cores were taken from the sediments and the top centimeter of each core was mixed into a combined sample. Four subsamples were then settled onto the bottom of 400-ml chambers made by gluing together the bottoms of acrylic Petri dishes. The amount of sediment in each chamber was less than 1.0 g dry weight to avoid CO_2 depletion. We found that up to 0.5 g of sediment could be used without a decline in the rate of photosynthesis per gram of sediment due to shading. Isotope was injected through a serum stopper and the chamber incubated in the pond (2 to 4 hours). After filtration of the algae and sediment at the end of the experiment, the samples were combusted by wet oxidation and the liberated CO_2 trapped and counted in a liquid scintillation cocktail.

TABLE 5-5 *Concentrations of Chlorophyll* a *and Phaeophyton in Pond Sediments, 1972**

Pond	Date	Sediment Depth (cm) 0 - 1		1 - 2		2 - 3		3 - 4		4 - 5	
		chl *a*	phaeo	chl *a*	phaeo	chl *a*	phaeo	chl *a*	phaeo	chl *a*	phaeo
B	20 May	59	39	22	129	24	96	2	18	93	141
B	7 July	188	119	75	150	60	160	40	240	22	160
B	3 Aug	199	291	80	330	84	171	62	177	22	170
B	14 Aug	96	199	-	-	36	267	-	-	10	292
B	21 Aug	194	233	-	-	-	-	-	-	42	188
C	20 May	12	131	5	73	2	192	0	94	5	44
C	27 Jun	60	105	40	200	50	150	30	205	10	60
C	7 July	20	30	0	160	0	270	0	310	0	335
C	31 July	54	247	8	241	27	207	0	168	6	107
C	7 Aug	157	151	0	277	0	235	10	214	0	249
C	17 Aug	101	186	-	-	48	238	-	-	19	154
C	24 Aug	65	195	-	-	79	233	-	-	66	279
D	20 May	11	120	13	92	7	58	9	51	0	178
D	27 Jun	65	150	50	160	45	150	20	80	30	150
D	7 July	10	130	0	120	0	300	0	355	0	300
D	31 July	90	145	10	182	-	-	-	-	5	197
D	7 Aug	41	226	0	190	0	135	0	96	0	97
D	14 Aug	54	136	-	-	44	128	-	-	18	74
D	24 Aug	74	105	-	-	56	132	-	-	26	81
E	20 May	38	72	18	201	30	255	3	238	0	309
E	19 July	120	300	75	250	30	235	45	210	20	200
E	3 Aug	76	238	45	236	72	197	71	260	36	128
E	14 Aug	63	280	71	239	-	-	-	-	12	317
E	21 Aug	129	143	-	-	71	181	-	-	38	217

*Data are expressed in μg pigment (g dry wt sediment)$^{-1}$.

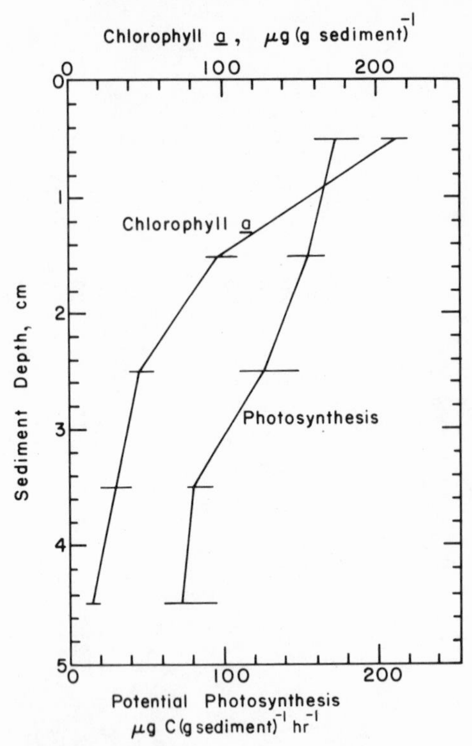

FIGURE 5-11. *Potential photosynthesis and chlorophyll* a *at different depths in the sediment of Pond B, July 1972.*

The data obtained in this way are estimates of potential production (Grontved 1962) rather than of true production per square meter. These values were converted to true production rates by the method suggested by Hunding (1971) which uses measurements of photosynthesis as a function of light intensities. These determinations, plus measures of the actual light extinction within the sediment, allow photosynthesis rates per square meter to be calculated.

Most of the primary production of the epipelic algae occurred in the top millimeter of sediment (Figure 5-12) because of the rapid extinction of light. The algae are abundant in the top 4 cm or so and are capable of photosynthesis when incubated in the light (Figure 5-11).

The seasonal curve of photosynthesis in Pond J in 1971 (Figure 5-10) is typical of the results. Rates of photosynthesis began to increase soon after the ponds thawed and warmed; the rates rose from 2 mg C m^{-2} hr^{-1} on 21 June to around 16 mg C m^{-2} hr^{-1} by mid-July. Following this peak, there was a steady decline throughout late July and August. Presumably photosynthesis continued until the ponds froze in mid-September. These

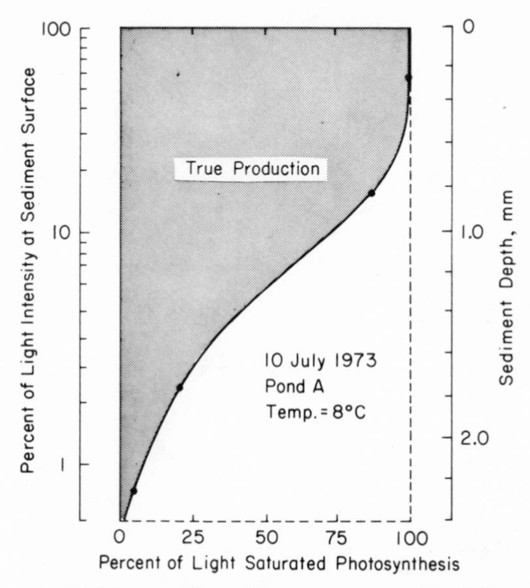

FIGURE 5-12. *Percentage of light-saturated photosynthesis at various depths in the sediment of a pond. The line also represents the relative light intensity at various depths; the shaded area represents estimated true productivity.*

mid-day photosynthesis rates were extrapolated to daily rates, based on measurements of the diurnal pattern of photosynthesis of these algae (Figure 5-13), and the daily rates summed (15 June to 1 September) for an estimate of annual production. These values ranged from 10.1 to 4.1 g C m^{-2} in Pond J, 1971 and in Pond A, 1973, respectively and are about 10 times greater than the plankton production of 0.6 to 0.9 g C m^{-2}.

The sediment algae are not so important in deeper arctic lakes. For example, in 1971 at nearby Ikroavik Lake (2.2 m deep) the total net epipelic fixation was only 2.3 g C m^{-2}, primarily because of the shorter ice-free period, lower water temperatures, and lower light intensity at the sediment surface. The plankton algae productivity was only slightly lower, about 2.2 g C m^{-2}. In still deeper Alaskan lakes, like Lake Schrader (average depth 30 meters), epipelic photosynthesis is probably insignificant because of the turbid water (Hobbie, personal communication). This was not the case in Char Lake in the Canadian Arctic which is both deep and very clear. Here, Kalff and Welch (1974) estimated a plankton production of 4.2 and a benthic microalgae production of 8.9 g C m^{-2}. This lake, however, has a moss-covered bottom (production of 8.4 g C).

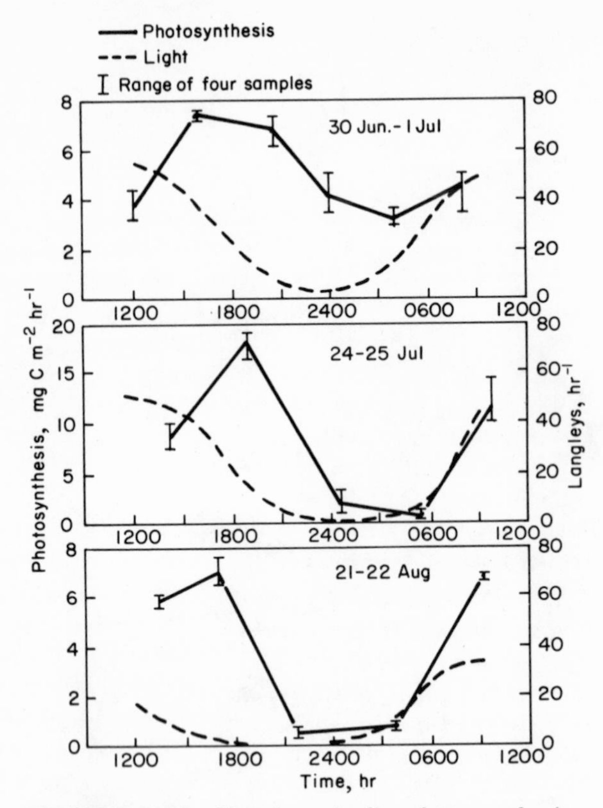

FIGURE 5-13. *Hourly epipelic photosynthesis rates and insolation, Pond B, taken on three days in 1972.*

In temperate regions only the shallow lakes have significant benthic photosynthesis. However, the combined total algal production is almost always greater than in the arctic ponds. In Marion Lake, British Columbia, epipelic algae contributed about 40 g C m^{-2} yr^{-1} to a total primary production of 66 g C m^{-2} yr^{-1} (Hargrave 1969). As another example, about 70% of the total primary production measured by Wetzel (1964) in Borax Lake was by benthic microalgae. Finally, in Lake Fureso, the production was 143 g C m^{-2} yr^{-1} for the epipelic algae but only 50 g C m^{-2} yr^{-1} for the plankton (Hunding 1971). All of these lakes are shallow, so significant amounts of light penetrate to the bottom. Thus, morphology seems to play an indirect role in determining the relative proportions of planktonic and benthic algal photosynthesis. Other, more direct controlling forces will be discussed in detail later.

FACTORS CONTROLLING ALGAE

Introduction

Primary production in any ecosystem depends on both the standing crop of producer organisms and their rate of net carbon fixation per unit biomass. The environmental factors controlling biomass and photosynthesis are interactive but their effects can be examined individually. The ultimate purpose of such experiments was to explain the highly characteristic daily and seasonal patterns of productivity and biomass (Figures 5-5, 5-6, 5-10, 5-13) seen for the phytoplankton and epipelic algae of these ponds. The primary biomass controls for plankton and epipelic algae appear to be zooplankton grazing and sediment mixing, respectively, while the important photosynthetic controls are phosphorus limitation, light intensity, and temperature.

The basic approach for assessing short-term effects was comparison of ^{14}C-productivity estimates of natural samples exposed to different amounts of light, different temperatures, and different nutrient concentrations. These 4-hour experiments were done at weekly intervals throughout the summer and occasionally over 24-hour diurnal cycles (details in Stanley 1976b). A variety of techniques were used for studying nutrient uptake directly including ^{32}P and ^{15}N tracers, analytical water chemistry, and cell extraction procedures. Longer term effects were evaluated by comparing biomass, productivity, and species composition changes in whole ponds and subpond enclosures before and after treatment with heat, nutrients, oil, etc. The magnitude of grazing effects was determined from experimental measurements of filtering and ingestion rates of the zooplankton and benthic herbivores and from experiments in which the zooplankton were removed from ponds.

Temperature

The generally cold climate of the Barrow site, with the resultant brief open-water season and low water temperatures, is undoubtedly a major cause of the low annual production levels of both phytoplankton and epipelic algae. For some reason, the responses of the pelagic and benthic photosynthetic processes to temperature were very different (Figure 5-14). The temperature optimum of the phytoplankton averaged about 15°C over the summer, but the optimum of the epipelic algae exceeded 20°C, the highest temperature tested. This difference can only rarely be of consequence since the ponds average only 5° to 9°C during the summer and seldom waım to above 15°C. However, differences in the photosynthetic

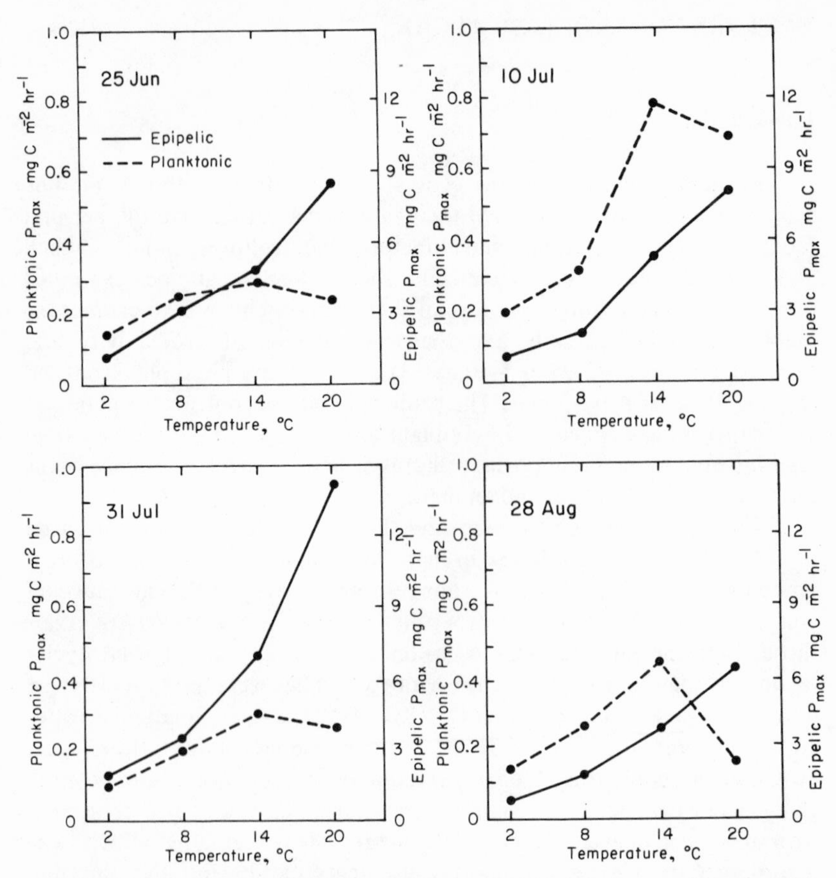

FIGURE 5-14. P_{max} *of phytoplankton and epipelic algae at four temperatures in Ponds A and B, 1973.*

response of the two groups to temperatures below the phytoplankton optimum probably are significant. Thus, the Q_{10} values between 2.5° and 12.5°C as calculated from the weekly P_{max} estimates averaged 3.0 (range 2.4 to 3.4) for the plankton algae and 2.2 (range 1.6 to 3.0) for the epipelic algae. This latter value is similar to the Q_{10} found for a benthic algal mat in two Antarctic ponds by Goldman et al. (1963) and is within the range of 2 to 4 considered typical for algae by Fogg (1965). The plankton values, on the other hand, are much higher than the 1.5 reported by Kalff (1969) for the plankton of two nearby Barrow ponds. His data were based on measurements at 10° and 20°C and would be an underestimation if the algae had a 15°C temperature optimum. In addition, Kalff employed a long incubation period (up to 40 hours) during which an unexplained decrease in photosynthesis rates took place.

These contrasting responses of planktonic and epipelic photosynthesis to temperature suggest adaptation of each group to its particular environment. The pond sediments are, in fact, warmer (up to 4°C on sunny days) than the overlying water; the sediment algae, with their higher optima, appear to take advantage of this fact. The planktonic algae, in contrast, appear to consist of species which photosynthesize more efficiently at the lower temperatures of the pond water but which are more strongly inhibited when temperatures are high. Since our Q_{10}'s are calculated from P_{max} values rather than from rate measurements at a single light intensity, the increased efficiency of planktonic photosynthesis at low temperatures is probably a result of increased enzyme activity in the dark reactions of photosynthesis. Low temperature optima, and presumably increased photosynthetic efficiency at low temperatures, have been seen by Goldman et al. (1963) in plankton in the Antarctic and by Bennett and Hobbie (1972) in a planktonic alga from Swedish Lapland. A direct comparison of P_{max} per unit carbon for similar arctic, subarctic, and temperate species would show the magnitude of this effect at low temperatures.

No clear seasonal trends in the Q_{10}'s or optimal temperatures were observed for the phytoplankton despite the species succession that occurred each year (Figure 5-2). This suggests that the temperature responses of the dominant phytoplankters were similar. However, we were not able to make measurements on individual species.

In general, then, normal diurnal and seasonal temperature fluctuations are capable of changing photosynthetic rates by as much as 50 to 75%. The relationship will normally be a direct one, since average pond temperatures are well below those for the algal optima. However, higher temperatures also stimulate other processes, such as respiration and herbivore grazing, which tend to reduce net photosynthesis or biomass. For this reason, the net effect of temperature increases on algal biomass in the ponds depends on whether photosynthesis can outpace the measured losses. For example, when temperature was increased 4°C in a sub-pond enclosure in Pond B in 1972, epipelic algal productivity showed no increase relative to the control sub-ponds and to Pond B itself.

Light

The hyperbolic form of photosynthesis-light curves for algae is well known (Vollenweider 1965, Steele 1962). Over a low illumination range the rate of photosynthesis varies almost linearly with light intensity (I), while at higher light intensities the rate rises more slowly and eventually reaches a maximum rate, P_{max}, at light saturation. The relationship can be described by

$$P = P_{max}(I)\ (I + I_{0.5})^{-1}$$

where P is the photosynthesis rate per unit biomass under a given set of temperature and light conditions, P_{max} is the light-saturated rate at the same temperature and $I_{0.5}$ is the light intensity at which P equals 0.5 P_{max}.

The half-saturation light intensity, $I_{0.5}$, can be interpreted as an integrated, inverse measure of photosynthetic efficiency, since it changes when either P_{max} or the slope of the photosynthesis-light curve at subsaturating illumination changes. The P_{max}, which is proportional to the rate of the dark reactions of photosynthesis, varies directly with temperature, as shown above, and inversely with the degree of nutrient limitation of the phytoplankton assemblage (Myers and Graham 1971, Thomas and Dodson 1972). It remains relatively constant, however, when light changes (Steele 1962, Myers 1970). On the other hand the slope of the curve at low light intensities, which is proportional to the ratio of chlorophyll to carbon in the cell and hence to the rate of the light reactions of photosynthesis, varies inversely with the light intensity ("sun-shade" light adaptation, Steemann Nielsen and Hansen 1961) and directly with

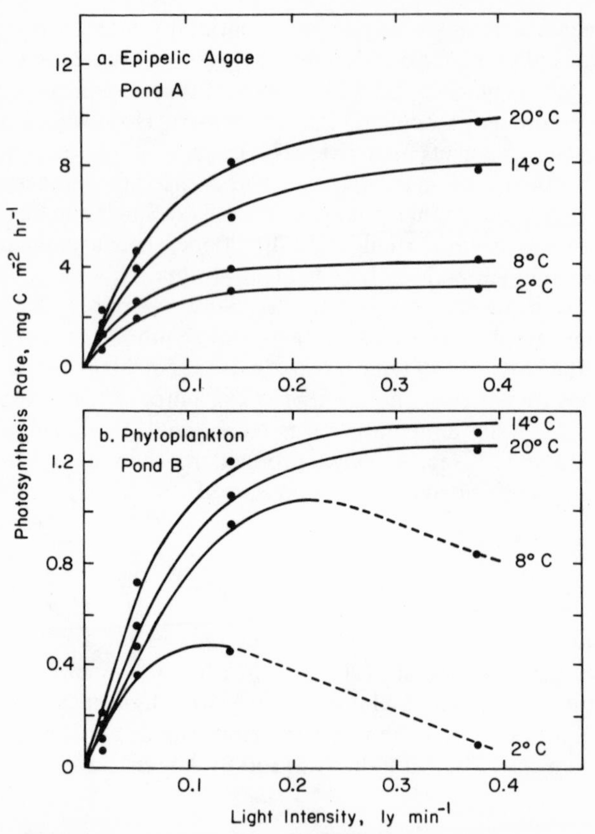

FIGURE 5-15. *A typical set of photosynthesis-light curves for the tundra pond algae, 3 July 1973.*

the nutrient status of the algae (Fuhs et al. 1972, Caperon and Meyer 1972). Hence, a decline in $I_{0.5}$ when P_{max} is constant is evidence for light adaptation, while an increase in $I_{0.5}$ with a decline in P_{max} is suggestive of increasing nutrient limitation.

Photosynthesis-light curves for the phytoplankton and epipelic algae (Figure 5-15) follow the typical hyperbolic form seen both in algal cultures (e.g., Rabinowitch 1951) and in some marine and freshwater phytoplankton assemblages (e.g., Vollenweider 1965, Ryther and Menzel 1959); the curves we mentioned do not show the high variability often encountered in temperate lakes. This tendency for algal assemblages from extreme environments to react like unialgal cultures, also noted by Goldman et al. (1963) for shallow Antarctic lakes, simplifies interpretation of field data considerably.

The $I_{0.5}$ is also a temperature dependent variable. It increases as much as 3-fold in the epipelic experiments (from 0.04 to 0.12 ly min^{-1} on 3 July) over a 10°C range (Figure 5-16). A comparable change was seen by Talling (1957) in the similar parameter, I_k (the intensity at which P equals 0.71 P_{max}), which he used as a measure of the onset of light saturation. As Talling pointed out, parameters such as I_k or $I_{0.5}$ should be expected to vary with temperature because of the nature of the light-photosynthesis curve. Theoretically, at low illumination the instantaneous photosynthesis rate of a given population rises linearly with light intensity and is independent of temperature. At higher illumination, photosynthesis increases more slowly and eventually reaches a light-saturated optimum, P_{max}, which is temperature dependent (Rabinowitch 1951). So as P_{max} changes with changing temperature, $I_{0.5}$ must also vary in order to maintain a constant slope in the lower portion of the light curve.

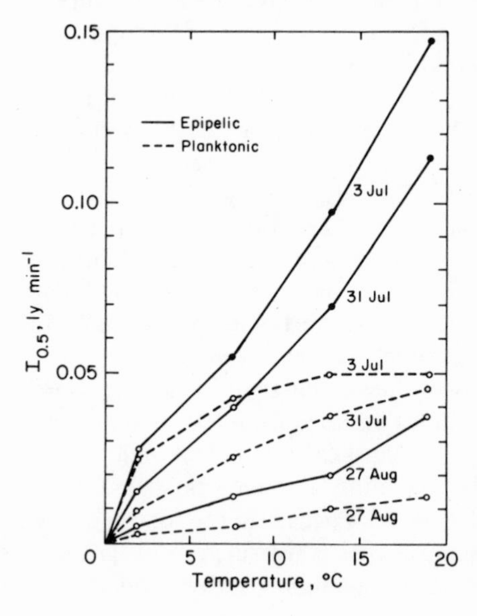

FIGURE 5-16. *Effect of temperature on the light half-saturation for photosynthesis ($I_{0.5}$) on three sampling dates in 1973.*

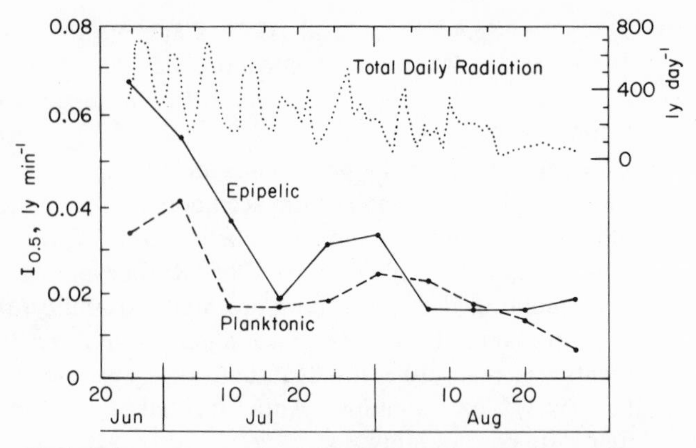

FIGURE 5-17. $I_{0.5}$ *at 8°C during the summer of 1973. The dotted line is the total daily radiation of the pond surface.*

The $I_{0.5}$ values for both the epipelic algae and phytoplankton (Figure 5-17), although they vary over the season, are not unusual when compared with other aquatic systems. The tundra pond values are slightly lower, for example, than the $I_{0.5}$ values of 0.04 to 2.0 ly min^{-1} calculated for a natural *Asterionella* population (Talling 1957). Photosynthesis in the pond populations became light saturated at 0.10 to 0.20 ly min^{-1}, placing them in the middle of a wide range of literature values summarized in Strickland (1960). For this reason, unusually high $I_{0.5}$ values would not appear to be contributing to the low seasonal productivity of the ponds.

The epipelic $I_{0.5}$ was higher than the planktonic $I_{0.5}$ at all experimental temperatures throughout most of 1973 (e.g., at 8°C, Figure 5-17), suggesting that the sediment algae are better adapted to high light intensities than are the phytoplankton. Similar results come from intertidal areas (Burkholder et al. 1965, Gargas 1971). However, it is our view that such comparisons may be invalid because of the necessary presence of sediment particles in the epipelic samples used for our ^{14}C measurements (see Stanley 1976a). At all times, some of the algae will be shaded by the particles. Thus, if shading effects by the sediment particles were significant, the average light intensities at the epipelic cell surface would be much lower than the light intensities incident on the whole sample and the calculated $I_{0.5}$ values would be too high. Attempts to separate the epipelic cells from the sediment were unsuccessful.

Both epipelic and planktonic $I_{0.5}$ values are highest in early summer and lowest in late August. Epipelic values declined from 0.06 ly min^{-1} on 25 June to about 0.02 ly min^{-1} by 27 August (at 8°C); planktonic values declined from 0.04 ly min^{-1} to 0.01 ly min^{-1} between the same dates (Figure 5-17). A roughly parallel decline also occurs in the total incident illumination following the summer solstice; this suggests a positive adaptive response of the algae to declining light as discussed above.

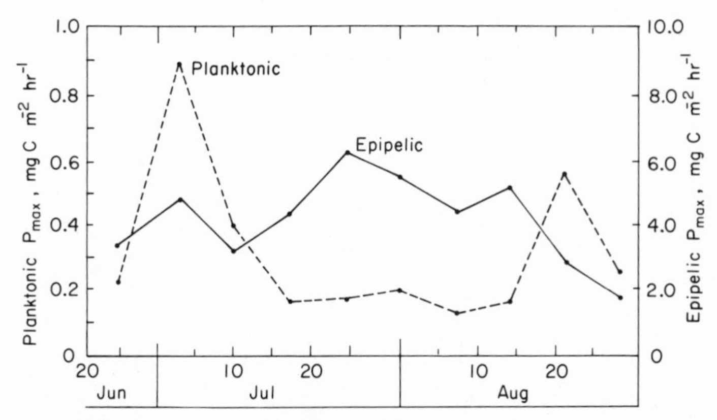

FIGURE 5-18. P_{max} *of planktonic and epipelic algae in tundra ponds at 8°C, 1973. (After Stanley 1974.)*

Gargas (1971) gave the same interpretation to a similar decrease in $I_{0.5}$ of estuarine epipelic algae in a Danish mudflat. Note, however, that we attribute the very rapid short-term decline in phytoplankton $I_{0.5}$ in early July to nutrient effects (see below).

The low productivity of the Barrow ponds is not due to low activities per unit of cell mass. Unfortunately, our evidence is not perfect as we have had to use the biomass data for 1971 (Figure 5-2, 5-10) with the P_{max} data for 1973 (Figure 5-18). An additional assumption that had to be made was that the photosynthetically active epipelic algae were one-sixth of the total biomass (B). Stanley (1976b) has predicted this based on models. The P_{max} B^{-1} ratios, (mg C)(mg algal C)$^{-1}$ day^{-1} range from 0.2 to 3.0 for phytoplankton and 0.4 to 1.0 for the epipelic algae. These values are as high as or even higher than the highest values tabulated by Berman and Pollingher (1974, their Table 6) from a variety of other studies. One reason for this high ratio in the plankton might be the high grazing rate. Usually, phytoplankton consist of a number of subpopulations, some of which are senescent, some actively dividing, and some growing slowly in an almost steady-state condition. Here, only the actively dividing subpopulation would survive.

The photosynthetic capacity of the phytoplankton, measured as P_{max} B^{-1}, varies somewhat over the season but in 9 out of 12 measurements it was between 0.4 and 1.0 (Figure 5-19) during 1971. This is expected from the close correspondence between the P_{max} m^{-2} (Figure 5-18) and the biomass for other years. The three excursions from the range given above occurred at the beginning of the first bloom, at the peak of the first bloom, and at the peak of the second bloom.

$P_{max}B^{-1}$ ratios for the epipelic algae are more difficult to estimate than for the phytoplankton because of methodological difficulties, but they too probably change little over the season. The total epipelic biomass in the top 5 cm of the sediment increases continuously over the summer

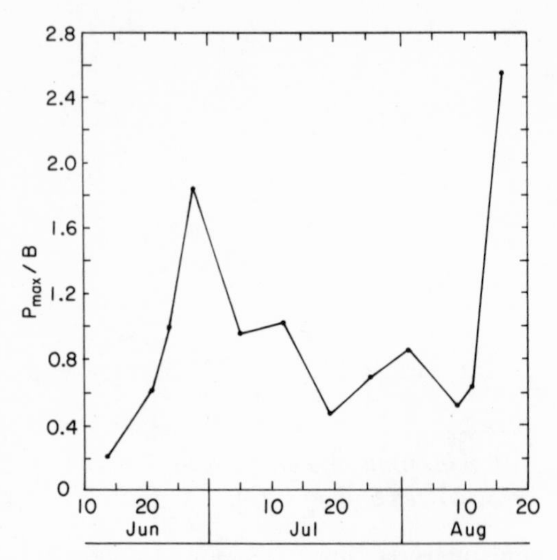

FIGURE 5-19. *Ratio of* P_{max} *to* B *(both in mg C) for phytoplankton at 8°C in Pond B, 1971.*

with no apparent July peak (Figure 5-10) as occurred for the 1973 P_{max} values (Figure 5-18). This suggests that the ratio of the epipelic P_{max} to B declines in late summer. However, no measurements could be made of the epipelic biomass actually photosynthesizing in the top several millimeters of sediment because it is not possible to section the coarse, unconsolidated sediments. If, as we think likely, these "surface" algae undergo cumulative burial into the dark underlying sediments so that the ratio of "surface" to "buried" algae decreases over the summer, then an apparent but incorrect decline in the ratio of P_{max} to B would result. In fact, in simulation runs of the benthic carbon-flow sub-model (see Chapter 10), the seasonal curve of "surface" algal biomass is identical in form to the P_{max} curve of Figure 5-18 (see Stanley 1976b).

The tundra pond phytoplankton are frequently inhibited by supraoptimal light intensities while even on the brightest of days the epipelic algae are undersaturated. This is illustrated in Figure 5-15, which shows the photosynthesis versus light curves, and in Figure 5-20, which gives the ratio of observed photosynthetic rates (P) at ambient light intensity to the light-saturated rate (P_{max}). A ratio of 1.0 in Figure 5-20 is equivalent to light-saturated photosynthesis; values less than 1.0 represent the degree of light undersaturation or oversaturation (inhibition) of photosynthesis. The epipelic algae were never inhibited and their photosynthesis was as much as 50% below saturation at 20°C, while at the same temperature the phytoplankton were around 80% light-saturated. At lower temperatures, however, light inhibition was common in the pelagic algae with the strongest inhibition (50% or greater) occurring at the lowest

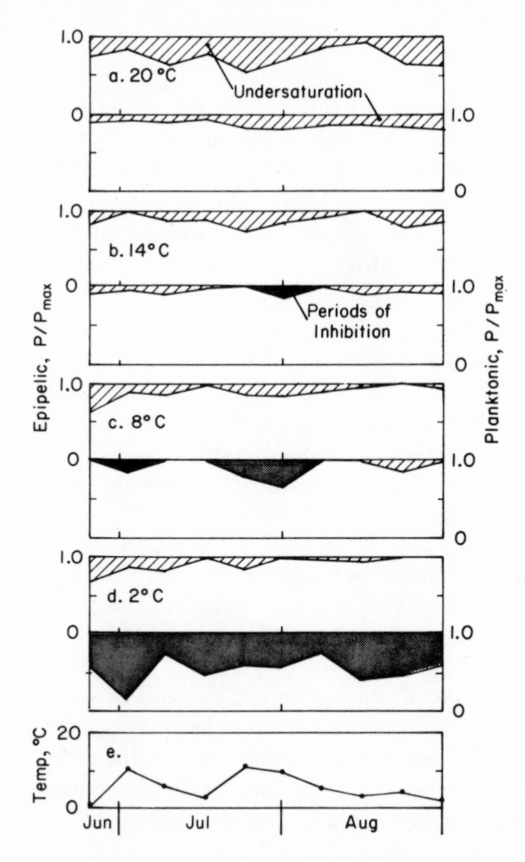

FIGURE 5-20. *Ratios of* P *to* P_{max} *for the phytoplanktonic and epipelic algae at four temperatures (a–d), 1973. Solid areas represent periods of inhibition; hatched areas represent undersaturation. Pond water daily maximum temperature (e) is given in the bottom panel. In each pair of panels in a–d, the upper panel represents epipelic and the lower planktonic.*

temperatures. Since pond water temperatures seldom rose over 10°C in 1973, it is clear that in that summer the phytoplankton were indeed light-inhibited during the middle portion of most days. The reasons for the differing responses of the benthic and pelagic algae to high illumination are unclear because of the difficulties in measuring true incident light intensities for the benthic group. Thus, the epipelic algal community may never reach inhibitory light intensities because of sediment shading of a part of the population.

The quantitative effect of photoinhibition on seasonal phytoplankton productivity is probably even greater than would be predicted from calculations of the duration of supraoptimal light intensities over the summer. Since photoinhibition apparently results from chlorophyll photo-oxidation (Yentsch and Ryther 1957), a process which is injurious to the cell, photosynthesis does not recover immediately when light intensity again falls below inhibitory levels. Instead, rates probably remain depressed for a period of time proportional both to temperature and to the absolute intensity and duration of the inhibitory illumination. In consequence, the diurnal photosynthetic maximum may not occur in the tundra ponds until midnight (Kalff 1969).

Our data suggest that temperature is indirectly responsible for the high degree of photosynthetic photoinhibition of arctic phytoplankton. Light levels during the Barrow summer (Chapter 3) are not appreciably higher than at lower latitudes nor, as mentioned above, are the $I_{0.5}$ values for the phytoplankton. What is important, then, is the inverse relationship between temperature and the minimum intensity for photoinhibition (the limiting optimum light intensity, Hutchinson 1957). A similar argument has been given by Sorokin and Konovalova (1973) for data on winter diatoms in the Japan Sea.

There is some evidence that the phytoplankton of these ponds are able to increase their photosynthetic efficiency in response to the declining light during a single day. This is similar to the seasonal response described earlier and to the "sun-shade" adaptations in marine plankton (Steemann Nielsen and Hansen 1961). The evidence is from Kalff's (1969) finding of a three-fold increase in I_{max} and a five-fold increase in P_{max} between noon and late evening in Barrow ponds. Our attempt to repeat Kalff's experiment for the phytoplankton on 10 July 1973 failed completely because a fog bank suddenly moved in and reduced the light to nearly zero (from 2100 to 0600). Data for the epipelic algae were better but no significant diurnal changes in P_{max} or in $I_{0.5}$ were seen (Figure 5-21).

The results of measurements of planktonic photosynthesis under constant temperature and light conditions raise the possibility of an endogenous rhythm in the planktonic P_{max}, despite the general observation (Doty 1959) that such rhythms decrease in amplitude towards the poles. Phytoplankton samples exposed to dim light or enriched with phosphate in carboys for 4 to 6 hours prior to the beginning of these constant light and

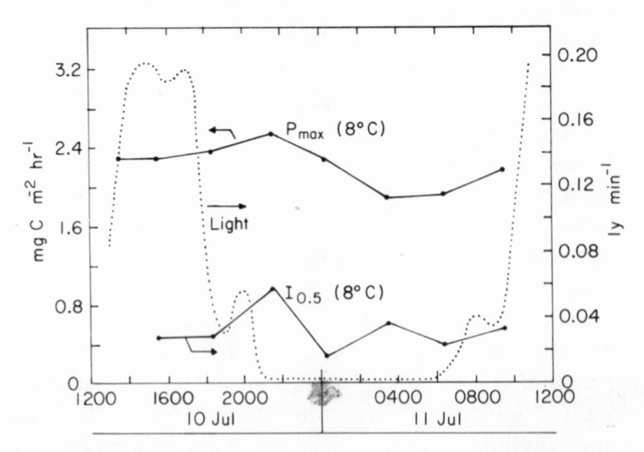

FIGURE 5-21. *Amount of photosynthetically available light (half of the incident radiation), rate of* P_{max} *for the epipelic algae, and* $I_{0.5}$ *on 10 and 11 July 1973, Pond B.*

temperature experiments showed a subsequent diel rhythm in rate with a peak at about 1800 and an amplitude of ±25% of the mean rate. The rhythm was suppressed, however, in samples taken directly from the pond or held in carboys exposed to normal oscillations in light intensity and temperature.

While endogenous rhythms in photosynthesis and other algal processes such as nutrient uptake (Chisholm 1974) may well occur in the Barrow phytoplankton and may be important in determining species composition (Stross et al. 1973, Stross and Pemrick 1973), their direct contribution to the low production levels seen in the Barrow ponds is probably negligible. Before such a rhythm could cause a large decrease in productivity, its amplitude must be large in comparison to light- and temperature-induced cycles in photosynthesis and be out of phase with them. A species having this rhythm, it would seem, would be quickly eliminated from the plankton assemblage. Furthermore, the usual absence of the rhythm in samples exposed to normal fluctuations in light and temperature and the frequent reversal of this suppression by addition of phosphate suggests that at most times the rhythm is insignificant because of nutrient limitation.

Nutrients

Pond studies at Barrow prior to the IBP program as well as preliminary chemical analyses early in this study suggested that nutrient limitation was probably an important control of photosynthetic rates. Kalff (1971), using 48-hour bottle bioassays, demonstrated both phosphorus limitation at times of peak phytoplankton productivity and ammonia limitation early in the season in water overlying unmelted ice and later during a period of very high photosynthesis. In addition, both N and P concentrations are low in tundra ponds (Figures 4-12, 4-16) in comparison with temperate ponds.

Our nitrogen data, viewed as a whole, indicate that the rates of supply of nitrogen in both the water column and sediments of the ponds usually exceed the average rates of nitrogen uptake by the algae. The mean rate of uptake of ammonia, the preferred form of N for these phytoplankton, was 0.29 μg N liter^{-1} hr^{-1} in 1970. This is about 1% of the average ammonia concentration in the pond water (32.8 μg N liter^{-1}). The rate of recycling of ammonia via biological ammonification in the water column was estimated from ^{15}N dilution procedures to range from 0.4 to 3.3 μg N liter^{-1} hr^{-1} in 1970 and 1971. Hence, ammonification processes alone could probably maintain a slightly higher rate of N uptake than is usually seen in the ponds. In addition, indirect evidence suggests that ammonia in the water column is also resupplied at a high rate from the sediments. For example, the addition of ammonia to bioassay bottles did not increase

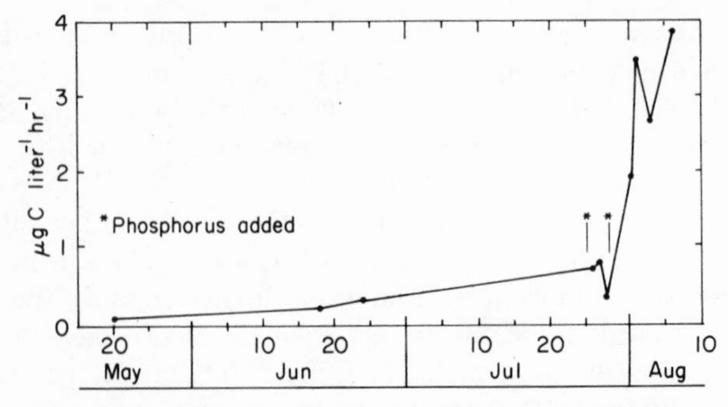

FIGURE 5-22. *Primary productivity of phytoplankton in Pond D before and after the addition of 46.5 g P on 25 July and 232 g P on 28 July 1970.*

ammonia uptake by the microflora as might have been expected if ammonification at the above-mentioned rates were the only means of supply. Secondly, in phosphorus fertilization experiments in Pond D, 1970, during which algal productivity and biomass increased (see below), the rates of ammonia uptake increased an order of magnitude (up to 2.1 to 3.5 μg N liter^{-1} hr^{-1}) relative to control ponds. Given that average NH_3-N concentrations in the interstitial water of the center of the ponds range from 2,000 to 3,000 μg liter^{-1}, it seems clear that the additional ammonia taken up in the absence of P-limitation came from the sediments. Finally, it seems unlikely that the epipelic algae, despite their higher biomass in comparison to the phytoplankton, would be N-limited in the presence of such high levels of ammonia in the sediments.

In contrast to the situation with nitrogen, a great deal of experimental evidence suggests that the pond algae are strongly limited by phosphorus. The most direct evidence comes from Pond D fertilization experiments in 1970 to 1972 (Figures 5-22 and 5-23). The first addition of phosphate (46.5 g P) to Pond D in 1970 resulted in a slow increase in phytoplankton productivity, and a second, much larger fertilization (232 g P, which resulted in a rise in concentration from 2.5 to 780 μg P liter^{-1}), completely depressed productivity for the next few days. After this, productivity rose to a peak (Figure 5-22) several times higher than that seen in other nearby ponds in that same summer. After phosphate enrichments in the same pond in July and August of 1971, epipelic algal productivity doubled within a month in contrast to the normal pattern of declining rates in control ponds during that period (Figure 5-23).

The Pond D fertilization experiments show that phosphorus limitation of algal photosynthesis in tundra ponds occurred throughout the growing season and not just during periods of peak productivity as Kalff's (1971) bioassays had indicated. All the 1970 enrichments to Pond D, as well as the first two in 1972, were made in mid- to late July when phytoplankton productivity in most ponds is normally very low.

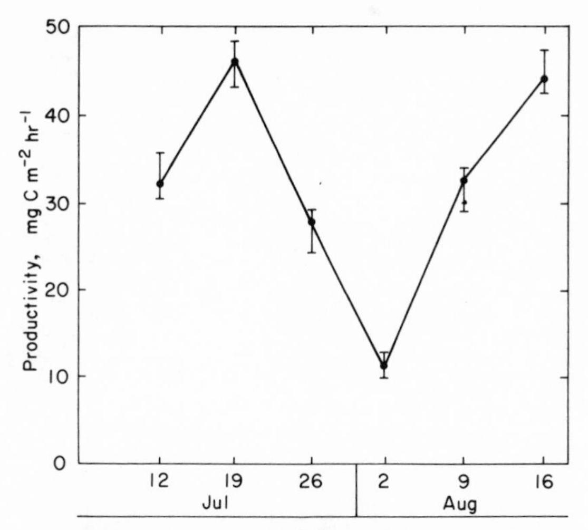

FIGURE 5-23. *Productivity of epipelic algae in Pond D after enrichments with P in July and August 1971. (After Stanley 1974.)*

Phosphorus limitation in both the sediments and water of the pond results from strong binding of phosphorus to iron in the sediments (see Chapter 4). The importance of the phosphate exchange reaction between the sediments and water as the rate-limiting step in phytoplankton productivity is seen in the correlation of primary production with phosphorus sorption indices for various Barrow ponds (Figure 4-31) and also in Pond B 1971 experiments where sub-ponds with plastic bottoms were used to eliminate sediment-water exchange processes. An extreme example of the effect of sediment binding capacity occurred in Pond G, a recently formed thermokarst pond with no accumulated pond sediments. Primary productivity in Pond G in late August 1970 was 77 μg C liter^{-1} hr^{-1} and the total phosphorus concentration was 593 μg P liter^{-1}. This contrasts with 26.2 μg total P liter^{-1} and 0.31 μg C liter^{-1} hr^{-1} fixed in the nearby Pond E at about the same date. In the Pond B sub-ponds with plastic bottoms, phosphate concentrations were more than double and phytoplankton biomass 10 times larger than other control sub-ponds and Pond B proper (Table 5-6, Figure 5-2). This presumably occurred because in the sub-ponds with no sediments the phosphate, recycled into the water via bacterial mineralization, zooplankton excretion, and soluble organic phosphate hydrolysis, was utilized for phytoplankton growth rather than sorbed to the sediments as in the controls.

Phosphorus is the principal cause of the spring phytoplankton bloom and decline. While light and temperature show no major increase or decline over the spring bloom period, there are drastic changes in the PO_4 concentrations in the water (Chapter 4, Figure 4-16). The $I_{0.5}$ values (Figure 5-17) and ratios of production to biomass (corrected for

TABLE 5-6 *Phosphorus Concentrations*[1] *and Phytoplankton Biomass*[2] *in Subponds B_1 (Control), B_2 and B_3 (Plastic Lined) of Pond B, 1970*

		DRP	DUP	PRP	PUP	TP	Total Algae
28 July	B_1	2.9	11.5	0.2	6.4	21.0	283
	B_2	2.1	11.5	0.2	5.6	19.4	485
	B_3	2.0	14.4	0.5	10.9	27.8	442
1 Aug.	B_1	3.2	11.6	0.3	4.3	19.4	325
	B_2	3.1	15.3	0.5	11.3	30.2	1538
	B_3	5.7	20.4	1.0	25.2	52.3	1720
5 Aug.	B_1	0.6	14.0	0.4	14.6	29.6	233
	B_2	1.8	18.2	0.2	9.9	30.1	749
	B_3	31.5	30.5	1.7	18.1	81.8	1198
10 Aug.	B_1	1.0	13.5	<0.1	9.5	24.0	256
	B_2	3.8	15.3	0.2	9.0	28.3	236
	B_3	35.4	12.6	1.0	24.2	73.2	445
12 Sept.	B_1	0.6	10.1	0.3	5.8	16.8	792
	B_2	1.8	11.3	0.7	9.3	23.1	1218
	B_3	16.9	15.1	4.8	21.0	57.8	3261

DRP = dissolved reactive phosphorus
DUP = dissolved unreactive phosphorus
PRP = particulate reactive phosphorus
PUP = particulate unreactive phosphorus
TP = Total phosphorus

[1] Phosphorus concentrations are expressed as μg P liter^{-1}.
[2] Phytoplankton biomass is expressed as 10^4 cells liter^{-1}.

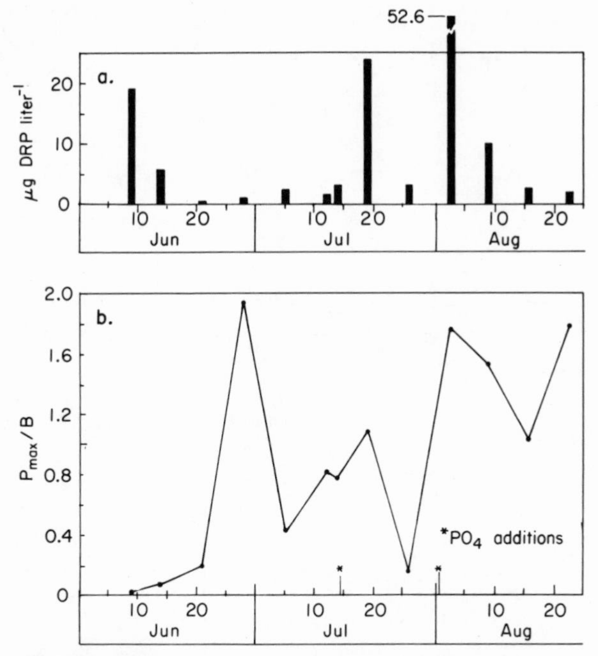

FIGURE 5-24. *a) Concentrations of dissolved reactive phosphorus in Pond D, 1971. b) Ratio of* P_{max} *to biomass of phytoplankton (8°C) in Pond D, 1971.*

temperature assuming a photosynthesis Q_{10} of 3.0) parallel productivity (Figure 5-19). Since the ratios of *P* to *B* often decline with increasing nutrient limitation, these patterns suggest that increasing PO_4 causes the spring increase in productivity. Soon, the increased algal production depresses PO_4 concentrations; because uptake rates for PO_4 are so much faster than transfer of PO_4 upwards from the sediment this nutrient limitation causes the decline in the bloom. Further evidence of limitation is seen in the fertilization experiments in Pond D, 1971 (Figure 5-24). Following addition of phosphate on 14 July, the *P* to *B* ratio increased quickly from 0.49 to 1.54 and then declined as the PO_4 concentration declined. The same response also occurred following fertilization on 1 August.

Despite these clear-cut demonstrations of general phosphorus limitation in the ponds, short-term (2- to 4-hour) ^{14}C bioassay tests for the effect of added phosphorus during 1972 and 1973 always gave negative results with both phytoplankton and epipelic algae samples (Figure 5-25). This was true regardless of the combination of temperature and light intensity used. Thus, the effects of phosphorus limitation, although important over the long term, are quantitatively insignificant over a short period in comparison to the effects of light and temperature.

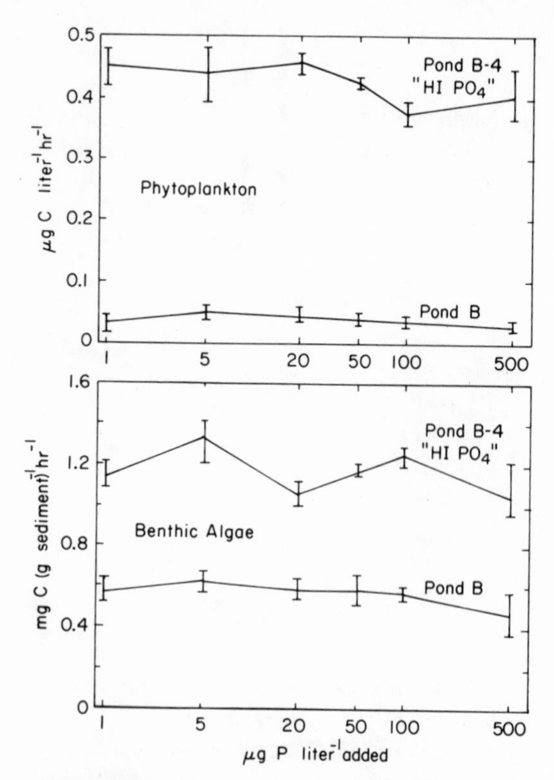

FIGURE 5-25. *Photosynthesis of planktonic and benthic algae from Pond B and from a phosphorus-enriched subpond (B-4) measured at different phosphate additions. The experiment was run on 16 August 1972 from 1300 to 1900 hours (Alaska Daylight Time).*

The contradictory results from the long- and short-term nutrient bioassays should be noted. Lack of response to nutrient addition in short-term bottle bioassays is not sufficient evidence, in itself, of lack of photosynthetic limitation by nutrients. The short-term ^{14}C assay does appear to have been used in this way in earlier studies (Ryther and Guillard 1959, Goldman 1963).

Although we attempted to carry out a detailed experimental analysis of the biologically important phosphorus fractions and their interactions in the ponds (Chapter 4, Prentki 1976), apparent contradictions remain and the specific mechanisms of phosphorus control on photosynthesis remain unidentified. For example, the phytoplankton at Barrow appear to take up phosphate into a functionally distinct "soluble" pool at a rate approximately 200 times that needed to sustain growth, despite the fact that they are P-limited. Only a very small fraction of the assimilated P is then transferred into the "bound," structural phosphorus pool and the remainder is excreted in an organic form. This organic "XP" fraction then either undergoes extracellular, enzymatic hydrolysis which again releases PO_4 or is sorbed onto water colloids which are in chemical equilibrium with XP (Lean 1973b). Whatever the adaptive value to the cells of this uptake-excretion-hydrolysis cycle (Prentki 1976), the very high rate-constant for PO_4 uptake in comparison to that of the sediment-water PO_4 exchange process obscures any relationship between photosynthesis and ambient PO_4 concentrations. The identification of the site of interaction between cellular-P and photosynthesis is thus exceedingly difficult.

One aspect of the PO_4 uptake data leads us to speculate on the reasons for the differences in our results between the short-term nutrient bioassays and long term enrichment experiments. Rates of phosphate uptake at ambient PO_4 concentrations appear to occur at, or very near, the V_{max} for PO_4 uptake, despite the demonstrated limitation of phosphorus on carbon fixation. This observation, together with the demonstrated ability of phytoplankton to undergo adaptive changes in the ratio of the phosphate uptake V_{max} to biomass (Stross and Pemrick 1973, Chisholm 1974) suggests that algal cells may respond to continuing phosphate limitation by reducing V_{max} for phosphate uptake or photosynthesis in a way analogous to light adaptation in photosynthesis. By so doing the cell would maximize P uptake per unit of enzyme for P uptake. This effect has been observed in the Barrow ponds where algae required a period of adaptation to higher phosphate concentrations before an increase in P_{max} or photosynthesis was observed. If so, such a cell could not show a photosynthetic response to phosphate addition until the V_{max} values had been increased by synthesis of additional enzymes for uptake. Perhaps this is the reason for our negative short-term bioassay results. Conversely, positive short-term bioassay results (Ryther and Guillard 1959) would only be seen in samples in which the PO_4 concentration had been reduced below levels permitting maximal rates of uptake shortly before the bioassay had begun so that an adaptive reduction in P_{max} had not yet taken place. Two distinct types of nutrient limitation follow from this hypothesis. The first occurs when the "instantaneous" measured P_{max} is at or below the maximum genetically set P_{max} under long-term saturating rates of supply of PO_4. The second occurs when rates of PO_4 uptake are at or below the "instantaneous" P_{max}. The former would correspond to Schindler's (1971) and O'Brien's (1972) nutrient control of algal biomass, the latter to nutrient control of photosynthesis (at least in terms of the time-scales over which a response can be detected).

Grazing

Daphnia, which comprises about 80% of the pond zooplankton biomass, has a high potential filtering capacity and, therefore, may exert a strong control on phytoplankton. *Daphnia middendorffiana,* the large dominant species, has a high filtering rate. At 11°C, it filters from 2 to 6 ml of pond water each hour depending on food density (Chisholm et al. 1975). At average *in situ* filtering rates and average population densities, *D. middendorffiana* is capable of filtering the entire pond volume within 2 days.

Two different calculation procedures show that the pond zooplankton probably ingest an average of 20% of the annual primary production in the ponds. The first assumes that the zooplankton production (growth and reproduction) amounts to about 1 mg C liter^{-1} yr^{-1} (Chapter 6) and that zooplankton assimilation efficiency is about 10%, while 50% of the amount assimilated is lost to respiration. The zooplankton would then ingest approximately 20 mg POC liter^{-1}. If the maximum volume or quantity of food (in mg C) that is ingested by *Daphnia* is unaffected by food particle size (McMahon and Rigler 1965), so that the small algae are ingested at a rate determined only by their volume proportion in the total pond seston (Chapter 6) which is 5% on the average, then 1 mg of algal C liter^{-1} yr^{-1} is lost to zooplankton grazing. Since the average annual phytoplankton production is about 5 mg liter^{-1}, 20% of pond primary productivity is lost to zooplankton. In the second calculation, total volumes of water filtered over several time periods were calculated from filtration rates per animal and from animal densities; total ingested sestonic carbon was then calculated from average particulate carbon concentrations during that time period. From this procedure the same approximate annual zooplankton consumption rate, 1 mg C liter^{-1}, was estimated. The magnitude of the grazing losses will be determined in large part by environmental conditions, which affect zooplankton survivorship, and by the quantity or quality of seston available in any year, which affects the assimilation efficiency of the zooplankton.

The strong grazing pressure may be hypothesized as an explanation for the apparent paradox that *P* to *B* ratios in the plankton of the ponds are very high relative to eutrophic systems despite the severe limitation of planktonic photosynthesis by phosphorus in the Barrow Ponds. In this hypothesis, the high rates of zooplankton ingestion are a means of cropping algal cells before they become "senescent", that is, before their *P* to *B* ratios begin to decline. This argument has been used to explain high turnover rates in bacterial microcosms grazed by a phagotrophic ciliate (isolated from the Barrow ponds) in comparison with systems with bacteria alone (Barsdate et al. 1974). However, as discussed earlier, it is likely in this case that the short-term changes in *P* to *B* ratios during algal blooms result from the sudden onset of phosphate limitation.

The studies of several ponds that contained added crude oil allowed further evaluation of the effect of zooplankton (details are given in Chapter 9). When oil was added to the entire pond (1970 and 1975), the zooplankton were killed while the algae survived. At a high dose rate of oil (1970), the algal production and biomass did not appreciably change from controls; at a lower dose rate (1975), there was an increase in algal biomass after a lag period.

Experiments in small subponds (240 liters) in 1976 tested the effect of removal of zooplankton with and without added oil (Figures 5-26, 5-27). The same effect was observed: after a lag of 2 weeks, oiled subponds recovered and had about the same primary productivity and algal biomass as did the controls. However, the pond containing no zooplankton had a 5-fold increase in primary production after 16 days and algal biomass also increased. Our conclusion is that zooplankton indeed are controlling algal biomass in the plankton of these ponds.

A similar question could be asked about control of sediment algae. However, epipelic algae make up only a small fraction of the total potential diet of benthic grazers. The sediment organic carbon is about 0.06% algae, 0.17% bacteria, and 99.77% detritus. The biomass of the epipelic algae ranges from 0.5 to 1.0 g C m^{-2} while the biomass of the sediment bacteria is approximately 3 g C m^{-2} but these organisms are very much diluted by the large amount of detrital carbon (1700 g C m^{-2}) in the top 5 cm of pond sediments. Consequently, unless the benthic grazers (primarily chironomids, oligochaetes, and micrometazoans) are highly selective feeders, they probably could not have a very significant effect on the epipelic algae.

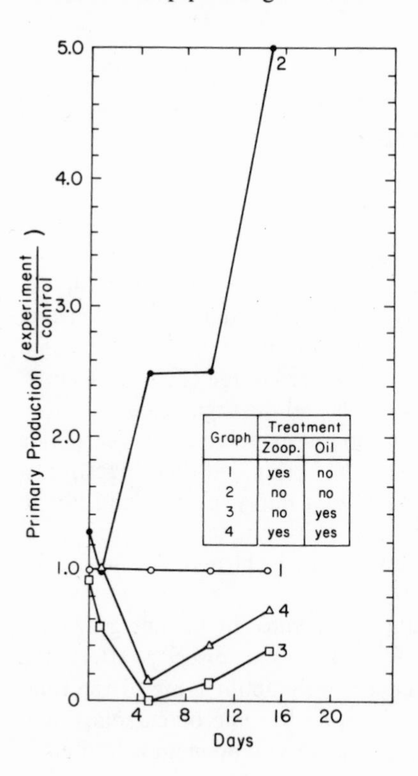

FIGURE 5-26. *Primary production (the ratio of experiment to control) in subponds after various treatments. (After Federle et al. 1979.)*

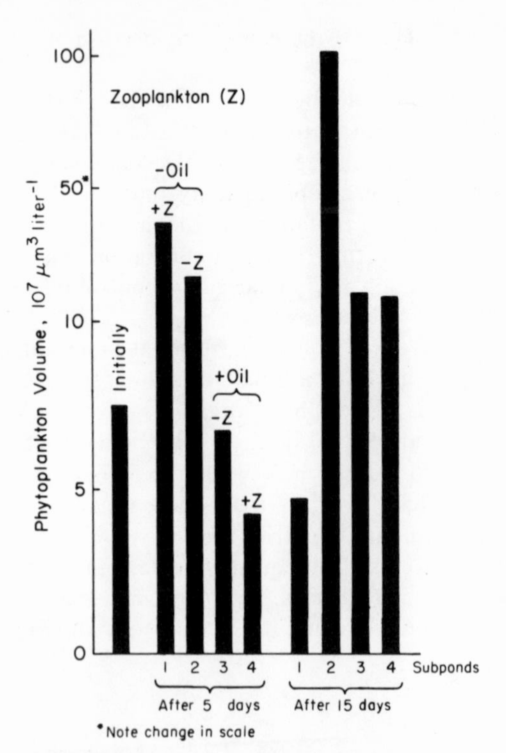

FIGURE 5-27. *Phytoplankton biomass (expressed as volume) in subponds, initially (Day 0), and after 5 and 15 days. Treatments are as described for Figure 5-26.*

Rates of ciliate feeding on algae have been measured, but for the chironomids and micrometazoans only an estimate can be made (Fenchel 1975). The ciliates and micrometazoans consume a total of no more than 8 mg algal C m^{-2} day^{-1}. Based on the measured value (325 mg C m^{-2} day^{-1}) of total ingestion by the chironomids and the organic composition of the sediment given above, the chironomid population in the ponds would consume only about 0.2 mg algal C m^{-2} day^{-1} if their ingestion is non-selective. If, as Kajak and Warda (1968) imply, chironomid larvae have a selectivity of 5:1 for epipelic algae over detritus, the algal consumption by the tundra pond population of chironomids would be approximately 1 mg algal C m^{-2} day^{-1}.

A large percentage of the total algae consumed by benthic grazers is "buried" epipelic algae, and therefore does not directly affect the productivity of the epipelic algae. Perhaps only about 1 mg of the total daily consumption of 9 mg algal C m^{-2} by micrometazoans and chironomids is "surface" epipelic algae since this component is limited to the upper 2 to 3 mm of sediment. This small consumption represents about 0.5% to 2% of the mid-summer biomass of "surface" epipelic algae and only 1% of their daily net productivity.

Sediment Disturbance

Animals living in the pond sediments exert a major control on epipelic productivity by physical disturbance of the surface sediments. They are largely responsible for the transfer of "surface" epipelic algae to "buried" epipelic algae, a transfer which has the same effect as grazing in that it removes a part of the biomass of photosynthesizing algae. Chironomid larvae and the tadpole shrimp, *Lepidurus arcticus,* create the most disturbance by their feeding activities in the surface sediments. Water currents and the release of gas bubbles from the sediments appear to be of secondary importance. The result is a very unstable substratum for the epipelic algae. Moss (1968), after comparing data from different benthic habitats, concluded that the relative stability of sediment, rock, and plant substrata is a key factor influencing the biomass and hence productivity of the benthic algae. Indeed, in mathematical simulations of epipelic algal productivity in the tundra ponds (Chapter 10 and Stanley 1976b), it was found that without the burial the algae at the sediment surface rapidly increased to a much higher biomass than is ever observed in the ponds.

Control of Species Composition

The phytoplankton species composition in the Barrow ponds changes dramatically with season in a manner which appears to be consistent from year to year. Although we can understand the factors controlling carbon fixation by considering the algae as a homogeneous group, it is possible that the seasonal changes in overall response are caused by changing species rather than by adaptation. For example, bloom species in the ponds might die out as phosphorus becomes less available. The change in the ratio of photosynthesis to biomass in the spring bloom might be a result of a change to species capable of tolerating reduced PO_4 rather than of starvation of an existing population.

In order to answer these questions about the plankton, we have developed ^{14}C track autoradiography techniques to study photosynthesis of individual species *in situ.* Preliminary results show different times of day for peak photosynthesis for *Rhodomonas* and *Chromulina;* this implies that these algae have different V_{max} and $I_{0.5}$ values (Figure 5-8). Whether these differences only play a part in niche separation or whether they can control overall rates is unknown. An example of the response we

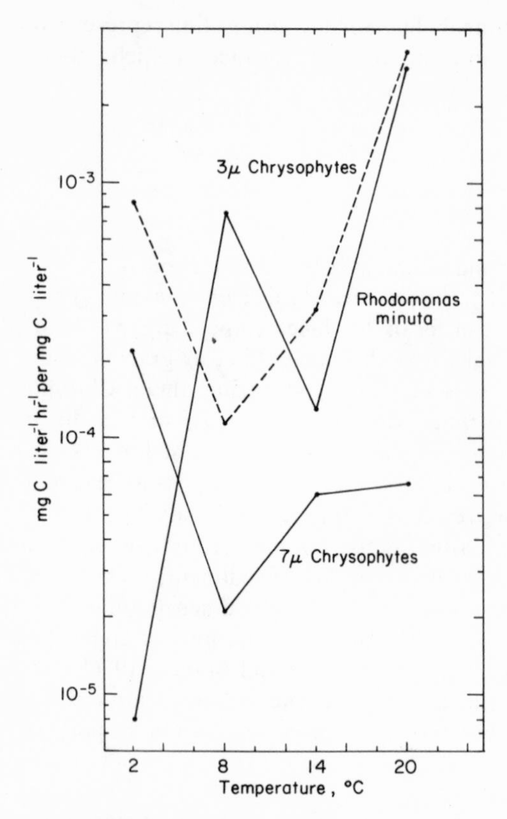

FIGURE 5-28. *The production to biomass ratios in three groups of plankton at four temperatures, 1972.*

expect to find is shown in Figure 5-28 where production to biomass ratio is plotted against temperature for three distinct organism types. With more data, such information could yield answers to questions of adaptations of phytoplankton populations and species dominance.

Another pathway of control of species composition leads from the zooplankton. As described earlier, we noticed that there was a change in species composition of the phytoplankton when oil was added to a pond (Figure 5-4). The same result, a replacement of *Rhodomonas* by *Uroglena,* was also found when the zooplankton were removed from subponds (Miller et al. 1978a, 1978b). Also, the *Rhodomonas* reappeared in Pond E, the 1970 oil treatment pond, at the same time that the zooplankton reappeared (1976). The mechanism for this apparent control is likely the grazing of the zooplankton which keeps the *Uroglena* at a low level. Another possibility is control through the regeneration of nutrients by zooplankton; in view of the strong control of nutrients by sediment processes, this should have little effect.

There is also some control mechanism that keeps diatoms from becoming important in the phytoplankton (they are present in the sediments). This mechanism could be the very low concentrations of silica in the water (almost always less than 0.5 mg liter^{-1}). Silica is present at low concentrations in the rainwater and runoff water (e.g., less than 0.1 mg liter^{-1} in the snowmelt in 1971) but does increase when water comes in contact with soil (e.g., 1.5 mg Si liter^{-1} in groundwater, 20 July 1970). Concentrations are low in the water, however, and it is possible that the sediment diatoms keep the silica at this low level. Papers by P. Kilham (1971) and S. Kilham (1975) have pointed out the control of diatom species by the silica concentrations.

Conclusions

The algae in the plankton are controlled by both zooplankton and nutrient supply. When zooplankton are removed, there is an immediate increase in both production and biomass of the algae. The same increase can be obtained by fertilization of a pond with phosphorus but the response takes many weeks. In nature, algae could increase if the rate of supply of nutrients increased and if the zooplankton increased very slowly. Usually, the phosphorus is tied up in the sediments and is released slowly.

The algae at the sediment surface live in a very nutrient-rich environment relative to that of the phytoplankton. However, they can only photosynthesize in a layer a few millimeters thick because of the rapid absorption of light by the sediments. Animals are not important in removing algae by grazing but do remove algae from the surface of the sediments by their activity which mixes algae below the euphotic zone. This is the main control of the epipelic algae and explains why there is no algal mat in these ponds.

Temperature also keeps algal production at a low level but the effects of raising the temperature are unclear because of the interaction of respiration and other processes. It is possible that the respiration of algae would increase faster than the gross photosynthesis and result in a lowered net photosynthesis at high temperatures. Also, the animals would respond to higher temperatures by increasing their grazing rate or their "sediment mixing" rate. Thus, the rates of individual processes would be speeded up by an increase in temperature but there may well be ecosystem compensation so that there is no overall effect of increased temperature.

ROOTED AQUATIC PLANTS*

Biology

There are three rooted aquatic plants found in the shallow ponds at Barrow: *Carex aquatilis, Arctophila fulva,* and *Ranunculus pallasii.* Only the first two are important in the ponds we have studied, although in the immediate area of the research site ponds can be seen that are completely covered by each of the three. *Carex aquatilis* is also one of the three dominant terrestrial plants and so was intensively studied in the terrestrial IBP study (see Brown et al. 1980). *Arctophila fulva,* on the other hand, is found only as an emergent in standing water.

Within the ponds we have studied, *C. aquatilis* is found in a pure stand in the shallow water less than approximately 15 cm in depth, while *A. fulva* is found in a pure stand in slightly deeper water (15 to 25 cm). In pond J in 1971, *C. aquatilis* occupied 32% of the pond area while *A. fulva* covered 13%. These percentages vary from pond to pond, of course, and the depths will also change with changes in water level of the pond. In some summers the entire *C. aquatilis* zone will become dry land.

Reproduction in the monocots at Barrow is usually vegetative (Johnson and Tieszen 1973). The result is that *C. aquatilis,* for example, produces a network of individual tillers connected by subsurface rhizomes (Shaver and Billings 1975). These authors define tiller as "the horizontal rhizome, stem, stembase and leaves originating at the point of insertion of the rhizome at the parent plant or 'mother tiller'." A typical part of this network (Figure 5-29) contains both spreading tillers, which produce almost all of the roots, and their daughter tillers called clumping tillers, which rarely produce roots or daughter tillers.

The life history of *Carex* is well known but we know little about *Arctophila.* Both *Carex* and *Arctophila* plants have almost complete aboveground die-back each year so that almost 100% of the living biomass is underground during the winter. Growth begins only when the snow and ice melt in mid-June. At this time, most surviving leaves are brown except for some with several centimeters of live material at the leaf base (Tieszen 1978a). These elongate rapidly at first (about 4 mm day^{-1}) and must draw on nutrient reserves from the stem base and rhizome as the roots are still frozen. In a typical *Carex* plant (Tieszen 1978a), two leaves had started growth the previous season; these attained their maximum length by 9 July after which they senesced. Other leaves (no. 3, 4, 5) exserted early in the season, grew for about 25 days, remained mature for another 20 days, and then senesced. Total leaf life for these three leaves averaged around 48 days. Finally, on terrestrial sites some tillers produced a sixth leaf later in

*C. P. McRoy

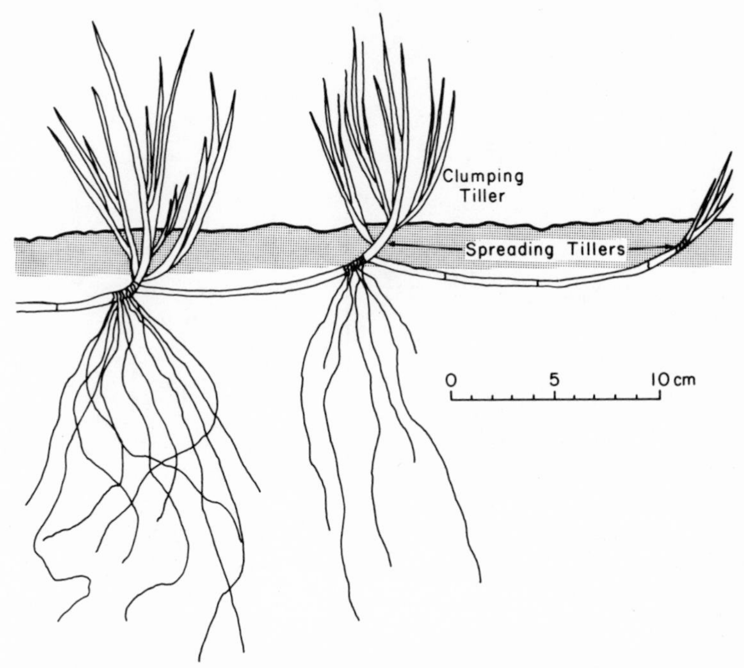

FIGURE 5-29. *Tiller types in* Carex aquatilis. *(After Shaver and Billings 1975.)*

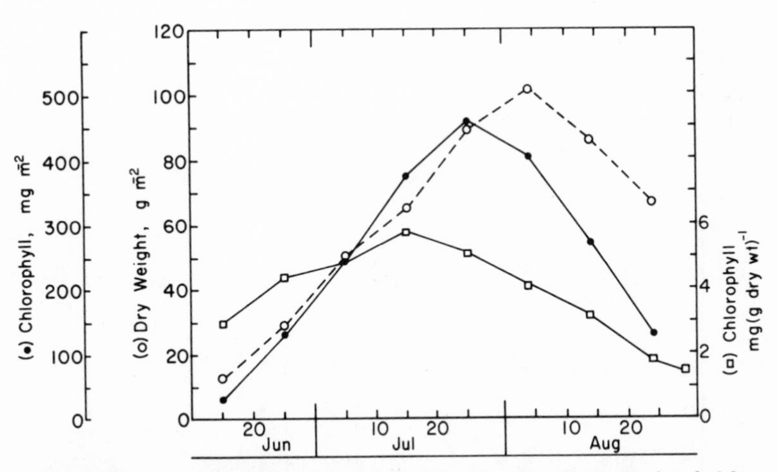

FIGURE 5-30. *Biomass of vascular plants, standing stock of chlorophyll, and concentration of chlorophyll near Barrow. (After Johnson and Tieszen 1973.)*

TABLE 5-7 *Vascular Plant Biomass in Barrow Ponds, 1971 and 1972**

I. *Carex aquatilis* (means of duplicate observations in Pond J), 1971

		21 Jun	28 Jun	5 Jul	12 Jul	19 Jul	5 Aug	9 Aug	16 Aug
1. Roots and rhizomes		-	3119	-	-	-	-	-	-
2. Submerged	Live	-	-	-	3	18	37	33	22
leaves and stems	Dead	-	-	-	106	81	100	145	102
	Total	-	-	159	109	99	137	178	124
3. Emergent	Live	-	-	-	39	47	89	83	50
leaves and stems	Dead	-	-	-	22	84	25	31	30
	Total	-	-	18	61	131	114	114	80
4. Whole plant	Live	-	-	-	42	65	126	116	72
leaves and stems	Dead	-	-	-	128	165	125	176	132
	Total	71	153	177	170	230	251	292	204

II. *Arctophila fulva* (single observations in Pond J), 1971

		21 Jun	28 Jun	5 Jul	12 Jul	19 Jul	5 Aug	9 Aug	16 Aug
1. Roots and rhizomes		-	31	-	-	-	-	-	-
2. Submerged	Live	-	-	-	6.7	12.4	5.3	8.2	-
leaves and stems	Dead	-	-	-	5.1	6.6	3.2	0	-
	Total	-	-	-	11.8	19.0	8.5	8.2	-
3. Emergent	Live	-	-	-	2.0	4.1	2.0	8.8	-
leaves and stems	Dead	-	-	-	0	0	0	0	-
	Total	-	-	-	2.0	4.1	2.0	8.8	-
4. Whole Plant	Live	-	-	-	8.7	16.5	7.3	17.0	-
leaves and stems	Dead	-	-	-	5.1	6.6	3.2	0	-
	Total	4.2	8.5	-	13.8	23.1	10.5	17.0	-

III. *Carex aquatilis* (from Pond J), 1972

			16 Jun	25 Jun	5 Jul	15 Jul	25 Jul	6 Aug	19 Aug	29 Aug
1.	Submerged	Live	-	-	-	12	12	18	-	-
	leaves and stems	Dead	-	-	-	89	47	56	-	-
		Total	-	-	-	101	59	74	-	-
2.	Emergent	Live	-	-	-	28	28	53	-	-
	leaves and stems	Dead	-	-	-	5	61	52	-	-
		Total	-	-	-	33	89	105	-	-
3.	Whole plant	Live	9	14	35	40	40	71	92	109
	leaves and stems	Dead	33	47	85	94	108	108	86	261
		Total	42	51	120	134	148	179	178	370
Number of Observations			10	12	12	12	12	12	12	8

IV. *Arctophila fulva* (from Pond L), 1972

			20 Jun	30 Jun	11 Jul	21 Jul	31 Jul	19 Aug	29 Aug
1.	Submerged	Live	-	-	-	32	37	58	58
	leaves and stems	Dead	-	-	-	13	23	30	43
		Total	-	-	-	45	60	88	101
2.	Emergent	Live	-	-	-	21	40	51	47
	leaves and stems	Dead	-	-	-	3	18	4	3
		Total	-	-	-	24	58	55	50
3.	Whole plant	Live	26	34	36	54	77	109	105
	leaves and stems	Dead	103	64	39	16	41	34	47
		Total	129	98	75	70	118	143	152
Number of observations			4	4	4	4	4	4	4

Source: McRoy and Leue 1973.
*Data are expressed as g dry wt m^{-2}.

the season but these did not complete their life cycle before senescence began in mid-August. *Carex* in the ponds actually produced a seventh leaf and the average number of leaves produced was 6.8 by 17 August 1972. These last leaves (5, 6 and 7) were still growing after the environmentally induced senescence had begun. In the late fall there is some dieback but the lower parts of the leaves remain green and will be productive next season. In the pond, *Carex* leaves had a maximum length of 190 mm which is about double the length of *Carex* in drier locations. Over the course of its 4- to 7-year life, a tiller may produce 20 to 25 leaves before dying aboveground (Shaver and Billings 1975).

As far as is known, the seasonal cycle of growth of *Arctophila fulva* is similar to that of *Carex*. The leaves are much wider than those of *Carex*, 8 to 11 cm versus 0.3 cm (Miller and Tieszen 1972), and the pure stands of *A. fulva* are the densest vegetation on the tundra. For example, the highest leaf area index (total area of leaves per unit area of ground) was 8.5 for a stand of *A. fulva* (Dennis et al. 1978). The usual maximum leaf area index in the wet meadow habitat at Barrow is 1.0 to 1.4 (Johnson and Tieszen 1973).

Biomass

The aboveground biomass of the plants in Pond J was measured by clipping and weighing (dried at 95°C for 24 hours). Peak biomass in 1971 was reached around 9 August in both the pond (Table 5-7) and in the dry tundra (Figure 5-30). These terrestrial data show very well the peak in chlorophyll that precedes the biomass peak by about 10 days. In fact, judging from these data the plants appear to reach their peak of photosynthetic capability in mid-July (chl(g dry wt)$^{-1}$). The data from 1972 (Figure 5-31) suggest that the *Carex* in Pond C continued to accumulate biomass until 29 August but we can not really say that these plants were basically different from the terrestrial plants. These biomass data agree well with the Pond C data for 1972 and 1973 (sampled in mid-August) when 219 and 89 g dry wt m^{-2} were found (Tieszen, Mandsager and Vetter unpublished data). This biomass represented 1100 and 370 tillers m^{-2}.

Arctophila fulva was collected from Pond J in 1971 and from Pond L in 1972. *Arctophila* was only a minor part of the total biomass in Pond J but was the dominant aquatic plant in Pond L. It did not occur at all in the main ponds we studied (B and C).

Most of the living biomass of these plants is in the belowground rhizomes and roots. If we take the single measurement of roots and rhizomes in Table 5-7 and assume that it does not change over the season, then 91 to 95% of the total plant material is underground for *Carex* and 57 to 78% for *Arctophila*. This agrees with the statements of Dennis and

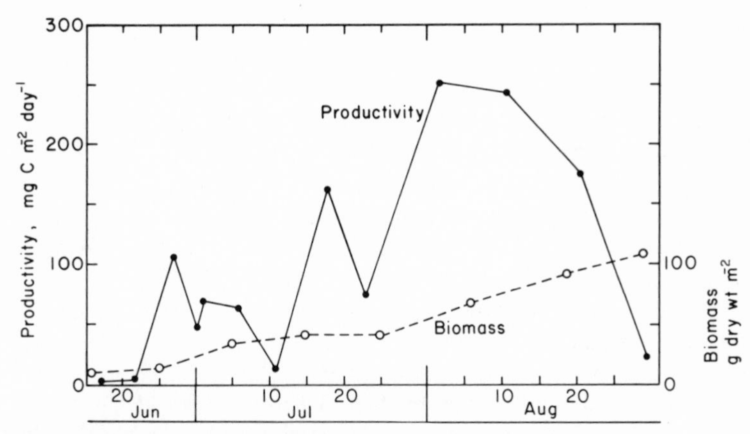

FIGURE 5-31. *Biomass and productivity of* Carex aquatilis *in Pond C, 1972, based on* ^{14}C *uptake.*

Johnson (1970) and Billings et al. (1978) that during the summer peak of aboveground graminoids at Barrow, 85 to 98% of the live vascular plant biomass is made up of roots and rhizomes. Shaver and Billings (1975) state that this high root-to-shoot ratio in arctic terrestrial plants is an effect of low nutrient levels, low soil oxygen levels, and high soil moisture content. All these factors act to decrease the effective uptake of water and nutrients by a unit of root surface.

This high ratio of root-to-shoot is also found in other aquatic plants so it is not a unique feature of arctic plants. Westlake (1968) reports that these ratios of roots to total biomass in a temperate reed swamp were 83% for *Phragmites*, 85% for *Equisetum*, and 90% for *Scirpus*.

Realistically, the root-rhizome weight we measured may be too high as it is likely that too much dead material was included. For example, Shaver et al. (1979) carefully dissected whole root systems from the tundra and found that 62% of the plant was root and rhizome (small roots may be lost). Dennis et al. (1978) consider that on the Barrow wet meadows there is a subsurface standing crop (live) of 534 to 620 g m^{-2} for an aboveground maximum standing crop of 80 to 130 g. The aboveground standing crop is up to twice this in the ponds (Table 5-7) so a value of 1000 to 1200 g dry wt m^{-2} is reasonable for belowground living matter.

Roots of *Carex* are found down to a depth of 25 cm in the sediment but most are concentrated in the 10 to 20 cm layers; about 85% of the total belowground biomass occurs in the top 10 cm (Shaver and Billings 1975). These authors report that roots live for 5 to 8 years but can elongate only for 2 to 3 years. In contrast to the above description, *Arctophila* has low amounts of root biomass. By analogy to similar terrestrial plants (Shaver and Billings 1975), such as *Eriophorum*, the roots likely live only 1 year.

Productivity

The primary production was measured in two ways. One was by measuring the accumulated aboveground standing crop at the end of the growing season. The other was by measuring the uptake of carbon-14 by plant parts incubated under water in clear or dark glass jars for 4 to 6 hours. After incubation, the dried plant material was combusted, the labeled carbon dioxide trapped in a strong base (NCS from Amersham-Searle), and the ^{14}C counted by liquid scintillation.

Based on the maximum aboveground biomass attained each year, two different projects estimated that the production of *Carex* in ponds was 290 to 370 g dry wt m^{-2} (Table 5-7) and 89 to 219 g dry wt m^{-2} (Tieszen, Mandsager, and Vetter unpublished data). Terrestrial production at Barrow is 80 to 130 g dry wt m^{-2} (Miller and Tieszen 1972) so these values are reasonable. All these values, however, refer only to aboveground biomass, and production data for belowground plant material cannot be obtained by a single harvesting procedure. Shaver and Billings (1975) suggest from longevity and standing stock measurements that 100 g dry wt m^{-2} is a likely value for root production in tundra plants. This value agrees with the data of Dennis and Johnson (1970).

The production of root biomass in the ponds is probably also equal to the aboveground production; a likely value is 140 to 180 g C m^{-2} yr^{-1} (388 to 500 g dry wt m^{-2}). The total root biomass could, therefore, be replaced every 3 years.

Productivity estimates based on ^{14}C uptake in the pond did not agree with the biomass measurements and so have only relative value (Figure 5-31). A rough integration of the curve of the photosynthesis per day gives 9.4 g C m^{-2} or about 20 g dry wt which is at least an order of magnitude too low. It is likely that the water contained too little carbon dioxide to permit maximum photosynthesis. For example, the highest photosynthesis rate in Figure 5-31 is equal to 0.27 mg C g^{-1} hr^{-1}. Yet the rates of carbon dioxide ^{14}C uptake measured by Tieszen and Johnson (1975) peaked at 5 to 9 mg C g^{-1} hr^{-1} for *Carex* in the air. When the leaves are incubated in 500 ml of pond water, there is only 0.1 to 2 mg C available for photosynthesis (Chapter 4) so obviously carbon soon became limiting.

Productivity of the plants on the terrestrial site has also been measured by continuous measurement of gas exchange using an IRGA (infrared gas analyzer). For the whole community, Tieszen (1978a) estimated a net incorporation of carbon dioxide carbon of 207 g C m^{-2} yr^{-1}. This agreed well with the 210 g estimated by the aerodynamic studies of Coyne and Kelley (1975). Tieszen (1978b) suggested that the carbon budget might be as follows: net production of shoots, 46 g C; respiration of shoots, 39 g C; net production of belowground rhizomes and roots, 61 g C; respiration belowground, 61 g C m^{-2} yr^{-1}.

It is difficult to know how closely the parallel can be drawn between production in the terrestrial and aquatic systems. Based on Tieszen's

(1978b) estimates of partitioning of carbon in terrestrial plants, the annual aboveground production of *Carex* in the pond is 140 to 180 g C m^{-2}, root production is equal to the aboveground (140 to 180), while the respiration is 308-396 g C m^{-2}, mostly in the roots, rhizomes, and stem bases. The total respiration above was estimated from a measure of total net shoot uptake of CO_2 minus the production above- and belowground (Tieszen 1978b). This amount of respiration is higher than the 20 to 50 g C m^{-2} we have measured for the entire pond (Chapters 4 and 8). Because of the discrepancy, our estimate must be taken as the best available but not necessarily the correct value.

The above estimates are for the plant stands. To estimate production on the basis of a square meter of the entire pond, the estimates were multiplied by 0.3.

Arctophila fulva production is quite low in the ponds we studied but can be much higher in ponds where the grass has completely overgrown the whole pond. From the data in Table 5-7, the aboveground production in Pond J was around 20 g dry wt m^{-2} (7.2 g C) in 1971 or 1.5 g C m^{-2} on a whole pond basis. In Pond L, where *A. fulva* covered the whole surface, the production was 150 g dry wt m^{-2} (54 g C). Root production is unknown but if, as suspected, this plant grows new roots each year, then the underground production will be 30 to 40 g dry wt m^{-2} (11 to 14 g C) in the stands of Pond J and a similar proportion for Pond L. In the ponds we are concentrating on in this report, Ponds A, B, C, and J, *Arctophila* is not important relative to *Carex* even though it does produce more than the phytoplankton.

Nutrient Content

The nutrient contents of *Carex* were determined as a part of a transect study of sites that extended from the driest tundra into the ponds (Tieszen, Mandsager, and Vetter unpublished data). After collection, the plant parts were separated into three groups (see Table 5-8), dried at 60 to 70°C for 1 to 2 days, pulverized with a Wiley Mill and then taken to the University of Alaska for analyses (details of methods in Van Cleve and Viereck 1972). Nitrogen was determined by titration of the ammonia released from a Kjeldahl digestion of the plant material. Phosphorus was determined by the molybdenum blue method on a perchloric acid digest while all the cations were measured by atomic absorption.

The nutrient content of *Carex* in these ponds (Table 5-8) and also in the terrestrial plants is as high or higher than that of non-cultivated plants from temperate areas (Harner and Harper 1973). This suggests "that tundra plants have adapted to a nutrient-poor environment by producing a relatively small biomass with high nutrient content . . ." (Chapin et al. 1975). In the transects, one set of plants from a wet polygon did contain

TABLE 5-8 *Nutrient Concentration in* Carex aquatilis *from Pond C Collected in the Second Week of August 1972**

	Ca	Fe	K	Mg	Mn	P	Zn	N
Youngest 3 leaves	.08	.02	1.21	.15	.01	.24	.01	2.95
Remainder of leaves	.13	.01	1.05	.11	.02	.12	.01	2.07
Stem base	.10	.01	1.67	.12	.03	.19	.01	1.64

Source: Tieszen (unpublished data).
*Data expressed as % dry wt.

only 0.05% P in the leaves and Ulrich and Gersper (1978) pointed out that this low a concentration can affect photosynthesis. Calcium (Table 5-8) is low in the new leaves but increases in older tissue. This could either be a result of its low mobility or a result of increase in secondary cell wall content. Chapin el al. (1975) show that calcium, magnesium, and iron all increase in concentration through the growing season (e.g., Ca in Figure 5-32).

Nitrogen is highest in the youngest leaves (Table 5-8) but this element, along with phosphorus and potassium, is actively retranslocated back to the roots and rhizomes late in the growing season. The nitrogen standing stock of *Carex* on a tundra site (Figure 5-32) actually peaked in mid-July and then was reduced to about 50% of the maximum by late August. The start of this decrease may have preceded the biomass peak. Exactly the same thing occurred in *Carex* and *Arctophila* from the ponds (Table 5-9). Chapin et al. (1975) reported that the ratios of end of season (24 August) standing stock to maximum standing stock were 0.47 for N, 0.45 for P, 0.36 for K, 1.0 for Ca, and 0.62 for biomass. Thus 53% of the N was retranslocated and even greater amounts (64%) of the K. Calcium was not retranslocated but 38% of the biomass was lost from the aboveground plant parts. A separate measurement (McCown 1978) showed that Barrow graminoids contain 30 to 40% non-structural carbohydrate at the time of maximum biomass so that it appears that these carbohydrates are also efficiently extracted from the senescing shoots.

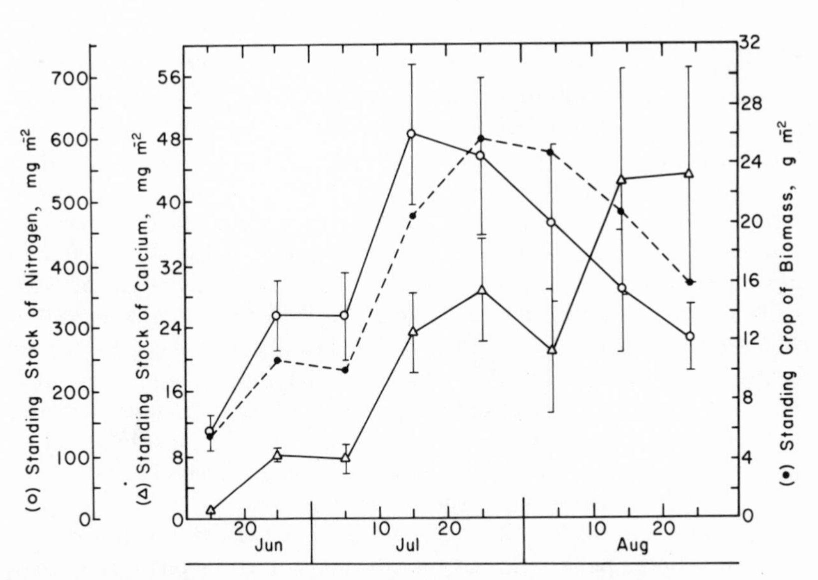

FIGURE 5-32. *Biomass and standing stocks of nitrogen and calcium in shoots of* Carex aquatilis *on tundra near Barrow. (After Chapin et al. 1975.)*

TABLE 5-9 *Nitrogen Content of Vascular Plants in Tundra Ponds, 1972**

	Carex aquatilis		*Arctophila fulva*	
Date	live	dead	live	dead
16 June	2.51	1.80		
20 June			2.23	1.74
25 June	2.46	1.63		
30 June			2.21	2.25
5 July	2.81	1.76		
11 July			2.58	2.04
15 July	3.21	1.87		
21 July			3.52	1.91
25 July	3.21	1.78		
31 July			3.13	1.76
6 August	2.99	1.75		
19 August	2.70	1.62	2.47	1.62
29 August	2.21	1.54	2.41	1.69
Mean	2.76	1.72	2.65	1.86

*Means of duplicates, expressed as % dry wt.

FACTORS CONTROLLING ROOTED AQUATIC PLANTS

Introduction

Numerous observations have shown that the stands of *Carex* and *Arctophila* in the ponds are the most productive of all the stands measured at Barrow (Bunnell et al. 1975). In addition, the two stands appear quite different because the terrestrial plants include standing dead leaves while the pond plants include only living leaves. Particularly after a number of years of low lemming density, the tundra appears brown for most of the growing season while the pond margins are always green. At the time of maximum standing crop on the tundra, the leaf area index ranges between 0.6 and 1.0 for living matter and from 0.0 to 1.5 for standing dead matter (Bunnell et al. 1975). Obviously the standing dead material is quite rapidly felled or otherwise removed in the ponds but the exact mechanism is unknown. Most likely the constant immersion of the stem bases increases the decomposition rate of this part of the plant. Another possibility is that the leaves growing in the pond are mechanically disrupted in some way by the formation of the ice. This process certainly appeared important in the destruction of *Carex* leaves that had overwintered twice (Chapter 8).

Carex in the ponds also has the advantage over the terrestrial *Carex* in that the ice preserves some green tissue. In the spring, the *Carex* in the ponds will have 5 or 6 cm of green tissue on each leaf that can begin immediate photosynthesis. We cannot quantify the importance of this but it certainly would help the pond plants and reduce the amount of carbohydrate translocated from the roots and rhizomes.

Light

The standing dead material absorbs a great deal of solar radiation in the canopy of the tundra. For this reason, the absence of standing dead in the ponds produces a better light climate for each leaf. The light climate is improved by each leaf being very erect so that it is close to perpendicular to the sun's rays. Dennis et al. (1978) report that 79% of the graminoid leaves at Barrow were inclined at 30° or greater and 77% of the *Carex aquatilis* leaves were inclined 60° or more. When the sun angle is low, and the average at Barrow is 25° on 21 June and 19° on 21 August, the combination of leaf angle and low sun angle results in almost complete absorption of all sunlight within the canopy (Caldwell et al. 1974). Bunnell et al. (1975) calculated that leaf area indices of 0.5 and 1.0 intercept 62 and 94% of the incoming solar radiation on 21 June; on 21 August they intercept 68 and 98%.

Early in the growing season, the rate of photosynthesis is clearly limited by the small amount of biomass. Accordingly, the plants are adapted to a rapid production of leaves immediately after the snowmelt by mobilizing stored carbohydrates from the roots and rhizomes and translocating them to the leaves. In terrestrial plants, this rapid early growth reaches 0.2 to 0.25 g (g dry wt)$^{-1}$ day^{-1} for the first week then falls to 0.03 g (g dry wt)$^{-1}$ day^{-1} after 10 days (Bunnell et al. 1975). The growth rate in the ponds was 0.07 g (g dry wt)$^{-1}$ day^{-1} at mid-season 1971 which compares very well with rates close to 0.06 g (g dry wt)$^{-1}$ day^{-1} measured in a *Carex rostrata* marsh in Minnesota (Bernard 1974).

Photosynthesis is a direct function of light in the Barrow tundra plants (Tieszen 1978a). The net photosynthesis of *C. aquatilis* was maximum at 15°C with severe inhibition above this temperature (Figure 5-33). Light saturation occurred at low light levels (about 0.1 ly min^{-1}) at 0°C and at lower temperatures but occurred at 0.24 ly min^{-1} at 15°C. These values are photosynthetically useful radiation (50% of incoming solar radiation). This saturation level is high relative to the usual insolation at Barrow because of the low solar angle and the high percentage of cloud cover during the summer. For example on 14 July (Figure 5-34a) the light level was saturating only around noontime. Other experiments show that the photosynthesis rates at optimum conditions were 18.5 mg CO_2 dm^{-2} hr^{-1} for *C. aquatilis* and 19.6 for *A. fulva.* Tieszen's studies also show that there is positive net photosynthesis (*Ps* greater than *R*) above an insolation of 0.01 cal cm^{-2} min^{-1}. Thus there is positive CO_2 uptake throughout most of the growing season. Only when the sun begins to set in August does the CO_2 uptake become negative at

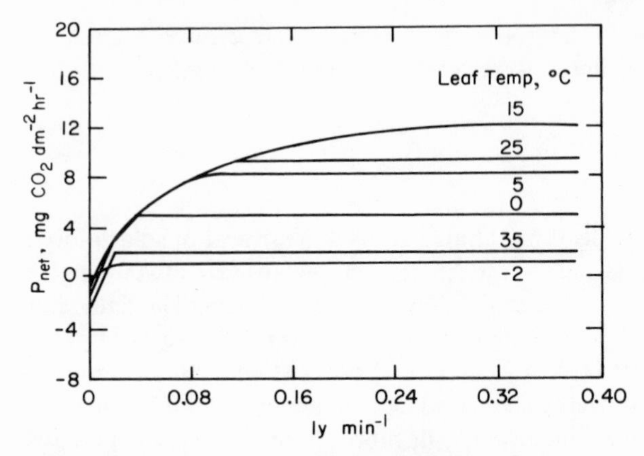

FIGURE 5-33. *Net photosynthesis of* Carex aquatilis *at various temperatures and radiation levels. The ly min^{-1} are for the 400 and 700 nm range (Tieszen, unpublished data.)*

night (Figure 5-34b). Final evidence for the limiting effect of light on vascular plant photosynthesis at Barrow is the direct relationship between daily solar radiation and primary production per day reported by Tieszen (1978a) for three species of sedges and grasses (including *Carex aquatilis*).

Temperature

The vascular plants at Barrow are well adapted to low temperatures. Photosynthesis, as described above, has a 15°C optimum and takes place even below −4°C (Tieszen 1978b). Another adaptation is significant respiration and translocation of carbohydrates at close to 0°C (Allessio and Tieszen 1973) and significant root elongation near 0°C (Shaver and Billings 1975). Other evidence for this adaptation to low temperature is the lack of relationship between daily temperature and daily photosynthesis in the tundra at Barrow (Webber 1978).

In summary, the vascular plants in the ponds are well adapted to local conditions of a short growing season and cold temperatures. Once free from snow and ice, their rate of growth per gram of plant is higher than temperate plants for a brief period and as high for most of the summer. While total plant photosynthesis would be greater at higher temperatures, the net photosynthesis may actually be lower due to the rapid increase in respiration. Indeed, the photosynthetic efficiency of 0.46% of the photosynthetically active radiation is respectable (Bunnell et al. 1975) and may be twice as high for the pond plants. Warming the tundra will have little effect on the net photosynthesis as generally the plants are operating below their light saturation on the part of the response curve where photosynthesis is insensitive to temperature (Figure 5-33).

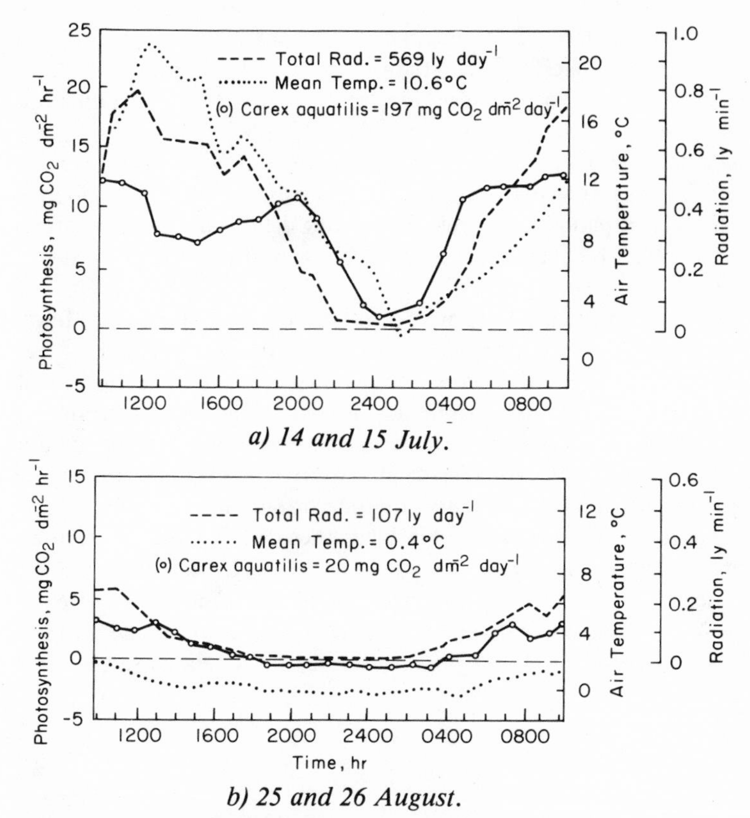

FIGURE 5-34. *Daily course of net CO_2 uptake by* Carex aquatilis *of chamber temperature, and of insolation. The hourly values for photosynthesis are means of five determinations (Tieszen, unpublished data.)*

Nutrients

Added nutrients had no immediate effects on plants, but over longer periods added nutrients did increase the plant biomass. Ten years after fertilization of terrestrial plots began at Barrow, the biomass had doubled (Dennis et al. 1978). The exact mode of action of nutrients was not clear, however, for Johnson and Tieszen (1973) found that fertilization of terrestrial plants did not affect either chlorophyll concentrations or photosynthetic and carboxylation rates per unit of biomass. Fertilization with either P alone or P + N did affect leaf area, canopy height and plant density. From these and other data, Bunnell et al. (1975) suggest that the effect of added phosphorus is not on photosynthesis but on the allocation of carbohydrates to leaf growth.

Under natural conditions, the plants are closely tuned to the prevailing climate and rate of nutrient supply. Therefore, added nutrients can only have effect after the plants have had time to change their photosynthetic capacity. The studies on terrestrial plants at Barrow showed that in the normal situation, the plants have a constant pool of carbohydrates. When nutrients are added, this pool is depleted slightly to produce more leaf tissue and the plants come to a new steady-state relationship with the environment. In the terrestrial system this increased growth is accompanied by increased standing dead and increased litter; the insulating effect of both of these may change soil temperatures, depth of thaw, etc.

The implication for the aquatic plants from these terrestrial findings is that high plant production in ponds is primarily a result of better nutrient conditions for the plants in the ponds than in the terrestrial sites. Yet, measurements of nutrients across transects indicate that concentrations of phosphorus are actually lower in the aquatic sediments than in terrestrial soils (Shaver et al. 1979). It is probable that there is increased availability of nutrients or a more rapid rate of supply in aquatic sediments than in soils. If this is true, then the low concentrations of nutrients in the sediments are caused by the uptake by the roots.

It is important to note that there may be differences between the interactions of phosphorus with terrestrial soils and with pond sediments. In ponds, the productivity of planktonic algae was related to the binding

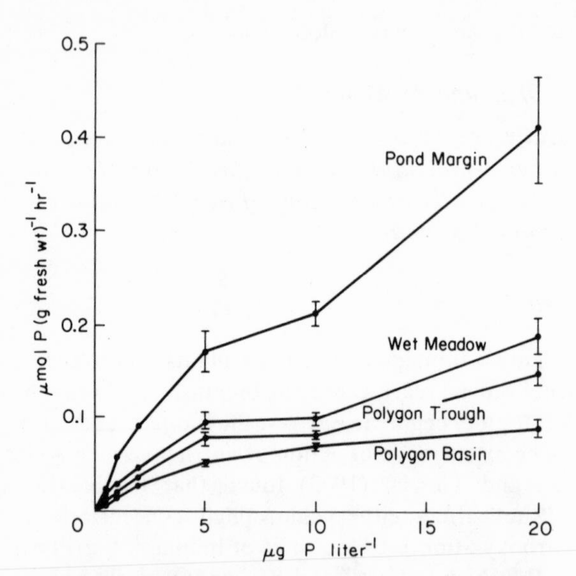

FIGURE 5-35. *Rate of P uptake by* Carex aquatilis *from different sites at four different concentrations of phosphate. Rates were measured at 10°C and each point is the average. (After Shaver et al. 1979.)*

capacity of the sediments (Figure 4-29) and not to the concentration of P in the water. This binding capacity is likely related to the supply rate. In contrast, in terrestrial soils at Barrow the production of vascular plants was correlated with the phosphorus concentrations (Ulrich and Gersper 1978, Bunnell et al. 1975). This conclusion, however, was based on standard methods of measuring available phosphorus and Chapin et al. (1978) argued that the available phosphorus in the soluble pool is really so small that it has a residence time of only 10 hours. They also argue that microbial lysis provides most of the available phosphorus. No matter which view is correct, there are differences between the reactions of the terrestrial soils and pond sediments.

Phosphorus

Chapin (1974), in his study of phosphate absorption in plants at low temperatures, found five adaptations: low temperature optima for root initiation, elongation, and production; large surface-to-volume ratios of roots; proportionately more nutrient absorbing tissue; higher phosphate absorption capacities at given temperatures; and less temperature sensitivity of the phosphate absorption system. One of the species he studied, *Carex aquatilis,* was collected from a pond margin at Barrow. Some of these adaptations, such as root growth at low temperatures and high root-to-shoot ratios, have already been discussed. The only measure of root diameter is given in Chapin (1974, his Figure 3). The diameters of *Carex aquatilis* roots were significantly smaller in the Barrow pond samples (soil temperature was 2°C) than in samples from the interior of Alaska where average soil temperature was 18°C. Chapin also found that the V_{max} of P uptake per unit of biomass was 4 times greater at Barrow than at the interior site.

Even within the Barrow area there were significant differences in P uptake by *Carex* roots from different sites (Figure 5-35). The relatively high V_{max} in the plants from the pond margin is further evidence that

TABLE 5-10 *Kinetic Parameters for Phosphate Uptake by* Carex *in Three Barrow Environments at 10°C*

Site	Uptake rate, μM (g fresh wt)$^{-1}$ hr^{-1}	V_{max} μM (g fresh wt)$^{-1}$ hr^{-1}	K_m (μM)
Pond C margin	0.212 ± 0.013	0.832 ± 0.195	21.6 ± 7.4
Wet meadow	0.099 ± 0.008	0.273 ± 0.042	11.2 ± 3.0
Polygon trough	0.082 ± 0.004	0.219 ± 0.029	11.0 ± 2.7

Source: Chapin and Bloom (1976).

nutrients are more abundant in this environment. The relatively high *K* for P uptake (a half-saturation constant) also is an indication of the relative richness of the pond sediment environment (Table 5-10); a low *K* is an indication of an uptake system that is effective at low concentrations (Wright and Hobbie 1966).

The Q_{10} for phosphate absorption is quite low, usually between 1.3 and 1.6. For *Carex aquatilis* Chapin and Bloom (1976) report 12 values whose average was 1.5. The temperature optimum for uptake was 30°C in this plant so adaptation does not include a decrease in this parameter. More important is the linearity between uptake of P and temperature which continues down to 1°C. Thus, roots are capable of absorbing phosphate as soon as the soil thaws. In contrast, Sutton's (1969) review shows that temperate plants are unable to absorb phosphate at low temperatures.

Nitrogen

There are no experimental data that prove a nitrogen limitation on the photosynthesis or growth of vascular plants in either the terrestrial or the aquatic environments. However, this does not mean that there is no limitation in the ponds. The only data we have which indicate that nitrogen is important are the very low concentrations of ammonia found in the stands of emergent plants (Figure 5-36). Obviously the plants keep the

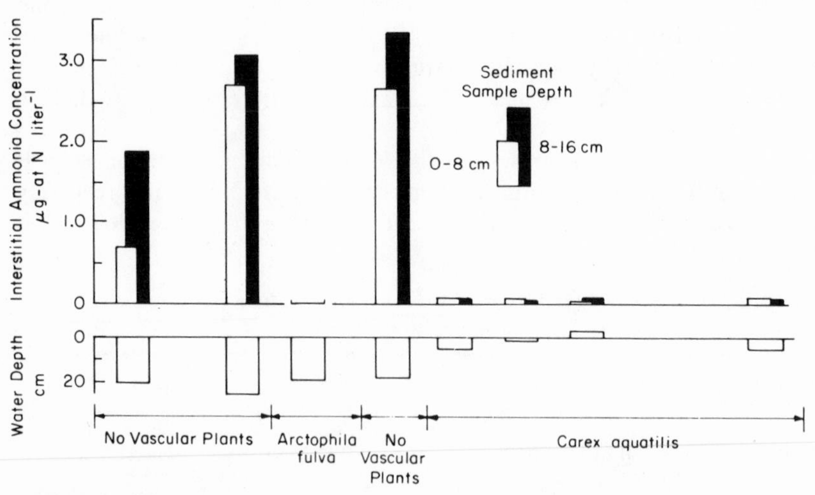

FIGURE 5-36. *Concentrations of interstitial ammonia (μg at N liter^{-1}) in the sediments of Pond J, 1973. The center of the pond, a region with no vascular plants, is at the left and the shore is at the right. (After Alexander and Barsdate 1975.)*

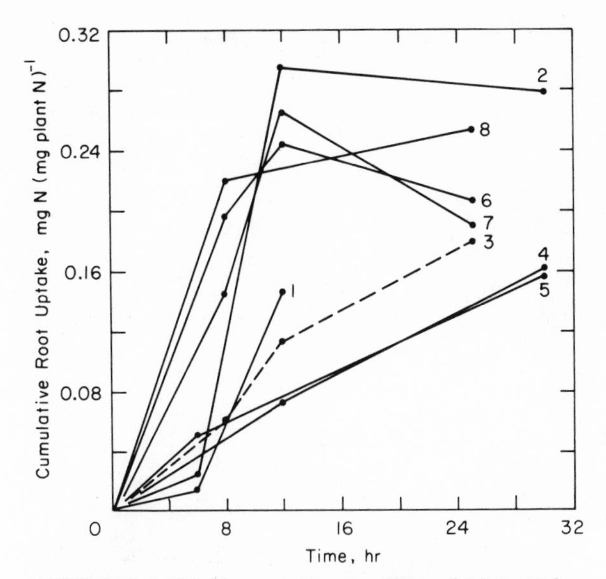

FIGURE 5-37. *Cumulative uptake of NH_3-N by* Carex *roots. Experiment 1 = 2.0 µg at N liter^{-1}, 2 = 3.3, 3 = 4.7, 4 = 6.0, 5 = 11.3, 6 = 15.3, 7 = 30.0 and 8 = 40.7. All experiments were run at 10°C in constant light. (After McRoy and Alexander 1975.)*

ammonia concentrations at a very low level while adjacent areas without plants have high concentrations.

In fact, the concentrations of ammonia in the sediments are so low that uptake rates are severely restricted (McRoy and Alexander 1975). These rates were measured with ^{15}N and with techniques described in Chapter 4. Whole plants were placed in partitioned containers such that the water surrounding the roots was isolated from that surrounding the leaves. The uptake by *Carex* roots was somewhat erratic but in general was linear over the first hours of the experiment (Figure 5-37). From data like these, the uptake rates at various concentrations could be plotted (Figure 5-38) and kinetic parameters calculated. The V_{max} for root uptake was 0.0275 µg N (µg plant N)$^{-1}$ hr^{-1} and the K (half-saturation constant) was between 117.6 and 175.0 µg NH_3-N liter^{-1}. The half-saturation values are 2 to 3 times greater than the concentrations of ammonia-nitrogen found in the sediments (53 to 75 µg N liter^{-1}, Chapter 4). In this region of the uptake curve in Figure 5-38 (4 to 6 µg-at liter^{-1}), the uptake is directly proportional to the concentration. Whether or not these low concentrations are at all limiting to photosynthesis or growth will depend upon the rate of supply of ammonia.

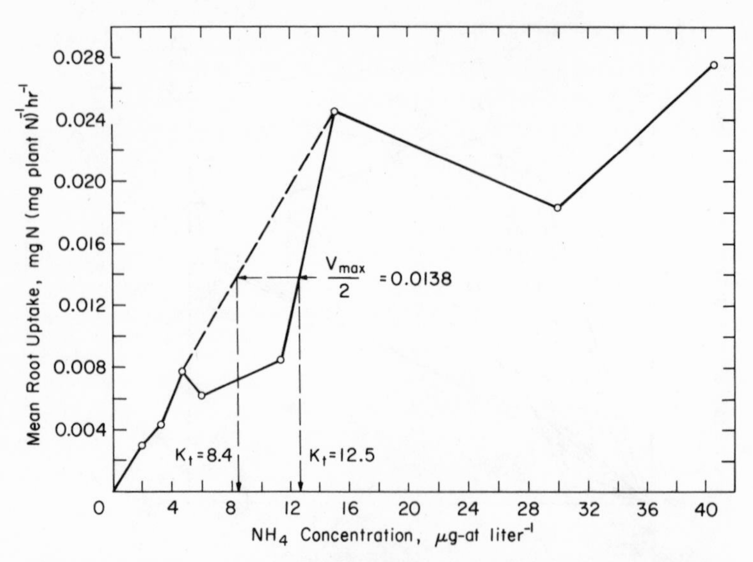

FIGURE 5-38. *Uptake of ammonia by* Carex *roots at different concentrations of ammonia. (After McRoy and Alexander 1975.)*

The NH_3-^{15}N that was taken up by the roots quickly moved to the leaves. In the same experiment described above for roots, the leaves also quickly became labeled. The V_{max} was 0.156 μg N (μg of plant N)$^{-1}$ hr^{-1} and the K was 56 μg liter^{-1}. This, of course, represents a transport from the roots. No ammonia was taken up by the leaves and stem and no $^{15}NH_3$ was lost from the plant by secretion. Thus, these results suggest that *Carex* plants obtain most of their nitrogen by means of ammonia absorption through the roots. This agrees with data from submerged plants in freshwater (Toetz 1971, 1973, McRoy and Goering unpublished observations).

Conclusions

In the Barrow ponds we studied, the most important primary producer was a sedge, *Carex aquatilis*. Adaptations, such as the ability to photosynthesize and translocate carbohydrates at low temperatures, enable this plant to virtually ignore the low temperatures and have as high primary productivity as temperate plants. Despite their low stature, the erect leaves of the rooted plants intercept nearly all of the low-angled solar radiation (average of 25° on 21 June).

The plants in the pond are actually more productive than *Carex aquatilis* plants on the tundra. One reason for this is the lack of standing dead leaves in the pond. Perhaps because of constant immersion, the dead leaves rapidly fall over. The water also gives some protection to the plants and may allow a slightly longer growing season for the aquatic plants. In addition, the green leaves and stems frozen into the ice are able to photosynthesize the next spring. Thus, these aquatic plants start the growing season with 5 or 6 cm of chlorophyllous tissue instead of the few millimeters that the terrestrial plants have. Another reason for the increased productivity may be an increased rate of nutrient supply. This is difficult to prove experimentally and the only evidence comes from studies of a series of terrestrial sites.

The field measurements described here show that *Carex aquatilis* populations growing in microhabitats (pond, meadow, ridge, basin, or trough) only a few decimeters from one another have different nutrient uptake and metabolic processes. When plants from these different populations were grown in a greenhouse, the nutrient uptake and metabolic differences were found to be genetically based (Shaver et al. 1979). These authors suggest that the genetic variability was present in a population of *Carex* that colonized a uniform, drained lake basin. As the microhabitats developed (when ice wedges and polygonal ground formed), there was competitive interaction and segregation into ecotypes which were maintained by vegetative reproduction.

SUMMARY

Phytoplankton

Algal biomass, measured by the amount of chlorophyll present and by the weight of cells, was very small in the water column of a pond. The chlorophyll rose from 0.1 μg liter^{-1} in early June to a peak of 0.8 to 1.2 μg liter^{-1} in late June or early July, then remained between 0.3 and 1.0 μg liter^{-1} until the end of August. Usually there was another peak, up to 2.0 μg liter^{-1}, at the end of August. The cell weight, determined from the direct count of the algae with an inverted microscope, was mostly between 200 and 400 μg wet wt liter^{-1} except for the late June—early July peak when the weight reached 800 μg.

The algae are all nannoplankton, mostly small flagellates less than 10 μm in length. There were usually several million cells per liter. The early season was dominated by chrysophytes (*Chromulina, Ochromonas*) but cryptophytes (*Cryptomonas, Rhodomonas, Chroomonas*) became dominant in early July and remained dominant for the rest of the summer. Some 105 different species of plankton algae were found in the ponds. The overwhelming importance of nannoplankton, of flagellates, and of chrysophytes and cryptophytes is typical of arctic ponds and lakes.

Detailed studies of the algal biomass every 3 hours for an entire day revealed that the total biomass changed as much as 3-fold in a few hours. The most likely explanation is that small flagellates (3 μm and 6 μm in length) moved down into the upper layers of the sediment during the low-light periods and moved back into the water column for the afternoon. Larger forms typical of plankton, such as *Rhodomonas minuta* and *Chromulina*, stayed in the water continually. Sediment algae were 60 times more abundant per milliliter than the plankton algae so there was an adequate reservoir of cells.

Phytoplankton primary productivity, measured by light and dark bottle uptake of $H^{14}CO_3$, paralleled the curves for chlorophyll and biomass. There was a peak in early July, a mid-summer low, and another peak at the end of August. The rates were reasonably high per unit of volume: 8 μg C liter^{-1} hr^{-1} in the early peak, 1 μg C liter^{-1} hr^{-1} in mid-summer, and up to 17 μg C liter^{-1} hr^{-1} at the end of August. However, the total volume of the system is so low that production was only about 1 g C m^{-2} yr^{-1}, one of the lowest ever measured. Detailed studies of productivity every 3 hours showed that the maximum occurred from early afternoon (1400) until mid-evening (2000). This coincided with the appearance of the 3 μm flagellates in the water column. On one day, the specific growth rate (μg C (μg C)$^{-1}$ hr^{-1}) of the 3 μm flagellates peaked at 0.04 hr^{-1} at 2000, the rate of the 6 μm flagellates and of *Rhodomonas* peaked at 0.03 hr^{-1} at 1400, and that of the *Chromulina* peaked at 0.02 hr^{-1} at 2300.

Epipelic Algae

In the sediments, the majority of the algae are attached to particles. Some 50% of the total biomass was chlorophytes (green algae) and 42% was cyanophytes (blue-green algae). Common algae were *Chlamydomonas, Closterium,* and *Aphanizomenon* (blue-green). Diatoms were also present. The cells are small, usually less than 10 μm in length; numbers ranged from 2 to 4 $\times 10^{10}$ m^{-2} (0.5 to 1.0 g C m^{-2}). Throughout the year, there was a gradual increase in biomass. Within the sediment, 50% of the biomass was in the top 1 cm; another 30% was at the 1-2 cm depth. Perhaps because of the animal activity which continually mixed the sediment, the algae did not form a mat at the surface.

Primary productivity was estimated with a ^{14}C technique in which a maximum potential photosynthesis rate was measured with a diluted sample and then an *in situ* rate calculated from the light-depth relationship. The depth of the 1% light level was only 2 mm but algae from greater depths were always able to photosynthesize when placed in light. The productivity began as soon as the ice melted and rose steadily from 2 mg C m^{-2} hr^{-1} on 21 June to 16 mg C m^{-2} hr^{-1} in mid-July. After this peak, the productivity steadily declined throughout the rest of the summer.

Total production was around 10 g C m^{-2} yr^{-1}, about 10 times the planktonic production.

Factors Controlling Algae

Planktonic algae have a temperature optimum for photosynthesis between 15° and 20°C while the epipelic algae have an optimum above 20° (likely above 25°). The Q_{10} response of photosynthesis is also slightly different as it is 3.0 for the plankton and 2.2 for the epipelic algae (2.5° to 12.5°C). In fact, if higher temperatures did occur, the algal productivity might not increase. Thus, we found that a 4°C increase in the temperature of an experimental pond did not change the productivity. The reason for the lack of a temperature effect may be that the rates of algal respiration and animal grazing also increase.

Light usually limits photosynthesis (Ps) in lakes but the ponds are so shallow that there is always more than enough light for the algae to have maximal rates of Ps. During June and the first half of July, there was even adequate light for net Ps to proceed all night. When there were no clouds or fog, the Ps rate at midnight was only 50% less than the rate at mid-day or mid-afternoon. Experiments relating Ps to light and temperature showed that the Ps increased to a maximum as the light increased (the curve was similar to a Michaelis-Menten saturation relationship). At low temperatures, 2° to 8°C, the Ps of the planktonic algae was inhibited at high light levels (>0.2 ly min^{-1}) while the Ps of epipelic algae was not inhibited. This may be a result of a slowed rate of repair of photo-damage to chlorophyll. The lack of inhibition of the epipelic algae may be a result of shading from sediment particles. It was impossible to separate the sediment from the algae so some algae was always wholly or partially shaded. Increasing the amount of light, therefore, will increase the Ps even if some algae might really be inhibited.

The half-saturation level of the Ps by light, $I_{0.5}$, is a useful value for comparison of the phytoplankton responses at different times of the year. In general, they fell throughout the year; this indicates that the algae respond to the lowered light levels by becoming more efficient. A very rapid decrease in $I_{0.5}$ in late June and early July may indicate a nutrient limitation. The specific growth rate (mg C fixed (mg algal C)$^{-1}$ day^{-1}) ranges from 0.2 to 3.0 for the phytoplankton and from 0.4 to 1.0 for the epipelic algae. These values, especially those for the plankton, are extremely high (reproduction several times per day). The actual populations do not change as rapidly so there must be an almost equally high removal rate. For the planktonic algae, the removal is due to grazing by the zooplankton; for the epipelic algae, the removal is due to downward mixing of the algae into the sediment.

The effect of nutrients on photosynthesis of the algae could not usually be demonstrated with the short-term ^{14}C experiments (less than 4 hours). However, there were dramatic changes in the Ps rates when nutrients were added to entire ponds. Nitrogen had no effect; as described in Chapter 4, the rate of ammonification in the plankton is likely high enough to supply all the needs of the algae and the sediments contain large quantities as well. In contrast, P concentration is controlled by the sediment. As described in the previous chapter, large amounts of phosphorus are adsorbed onto iron hydroxides in the sediment; the equilibrium allows only a few μg P liter^{-1} to exist in the water column. A relatively high rate of input of P in the spring allows the algae to grow rapidly and bloom. The input soon ceases as runoff stops and the algae rapidly take up the P faster than it can be resupplied by the sediment. As a result, concentrations fall and the algal bloom ends.

Additional evidence for the importance of phosphorus in controlling the primary productivity comes from the fertilization of whole ponds. When phosphorus was added there was actually a decrease in productivity (inhibition) followed several days later by a rise to peaks several times higher than those of productivity in control ponds. Epipelic algae also responded to fertilization with P; their productivity doubled within a month.

Ponds where the iron-P trapping mechanism was not well established, such·as newly formed thaw ponds with no accumulated sediments, had 20 times the P concentration and 200 times the primary productivity of control ponds at the same date.

There are still many remaining problems of the interactions between algae and nutrients. For example, what is responsible for the rapid uptake of phosphorus and its release as a low molecular weight organic compound (XP)? Do the algae do this? The bacteria? Is it in a form usable by algae?

Grazing by zooplankton could be another mechanism controlling the algae in the pond. The dominant zooplankton was a large *Daphnia*; each animal is capable of filtering 2-6 ml of water per hour at 11°C. The whole population probably filters all of the pond water every 2 days (range, 0.5 to 12 days). Two different calculations, one a back-calculation from the zooplankton production and the other based on measurements of zooplankton filtering rates under different conditions of temperature and food density, indicate that about 20% of the total algal productivity is grazed by zooplankton. At certain times of the year the grazing pressure is intense, and so zooplankton could also be very important as a control. When the zooplankton were removed from a pond, the biomass and primary productivity of the algae increased but this increase was much less than expected. Evidently other factors immediately became limiting. In addition to these small changes, the zooplankton removal caused one of the dominant species, *Rhodomonas minuta*, to be replaced by *Uroglena*, but it is not known whether the replacement was a result of the release of grazing pressure which had kept the *Uroglena* at a low level or whether it was caused by some other interaction between the *Daphnia* and the *Rhodomonas*.

In the sediments, the algae are grazed by protozoans, chironomid larvae, and other larger animals such as oligochaetes and nematodes. Whether or not this grazing is an important control of these algae depends upon the degree of selectivity these animals have because the sediments are composed of about 0.06% algae, 0.17% bacteria, and 99.7% detritus (all are percent of total organic matter). The measured feeding rate of protozoans and micrometazoans was a maximum of 8 mg algal C m^{-2} day^{-1} out of a total algal biomass of 500 to 1000 mg m^{-2}. Chironomid feeding is less selective and totalled 325 mg C m^{-2} day^{-1}. If there were no selectivity, this would equal 0.2 mg algal C; if there were a 5-fold selection of algae over detritus, the consumption would be 1 mg m^{-2} day^{-1}. Thus, total consumption may be about 9 mg algal C m^{-2} day^{-1} which is about 10% of daily production.

In some shallow ponds, the benthic algae have an extremely high productivity because they form thick mats. This does not happen in the tundra ponds because of the continued mixing of the sediments by the chironomid larvae. This mixing of the algae down below the lighted zone is likely the dominant control in view of the abundant nutrients in the pond sediments.

Aside from the example of the zooplankton-*Rhodomonas* link mentioned above, there was no evidence about what controls algal species composition. Autoradiography did show that *Chromulina* and *Rhodomonas*, which were present at the same time in the plankton, have photosynthetic peaks at different times of the day; thus, there is always physiological diversity present in the plankton and different species can take advantage of seasonally changing conditions. Diatoms may have been kept out of the plankton by the low concentration of silica, always less than 0.5 mg $liter^{-1}$.

Biology of Rooted Aquatic Plants

In the Barrow ponds, the dominant higher plant is the sedge *Carex aquatilis*. This plant is actually amphibious as it grows on the tundra as well. It occurs as a bed around the edge of the pond out to a depth of 15 cm. A grass, *Arctophila fulva*, occurs only in ponds and is found in 15 to 25 cm of water. The percent of the surface area covered by plants varies from pond to pond; in the "typical" pond we studied intensively, *C. aquatilis* covered 32% and *A. fulva* covered 13%.

These plants reproduce vegetatively and so have a network of horizontal rhizomes and stems (the tillers) connecting mother and daughter plants. Every fall there is almost complete die-back of the aboveground leaves; growth begins from the stem base immediately after snow and ice melt in mid-June. Even though the roots may still be frozen, the leaves elongate rapidly in late June and the two original leaves may attain their maximal length by 9 July and then begin to senesce. Other leaves appear a little later in the season, grow for about 25 days, remain mature for about 20 days, then senesce. Each plant produces around five leaves each year and 20 to 30 during its lifetime of 4 to 7 years.

In a typical pond, the aboveground biomass of *C. aquatilis* was 170 to 230 g dry wt m^{-2} while that of *A. fulva* was 10 to 23 (July and August). Biomass reached a maximum in the first week in August and declined rapidly thereafter. However, it is typical of tundra plants that most of the biomass is belowground; accordingly, some 3000 g dry wt m^{-2} of roots were found for *Carex* (90 to 95% of total biomass). *Arctophila* has smaller root systems and a biomass of 31 g of roots was found (57 to 78% of total biomass). The high ratio of root-to-shoot is an effect of low nutrient levels, low soil oxygen levels, and high soil moisture content; it is typical of other plants found in temperate swamps. *Carex* roots live for 5 to 8 years while *Arctophila* roots live one year.

The aboveground *Carex* production in ponds ranged from 140 to 180 g C m^{-2} yr^{-1}, which is slightly higher than production in the terrestrial tundra. Root production is likely about the same amount; the total quantity equals 308 to 396 g C m^{-2} yr^{-1}. Additional carbon is used in root respiration; this is probably around 320 g C m^{-2} yr^{-1}. These data are for plant beds; total pond values are 1/3 of the above.

Arctophila production was not too important in the intensively studied ponds as it ranged from 18 to 21 g C m^{-2} yr^{-1} in the plant beds.

The nutrient content of *Carex* is as high as or higher than that for similar temperate plants. The content did change throughout the growing season, however, and the percent of nitrogen, phosphorus, and potassium reached a peak in mid-July, well before the peak of biomass. By late August, a large quantity of these nutrients had been retranslocated back to the roots and rhizomes for storage. Thus, by 29 August the N content of *Carex* was 6 to 8% of the content at maximum (both as percent of dry weight). Some structural carbohydrates may also be removed and translocated. Calcium does not change in the plants because it becomes a part of the cell wall.

Control of Rooted Plants

The production of plants in the stands within the ponds is actually higher than production in terrestrial sites. One reason for this is the relative lack of standing dead leaves in the pond plants. As a result, the plants of the tundra appear brown while the plant beds in the ponds are green and the light climate for each leaf is much better. The cause of the lack of standing dead material may be an increased rate of decomposition of the leaf parts that are in contact with the water or mechanical damage by ice.

The canopy of leaves is very effective in absorbing nearly all of the solar radiation. Each leaf is erect and is angled so that it is close to perpendicular to the sun's rays (it is 85% from the horizontal for *C. aquatilis*). The sun angle is low, only 21° to 25° during the summer.

Early in the growing season, the rate of photosynthesis is clearly limited by the small amount of green plant material. Accordingly, the plants are adapted to a rapid production of leaves immediately after the snowmelt by moving stored carbohydrates and nutrients from roots and rhizomes into leaves. This early growth rate can be as high as 0.25 g (g of plant)$^{-1}$ day^{-1} but falls to 0.07 g (g of plant)$^{-1}$ day^{-1} in mid-season. Plants in the pond may also get a jump on the terrestrial plants because the ice preserves some green tissue over the winter. In the spring, *Carex* in the ponds has 5 to 6 cm of green tissue on each leaf, whereas terrestrial plants will have only 2 cm.

The maximum photosynthesis rate of *Carex* occurs at 15°C but positive net photosynthesis occurs as low as 12°C. The Ps vs. light curve is saturated at a relatively high level so that on a typical day the Ps is saturated only briefly, around noon. There is continuous positive net photosynthesis from mid-June until early August when the sun begins to set. Thus, photosynthesis of the plants is directly related to the solar radiation throughout the summer.

The *Carex* and *Arctophila* are similar to other tundra plants in that they can function at low temperatures. Even at 0°C, roots grow, net photosynthesis occurs, and translocation of carbohydrates takes place. The low temperatures also reduce respiration so that the photosynthesis efficiency of the terrestrial tundra plants was 0.46% and that of the aquatic plants was even higher. These values are about the same as for temperate plants.

Experiments with terrestrial plants of the Barrow tundra suggest that phosphorus concentrations in the soils limit the plant production. Yet, the concentrations of phosphorus are lowest in the pond sediments where the productivity of the plants is the highest. It appears that the phosphorus in the ponds is more available than that in the soils and, indeed, the binding process is very different in these two systems. Thus, plant production in the ponds is higher than that in the terrestrial system in part because of greater availability of phosphorus. Improved light conditions for the pond plants will also aid in higher productivity rates.

When compared with temperate plants, phosphate absorption of arctic plants is faster at low temperatures. This is, in part, tied to the high amounts of roots in arctic plants and to root elongation at low temperatures. At Barrow, plants from the ponds had a high V_{max} and a high K for phosphate absorption when compared to terrestrial plants. These indicate that the phosphorus is more available in the ponds.

Nitrogen may also limit production of the plants, but the evidence is indirect. One type of evidence is the extremely low concentration of ammonia within the sediments of the plant beds compared with the very high amounts outside the plant beds (0.05 to 0.07 vs. 3 mg NH_3-N liter^{-1}). Also, the half-saturation values for ammonia uptake by the roots were 2 to 3 times the concentrations so the transport systems are undersaturated. The plants are able to operate very effectively with these low concentrations of ammonia and we conclude that there is a close balance between need and rates of supply. It was not possible to decide whether or not there was an actual limitation by nitrogen, but it appears that the plants are operating at close to the maximum rates of supply of ammonia.

Greenhouse experiments with *Carex aquatilis* show that there are genetic differences in metabolism and nutrient uptake between populations located only a few decimeters apart (in pond, ridge, trough, or meadow). These ecotypes have developed by competitive interaction which took place as the ice wedges and polygonal ground grew and caused the microhabitats to develop.

6

Zooplankton

R. G. Stross, M. C. Miller, and R. J. Daley

COMMUNITIES, LIFE CYCLES, AND PRODUCTION *

Community Composition and Life Cycles

Crustaceans dominate the zooplankton of polygon center ponds and also of pools in nearby troughs which form between polygons. Rotifers are present but rare. The zooplankton species found at Barrow (Table 6-1, Figure 6-1) are commonly found throughout northern Alaska, as shown by collections from the Colville River to the east (Reed 1962) and from Cape Thompson to the west (Hilliard and Tash 1966).

Despite the similarity of species over the entire region, there are local differences in the community structure between the zooplankton of these shallow waters and the zooplankton of nearby lakes. In the shallow ponds and trough pools the Crustacea are quite large and often heavily pigmented. For example, the fairyshrimp frequently reach 15 mm and the *Heterocope* is the largest freshwater copepod. The species of the genus *Daphnia, D. middendorffiana* and *D. pulex,* are two of the largest known. In contrast, smaller forms of zooplankton are found in lakes containing planktivorous fish. Ikroavik and Sungoroak Lakes, for example, have no fairyshrimp but do have a small, transparent *Daphnia* (*D. longiremis*). A third large lake, Imikpuk, has no fish; it contains the same species of large Crustacea as the ponds. From this evidence, it is likely that size-selective predation by fish may prevent the larger species of zooplankton from inhabiting deep lakes.

In addition to the differences in zooplankton species between lakes and ponds, there are also differences between the ponds and trough pools at the IBP site (Figure 6-1). Calanoid copepods (*Diaptomus* and *Heterocope*) and *D. middendorffiana* are present in the ponds but not in the trough pools; fairyshrimps are present in both but are rare in pools. *Daphnia pulex,* present only in the troughs, is somewhat smaller than *D. middendorffiana* and has only a small amount of pigmentation (Edmondson 1955). Cyclopoid copepods (*Cyclops*) are abundant in both habitats. These tundra ponds are as species-rich as alpine ponds at mid-latitudes (Dodson 1974, Sprules 1972).

*R. Stross

TABLE 6-1 *Species of Crustacea in the Plankton and Their Life Cycles in a Typical Polygon Pool (IBP Pond C) on the Tundra near Barrow*

Species	Overwintering stage	Start of reproduction
Copepoda		
Cyclops vernalis Fischer	copepodid (pre-adult)	June 15
Cyclops magnus Marsh	copepodid (pre-adult)	June 15
Diaptomus bacillifer Kolbel	embryo (early)	July 20
Diaptomus glacialis Lilljeborg	embryo (early)	late July
Diaptomus alaskaensis M. S. Wilson	--	--
Heterocope septentrionalis Juday and Matthowski	embryo	late July-August
Branchiopoda		
Branchinecta paludosa O.F. Miller	embryo (early)	July 20
Polyartemiella hazeni Murdoch	--	--
Daphnia middendorffiana Fischer	embryo (early)	July 15

All Crustaceans except for the *Daphnia* are monovoltine (one generation per year); offspring of the overwintering generation enter diapause to survive through the next winter. Usually, the embryonic stages enter diapause, but in the cyclopoid copepods a copepodid (pre-adult) instar overwinters. (The precise overwintering instar was not determined.) In some species, such as *Cyclops strenuus*, nearly all copepodid instars have the ability to encyst (Elgmork 1959, Szlauer 1963). The advanced stage of development of the overwintering cyclopoids enables them to reach adulthood and reproduce shortly after the ice melts in June.

The *Daphnia* are the only taxa at Barrow to produce a generation that does not overwinter. The sequence of events in the *Daphnia* life cycle follows: first, the overwintering embryos (formerly called ephippial eggs) hatch shortly after the ice thaws; second, the hatchlings, which are all female, reach maturity and in mid-July release a brood of young (Stross 1969, Stross and Kangas 1969); third, after this brood of young is produced, the females continue to reproduce but now produce diapausing embryos (Edmondson 1955, Stross 1969). If food is exceptionally abundant, the brood of young (called young-of-the-year) will also produce diapausing embryos. Only females are known for *D. middendorffiana* but

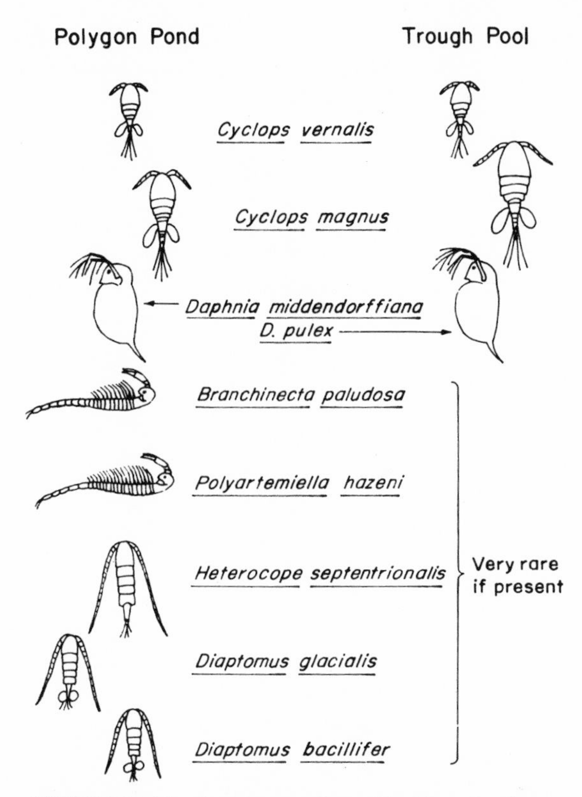

FIGURE 6-1. *Crustacea found in the plankton of ponds and trough pools at Barrow, Alaska.*

in *D. pulex* a small number (4%) of the young-of-the-year may be infertile males (Brooks 1957).

Stross (1969) has shown that the switch from production of young to production of diapausing embryos is facultative and under environmental control. In an experiment with constant light and constant temperature (12°C) in the laboratory, the females that hatched from the overwintering eggs produced only young; at 20 hours or less of light per day only diapausing embryos were produced but this switch occurred only in animals which had already produced a brood of young. The critical photoperiod determined in the laboratory of L22:D2 may well be acting in the populations of lakes too, as these animals made the switch in late August when the ratio was about L22:D2. On the other hand, populations in pools and ponds make the switch in mid-July (Stross and Kangas 1969). This earlier switch may be caused by the temperature regime of the ponds which is warmer and more variable than that of the lakes. Thus, pond temperatures may change as much as 10° in a day and may fall to 5°C on any day of the summer. These low temperatures, below 10°C, may act as some sort of physiological "dark interval." Other arthropods have a similar interval timer that determines when they can respond to a photoperiodic induction (Lees 1960); in the *Daphnia,* the production of a brood of young may also be an interval timer.

From the foregoing account it is clear that the initial density of zooplankton populations in spring is determined by the number of overwintering animals that emerge. An upper limit to density is thus placed on all monovoltine species and therefore on the annual production. The *Daphnia* have an additional production capability from the young produced by the overwintering generation.

Standing Crops

The Crustacea were collected by pouring 8, 10, or 12 liters of water through a No. 25 mesh plankton net. The net could not be towed in the pond because it would have disturbed the sediment; instead, 1- or 3-liter water samples were dipped from the pond with a plastic bucket mounted on a long pole. This procedure may have resulted in underestimates of the fast moving animals such as the calanoid copepods and of the bottom huggers such as the fairyshrimps. Triplicate samples were taken at one or more stations, depending on the pond being sampled.

The distribution of the zooplankton within a pond was not random. Fairyshrimps and *Daphnia* were frequently clumped, either as a dense swarm (*Daphnia*) or as a reproductive group (fairyshrimps). The two species of fairyshrimps were segregated one from the other in the mating swarms and the individuals or attached pairs often hovered in rows. Obviously, there was poor sampling precision for these forms. The

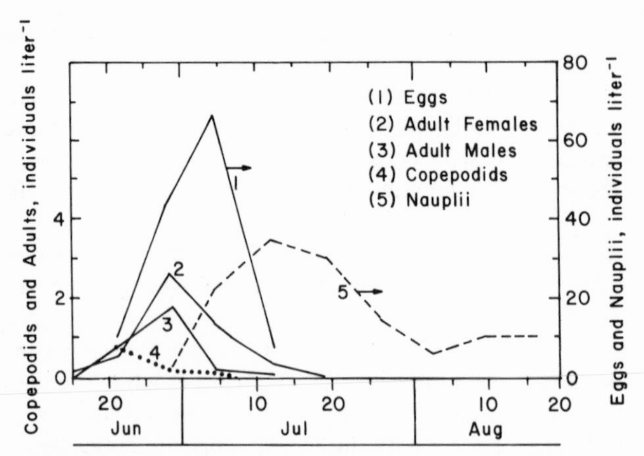

FIGURE 6-2. *Densities of various stages of* Cyclops vernalis *in Pond C, 1971. Sample points are means of four to eight 10-liter samples.*

precision was better for the copepods, with only a small variance between samples at the same station. The variance was greater between stations, however, and there was also a tendency for the copepodids to cluster at the margins of the ponds in the spring.

In a later study of *Daphnia* distribution, Haney and Buchanan (personal communication) showed a correlation between the location of most of the *D. middendorffiana* in a pond and the sun on clear or hazy days. The population seemed to move toward the margin of the pond farthest from the sun and in midsummer actually circled the entire pond margin over 24 hours.

The cyclopoid copepods were numerically more abundant than other zooplankton forms, especially early in the summer (Figure 6-2). As the ice melted in 1971, the late juvenile and adult stages of *Cyclops vernalis* and *C. magnus* appeared in the water. By mid-June the adults of *C. vernalis* were carrying egg sacs and the first nauplii were seen on 21 June. By the last week in June, most of the overwintering animals had been transformed into adults which reached a maximum of 2.5 females and 1.7 males liter^{-1}. The maximum number of nauplii, 35 liter^{-1}, were found on 13 July. After 25 July no more than 10 liter^{-1} were collected. The copepodids began to appear in early July and were most abundant during the first third of July. From late July to mid-August, there was approximately 1.0 animal liter^{-1}. Other species of cyclopoid copepods were present, but the 10-liter samples were not large enough to permit quantitative estimates. Although the copepodids of all the cyclopoid species may enter diapause in the sediment (Elgmork 1959, Szlauer 1963) their entry was not examined in the ponds at Barrow.

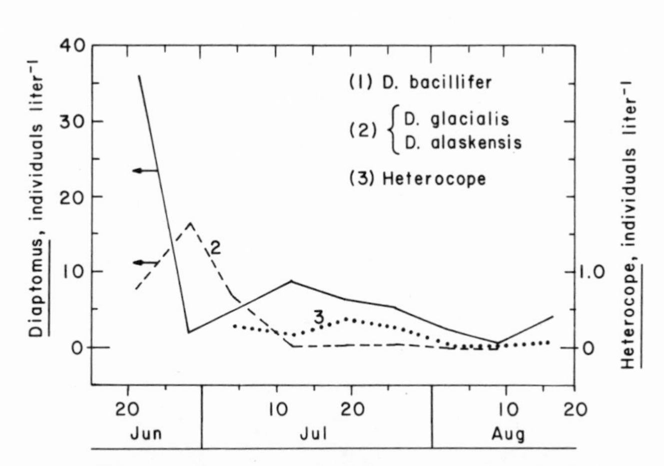

FIGURE 6-3. *Densities of all stages combined of three species of calanoid copepods from Pond C, 1971. (Same samples as Figure 6-2.)*

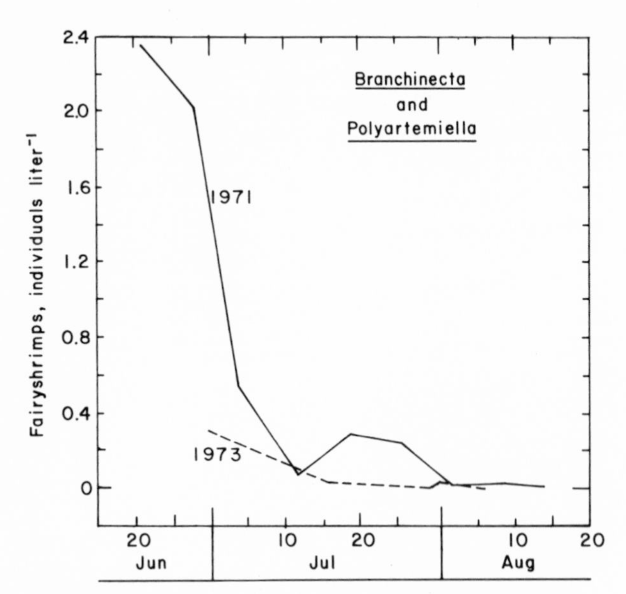

FIGURE 6-4. *Densities of fairyshrimps* Branchinecta paludosa *and* Polyartemiella hazeni *in Pond C, 1971. (Same samples as Figure 6-2.)*

The calanoid copepods overwinter as embryos. They hatched within a week of the pond thaw (Figure 6-3) and by 19 June there were two sizes of calanoid nauplii present. The most abundant species in the study ponds was *Diaptomus bacillifer* at about 35 liter^{-1}. It was also the smallest. The medium-sized *D. glacialis* had a density of 8 liter^{-1} while the largest calanoid, the predaceous *Heterocope septentrionalis*, was not detected until 3 July when 2.5 liter^{-1} were recorded. Densities of all populations decreased as the individuals developed, although after 15 July numbers stabilized at 3 liter^{-1} for *D. bacillifer*, less than 1 liter^{-1} for *D. glacialis*, and about 2 liter^{-1} for *Heterocope*.

The fairyshrimp are very large in comparison with the other zooplankton. For example, their mass may be 10 times that of a large daphnid. As a result, they were the dominant species in biomass in 1971 even though the numbers were low (Figure 6-4). The two species, *Branchinecta* and *Polyartemiella*, hatched at about the same time as the cyclopoid copepods emerged and then reached a maximum density of 2.3 nauplii liter^{-1} on 21 June. The fairyshrimp numbers then declined to about 0.3 liter^{-1} in early July and by early August they were too rare to appear in the samples. During mating, which took place in the 3rd week in July, the fairyshrimp of each species moved to one place in the pond, formed pairs, and hovered in large rafts immediately beneath the surface. The fairyshrimp numbers declined throughout the 3 years of the study; in

1973 there was less than 0.05 animal liter^{-1} in the intensively studied ponds after the first sampling date in June.

Young *Daphnia* appeared in the water several weeks after the ponds melted (Figure 6-5); hatching occurred on 20 June in 1971 and on 25 and 28 June in 1972 and 1973. After hatching in late June, the number of *Daphnia* should be constant or decrease until young-of-the-year begin to appear in mid-July. Sampling errors cause fluctuations in the population estimates, however, and in some years (e.g., 1973 in Figure 6-5) there is an apparent increase. After mid-July, the young-of-the-year hatch and the total numbers should increase. In 1971, the brood size was 1.2 embryos per female, which should have resulted in 2.2 to 4.8 young liter^{-1}. Instead only 0.5 young liter^{-1} were found (Figure 6-5). As discussed later, this mortality is likely due to the predaceous copepod, *Heterocope*, and occurs each year. As a result, the young-of-the-year never contribute much to the total population of *Daphnia*. This observation leads to the conclusion that the summer abundance of *Daphnia* is determined exclusively by the size and hatching success of the overwintering population.

Reproduction in the Crustacea

Production of eggs by *Cyclops* began in mid-June. In fact, gravid *Cyclops vernalis* were present on 14 June 1971 while there was still some ice floating in the pond.

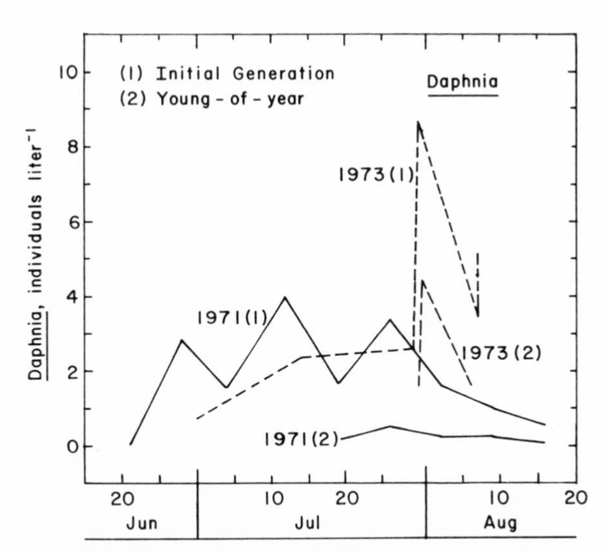

FIGURE 6-5. *Density of* Daphnia middendorffiana *for 1971 and 1973 of both the overwintering and young-of-year generations.*

TABLE 6-2 *Egg Production in Species of Crustacea in Pond C, 1971*

Species	Mean eggs (gravid female)$^{-1}$	Total eggs female^{-1}	Broods female^{-1} yr^{-1}	Egg production liter^{-1} yr^{-1}
Daphnia middendorffiana				
a. animals hatched from overwintering eggs and produce:				
1. eggs that immediately hatch	1.2	1.1	1	4.3
2. resting eggs	1.2**	3.6	3*	6.8
b. animals hatched from a-1 above will produce resting eggs	1.2**	1.2	1*	0.3
Branchinecta and *Polyartemiella*	5.0	15.0	continuous	0.8
Diaptomus bacillifer	14.0	11.6	1*	17.4
D. glacialis	74.0	74.0	1*	3.7
Heterocope septentrionalis	?	?	?	?
Cyclops vernalis	35.5	142.0	4*	126.6

*Estimated
**Determined in June 1972

The total production of *Cyclops* eggs was estimated from the duration of embryonic development, the numbers of gravid females, and the number of eggs per female at each sampling (Table 6-2). An estimate of the maximum development time came from measurements in which gravid females were placed in glass tubes 2.5 x 20 cm and held in the pond. Unfortunately, the hatching success of broods of eggs other than the first brood was low and an accurate development time could not be measured. The measured development time of 1 week at approximately 10°C is conservative and agrees with the literature.

Despite the large number of eggs produced by *Cyclops* each year (Table 6-2), only a few nauplii survive. For example, from the 126.6 eggs liter^{-1} produced in 1971, only 10 nauplii liter^{-1} were found in the pond during August. The loss could have been caused by sampling error, by invertebrate predation, or by failure to hatch. In fact, hatching success was poor in experiments. Gravid females ($n = 20$) with about 60 eggs in each of the two egg sacs were collected from the pond and placed in glass tubes. They produced a mean of approximately 40 nauplii from the first brood and about 4 and 10 nauplii, respectively, from the two later broods.

Egg production of the fairyshrimp was calculated (Table 6-2) but because the embryos enter diapause, the hatching rates were not experimentally measured. It is known that animals in the field almost always have fewer eggs than their potential maximum. To measure this maximum, Kangas (1972) confined both species in food-enriched cultures at 5.5°C and found a mean brood size for *Branchinecta* of 22.5 and for *Polyartemiella* of 14.3 eggs per female. In the same year (1969), the pond-grown animals had brood sizes of 6.7 for *Branchinecta* and 9.3 for *Polyartemiella*. In a pond adjacent to Pond C in 1969 and 1973, the brood size was 9.0 and 14.9 for *Branchinecta* and 4.2 and 11.8 for *Polyartemiella*. In other words, it appears that the egg productivity varied from year to year. Kangas also found that there was a 4- or 5-day interval from the start of one brood to the beginning of the next.

In the Barrow ponds, both species of *Daphnia* produce a single brood of young before shifting to the production of resting eggs. This single brood of developing embryos, visible through the translucent carapace of the mother, may be as numerous as 20 in *D. pulex* and 12 to 14 in *D. middendorffiana*. These are field measurements only and are not the maximum brood sizes reported for these species. In contrast to these broods of young, only two resting eggs are produced at a time.

In Pond C, where the only *Daphnia* present is *D. middendorffiana*, there was an increase both in abundance (Figure 6-5) and in the mean brood size (Figure 6-6) from 1971 to 1973. At the same time the mean size of the adults also increased. These data are summarized in Table 6-3 along with the computed coefficients for the equation $Y = aX^b$ where Y is the number of embryos in the brood and X is the length of the adult. The estimated values, based on least squares regression, are 12.0 and 0.01 for a

TABLE 6-3 *Least Squares Linear Regression Coefficients for Brood Size vs. Length of* Daphnia middendorffiana *from Barrow, Alaska*

Year	Pond	Brood size mean ± S.E.		Length (mm) mean ± S.E.		Coeff. of correlation (*r*)	Slope coeff. (*b*)	Y-intercept (*a*)
1973	C	2.86	0.01	4.15	0.2	0.97	13.6	0.00
1977	C	2.53	0.02	1.02	0.1	*	-	-
	A	2.66	0.01	1.53	0.1	0.98	9.1	0.03
	B	2.57	0.03	1.54	0.1	0.94	12.4	0.01
	D	2.88	0.02	4.85	0.1	0.94	10.7	0.01

*Only two non-zero brood sizes were observed.

and *b*. In comparison, the exponent, *b*, for *D. pulex* in the trough ponds was roughly 5.0. Thus, despite the variability from year to year, the increase in number of each brood for an increase of 1 mm body length was lower in *D. pulex* than in *D. middendorffiana*.

During the 3 years of study in Pond C, the biomass of fairyshrimps decreased steadily while that of *Daphnia middendorffiana* peaked in 1973 (Table 6-4). The increase in brood size of the *Daphnia* suggests that food supply for the *Daphnia* had increased, a deduction supported by a number of previous studies (e.g., those on *D. magna* by Green (1954, 1956)). We tested this with young *Daphnia* cultured in pond water with extra sestonic particles added from the same pond. The particles were concentrated from the pond water with an Amicon macromolecular filter. Seven *Daphnia* were placed in 750 ml of water and fresh particle suspensions were added daily for 11 days. At the end of the experiment, control flasks with no added food contained animals with an average of 1.0 embryos per brood

TABLE 6-4 Daphnia *Brood Size and Biomass of* Daphnia *Contrasted with the Biomass for the Other Crustacea in Pond D*

Year	Sample dates	Mean brood size of *Daphnia*	Biomass *Daphnia* only μg C liter^{-1}	Biomass *Daphnia* + fairyshrimps +*Diaptomus bacillifer* μg C liter^{-1}	Biomass Fairyshrimps μg C liter^{-1}
1971	7/5, 7/12	1.18	20.6	83.2	54.4
1972	7/3, 7/13	3.50	264.0	312.9	3.2
1973	7/16	4.15	73.5	99.8	0

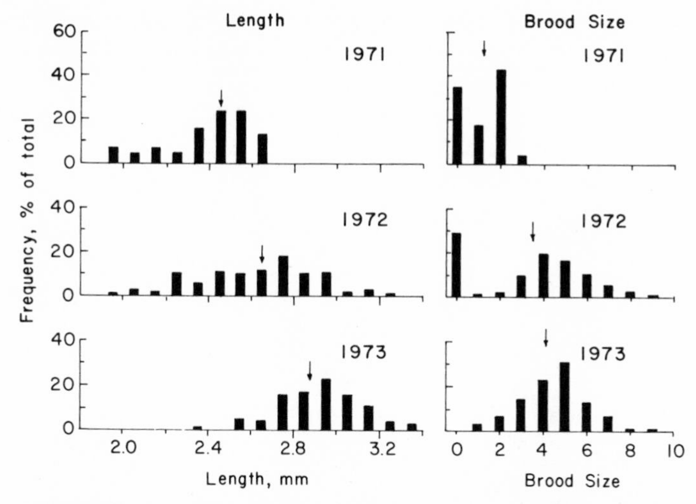

FIGURE 6-6. *Size (length) frequency distributions of the overwintering generation of* D. middendorffiana *while gravid with the first brood. Frequency distributions are also shown for each year. Means indicated by arrows.*

while animals grown at 3 times the natural food supply had 3.2 embryos per brood. Above this food concentration, the carapaces and appendages of the *Daphnia* became covered with peritrich protozoans which interfered with movements. As a result, at 5 times the control concentrations of food, the brood size decreased to 1.6. Thus, food concentration did affect the brood size.

It was less easy to interpret data on *D. pulex* from a trough pond (Near Ditch). In 1972, a dry summer, the pond volume decreased greatly with a resulting increase in numbers of animals per liter. *Daphnia* which had hatched from the overwintering embryos were 1.4 to 2.8 mm long at maturity and had small broods (less than 11). In contrast, the animals were sparse the next summer but were larger (2.4 to 2.8 mm) and had larger broods (9 to 18). Despite the large differences in length and brood size, the constants for the equation describing the relationship between length and brood size remained the same.

Biomass

The carbon content of various life stages and sizes of the important species was measured so that numbers could be converted to carbon. Early in the study the carbon content for individuals or groups was measured with a dichromate oxidation (Strickland and Parsons 1968). Later, the animals were combusted and the CO_2 measured with an IR gas analyzer.

TABLE 6-5 *Mean Weights of Eggs and of the Maximum Size of Adults, and Specific Growth Rates (g day^{-1}) for Crustacea in the Plankton of an Arctic Pond* [a]

	Egg mass µg C calc.	Egg mass µg C obs.	Growth rate *g*	Maximum adult mass calc.	Maximum adult mass obs.	Growth period
Daphnia middendorffiana	[b]5.6	8.5	0.10	45	41	6/24 to 7/15
	[c]5.6	3.7	0.10			7/15 to 8/6
Polyartemiella	5.9	3.7	0.12	402	416	6/15 to 7/19
Branchinecta	6.1	9.7	0.12	574	307	6/15 to 7/19
Heterocope	2.8	2.0	0.09	43	84	6/15 to 7/19
Diaptomus glacialis	2.0	0.5	0.12	30	25	6/15 to 7/19
D. bacillifer	0.16	0.20	0.10	4.8	11	6/15 to 7/19
Cyclops strenuus		0.20	0.21	13.3	16.2	6/15 to 7/5
C. vernalis		0.20	0.20	10.9	13.3	6/15 to 7/5
C. magnuus		0.40	0.20	21.8	25.0	6/15 to 7/5

a. Calculated refers to the value from regressions based on measured carbon content; observed values are means of field data.
b. Animals hatched from overwintering embryos.
c. Young-of-the-year animals.

A mean of 3.7 µg C per egg was determined for the non-diapausing *D. middendorffiana*; this was similar to the weight of the diapausing eggs. There was no weight difference between the two kinds of eggs of *D. magna* (Green 1956). The mean size of the *Daphnia* was 41 µg C (range was 33 to 70 µg). For our calculations, we assumed that the overwintering generation of *Daphnia* stopped growing in mid-July when the biomass seems to approach an equilibrium (Winberg 1971) (Figure 6-5), even though this species is indeterminate in growth. The fairyshrimp *Polyartemiella* averaged 416 µg C per animal and *Branchinecta* 307 µg C.

Growth Rates

Coefficients of growth and maximum size attained by each of the major species were calculated from regressions. For convenience, the growth of all species of Crustacea was assumed to have both a predictable interval and an upper limit (Table 6-5). Daily rates of instantaneous growth were estimated to range from 0.09 for *Heterocope* to 0.20 for the cyclopoid copepods.

Mortality

Estimates of mortality were made from data on population size over time. The largest loss in 1971 was of the *Cyclops* embryos or nauplii. As already mentioned, only 35 out of 127 eggs produced per liter gave rise to nauplii, and the first hatch was much more successful than later hatches.

Mortality rates of fairyshrimps and *Daphnia* were high in 1971. In fact, the loss of young fairyshrimp was so high early in the season that it resembled an exponential function (Figure 6-4). The loss of *Daphnia* was always largest in the young-of-the-year generation where 85 to 90% of the newborn did not survive. Much of this loss may be the result of predation by *Heterocope*, the large copepod.

We conclude that most of the loss from the populations occurred in the young zooplankton stages. The adults all die at or before freezeup but the exact time is unknown. The percentage survival of the overwintering embryos is also unknown.

Predation

There are no fish and the feeding of phalaropes is not important; therefore, most of the predation on Crustacea is by other Crustacea. A number of experiments were carried out by placing various numbers of prey in small jars along with a fixed number of predators. The jars were incubated on the bottom of a pond for a number of days. The data give possible relationships among these animals but can not give quantitative relationships because the conditions are artificial. For example, Strickler and Bal (1973) have shown that some copepods find their prey by following turbulent wakes. Placing the animals inside a bottle may drastically change the natural turbulence and may well give predators an advantage. A measure of the possible intensity of predation is the coefficient, K (liters time^{-1}), calculated from

$$\text{Prey remaining} = (\text{Prey available at } T_o)e^{-KTX}$$

where T is time and X is number of predators liter^{-1}.

From 16 to 21 July 1972, young *Daphnia* were eaten by *Heterocope* (five stage V copepodids or adults liter^{-1}) with K of 0.69 liter—this is a daily K of 0.14. Adult *Daphnia* were killed at a daily K of 0.09 liter. From 23 to 31 July, both young *Daphnia* (slightly larger now) and fairyshrimp were killed at a daily K of 0.09.

Experiments by Dodson (1975) in 1973 tested more of the possible feeding relationships within the food web (Figure 6-7). These experiments were run in 220 ml of filtered (93-μm mesh net) pond water for 48 hours. Data for all copepod nauplii were combined so the nauplii of *Cyclops vernalis* also includes those of *C. strenuus, C. languidoides,* and *Diaptomus alaskaensis*.

The value of K for the interactions (Table 6-6) confirms the voracity of *Heterocope* for *Daphnia* young and indicates that they feed equally well on *Cyclops* nauplii. Adult *Cyclops vernalis* had a high K for their own

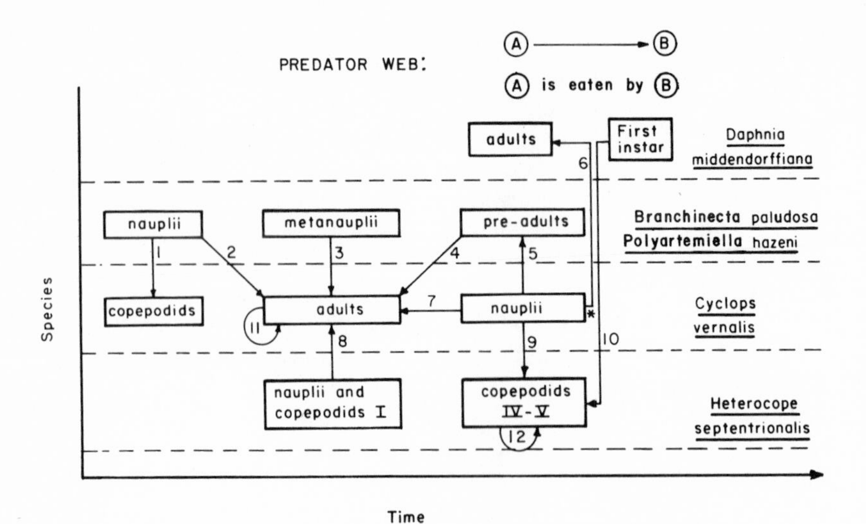

FIGURE 6-7. *Predatory web of the planktonic Crustacea in Pond C as constructed by Dodson (1975).*

nauplii and for the young *Heterocope*. All stages of the fairyshrimps were eaten by the *Cyclops*, which seems to be a very versatile predator. We do not know whether *Heterocope* prefers *Daphnia* young or *Cyclops* nauplii. The nauplii have become rare by the time the young *Daphnia* hatch and appear in the last 2 weeks of July.

Some species interactions gave very low K values. For example, both the fairyshrimp and *Daphnia* accidentally trapped *Cyclops* nauplii (interactions 5 and 6). Others, such as *C. strenuus* and *Diaptomus alaskaensis*, were not predatory.

These experiments indicate the potential importance of predation in controlling zooplankton production and perhaps even species composition. From these experiments and from our observations on the *Heterocope*-*Daphnia* interaction, it appears that *Heterocope*'s success could depend upon abundance of the young-of-the-year *Daphnia* in the July hatch.

Production

Estimation of production by the Crustacea is simplified in the Barrow ponds because there is, with one exception, but a single generation and because reproduction is restricted to specific periods. Here, we define production to be the sum of biomass of eggs and individuals surviving at the end of the summer and the biomass of the same lost throughout the summer. In general form, production (μg C liter^{-1}) is

$$P=\Sigma\left\{(N_t-N_{t+1})(M_t e^{(gi/2)})\right\}+N_T M_T+N_E M_E \qquad (1)$$

TABLE 6-6 *Size of Predator and Prey Plus the Predation Coefficient for the Interactions Shown in Figure 6-5*

Date	Predator length (mm)	Prey length (mm)	K value[1] (220 ml)	Interaction[2]
June 20	1.08 - 1.17	1.23	0.0040[3]	1
June 20	1.50	1.23	.0145	2
June 26	1.50	1.67 - 3.42	.0101	3
July 3	1.50	2.53 - 4.16	.0051	4
July 6	5.33 - 6.66	0.28 - 0.80	.0225	5
July 12	2.32 - 2.92	.32 - 1.08	.0205	6
June 30 - July 3	1.50	.28 - .68	.100	7
June 26 - 30	1.50	.66 - .80	.0722	8
July 4 - 12	2.20 - 2.66	.28 - 1.08	.250	9
July 12	2.32 - 2.92	.32 - 1.08	.236	10
	1.50	1.50	.0026	11
July 13	2.25 - 2.66	2.25 - 2.66	.0144	12

1. K = predation coefficient, liters day^{-1}.
2. Interactions are those shown in Figure 6-5.
3. An underestimate, because at least half the copepodites were the non-predaceous *C. strenuus.*

Source: Modified from Dodson (1975).

where N is numbers liter^{-1}, M is μg C per individual or egg, g is the instantaneous growth rate, i is interval (days), T refers to survivors and E to eggs produced.

In the above formula, $(N_t - N_{t+1})$ is the mortality over a given time period. Multiplication of mortality by the geometric mean of the biomass (M_t) and by $e^{(gi/2)}$ estimates the production of the animals lost by mortality during that interval. The summation of production for each of the intervals estimates total production of all animals not surviving to the end of the summer. The production of animals that did survive to adulthood is $N_T M_T$ and the production of eggs is $N_E M_E$.

The above formula holds for monovoltine fairyshrimps and calanoid copepods that diapause as embryos. For *Daphnia* and cyclopoid copepods, subscripts are used for overwintering (A) and young-of-the-year (J) generations.

The cyclopoid copepods are monovoltine but enter diapause as near adults. Thus, the J category shows much of the production. Also, the E subscript now indicates the copepod stage.

The total production of zooplankton in Pond C was approximately 1 mg C liter^{-1} yr^{-1} (Table 6-7) or 200 mg C m^{-2} assuming an average depth of 20 cm. Since sampling errors are unknown, and because the range of estimates is not large (820 to 1400 μg C liter^{-1} yr^{-1}), we assume that total zooplankton production was about the same each year. Estimates for other ponds are given in Figure 6-8.

TABLE 6-7 *Annual Production of Crustacea in Pond C*[1,2]

	1971	1972	1973
Cyclops vernalis	80.4	20.4	76.1
Diaptomus bacillifer	159.2	10.3	157.7
D. glacialis	122.0	180.2	9.5
Heterocope	21.8	32.9	24.0
Fairyshrimps[3]	811.6	122.1	106.2
Daphnia	203.1	821.2	447.1
Total	1398.1	1187.1	820.6

[1] μg C liter^{-1}.
[2] Except for *Daphnia* all annual rates are exclusive of the production of diapausing embryos. Since the initial stock each year is not subtracted, each estimate in effect assumes 100% survival of the overwintering individuals.
[3] 95% or more *Polyartemiella*.

However, in Pond C the distribution of this production did vary between the groups; the fairyshrimps had the highest production in 1971 while *Daphnia* were the highest in 1972 and 1973. The very high value for *Daphnia* in 1972 is real but somewhat misleading as the young and adults were concentrated by the low water levels of late July. There may also have been an unusual second brood of young in 1972 as a result of minimum July water temperatures of 10°C or higher during a critical interval. On the other hand, the *Cyclops* and fairyshrimp that have most of their production early in the year, when water levels were high, had their apparent production (on a per liter basis) lowered as a result of the greater pond volume at that time.

Daphnia Feeding, Growth, and Reproduction

Filtering rates were determined with the short-term feeding method (Nauwerck 1959) using ^{14}C-labeled *Chlamydomonas reinhardtii* (see

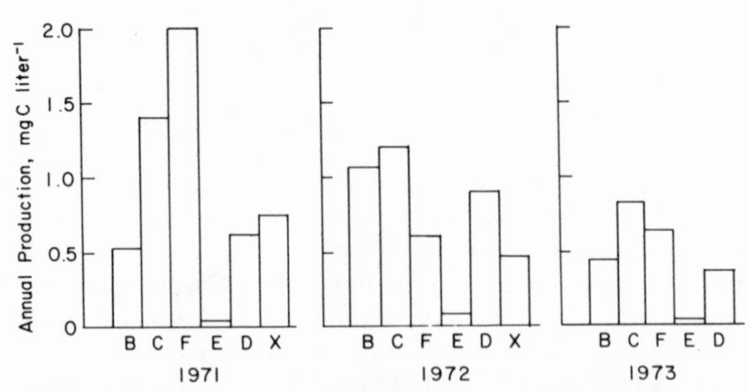

FIGURE 6-8. *Annual production of the planktonic Crustacea in ponds for the years 1971, 1972 and 1973.*

Chisholm et al. 1975, for details). Incubation times were kept short so that there was no loss of label from defecation or excretion. The labeled algae and *Daphnia* were added to pond water which had been filtered through a 64-μm mesh net. Final numbers of particles in the incubation medium were determined with a Coulter Model B Electronic Cell Counter. The amount of radioactivity both added and taken up was measured in a liquid scintillation counter after digestion of the organisms with Protosol and maceration with a tissue homogenizer.

In *D. middendorffiana*, the relationships of the feeding rate to food concentration, to body length, and to temperature showed generally the same pattern found for other *Daphnia* species. Maximum filtering rate for all size classes of *D. middendorffiana* at Barrow was observed at 12°C. In other species of *Daphnia*, the maximum filtering rate was reported at higher temperatures (McMahon and Rigler 1965, Burns and Rigler 1967).

Before the experiments, the freshly caught animals were held overnight in constant dim light. Aside from this, no attempt was made to acclimate the animals because the test temperatures (roughly 5°, 10°, 15°C) were within the range of the natural temperature changes on clear summer days. Yet acclimation may be important for animals grown at a constant temperature because one species, *D. rosea*, has a temperature optimum for filtering rate that was entirely a function of the temperature (12° to 20°C) at which the animals were grown (Kibby 1971).

The filtering rate of *D. middendorffiana* is slightly higher than rates for other species of *Daphnia*. The relationship of this rate, F (ml per animal hr^{-1}), to body length L (mm), can be expressed as a power

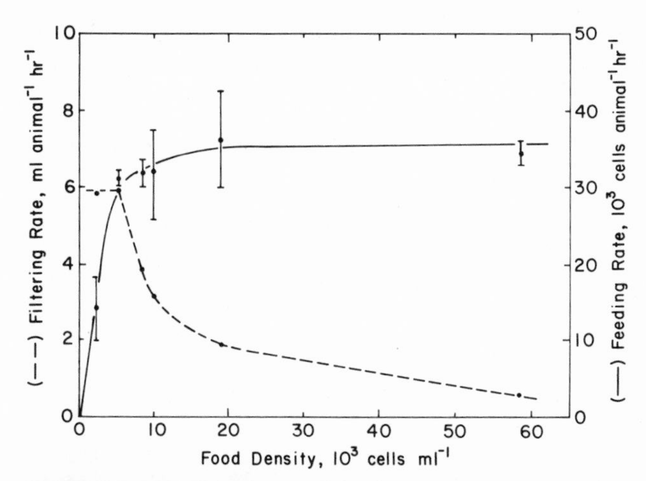

FIGURE 6-9. *Filtering and feeding rates for individuals of* Daphnia middendorffiana *as determined by feeding natural suspensions of particulates containing labeled cells of* Chlamydomonas reinhardtii.

TABLE 6-8 *Estimated Rates of Filtering, Ingestion, Assimilation, Growth, and Reproduction for* Daphnia middendorffiana *in a typical Tundra Pond at Barrow, Alaska on a Clear Day in Mid-July*

Time of Day		0600	1300	1800	2400	Daily Mean
Temp (°C)		5	11	15	11	10
Activity Coeff.		1.0	2.0	1.0	2.0	1.5
Filtering Rate (ml hr^{-1})						
Length (mm)	1.5	.37	1.46	0.67	1.46	0.99
	2.0	.50	3.41	1.48	3.41	2.20
	2.5	.63	6.58	2.74	6.58	4.13
	3.0	.76	11.27	4.52	11.27	6.95

Mean Daily Ingestion Rate (μg C hr^{-1})					
Length (mm)		1.5	2.0	2.5	3.0
Food	50 μg C $liter^{-1}$	0.05	.11	.21	.35
	60 μg C $liter^{-1}$	.06	.13	.25	.42
	75 μg C $liter^{-1}$	.07	.16	.31	.52
	100 μg C $liter^{-1}$	.10	.22	.41	.70
Carbon Content $(ind^{a})^{-1}$ (μg C)		7.56	17.92	35.0	60.5
Respiratory Rate (crude) 10°C mean (μg C hr^{-1})		.03	.06	.11	.17

Assimilation (μg C hr^{-1})[b]					
Length (mm)		1.5	2.0	2.5	3.0
Food	50 μg C $liter^{-1}$	.035	.077	.147	.245
	60 μg C $liter^{-1}$	.042	.091	.175	.294
	75 μg C $liter^{-1}$	.049	.112	.217	.364
	100 μg C $liter^{-1}$	.070	.154	.287	.490

Growth Rate[c]					
Length (mm)		1.5	2.0	2.5	3.0
	50 μg C $liter^{-1}$	.016	.023	.025	.029
	60 μg C $liter^{-1}$	.037	.067	.044	.048
	75 μg C $liter^{-1}$	.059	.067	.071	.074
	100 μg C $liter^{-1}$	.120	.120	.114	.120

Eggs day^{-1} [d]					
Egg Size		2.94	2.94	3.73	4.5
Food	50 μg C $liter^{-1}$	.04	.14	.23	0.40
	60μg C $liter^{-1}$	.06	.25	.42	.66
	75 μg C $liter^{-1}$	.16	.42	.69	1.03
	100 μg C $liter^{-1}$	.32	.76	1.14	1.71

a. μg C ind = 2.24 (length, mm)^3
b. efficiency = 0.7
c. biomass, $m = M_o \exp(gt)$, where M_o is egg mass
d. eggs day^{-1} = (assimilate day^{-1}) $(0.2 + 157.2\,(0.0079 \text{ length (mm)})^{1.14})$

function equation, $F=0.458\ L^{3.17}$ at 11°C and $F=0.458\ L^{1.65}$ at 5°C (Chisholm et al. 1975 and Table 6-8). Because both the coefficient and the exponents are higher than those for other species, an individual *D. middendorffiana* will filter more water than individuals of other species of the same size.

The feeding rates increased hyperbolically with increasing food concentration (Figure 6-9) and reached saturation at a food concentration of 37,000 particles ml^{-1} and a filtration rate of 8.0 ml hr^{-1} (for a 2.6-mm animal). This saturated feeding rate of the adults decreased considerably over the summer; at 11°C on 5 July the rate was 265,000±76,000. Since these rates were measured with the ^{14}C-labeled *Chlamydomonas* plus the natural assemblages of particles, and thus included a wide range of sizes, the saturation values can not be compared with other measurements where monospecific suspensions were used.

The laboratory feeding measurements were carried out with the natural assemblage of particles less than 64 μm still present. This is closer to natural conditions than most other experiments where pure cultures of algae are used, but still leaves the problem of how to extrapolate the laboratory data back to field conditions. One approach is to measure the amount of the particulate organic carbon (POC) both in the pond and in the experiments. We assumed that the 10,000 particles ml^{-1} measured in the experiments were spheres with an average diameter of 10 μm. From the usual assumptions about phytoplankton density and carbon content, we calculated that 10,000 particles contained 0.2 μg C. In the field, the POC ranged from 0.2 to 0.8 μg C ml^{-1}, so it is reasonable that the *Daphnia* in the pond were always feeding at or near their maximum rate (Figure 6-9). Additional evidence that the experimentally measured filtering rates simulate nature comes from measurements by J. Haney (personal communication) using ^{32}P-labeled yeast cells (3 μm) added to unfiltered pond water. The results agreed with the estimates given above and in Table 6-8.

If we accept the filtering rates above as close to natural rates, then the effect of food density, temperature, and length of animal on growth and reproduction could be calculated if the assimilation efficiency were known. Literature estimates of assimilation efficiency vary widely and we made no independent measurements. Our arbitrary assumption of 70% efficiency is useful for illustrating the various interactions, but the results of the calculations (Table 6-8) have only comparative value. In the table, the ingestion rate per individual equaled $2.24\ L^3$ where L is length (mm). The respiration rate was a mean of our measurements in a Gilson Respirometer (4.0 to 8.0 μl O_2 (mg C)$^{-1}$ hr^{-1} at 10°C). These rates agree in general with those measured by other investigators although our precision was not as good as hoped for.

From these crude calculations (Table 6-8), we estimate that the smallest animal (1.5 mm) can barely grow at a food density of 50 μg C $liter^{-1}$. At 60 μg C the growth rate is doubled but still is very low. At 100 μg C $liter^{-1}$ the individual may grow at a rate comparable to that actually observed at Barrow (Table 6-5).

Egg production is also related to food concentration and to size of the individual in much the same way as growth is related. When food is scarce, only the larger individuals can accumulate sufficient energy to reproduce within the time allotted. In *Daphnia* populations of temperate waters, the time between broods is about 6 days at 10°C. In this arctic population, 6 days would allow the smallest individual living on 50 μg C liter^{-1} to eat enough to produce about one-fourth of an egg. In contrast, the largest animal (3.0 mm) might produce 10 eggs on the same food level. It must be remembered that the Barrow *Daphnia* have but a single brood of young.

One indication of the inaccuracy of the calculations in Table 6-8 is that the reproductive potential of a 3.0-mm *Daphnia* is estimated to be nearly 2 times that of a 2.5-mm *Daphnia* rather than the 4 to 5 times found in the field. A more detailed model has been constructed which more closely approximates the field conditions (Stross et al. 1979).

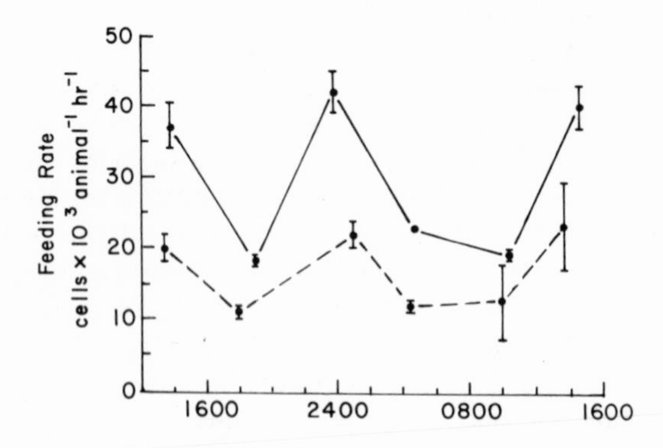

a) Feeding rates at 11°C on two different days.

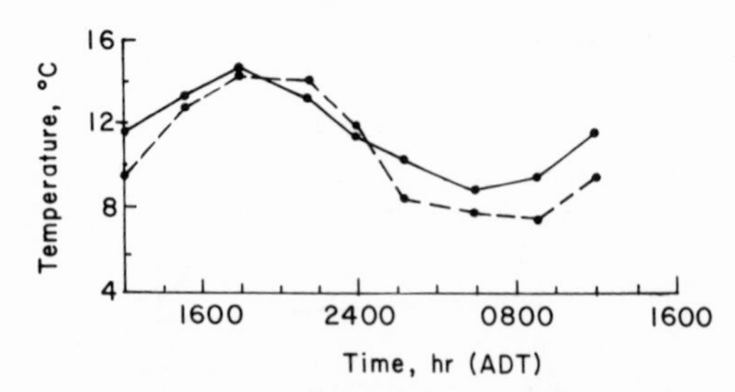

b) Temperature oscillations in the pond on day preceding removal of the Daphnia *to the laboratory.*

FIGURE 6-10. *Endogenous oscillations in feeding activity of* D. middendorffiana *in a constant environment. (After Chisholm et al. 1975.)*

Coincidentally, a particulate concentration of 100 μg C liter^{-1}, which is the saturation level for filtering, is near the actual amount of living biomass in the ponds. However, there is also a much larger component of detrital-like particles in suspension which dilutes the living material. The quantity of detritus ranges from 250 to 800 μg C liter^{-1}. Although no actual calculations have been made, the effect of dilution can be accounted for in the calculation by lowering the assimilation efficiency. However, if the assimilation efficiency were reduced from 70% to 14%, then 500 μg C liter^{-1} would be needed to produce the same energy for the animal as 100 μg C liter^{-1} of living biomass. We could not go further with the question of how zooplankton handle low quality detritus; it is an unknown area in our understanding of this or any other system.

When *Daphnia* were held under conditions of constant light and temperature, they had peaks of feeding activity at 1400 and 2400 (Chisholm et al. 1975 and Figure 6-10a). It may be significant that these are the same times that the water temperature of the pond, measured on a clear day (Figure 6-10b), passes through the temperature optimum for the feeding of this species. These peaks could be caused by oscillations in metabolic activity similar to those found by Ringleberg and Servass (1971) for the phototactic response in *Daphnia*. Daily rhythms are also known for arctic algae (Müller-Haeckel 1970), invertebrates (Mendl and Müller 1970) and fishes (Müller 1970) and double activity peaks have been found in copepods (Spindler 1971a, 1971b) for egg laying and molting. There is also the possibility, suggested by Remmert (1969), that two discrete populations are present, with one active only in the daylight and one active at night. It is difficult to assess the importance of the daily rhythms at Barrow because clear days are so rare (as described in Chapter 2, cloud cover averages 80 to 90% during July and August).

Synchrony

The Crustacea in each of these ponds exhibit a certain degree of seasonal synchrony in their major life cycle events such as hatching and molting. This was measured in a 1971 experiment in which overwintering embryos (ephippia), collected from a trough pool (*D. pulex*) and from Pond B (*D. middendorffiana*) on 15 June, were incubated in glass tubes in water from their own pond under natural light and temperatures. A parallel set was also incubated under constant light and a constant temperature of 10°C. All tubes were inspected six times each day during the hatch and the young removed.

A seasonal synchrony for each species was certainly evident but a daily synchrony of hatching was not demonstrated (Table 6-9). The best example is for the *D. pulex* embryos incubated under natural conditions, whose hatch was 75% completed within the first 24 hours of the total

TABLE 6-9 *Hatching Times of* Daphnia *in Tundra Ponds**

		D. pulex				*D. middendorffiana*	
	Hour of	Rack 1		Rack 2		Rack 1 (only)	
Date	examination	#	Cum %	#	Cum %	#	Cum %
6/20	0400	0		0		0	
	0800	0		0		2	
	1200	0		0		2	
	1600	12		3		5	
	2000	19		8		11	
	2400	12	71.7	4	57.9	11	38.7
6/21	0400	1		2		4	
	0800	3		1		7	
	1200	0		1		6	
	1600	6		5		6	
	2000	4		1		7	
	2400	0	95.0	0	96.1	4	81.0
6/22	0400	-		-		-	
	0800	-		-		-	
	1200	3		0		14	
	1600	0		0		0	
	2000	0		0		0	
	2400	0	100.0	0	100.0	0	99.0
6/23-26		0		0		10	
Total hatch		60		26		81	
% of Viable embryos		80.0%		67.5%		89.9%	
% of Egg cases		23.0%		18.7%		42.1%	

*Egg pods of overwintering embryos of *D. pulex* and *D. middendorffiana* were collected on 15 June 1971 and were incubated in glass tubes (25 x 200 mm), five pods per tube. Racks of tubes were held at ambient temperature and natural light levels of Near Ditch.

hatching interval and 95% complete in the first 36 hours. Under the same conditions, the *D. middendorffiana* hatch was more spread out with about 50% completion in the first 24 hours and near 100% by 72 hours. The daily patterns of hatching were also different for the two species. Thus, *D. pulex* hatching took place mostly in the warm half of the first day (the pond water reached maximum temperature at 1800 hours) and the pattern repeated on the second day. In contrast, *D. middendorffiana* emerged at approximately a constant rate over the entire 24 hours. The same pattern for *D. middendorffiana* was seen in the embryos held under constant light and temperature.

Later in the same summer, there was a similar synchrony in the time of appearance of non-diapausing embryos in the brood pouches of the females. Hatching of the embryos was likewise synchronous (Figure 6-11). For example, hatching was mostly complete within 48 hours in Pond B and

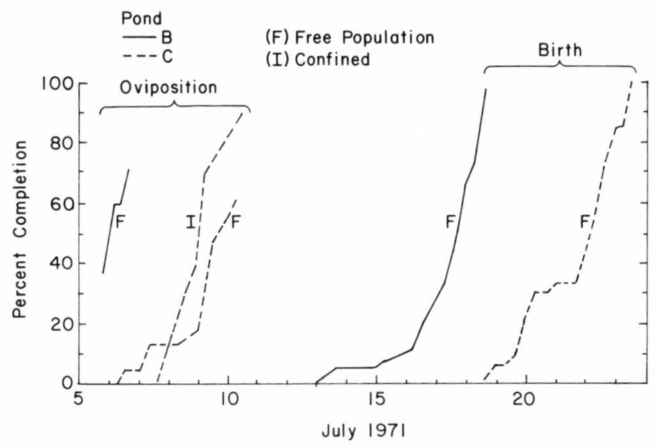

FIGURE 6-11. *Percent completion of release of eggs into brood pouches (oviposition) and of birth of young* Daphnia middendorffiana *in Ponds B and C, July 1971.*

70% complete in 48 hours in Pond C *Daphnia*. Thus synchrony in the population appeared to be retained at least through the first adult instar.

Synchrony was not equally sharp in all years of observation, however. In 1973 a sudden cool period occurred after the females had begun to oviposit. The low temperatures slowed the oviposition process and resulted in a spreading out in time of both the ovipositioning of eggs and later hatching of the young.

The hypothesis that molting was under the control of an endogenous circadian rhythm was also tested. Measurements were carried out in the field in chambers in which day length and temperatures were controlled. Sample populations of *Daphnia* were observed at intervals of a few hours (1971) or were continuously monitored in the ZEHMAC (1972, 1973). This apparatus was a bottomless cage of *Daphnia* that revolved on a circular track once every 24 hours. Cast exuviae, dead animals and egg pods (ephippia) fell to the track surface. The cage, equipped with a roller that fitted into the rear wall, pressed them against the bottom as it passed.

There was a distinct rhythmicity in molting. Under natural conditions a peak in molting intensity during the last 8 hours of a day was followed by a second peak during the first 8 hours of the day following. This bimodality seemed similar to that observed for oscillations in feeding intensity (see above). Only the experimental conditions of constant light and constant temperature or constant light and natural temperature gave a response similar to the results under natural conditions. A treatment of constant temperature plus natural light resulted in a near-uniform spread of molting.

In 1973, an interesting synchrony occurred in the timing of the molt to the first adult instar of *Daphnia*. In one of the two monitors operating under natural conditions, nearly half of the animals molted within 1 hour around 0100. The second monitor also showed a similar burst of molting but it occurred 12 hours later (1300).

CONTROL OF ZOOPLANKTON PRODUCTION (I)*

General Considerations

Populations of Crustacea in arctic pond ecosystems exemplify the antithesis of the steady state. Each population spends only a small part of the year in the active phase of the life cycle and the remainder of the year in diapause. The annual cycles are mostly monovoltine. The switch from diapause to active phase each spring is nearly synchronous. Stable age distributions, demanded of populations in the steady state, are therefore not possible. The populations may be viewed as batch cultures (Smith 1952, 1963) in which endogenous constraints of the genotype rather than resources prevent attainment of the steady state.

An analysis of production and its control was carried out under the principles set forth by Slobodkin (1959, 1960). The biotope is viewed through the eyes of the dominant planktonic consumer, *Daphnia middendorffiana*. The questions seemingly most pertinent to both the answers available and the uniqueness of the arctic community are: (1) How are *Daphnia* populations able to dominate the assemblage? (2) What determines the number of generations each year? (3) What controls reproductive effort of each generation? (4) What, if anything, prevents the maximum yield from each generation?

Polygon ponds near Barrow and elsewhere in arctic Alaska are populated with Crustacea that feed by filtering particulates from the water. The large size of both the dominant species, *D. middendorffiana*, and the fairyshrimps, in combination with the absence of vertebrate predators, suggests that large size could be an adjustment in the face of competition (Hrbáček 1962, Brooks and Dodson 1965). An alternative argument presented by Dodson (1972, 1974) attributes the large size to the selective action of invertebrate predators.

A second type of small pond at Barrow, the trough pond, contains *Daphnia pulex*. Fairyshrimps are present but exceedingly rare. *D. pulex* is absent from polygon ponds, presumably as a result of inferior competitive ability or as a result of greater vulnerability to predators. This species may be less able to satisfy its food needs at densities of pond particles found in the polygon ponds. Both the filtering power and its relative growth with body size are significantly smaller in *D. pulex*. Filtering rate F equals aL^b

*R. Stross

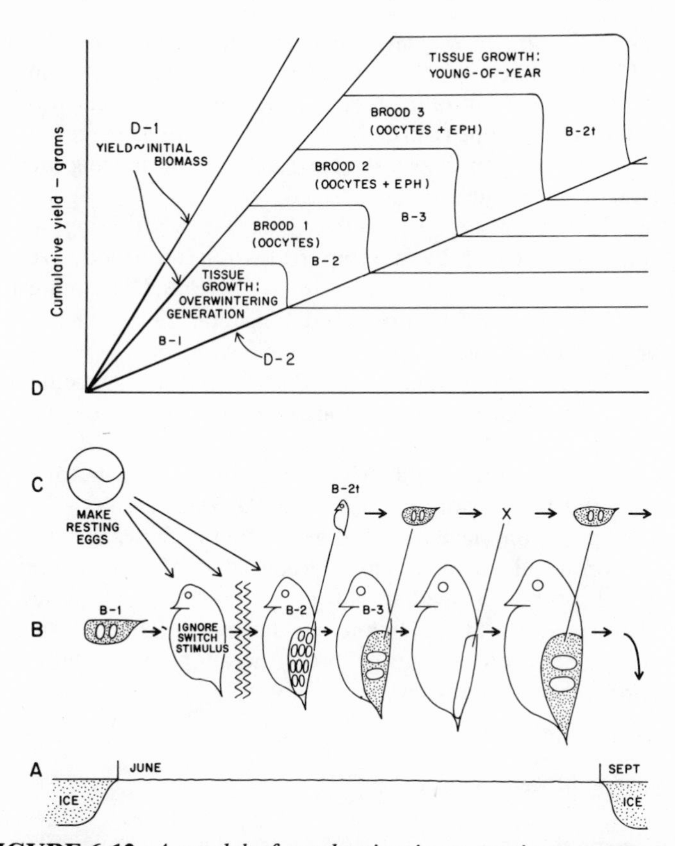

FIGURE 6-12. *A model of production in an arctic consumer and its control. The model attempts to summarize the components of production in a dominant species, such as* Daphnia middendorffiana, *and to illustrate how production may be restricted by a variety of agents within the organism, or population, and by the external environment. A—Time limitation placed on the interval of potential productivity by temperature. The interval is the ice-free interval from mid-June to September. B—The seasonal cycle in* Daphnia, *starting with the hatching of the overwintering embryos (B-1). After one brood of young (B-2), reproduction switches to the eggs that enter a diapause as embryos (B-3). C—The switch in* Daphnia *reproduction which limits the number of young produced to the equivalent of one brood is controlled intrinsically by the environment. D—Cumulative yield based on each source of product, namely the overwintering and the young-of-year generation. Two major forces are at work in determining cumulative yield. In a mortality-free environment (D-1) actual yield seems to be determined by the biomass present. That is, the larger the population the larger the yield. Explanations are provided in the text. In actuality, mortality (D-2) reduces the potential yield. See text for further discussion.*

where L is length in mm. In *D. pulex* the exponent b is between 2.6 and 2.8 (Burns 1969) while in *D. middendorffiana* the exponent is approximately 3.0 at the optimum temperature; this is suggestive of a regression with a steeper slope (Chisholm et al. 1975). Other species, such as *D. rosea*, are also likely to show a steeper regression of filtering rate on body length (Burns and Rigler 1967).

The steep regression for the relative increase of filtering in *D. middendorffiana* is matched by a steep regression of brood size on maternal length. Brood size in *Daphnia* is closely correlated with maternal length (Green 1954, Hall 1964, Stross and Chisholm 1975). There is a faster relative increase of fecundity in *D. middendorffiana*, the species from the more oligotrophic polygon ponds. Both feeding and reproductive adjustments in *D. middendorffiana* are consistent with existence in sparse food environments.

The life cycle of *Daphnia* could also be a significant contributor to its dominance over the fairyshrimp species. The *Daphnia*, unlike all other Crustacea in the plankton, have a second generation each year (Figure 6-12). The single brood of young which is produced by the overwintering generation could permit a feeding advantage over other species of Crustacea. The mechanism is unknown, however. The alternative argument, that invertebrate predation forces the allopatry of these two species of *Daphnia*, has been advanced by Dodson (1975).

Control of Number of Generations

Annual production rate can be directly proportional to the number of generations when rates of birth and survival are constant. An experimental basis for the determination of the control of the number of generations of *Daphnia* at Barrow each year has been provided (Stross 1969). The effect of a restricted number of generations on productivity has already been discussed (Stross and Kangas 1969). A primary question is whether the environment or some intrinsic quality of the species regulates the number of generations in the planktonic Crustacea.

The fairyshrimps are widely known to be restricted to a single generation, with an obligatory arrest appearing in the embryos (Broch 1965). Life cycle control in the copepods at Barrow is uninvestigated, although it is likely that cyclopoid and calanoid species may function differently. The cyclopoids enter diapause in late juvenile (copepodid) stages and do so near the summer solstice while the sun is continuously above the horizon. The species of *Cyclops* in the ponds at Barrow could be long-day inductive if they function as suggested by the work of Spindler (1971a, 1971b).

Whether control is obligatory or facultative, the fairyshrimps and copepods have but a single generation each year. Thus, while the ice-free season is admittedly short the populations fail even to fill that time interval with reproduction. Surely there can be no direct connection between available length of the season at Barrow and productivity in tundra ponds.

In the case of the two species of pond *Daphnia* the number of generations each year is known to be under the influence of environment (Stross 1969). In the laboratory multiple broods of young are produced at constant temperature and long-day photoperiods. In the field the first generation overwinters as embryos (ephippial eggs) (Edmondson 1955, Stross and Kangas 1969). The overwintering generation produces but one brood of young. It then produces the resting or ephippial eggs for the remainder of the season. Two or three broods of resting embryos are likely if the adult survives. The phenological pattern applies to both species of *Daphnia* in Barrow ponds.

In the laboratory the overwintering generation of *D. pulex* was shown to be capable of producing broods of young throughout life (Stross 1969). The controlling stimulus at constant temperature was day length. At a photoperiod of L22:D2, 50% of the broods were diapausing; this is the longest critical photoperiod on record (Stross 1969). The switch in the field was postulated to involve temperature but as a token stimulus that interacted with photoperiod. Later experiments with *D. middendorffiana* with temperature held constant under a field (natural) light regimen have shown that a temperature of 10°C allows successive broods of young to be produced.

In other words, low summer temperatures in ponds at Barrow may not limit production directly. The temperature, particularly the oscillating temperatures, acts in conjunction with photoperiod to regulate the reproductive polymorphism of the dominant species. The presence of sufficient degree-days to allow for more broods of young has been discussed (Stross and Kangas 1969). The question is whether the switching mechanism that determines broods of young vs. broods of resting embryos can be modified to allow more than one brood of young.

The first brood of young *Daphnia* cannot be stopped. It is the result of an internal switching mechanism analogous to that in aphids (Lees 1965). It prevents the environment from triggering the transformation until after one brood of young is released. However, the reproductive pattern in *Daphnia* at Barrow could represent an adaptation. Olofsson (1918) reported populations of *Daphnia* on Spitzbergen (Svalbard) that have one, two, or no broods of young before starting to release diapausing embryos. However, Meijering (personal communication) has found that the annual cycle on Svalbard is the same as that at Barrow.

Reproductive Effort

The realization is that annual production is limited by the density and fecundity of the overwintering generation. The studies described in the previous section show that reproductive effort in both *Daphnia* and the fairyshrimps is resource-limited. Evidence for food limitation in fairyshrimps was provided in measurements of fecundity. Kangas (1972) showed that females of *Branchinecta paludosa* and *Polyartemiella hazeni* in the laboratory released more eggs that did their counterparts in the ponds near Barrow.

An elaborate analysis of growth and reproduction was performed on *D. middendorffiana*. Results from field observation and from experiment and model simulation of the *Daphnia* populations in Barrow ponds clearly show the limiting effect of food on growth and fecundity.

A number of biological aspects of arctic *Daphnia* simplify the analysis of growth and reproduction. First, embryonic development is activated synchronously in spring, and synchrony is retained in the population through much of the summer. Second, there is a single brood or clutch of young carried in a brood pouch. Third, the whole population is female. Fourth, clutch or brood size is proportional to length. A characteristic of the *Daphnia* populations at Barrow is the large amount of size variability, similar to that found in certain aphid populations (Taylor 1975).

Field observations show year-to-year changes in both mean female length and in mean number of embryos per brood. In three summers in one pond (Pond C), mean brood size increased from 1.2 to 4.2 embryos per female. Mean length of the females also increased. However, the regression of brood size on female length showed no significant change. The major increase over the three-year period was in size of the mothers, hence in a larger fecundity per female. A simulation model indicates that an increased food supply or a larger size at hatching allowed the females to grow to a larger size. That is, the larger individuals acquire more food and grow to a larger size, hence produce more young. (In 1977 the length and brood size were once again at the 1971 values in Pond C.)

Support for the function of initial size on adult length was provided by SUNDAY, the simulation model (Stross et al. 1979). A regression of brood size on maternal length was made to simulate the data observed in field populations. In the simulation larger adults and a larger brood were obtainable either by increasing the food concentration or by increasing the initial size of the newborn (a sample output is shown in Figure 6-13).

The sensitivity of the SUNDAY model to environment is illustrated with two sample problems. In the first problem the model was asked to determine the effect of shifting the phase of the endogenous oscillation in metabolic activity. Incorporated into the model was the knowledge that a feeding activity may be maximum twice each day in Pond C and also that maximum activity is twice the minimum (Chisholm et al. 1975). A daily temperature oscillation was employed that corresponds to observed values on clear summer days (5° to 15°C).

The strategy of field populations of *Daphnia* was thereby suggested. As a result of having two daily peaks of activity each day with a maximum

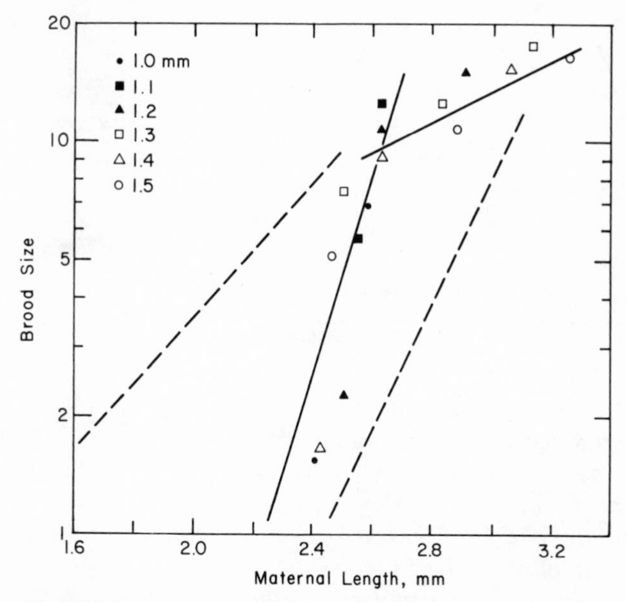

FIGURE 6-13. *Simulation of the number of embryos in the brood pouch of instar 6 as a function of length of the female in instar 5 of a* Daphnia *population (solid line). Simulation is from a* SUNDAY Daphnia *population fed at various concentrations of food and beginning at various initial lengths. Note the biphasic slope which flexes at the length where growth efficiency reaches its maximum. The regressions for two real populations from arctic pools are shown for comparison (dashed line). The left is from* D. pulex, *the right from* D. middendorffiana, *which is also biphasic under experimental conditions. (After Stross et al. 1979.)*

activity near noon and again at midnight, the reproductive rate was doubled. There was only a small increase in length of the female (Figure 6-14). Haney and Buchanan (personal communication) in direct field measurements have shown a daily oscillation in feeding activity that is out of phase with temperature. They found only one maximum each day, although their measurements were in different ponds from those of Chisholm et al. (1975).

A part of the increased reproductive effort was attributed to a faster rate of food ingestion and a slower rate of respiration. The so-called McLaren (1963, 1974) effect was tested in a second modeling exercise. When night temperatures were set at a constant 14° or 15°C and prevented from dropping, growth was much reduced and the individuals failed to

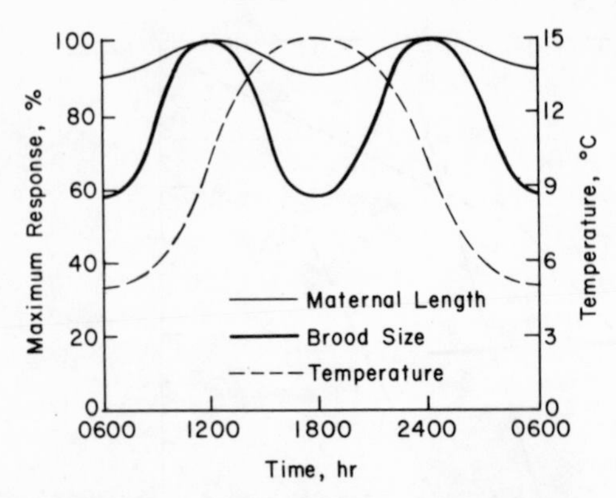

FIGURE 6-14. *Simulation of brood size and maternal length of* Daphnia middendorffiana *by the model* SUNDAY. *Relative length and brood size are shown to depend on the phase of an endogenous rhythm of activity relative to the phase of a daily oscillation in temperature. Length, and especially brood size, are maximum when the phase of peak activity coincides with the time of day when pond temperatures are intermediate to the extremes. At these times the temperature is at the biological optimum for the species, as determined by Chisholm et al. 1975. (After Stross et al. 1979.)*

reach reproductive length. When temperatures were allowed to oscillate, the same food concentrations permitted attainment of large maternal lengths and large brood sizes (Stross et al. 1979).

Available food influences both adult length and brood size in *Daphnia*. They in turn reflect the concentration of food available and the initial length of the individual. Therefore the increased reproductive effort in Pond C over a three-year interval suggests either an increase in food availability or a larger size at birth. Despite the increase, mean brood size was less than saturation as shown experimentally (see below).

Model simulation of reproductive characteristics typical of a real arctic population reveals the rapid increase in brood size with increase in maternal length (Figure 6-13). Three conditions contribute. First, a larger intake of food is accompanied by, second, the more efficient conversion of foodstuffs to growth. The third factor is the larger proportion of the growth fraction that goes into synthesis of gametes. The reader will note that the simulated regression of brood size on maternal length is biphasic. This has been shown in real populations of *D. middendorffiana* at Barrow but not in *D. pulex*.

The enhanced growth efficiency plays an important part in a "regulatory mechanism" of birth rates (Stross and Chisholm 1975). A population with individuals of differing rates of food intake can bring about a disproportionate allocation of food resources. The inequality of food distribution compounded by an increased growth efficiency allows some members of the population to reproduce while others enjoy only a marginal maintenance intake (Stross, et al. 1979). The important point here is that reproductive effort reflects both food availability and the efficiency with which the food is utilized. The concept of an efficient predator with a non-linear efficiency (Slobodkin 1960, 1974) emphasizes the significance of balance between consumer and the concentration of food. The model SUNDAY dramatizes that relationship for *Daphnia* in arctic ponds.

Simulation of *Daphnia* is based on a carbon equivalent. Computed growth efficiency which is the ratio of accumulated biomass to food ingested shows the nearly 3-fold increase that results from the appropriate ratio between consumer density and food concentration (Figure 6-15). For example,at a food density of 3,000 cells ml^{-1} the ratio of yield at the end of the first pregnancy to assimilation since birth is about 3%. At 7,000 cells it has increased to 17% in the model. Figure 6-15 depicts the relationship of efficiency to initial length at various food concentrations.

Another kind of delicate relationship in arctic *Daphnia* is indicated by output from the simulation model. The growth efficiency of an adult

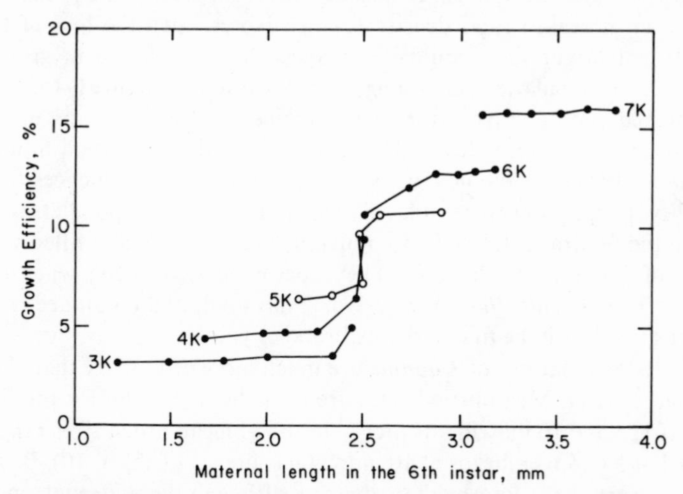

FIGURE 6-15. *Simulated growth efficiency of* Daphnia *populations as a function of length of the female in instar 6. Each point represents a population begun at one initial length. Each line connects a series of six groups of animals which were all fed the same density of food particles (given in thousands of cells ml^{-1} so that 3K equals 3000 cells ml^{-1}).*

female may be related to her initial size. At 6,000 cells ml^{-1} growth efficiency increases from 10 to 16% when the initial (feeding power) length is increased from 1.0 to 1.4 mm. The minimum length for maximum efficiency decreases as food concentration increases (Figure 6-15).

The buffering effect of a differential growth efficiency becomes evident. When density approaches saturation, the population will have only some of its individuals within the critical range of initial length. These individuals will reproduce, while the major portion of the population can gather only sufficient food for the maintenance requirement. In effect, at high density the bulk of the population gives a part of its ration to the few large individuals. These in turn will, with increased growth efficiency, produce gametes. Thus in *Daphnia*, as in species with more elaborate behaviors, a mechanism exists for stabilizing birth rates as the population saturates its food resource. Viewed from outside the population, the large *Daphnia* have a disproportionate effect on food resources for, say, a potential competitor such as the fairyshrimps. Internally, however, the population may be said to distribute its food resources unequally into a size hierarchy that functions to "regulate" birth rate. The large variability noted in the size of *Daphnia* (see previous section) and aphids (Taylor 1975) could represent a selective adjustment, one function of which is to stabilize birth rates at high densities.

Predation that is selective for the larger sizes may prevent such a system from functioning, however. The absence of vertebrates in tundra pools may be a factor in the observed size patterns. If vertebrate predation results in an increased food density for herbivores, then the loss of the more efficient larger sizes could be balanced by an increase in growth efficiency of the population remaining, as is illustrated in Figure 6-15.

Observed reproductive efforts of *Daphnia* and fairyshrimps in the ponds at Barrow were less than expected and infer food-limited populations. To determine how density of the population influences food availability, populations were isolated in small ponds. These pondlets were carved in the tundra turf, lined with polyethylene sheeting, and filled with 10 liters of water from Pond C. The experiment was begun when the *Daphnia* (*D. middendorffiana*) were 1 or 2 days old, and continued until they were gravid with the first and only brood of young.

Isolated populations of *Daphnia* are much more productive than their pond counterparts. Mean brood sizes are near the maximum for the first brood. At 1, 5 and 10 individuals $liter^{-1}$ initially, mean brood sizes ranged from 10.4 at 5 to 5.6 embryos at 10 individuals $liter^{-1}$ (Table 6-10). Brood size was largest at the intermediate density, although the maternal length was similar at both 1 and 5 individuals. Attention is drawn to a large mortality in both pondlets at the intermediate density.

The average number of embryos per brood was 2.5 times larger than those of *Daphnia* in Pond C although the mean maternal length was the same (2.9 mm). Brood size in the isolated population was reduced to that

TABLE 6-10 *Survival, Growth, and Reproductive Effort of* Daphnia middendorffiana *when Grown in 10-liter Volume Ponds**

Density ($liter^{-1}$)		Mean final	Maternal biomass	Mean brood	Total biomass
Initial	Final	length (mm)	(μg C $liter^{-1}$)†	size	(μg C $liter^{-1}$)†
1.0	1.0	2.89	62.4	7.9	98.8
5.0	2.2	2.88	117.7	10.4	209.2
10.0	9.6	2.72	413.9	5.6	579.0

*The animals were isolated from Pond C within 48 hours of birth on 28 June 1973 and cultured until 25 July when the females were gravid with first brood.
†Carbon content was estimated from the regression $C = 2.24\ L^{3.0}$ where "L" is length in mm; each egg was assumed to contain 4.0 μg C although this is to some degree a function of maternal size.

TABLE 6-11 *The Influence of Various Densities of Fairyshrimps on Growth and Reproductive Effort of* D. middendorffiana*

Daphnia				Fairyshrimps			
Final density	Length (mm)	Eggs per brood	Biomass (μg C $liter^{-1}$)	Density ($liter^{-1}$) Initial	Final	Biomass (μg C $liter^{-1}$)	Total biomass (μg C $liter^{-1}$)
4.0	2.74	7.2	184	0.5	0.1	35	335
5.0	2.79	8.1	243	2.0	1.2	420	825
4.7	2.67	5.2	200	5.0	3.9	1365	1663
3.5	2.66	4.3	148	10.0	7.7	2695	2903

*Initially each pond contained 5.0 *Daphnia* $liter^{-1}$ (see Table **6-10** for added details). Final biomass of fairyshrimps, primarily *Polyartemiella*, assumed to be 350.0 μg C.

of broods in Pond C, but only when the initial density was 10 *Daphnia* liter^{-1}. At the same time total biomass was 579 μg C liter^{-1} in the isolates and only 267 μg C liter^{-1} in Pond C.

The upper limit of brood size for the first brood of *D. middendorffiana* is roughly 10, much smaller than the limit of 20 or so for *D. pulex*. Brood size in the natural populations with a mean of 4.5 was therefore never at the reproductive limit. Although indirect, this is evidence for food limitation *in situ*. Similar evidence comes from feeding of fairyshrimps. Kangas (1972) showed that both species of fairyshrimps were capable of a larger fecundity than was attained by the females in Ponds B and C.

Now, if the brood size is a valid indicator of food limitation, the mean brood size of *Daphnia* shows that a limitation was present in early July and the brood size of fairyshrimps shows that food limitation was also present later in July when these animals produce their eggs.

Another set of experiments in the tundra pondlets indicated that more food was available in the pondlets than in the control ponds. In these experiments, 5 *Daphnia* liter^{-1} were grown with either 0.5, 2, 5 or 10 fairyshrimp liter^{-1}. The results were similar to those from the previous experiment in that brood size increased at first to 8.1 as the number of fairyshrimps were increased but it then decreased to 4.3 at the maximum density of fairyshrimps (Table 6-11). *Daphnia* biomass showed the same pattern but the changes can be accounted for by the lower numbers of surviving *Daphnia* at the high fairyshrimp densities. Again, the conclusion must be that Pond C water in isolated pondlets can support a larger biomass than is observed in the ponds.

Model, observation, and experiment all show that reproductive effort in Pond C may be food-limited. Production rate for the principal generation is therefore food-limited. The "nested" pattern begins to emerge. First of all, annual production in a body of water that is frozen for 9 months each year must indeed be temperature-limited. However, the populations that are responsible for the synthesis of tissues have apparently undergone an adjustment. Their life cycles are restricted to essentially a single generation. Thus the number of generations, not temperature, exerts direct control on annual productivity. Restriction to a single generation is probably controlled intrinsically in the fairyshrimps and is controlled environmentally in the *Daphnia* populations. In the latter, temperature is involved but only as a cue in converting photoperiod to "short day." Temperature therefore presents no direct limit to the duration of the productive period. Indeed it is possible to show that extra degree-days are available (Stross and Kangas 1969).

Food limitation acts to control productivity, but within the principal generation only. The evidence is that brood size is limited by food concentration. There may be a compensatory system indicated in the population, however. In simulation, a given number of young is more efficiently produced by a part of a food-limited population.

Predation

Food may be limiting fecundity, and predation seems further to restrict maximum population density. Predation can also restrict annual productivity by removing the young before they reach full size. The predator *Heterocope* seems quite effective in removing the young-of-the-year generation. Yield and its efficiency are known to be related to age of the individual at harvest (Slobodkin 1959, 1960). Mortality at an early age, especially in the young-of-the-year *Daphnia*, may reduce the yield. Most of the young are removed, presumably by *Heterocope*, within the first 48 hours, long before the individuals have reached adult size (see previous section). The argument that predation reduces annual productivity can be supported since yields from populations at densities larger than existed in the ponds were obtained in the pondlets.

Other evidence indicates that *Heterocope* predation may have reduced *Daphnia* production. In a correlation of the densities of the two populations from five ponds for 3 years, the production of the two were inversely proportional (Figure 6-16). The inverse relationship was easily detected when the annual yields for the ponds on the IBP site were compared. A separate regression for each of the 3 years was necessary because the annual yields are on a unit-volume basis and were uncorrected for evaporation. A drought in 1972 was an especially effective concentrator of the pond volume, hence the yield was apparently larger. Slopes ranged from −6.4 to −11.2. To achieve such a consistent range, three data points in 1973 were omitted as indicated (Figure 6-16). The data were less comprehensible for the year (1973) when fewest samples per data point were available. The regression slopes could express the 6 or 11 units of *Daphnia* necessary to produce one unit of *Heterocope*. The relationship is likely to be coincidental, however.

Dodson (1975, see Figure 6-7) has shown that other predator-prey relationships exist in the polygon pools at Barrow. All populations could suffer a restricted production from predation, while conceivably having a fecundity limited by food concentration. The point is both significant and vital to the coexistence of controls by predator selection and by size efficiency.

Food Production and Community

Trophic level interactions are likely to modify quantity and quality of food available to the zooplankton in tundra ponds. The *Daphnia* population in pondlets was more fecund than was its counterpart in the original pond, but only at an intermediate density. At least two explanations are possible. One is that nutritive quality is diluted by the detritus in suspension. When the detritus is heavily cropped, it is re-

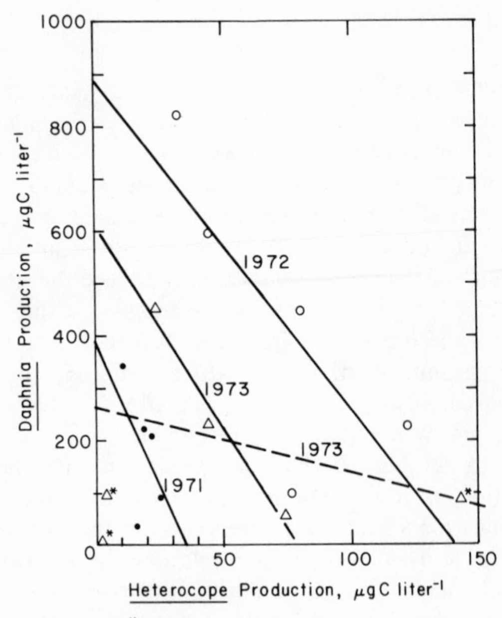

FIGURE 6-16. *Linear regressions of the annual rate of* Daphnia *production on annual rate of* Heterocope *production for each of five or six ponds. Each year was treated separately, a procedure made necessary by the annual variability in pond volume. The concentration as a result of the 1972 drought year is particularly striking.*

generated, but more slowly than the growth of bacteria and algae; the result is that cropping increases the food quality. This hypothesis is discussed in the next section.

An alternative explanation is that the production of algae, or bacterial food, is increased by phosphate excreted by zooplankton. Almost overwhelming evidence now describes algal growth rate in a nutrient-limited condition to be proportional to the rate of active uptake of the limiting resource. Uptake rate and growth rate describe a hyperbola with respect to concentration (Caperon 1968, Dugdale 1967, Eppley and Thomas 1969, Rhee 1973, among others). Judging from half-saturation constants for phosphate and nitrogen (Eppley et al. 1969, Carpenter and Guillard 1971, Rhee 1973, Chisholm 1974), ambient phosphate concentrations in the pond (Chapter 4, Prentki 1976) and bioassay experiments (Stross 1975) show that phosphate is limiting. Growth rates are likely to be increased in proportion to an increase in ambient concentrations. Measured uptake rates in the Barrow ponds ranged from 13 to 320 μg P liter day^{-1} (Chapter 5), indicating a large capacity despite a residence time only slightly longer than has been described for more eutrophic lakes (Rigler 1956, 1964, Lean 1973b).

Excretion of phosphate by zooplankton is an established fact. At least 40% of the phosphate excreted in the Barrow ponds is reactive phosphate (SRP), a value that agrees with the literature (see Chapter 4). In fact, when computed, zooplankton excretion could supply the growth needs of the phytoplankton in the ponds. Each *Daphnia*, in filtering 5.0 ml hr^{-1} (Chisholm et al. 1975) can increase the ambient concentration by 4% hr^{-1} if it excretes 90% of ingested phosphorus, half of which is SRP. However, if available phosphorus is only 10 to 20% of the total, actual increase attributable to each individual is conservatively 20% day^{-1}. Thus, if the *Daphnia* can be increased from one to five individuals per liter and ambient concentration of reactive phosphorus can be doubled, the increased grazer density would still sweep only half the water volume free of algal cells (69% of the water may be swept free of cells while maintaining steady density if all cells divide daily). Clearly, a stimulating effect on algal growth that can result from excretion of phosphate by the *Daphnia* grazing on the food supply can be a simple explanation for maximum fecundity in *Daphnia* at intermediate concentrations of animals.

Perturbations

One planned disturbance revealed something about the stability of zooplankton in the Barrow ponds. A deliberate addition of crude oil to one of the ponds (Pond E) in 1970 destroyed the Crustacea and prevented reproduction until at least the summer of 1973. Yet each spring a substantial number of naupliar stages of copepods and fairyshrimps were discovered in the pond. If the inoculation is from adjacent ponds via the annual flood water from melting snows, then a substantial alteration of each pond is possible. The fact that some degree of differentiation in relative abundances is possible yet does not occur, suggests a functional relationship of species to characteristics of the environment in each pond. Control may be functional and based on the biological characteristics of each species. That is not a new idea but intended merely to illuminate the experimental potential of arctic ponds.

Conclusions

The community in arctic ponds contains a stable assemblage of consumers. Reproductive and other ecologically pertinent traits of the dominant species seem similar to those of so-called equilibrium species

(MacArthur 1960). Factors that may control productivity of the planktonic consumers are amenable to analysis. Questions regarding production control, however, seem to be simpler than the questions regarding density regulation.

Productivity of zooplankton may be controlled by a combination of environmental factors and intrinsic properties of organisms adapted to an arctic environment. Intrinsic properties restrict the number of generations each year, either with intervention from the environment, as in the case of *Daphnia*, or without, as in the case of fairyshrimps. Thus while the pond is frozen most of the year, an adapted life cycle seems not to use fully even those months of a growing season that are available.

Within the single generation, however, food resources and not temperature may restrict reproductive effort. The availability of food may be coupled to the density of consumers in experimental pondlets. Densities are sufficiently low that a direct stimulation in growth of algae (that might result from excretion) is probable. A decreased dilution effect resulting from removal of detritus is an alternative possibility.

The capacity of the consumers to stimulate the food supply makes each individual vital to the effort. Young and juvenile mortality may have a dual effect. Elimination of an individual before it completes growth prevents its maximum yield. Its loss may also lower productivity if its excretion is essential to maintain the necessary phosphate concentrations for growth of algae. That is, given a single generation in which food limits reproduction, the presence of an infant-killer such as *Heterocope* can further reduce production.

Temperature must ultimately be significant in controlling productivity within arctic communities. Its immediate effect cannot be discovered, however. Food resources and predation in combination with intrinsic qualities of the genotype are the obvious immediate determinants.

CONTROL OF ZOOPLANKTON PRODUCTION (II) *

Introduction

The preceding section examined the control of zooplankton by looking only at their responses, such as growth, length, and brood size. When information on the composition of the food is added, then it appears that this composition is exerting some amount of control on the growth and reproduction of the zooplankton. From this we have formulated the hypothesis that it is the relative amount of high-quality food, such as algae and bacteria, in the total available food that controls zooplankton production. The idea that bacteria and detritus are important to

*M. C. Miller and R. J. Daley

zooplankton nutrition is, of course, an old one and dates back at least to Naumann (1921, quoted in Saunders 1969). More recently Nauwerck (1963) and Saunders (1969, 1972) have championed this idea. It has been happily ignored by most aquatic scientists, in part because of the difficulties of quantifying the nannoplankton and bacteria. As techniques are now available for counting both of these, perhaps the relationships among zooplankton, algae, bacteria and detritus will receive detailed study.

In this presentation, the word seston is synonymous with particulate organic carbon (POC). Seston is made up of algae, bacteria, and detritus (which is defined as non-living particulate matter). Operationally, it is impossible to separate attached bacteria from the detritus. In this case it does not matter because less than 20% of the total bacteria were attached to particles (Rublee personal communication).

The available evidence pointing to the importance of detritus and bacteria for zooplankton nutrition comes from three different approaches. The first is an indirect approach that examines the amount of phytoplankton available as food. Juday (1942) and Nauwerck (1963) found inverted biomass pyramids in lakes where zooplankton biomass is much greater than phytoplankton. For example, in Lake Erken the phytoplankton to zooplankton ratio averaged 1:7 (Nauwerck 1963). Even when Nauwerck measured the algal production, it was not enough to produce the zooplankton. Accordingly, he believed that detritus and bacteria were important as food.

The second approach, also an indirect one, stresses the abundance of detritus. In Lake Erken, the ratio of detritus to living matter is about 10:1 (Hobbie personal communication) but in other eutrophic lakes it may be 1:1. In Frains Lake, Michigan, Saunders (1969) found that the ratio ranged from 2.7 to 14.2 (no bacteria were included). In the Barrow ponds, the size distribution of the seston, measured with a Coulter Counter, is similar to that of oceanic seston (Sheldon and Parsons 1967, Sheldon 1972). Most of the particles are less than 25 μm in diameter and are therefore in the size range of particles taken by zooplankton. Given that most of the particles of seston are detritus, the zooplankton must collect more detritus than anything else. In small lakes and reservoirs the detritus can be so important that it is the base of most of the food chain (Wright 1959).

The third approach investigates the actual assimilation rates of algae, bacteria, and detritus by zooplankton. Saunders (1969, 1972) found in one series of experiments that their assimilation of natural algae (*Rhodomonas* and *Cryptomonas*) was 52 to 88%, bacteria was 13 to 52%, and detritus was 3 to 14% of the amount ingested. These experiments have the basic problem that the detritus used was an artificial one that may not be similar to natural detritus. It was, if anything, likely to be more nutritious so the assimilation value above may be too high.

From his studies, Saunders (1969, 1972) concluded that not only was detritus a food for zooplankton but it also acted as a biological buffer that evened out the fluctuations in food supply. He states (1969) that ". . . low calorie detritus in large amounts . . . acting as a dilutant, . . . may also dampen the assimilation of high quality algae, particularly when the seston is present in quantities approaching the incipient limiting concentration for feeding. This dampening would tend to suppress the growth rate of an exploding zooplankton population and perhaps permit the phytoplankton community to adjust to a high rate of grazing."

Zooplankton Feeding and Seston

As Saunders states above, it is important to establish that the seston concentration in the tundra ponds is above the limiting concentration for zooplankton feeding. We were not able to determine what this limiting concentration was, using the natural seston. However, we did determine that dilution of the water sample, which reduced the concentration of seston, increased the filtration rate of *Daphnia*. The seston in these ponds is made up of only 5 to 13% living matter (algae and bacteria) and is always greater than 200 μg C liter^{-1}. In Pond B in 1971, the detritus was 300 to 2000 μg liter^{-1} while the algal biomass was 0.6 to 15.0 μg C liter^{-1}.

The ingestion rate of *Daphnia middendorffiana* rises with increasing food concentration then reaches a constant, saturated rate at about 10^4 *Chlamydomonas* cells liter^{-1} (Figure 6-9, Chisholm et al. 1975). The point at which the rate became constant is a little less than 450 μg C liter^{-1}. (The experiment was carried out with natural pond water containing 200 μg detrital C liter^{-1}. The *Chlamydomonas* cells were 5% carbon and had a volume of 500 μm^3 cell^{-1} to give 250 μg C liter^{-1}.) This is also the point above which the volume cleared per hour drops sharply from 8 ml or so down to 7 ml (Chisholm et al. 1975). It appears that above this point the filtering apparatus is delivering all the food to the food groove that the *Daphnia* can handle. Thus, the energy intake of the animal is limited by the quality of the food. If the filtered seston is of poor quality, that is, made up mostly of unusable detritus, then the animal may not be able to obtain enough food for growth despite the abundance of seston and the presence of algae. It should be stressed that we imply in this section that carbon or energy is limiting the zooplankton growth. In reality it may just as well be digestibility or even nitrogen content. Sick (1970) proved that growth and reproduction in the brine shrimp, *Artemia*, correlated best with the nitrogen content of the different algae they were fed.

The concentration at which saturation of the feeding apparatus begins, around 450 μg C liter^{-1}, is about the same for *D. middendorffiana* as for other large zooplankters. For example, Bell and Ward (1968) found a constant ingestion rate in *D. pulex* when the food was above 200 μg C liter^{-1}. Marshall (1973) reported that *Calanus* produced eggs at a food concentration from 270 to 300 μg C liter^{-1}. In *D. rosea*, Burns and Rigler (1967) found that saturation began at about 190 μg yeast C liter^{-1} Finally, McMahon and Rigler (1965) found that saturation began at 135 to 660 μg C liter^{-1} when yeast, *Chlorella*, or *Escherichia coli* were used.

Production and Seston

There was an inverse relationship between the mean concentration of seston and the sum of the maximum biomass attained by each species of zooplankton (Figure 6-17). The correlation coefficient of 0.946 (n=8) is

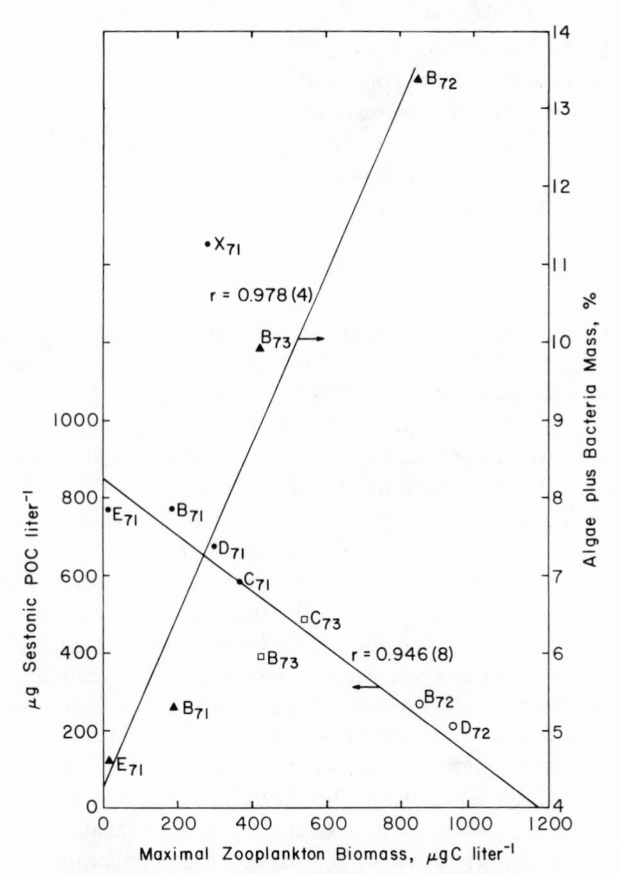

FIGURE 6-17. *Sestonic POC and percent algae and bacteria of POC as a function of zooplankton density, 1971 to 1973. The value for Pond X in 1971 was omitted from the regression calculation as it was heavily fertilized with phosphorus.*

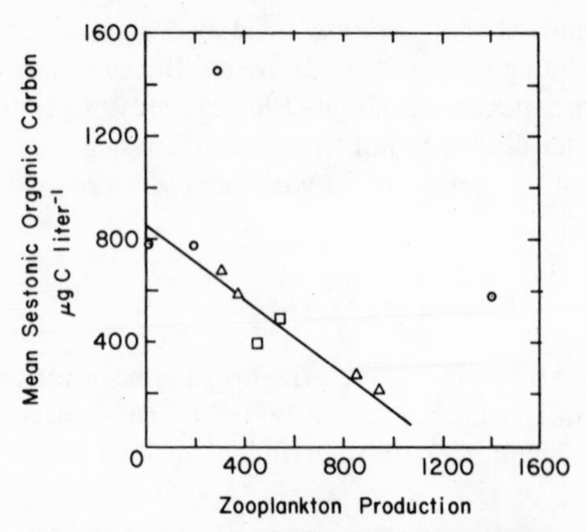

FIGURE 6-18. *The maximal biomass of zooplankton and the total zooplankton production plotted against the mean sestonic organic carbon during June, July and August for ponds in 1971 (circles), 1972 (squares) and 1973 (triangles).*

highly significant. Zooplankton biomass was the smallest in 1971, the year the algae and bacteria made up only 4.6 to 5.3% of the total seston. In contrast, the biomass was highest in 1972 when the seston was 13.4% algae and bacteria. Pond E had been subjected to an oil spill in 1970 which killed all the zooplankton. Biomass was very low in 1971 and consisted only of *Cyclops* sp. and fairyshrimp. Although the differences are not significant, Pond E did have a high seston concentration and a low ratio of living matter to detritus.

The relationship between the seston concentration and zooplankton production (Figure 6-18) is not as good a fit to a straight line as was the biomass to seston relationship, but the trend is obviously the same. Thus, the high seston concentrations are present only when both zooplankton biomass and production are low. At the same time, the absolute amount of living biomass remains constant and the living material becomes a higher percentage of the seston (Figure 6-17). One explanation is that grazing by zooplankton will reduce both living and dead material but the living material is replenished rapidly, so in a steady-state its percentage of the total seston rises. From this, we might expect that reduction of the zooplankton would result in higher seston concentrations and a lower living to detritus ratio; this occurred in Pond E following the oil spill. If a better quality food results from the increase in percentage of algae and bacteria, then there would be a positive feedback with better quality food leading to higher production leading to still better quality food, etc.

Another explanation of the inverse relationship between zooplankton and seston is that the detritus portion of the seston is less abundant in some years than in others for unknown reasons (perhaps less wind). The algae and bacteria grow as usual so that their portion of the seston increases and the improvement in food quality produces more zooplankton.

Experiments

Additional experiments by Miller and Federle in 1976 and 1977 help to explain the results of the pondlet experiments. In these pondlets, increased numbers of *Daphnia* and fairyshrimp increased *Daphnia* production up to a certain limit above which the additional numbers decreased production (Tables 6-10, 6-11). The 1976 and 1977 experiments made use of 20-liter containers of pond water incubated in Pond C. Additions of *Daphnia* and fairyshrimp at natural concentrations did not change the amount of POC, the rate of primary productivity, the bacterial density, or the bacterial activity (acetate-^{14}C uptake) from rates and levels in control chambers with no zooplankton. However, zooplankton additions at 5 times natural concentrations increased the POC and the bacterial activity.

The results indicate that in these chambers the POC is not controlled by zooplankton grazing at natural densities of animals. Only at the five times natural level of animals was the POC affected and there the high POC came from fecal material. From these and other experiments (the bacteria-protozoa-PO_4 cycling described in Barsdate et al. 1974), it appears likely that increased zooplankton production when crowded comes from increased rates of nutrient cycling.

However, conditions in the chambers are highly artificial. In the natural ponds, the zooplankton do not change the POC concentrations appreciably and can not affect the PO_4 cycling which is so dominated by sediment-water interactions.

It still is reasonable to us that the seston-POC-food quality interaction is affecting the zooplankton productivity. Control of the seston concentration must come from an external agent such as wind currents.

SUMMARY

The ponds contain calanoid and cyclopoid copepods, two species of fairyshrimp, and *Daphnia middendorffiana*. Rotifers are present but rare. All the crustaceans except for *Daphnia* have a single brood per year; the offspring of the overwintering generation enter diapause and overwinter in this form. The embryos of calanoid copepods and of the fairyshrimp overwinter, but in the case of the cyclopoid copepods, a copepodid stage (pre-adult) enters diapause. *Daphnia* hatch from overwintering eggs (ephippial eggs) and the hatchlings grow and produce a brood of young. Next, the same animals switch to production of the ephippial eggs. It is likely that the single brood of young does not survive to reproduce. Only parthenogenetic females of *D. middendorffiana* have been found at Barrow. The switch to ephippial egg production is likely controlled both by photoperiod and by temperature.

The life cycle of the crustaceans implies that an upper limit is set on the number of animals that grow in a given year by the number of diapausing animals that survived the winter.

The first animals to hatch in the spring are the cyclopoid copepods (*Cyclops vernalis* and *C. magnus*). Copepodid and adult forms were present almost as soon as any water at all appeared around the edge of the ponds; by mid-June, egg sacs and even nauplii were present. The egg crop reached its peak in early July when there were about 2.5 females, 1.7 males, and 65 eggs liter^{-1}. A maximum of 35 nauplii liter^{-1} were seen in late June but in late July and in August there were only 10 animals liter^{-1}

Calanoid copepods were more abundant than the cyclopoids. Soon after they hatched in mid-June there were up to 40 liter^{-1} of *Diaptomus bacillifer* and *D. glacialis*. After 15 July the numbers stabilized at about 4 *Diaptomus* liter^{-1} and 2 *Heterocope septentrionalis*, a predaceous form. Fairyshrimp are so large that they had the greatest quantity of biomass of any zooplankton group even though their numbers were usually low. The two genera, *Branchinecta* and *Polyartemiella*, hatched in mid-June and soon up to 2 nauplii liter^{-1} were present. By early July there were less than 0.3 liter^{-1} In some years the numbers were so low that *Daphnia* became the dominant organism in the plankton.

Daphnia hatch in late June and reproduce in early- to mid-July. Between 2 and 5 animals liter^{-1} are present much of the summer but only about 0.5 new animals liter^{-1} appear from the single brood. It is likely that most of the young of this summer brood are eaten by *Heterocope*.

In general, the zooplankton egg production was below the maximum possible production for each species. For example, when grown in the laboratory, the fairyshrimp produced 1.5 to 2 times more eggs than in the field. Even when large numbers of eggs were produced, the resulting number of nauplii or young was small. For example, *Cyclops* produced 126 eggs liter^{-1} in one year but only 7.5 nauplii liter^{-1} resulted. Production by invertebrates is important but this can not be separated from a low rate of hatching. *Daphnia* has the ability to produce 12 eggs per brood but the average brood sizes were 2 to 4. Larger animals had larger broods; this was controlled by the food supply. An experimental 3-fold increase in the food concentration for *Daphnia* resulted in a 3-fold increase in brood size.

Because there are no fish in the ponds, most of the predation on crustaceans is by other zooplankton. In artificial experiments in small containers, an adult or copepodid stage of *Heterocope* killed young *Daphnia* with a daily coefficient of 0.14 liter $^{-1}$. The coefficient for adult *Daphnia* was 0.09. *Heterocope* also feed on *Cyclops* nauplii while adult *Cyclops* feed on their own nauplii, on young *Heterocope*, and on fairyshrimp. In the field, the *Daphnia* which first hatch from the overwintering eggs are relatively immune from predation because the *Cyclops* adults are small and because the *Heterocope* are still in their naupliar stage. However, by mid-July when the single brood of *Daphnia* hatches, the *Heterocope* are adult and appear to eat nearly the entire brood. It is possible that the large size of *Daphnia* and other crustaceans is an adaptation to minimize vulnerability to invertebrate predation. Another possibility is that the large size results from competition for food.

The production of zooplankton was calculated as the biomass of individuals surviving to the end of the summer plus the loss from mortality plus the mass of eggs produced. The annual production was close to 1 mg C liter^{-1} or 200 mg C m^{-2}. In one pond, the fairyshrimp contributed 0.8, 0.1, and 0.1 mg C in each of three separate years while *Daphnia* contributed 0.2, 0.8, and 0.5 mg C in the same years.

Daphnia filtration rate was at a maximum at 12°C, much lower than the maximum temperature for temperate species. When compared with individuals of the same size of other species of *Daphnia*, *D. middendorffiana* filters slightly faster. A 2.6-mm-long animal has a filtration rate of 8.0 ml hr^{-1} at the saturation level of 37,000 particles ml^{-1} (the experiment was run at 11°C). In the pond, the *Daphnia* are believed to feed at or near their maximum rate.

Because the relationship between size of animal and filtering rate is a power function, the velocities increase rapidly as the animals grow. If an efficiency of assimilation of 70% is assumed (no measurements were made), then the growth rate of animals can be related to food concentration and animal size. Thus, a 1.5-mm animal can barely grow at 50 μg POC liter^{-1} while a 3.0-mm animal will not only grow well but will also produce 10 eggs. Thus, large animals have a decided advantage over small animals at all stages of growth and only large animals can hope to accumulate enough stored energy to reproduce within the short summer. Animals in the ponds follow this general rule too; the reproductive potential of 3.0-mm *Daphnia* was 4 to 5 times that of a 2.5-mm *Daphnia*.

In the ponds, the zooplankton have the additional problem of coping with large quantities of non-living particles. Some of these may well be edible but most are likely resistant to breakdown. Thus, there was about 100 μg living C liter^{-1} and 250-800 μg non-living C liter^{-1} in a typical pond. Nothing is known of the effect of these non-living particles on *Daphnia*; at one extreme, they might digest some or most of them while at the other extreme the particles would clog the feeding system and actually reduce the efficiency of assimilation.

D. middendorffiana possesses an endogenous rhythm of feeding activity. Animals held under constant light and temperature had peaks of feeding at 1400 and 2400. This could be a strategy for optimizing feeding, for these are the same hours that the water temperature in the natural pond on a clear day is optimum for feeding for this species.

The population of *Daphnia* hatch from the overwintering eggs in a single burst which lasts for 2 or 3 days. This synchrony is retained throughout the lifetime of the population as the single brood of eggs all appeared within a 4-day period and these eggs also hatched in a rough synchrony. There was also synchrony of molting within a part of a single day but a circadian rhythm could not be demonstrated.

7

Macrobenthos

M. Butler, M.C. Miller, and S. Mozley

INTRODUCTION TO ARCTIC BENTHOS *

The larger benthic animals, which are mostly insects although snails and worms do occur, are extremely important to the ecology of the ponds. Not only are their biomass and productivity larger than those of any other group of animals, but they also continually affect the structure of the sediments by their feeding and burrowing. In one experiment, for example, chironomids (at twice natural density) mixed and moved the sediments so much that sand grains placed at the surface were completely covered in 10 days and 10% had reached a depth of 3 cm. This movement, plus the circulation of water through their burrows, mixes oxygen into the sediments and promotes the exchange of dissolved material. The benthic animals also eat zooplankton, microfauna, benthic algae, and each other and are, in turn, a food for insectivorous birds and even zooplankton.

Macrobenthos of the ponds were investigated in two periods by different investigators. In 1971 and 1972, D. M. Bierle collected extensive data on distribution and dynamics, population densities, feeding, growth, respiration, and production. Most data were collected in Pond J and on larvae of the genus *Chironomus*, but comparative estimates of total benthic secondary production were obtained for three other ponds, B, D, and E. Unfortunately, the abundance of small forms and early larval instars was underestimated and species were not adequately distinguished, with the result that many population data represented mixtures of animals with different life cycles. Some estimates were too low, such as larval densities, spatial variance of larval populations, densities of emerging adults, and length of larval lifespans, but estimates of larval and adult biomass seem to have been too high. On the other hand, many data collected in the earlier years on respiration, growth, larval feeding, trends in directly measured total biomass, and occurrence of benthos or their terrestrial life stages in bird guts were apparently valid. The present authors collected further data on species composition, density, age structure, emergence, and within- and between-pond variance in 1975, 1976, and 1977. A detailed description of these studies will be included in a

*S. Mozley and M. Butler

Ph. D. thesis at the University of Michigan by M. Butler. We are grateful for Dr. Bierle's permission to use his data and samples in this account.

Benthic Studies

Research on arctic macrobenthos has been sparse and primarily descriptive. The earliest detailed work was Andersen's (1946) study of several shallow lakes in Greenland. He pointed out the relatively great importance of soft-bottom benthos and particularly the Chironomidae in northern waters. He observed large populations but comparatively few species, primarily Chironomidae in the groups Tanytarsini, *Chironomus, Phaenopsectra, Psectrocladius, Cricotopus, Corynoneura,* and *Procladius*, and related their abundance and distribution to environmental features such as dissolved oxygen over winter, macrophytes, development of dissolved H_2S, and depth of water. He mentioned also Acari, Turbellaria and *Lepidurus*. He believed most Chironomidae to have one-year life cycles. A brief survey of lakes in the mid-1950's (Livingstone et al. 1958) also emphasized the low diversity and moderate standing crops of benthos and littoral insects and the importance of chironomids in the arctic standing waters, but provided little direct data on their ecology. They found an unusual Tanytarsini preserved in lake sediments, *Corynocera* (or *Dryadotanytarsus*) and remarked on the very high density of chironomid remains in some lakes. Their higher standing crops of macrobenthos were similar to those in Barrow ponds.

The classic studies of Scholander et al. (1953) on physiology of arctic animals showed how *Chironomus* larvae from Barrow ponds were able to freeze and return to activity with the subsequent thaw. Oliver (1968) and Danks and Oliver (1972), working in the Lake Hazen area of extreme northeast Canada, showed that metamorphosis and emergence of Chironomidae adults were extremely synchronous but varied between ponds and to a lesser extent between habitats within ponds. The Chironomidae in those studies were in the same genera but were often different species than the Barrow midges. Oliver (1968) summarized a number of other features of arctic Chironomidae, including the occurrence of copulation on surfaces rather than in aerial swarms in several species, and mechanisms controlling emergence, oogenesis, and long life cycles. Oliver found that water temperature regulated the time of emergence in shallow environments. In Cape Thompson ponds (Watson et al. 1966a), macrobenthos densities were high (38,000 m^{-2}) and chironomids were the most numerous taxon, followed by oligochaetes. Turbellaria, Polychaeta, Isopoda, Amphipoda, Acari, and the insect orders Plecoptera, Trichoptera, Ephemeroptera, and Coleoptera were well represented, especially in beds of the macrophyte *Arctophila*. Unfortunately they were unable to give more taxonomic detail.

Most species lists of arctic or cold-subarctic insects are based on collections of adults and do not discriminate among terrestrial, semi-terrestrial and truly aquatic forms (Watson et al. 1966b; McAlpine 1965; Rickard and Harmston 1972). The benthic studies of Holmquist (1973) have focused on coastal lakes and the presence or absence of marine relicts among the macrobenthos.

The Char Lake IBP study emphasized dynamics, production, and controls on macrobenthos, particularly Chironomidae (Welch 1973, 1976, Welch and Kalff 1974, Lasenby and Langford 1972). Chironomids shared dominance with *Mysis relicta* in Char Lake but there were few species; Orthocladiinae were dominant and the variable ice cover occasionally interfered with annual reproduction. Char Lake conditions represented a more extreme polar environment than those in Barrow. Secondary productivity in Char Lake was extremely low because of low primary productivity (Welch 1976). It was also found that chironomids have longer lifespans in the Arctic than in temperate latitudes, two or three years in the Char Lake species.

Barrow thaw ponds contrast with other investigated arctic habitats in several ways. The species of Chironomidae are somewhat different from those in other areas, although genera are similar. Some of the statements and suppositions of Livingstone et al. (1958) about the absence of dytiscid beetles and widespread dominance of *Corynocera* in the Arctic are not borne out. Lifespans of benthic animals are longer than earlier authors report, and growth is slow in terms of calendar years. However, development is rapid in view of the low summer temperatures and when compared with daily rates at lower latitudes. A sum of less than 1 year of ice-free days is required for maturation in all the smaller species.

Species Composition

The principal organisms were collected from core samples of sediments as well as by sweeps of emergent vegetation, by submerged pitfalls, by quantitative entrapment of emerging adult insects, and by some aerial netting. The list (Table 7-1) is remarkable in several ways. Virtually every taxon contains fewer species than would be expected in a temperate pond and many taxa are missing altogether. There are no Ephemeroptera, Odonata, Hemiptera or Megaloptera, no Hirudinea, lumbriculid or naidid Oligochaeta, no Pelecypoda, no Amphipoda, Isopoda, or Decapoda. Mosquitoes are absent in contrast to their notorious abundances in thaw ponds and pools farther inland. Indeed, there are few dipterans of any sort besides the abundant and (comparatively) diverse Chironomidae. The numbers of species of Coleoptera, Gastropoda, Trichoptera, and Acari are all lower than in temperate waters. Even Chironomidae are not as rich in species here as they are in permanent temperate ponds.

TABLE 7-1 *Macrobenthos Collected From Tundra Ponds Near Barrow, Alaska*

Turbellaria	1 sp.
Annelida	
Oligochaeta	
Tubificidae	*Tubifex* sp.
Enchytraeidae	*Propappus* sp.
Hydracarina	
Lebertiidae	*Lebertia* sp.
Gastropoda	
Physidae	*Physa* sp.
Insecta	
Plecoptera	*Nemoura arctica*
Coleoptera	
Dytiscidae	*Agabus* sp.
	Hydroporus sp.
Diptera	
Chironomidae	
Podonominae	*Trichotanypus alaskensis*
Tanypodinae	*Procladius prolongatus*
	P. vesus
	Derotanypus alaskensis
	D. aclines
Chironominae	
Chironomini	*Chironomus pilicornis*
	C. hyperboreus
	C. riparius
	C. (Camptochironomus) grandivalva
	Stictochironomus sp.
	Dicrotendipes lobiger
	Cryptochironomus sp.
Tanytarsini	*Tanytarsus inequalis*
	T. gregarius gr. sp. 2
	Paratanytarsus penicillatus
	Constempellina sp.
	Cladotanytarsus sp.
Orthocladiinae	
Orthocladini	*Psectrocladius psilopterus* gr. spp. (at least 2)
	P. dilatatus gr. sp.
	Cricotopus tibialis
	C. nr. *perniger*
	Cricotopus sp. 3
	Cricotopus sp. 4
	Orthocladius (Eudactylocladius) sp.
	O. (Pogonocladius) sp.

TABLE 7-1 (Continued)

	Metriocnemini	*Corynoneura* spp. (several)
		Metriocnemus spp. (at least 2)
		Pseudosmittia gr. sp.
		Parakiefferiella gracillima
		P. nigra
		Limnophyes sp.
		Lapposmittia sp.
		Mesosmittia sp.
		Bryophaenocladius sp.

Individual species of animals tend to occur primarily either in the soft sediments in pond centers or among stems and leaves of *Carex* and *Arctophila* around the edge. These two habitats and the animals associated with them are represented in Figures 7-1 and 7-2. In the centers, Chironomidae, led by *Chironomus* sp., *Tanytarsini* spp., and *Procladius*, make up 75 to 95% of the biomass while Oligochaeta contribute almost all the remainder (Figure 7-3). Tubificidae account for most worm abundance but the rarely encountered Enchytraeidae are very large forms and may be locally important. At the edges among plants, other insects and the snail *Physa* account for much of the biomass since their weight per individual is high.

Distribution of chironomid species between macrophytes and pond centers is best illustrated by data from emergence traps (Figure 7-4). *Trichotanypus* and most Orthocladiinae were associated with the *Carex* plants. Emergence of *Corynoneura* into all traps probably reflects redistribution of active prepupae or pupae, as benthic samples showed larvae of this species to be restricted to the vegetated pond margins. Other subfamilies had species-specific habitat preferences, with clear distinctions seen even within genera. *Chironomus pilicornis, Procladius vesus,* and *Tanytarsus inaequalis* were all most abundant in pond center traps, while *C. riparius, P. prolongatus,* and *T. gregarius* gr. sp. 2 occurred equally or predominately in "*Carex*" traps.

Most macrobenthos in the ponds are detritivores and most of these are deposit feeders. As noted, however, most species we found were chironomids; the trophic structure of these will be considered later in the chapter. The remaining macrobenthos species have a variety of feeding mechanisms.

All the Oligochaeta at Barrow are deposit feeders that always feed below the surface. *Nemoura* and *Tipula* may be shredders or gatherers of fine detritus among the macrophytes. *Micrasema* probably feeds on living plants (Wiggins 1977) while *Asynarchus*, another caddis fly, is omnivorous and even feeds on Turbellaria, dead *Nemoura*, and other *Asynarchus*. The dytiscid beetles are predators both as adults and as larvae. The mite

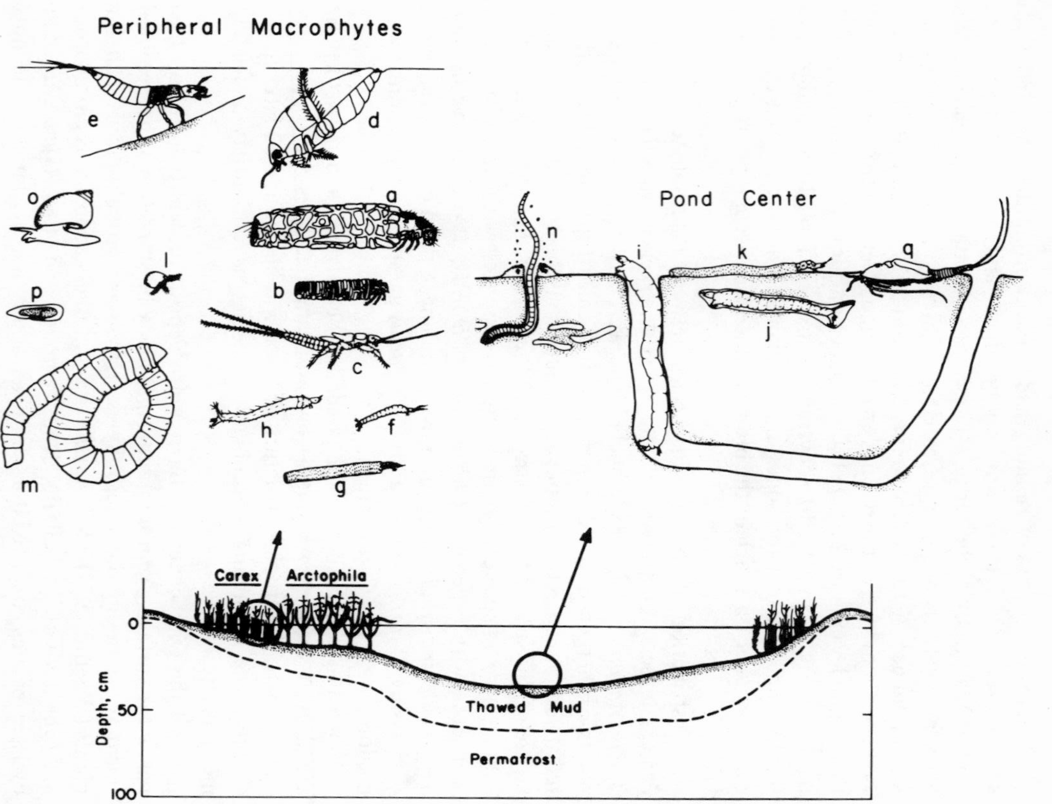

FIGURE 7-1. *Habitats and principal taxa of benthic macroinvertebrates in a thaw pond.*

a *Limnephilus* sp.
b *Micrasema scissum*
c *Nemoura arctica*
d *Agabus* sp. (adult)
e *Hydroporus* sp. (larva)
f *Corynoneura* sp.
g *Paratanytarsus penicillatus*
h *Trichotanypus alaskensis*
i *Chironomus pilicornis*
j *Procladius vesus*
k *Tanytarsus inaequalis*
l *Libertia* sp.
m *Propappus* sp.
n *Tubifex* sp.
o *Physa* sp.
p *Turbellaria*
q *Lepidurus arctica*

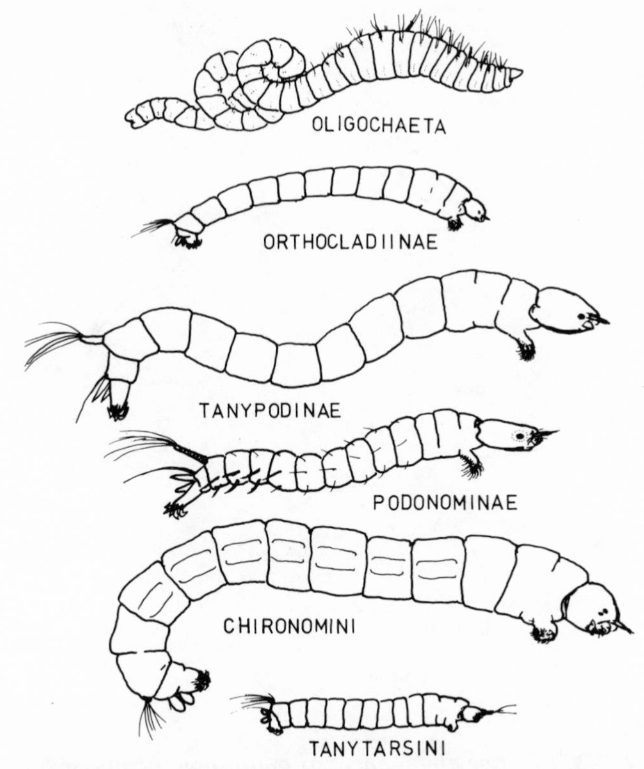

FIGURE 7-2. *Representative chironomid larvae and a tubificid oligochaete from a Barrow tundra pond.*

Lebertia is predaceous but also is parasitic on emerging adult insects. The snail *Physa* grazes on attached algae, microfauna, and detritus on the macrophytes or sediment surface. *Lepidurus* is described in detail later.

CHIRONOMIDAE STUDIES *

Ponds as Midge Habitats

There are two kinds of habitats for midges in the ponds, soft organic muds and emergent vegetation (*Carex* and *Arctophila*). Macrophyte-free sediments can be further subdivided into fine, unconsolidated sediments in the centers of the larger ponds and the irregularly spaced peaty sediments, closer to shore. The soft sediments represent deposits of finer detritus which are eroded from peripheral aquatic and terrestrial plant beds by

*M. Butler and S. Mozley

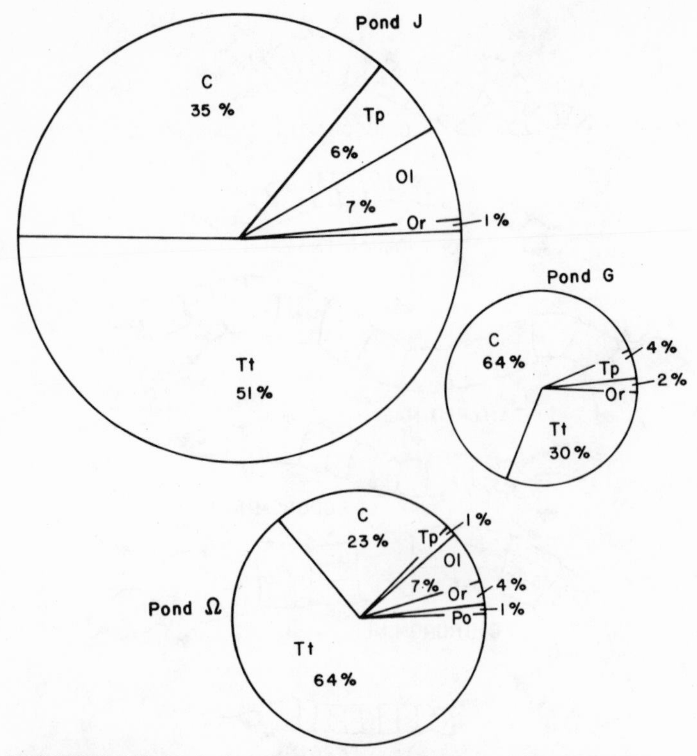

FIGURE 7-3. *Relative abundances of major macrobenthic invertebrate taxa in the central sediments of Ponds J, G and Ω averaged over 1975 and 1976. C—Chironomini, Tt—Tanytarsini, Tp—Tanypodinae, Or—Orthocladiinae, Po—Podonominae, Ol—Oligochaeta.*

wind-generated currents. Peaty areas appear to be former stands of plants with rhizomes still present but no longer growing. In the emergent vegetation habitat, the water is shallower than in the pond centers and circulates less. As a result, the water temperature at midday may be several degrees warmer in the plant stands than in the pond centers. Because of almost constant winds, complete exposure of the ponds, and low water temperatures, there is no depletion of oxygen above the sediments in the center although there may be a slight decrease among the plants on especially warm days. Sediments in the center of Pond J have a median grain size (wet sieved) of 500 μm and 95% of the particles were larger than 6.2 μm. The sediments in Pond J lose 68% of their dry weight on ignition, but other ponds have up to 84% organic matter. Their water content is very high and the interface between sediments and water is often indeterminate. There are slight differences in fauna between the types of sediments.

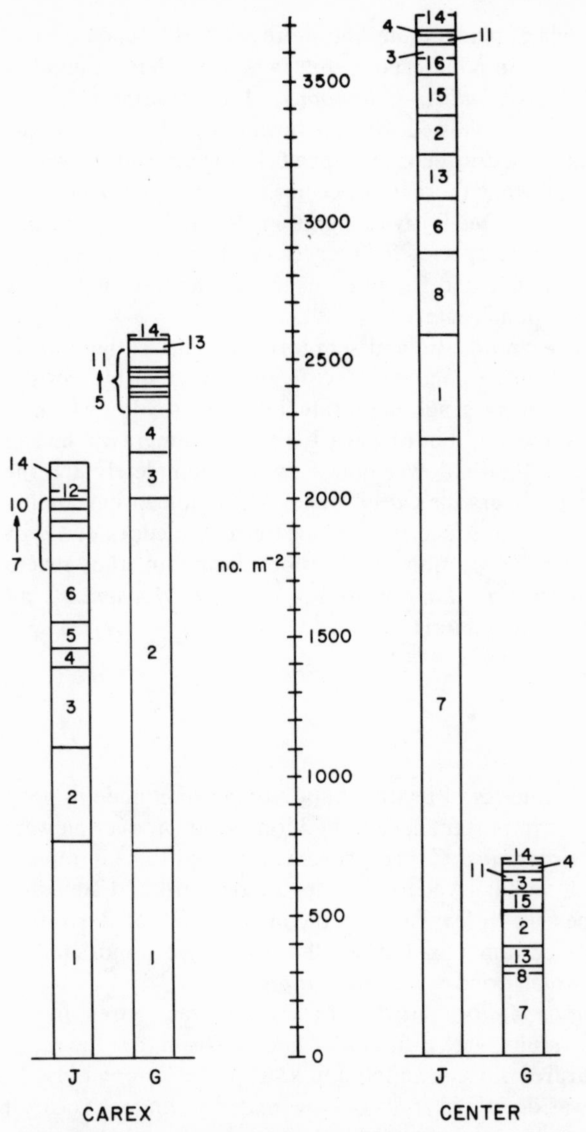

FIGURE 7-4. *Total emergence of Chironomidae from two habitats of Ponds G and J during 1976.*

1. *Paratanytarsus penicillatus*
2. *Corynoneura* spp.
3. Other Orthocladiinae spp.
4. *Trichotanypus alaskensis*
5. *Derotanypus alaskensis*
6. *Tanytarsus gregarius,* group sp. 2
7. *Tanytarsus inaequalis*
8. *Constempellina* sp.
9. *Psectrocladius* sp.
10. *Stictochironomus rosenschoeldi*
11. *Procladius prolongatus*
12. *Cricotopus tibialis*
13. *Cladotanytarsus* sp.
14. Miscellaneous spp.
15. *Chironomus pilicornis*
16. *Procladius vesus*

Sediments in stands of plants along the shore support a fauna similar to that in peaty sediments in pond centers, but in addition have a number of Orthocladiinae (*Psectrocladius, Cricotopus*). Larval samples along transects in Pond J in 1971 showed a trend toward smaller sizes at the edges. Chironomids in the *Carex* beds averaged 0.14 mg dry weight, with a range of 0.06 to 0.26 mg, while those in the center ranged from 0.23 to 0.58 mg and had a mean of 0.46. Qualitative samples in 1975 indicated that size differences corresponded to species differences, with many *Cricotopus, Psectrocladius, Corynoneura* and Tanytarsini in the shallows but many large *Chironomus* in the pond center.

The boundary between aquatic and semiterrestrial marsh habitats is sometimes sharp, as when the pond bank is slightly undercut by erosion, but more often the shallows grade smoothly into wet tundra which is submerged just after snow melt but dries early in the summer. By the end of summer in dry years, the shallower ponds dry out completely and the sediments crack. In most years the *Carex* stands are no longer in standing water by freezeup time. Emergence traps placed near the edges of *Carex* beds caught a greater proportion of Orthocladiinae in the genera *Limnophyes, Pseudosmittia, Lapposmittia,* and *Metriocnemus,* all suggestive of semiterrestrial habitats.

Species

All quantitative estimates of benthic populations were made in pond centers, where the most important species by biomass or production were those in the genus *Chironomus. C. hyperboreus,* reported to be dominant in arctic and subarctic ponds (Andersen 1946; Lindegaard and Jonasson 1975), occurred in the area but was rare in the larger ponds at Barrow. It appeared to be most common in the smaller ice-wedge ponds and in flooded grassy areas which dried out in late summer.

The most common *Chironomus* in the ponds was reared in the laboratory and many adults were available for comparison, but the range of morphological features did not match any known species precisely. *C. pilicornis* (Townes) was described from Barrow material, and was nearest the dominant *Chironomus* in pond centers, so we have used that name here. Details of size, the shape of the superior volsella and antennal ratios differed between *Chironomus* collected in early and late July, and emergence periods were distinct for the two forms. Larval chromosomes have not been compared so we do not know whether these two are distinct from each other, or whether one or two new species is involved. The other common species of *Chironomus, C. riparius,* was identifiable as larvae and adults.

Tanytarsus gregarius gr. sp. 2 is definitely a new species. The male genitalia place it in the *T. gregarius* group (Reiss and Fittkau 1971), but its brachyptery, shortened mid- and hind legs, and reduced antennal plume separate it clearly from all other known *Tanytarsus*. It is a member of one of at least four species pairs of Chironomidae in which one has normal antennae and the other a reduced plume. In each case the reduced-plume form cannot be identified in the literature and the two forms have temporally distinct emergence periods. Oliver (1968) and Wülker (1959) have noted similar trends in other chironomids in extreme environments.

In some cases, especially *Metriocnemus, Psectrocladius* and several other Orthocladiinae, the species definitions and available keys are inadequate. In others, Barrow pond forms are obviously new species, e.g., *Cladotanytarsus, Cricotopus* and the aforementioned *Tanytarsus*. Other species have type localities in Barrow: *Trichotanypus alaskensis* and *Derotanypus alaskensis*. Barrow tundra ponds clearly have a fauna distinct from that of previously studied areas.

Emergence of Adults

In 1975 adult midges were trapped in floating polyethylene pyramid traps, each covering 0.1 m^2 of pond surface. In 1976 inverted funnel traps were constructed of clear polycarbonate plastic. These funnel traps were in contact with the bottom and enclosed 0.05 m^2 of sediment; this eliminated problems of trap avoidance or attraction. Traps were cleared at 2- or 3-day intervals from the earliest emergence (usually *Trichotanypus* in late June), until mid-August when emergence of all species was virtually complete. Both cast pupal skins and adults were collected.

Comparisons of numbers of pupal skins with adults revealed that floating traps introduced greater errors than bottom-resting traps. In floating traps, the numbers of each fluctuated widely but sometimes skins and other times adults were more numerous. In bottom-resting traps the fluctuations were smaller and usually the number of adults exceeded the number of pupal skins of each species. The greatest fluctuations in the number of pupal skins occurred after the longest intervals between trap clearing; therefore, skins are likely lost by sinking. Some adults were lost or damaged by predators (probably mites, beetles, and perhaps *Heterocope*) and by the act of clearing the traps; however, such losses were greatly reduced in 1976 by the improved traps.

All species had a single emergence each year in the ponds, and most, especially the Tanytarsini and *Trichotanypus,* showed high synchrony (Figure 7-5). Species emerging early in the season tended to show greater synchrony than those emerging later. If the populations are synchronized by thermal cues at or soon after the time of thaw as suggested by Danks and Oliver (1972), then later emerging species would be expected to show progressively decreasing synchrony. Another factor which may reduce population synchrony is parasitism. During the 1976 emergence of

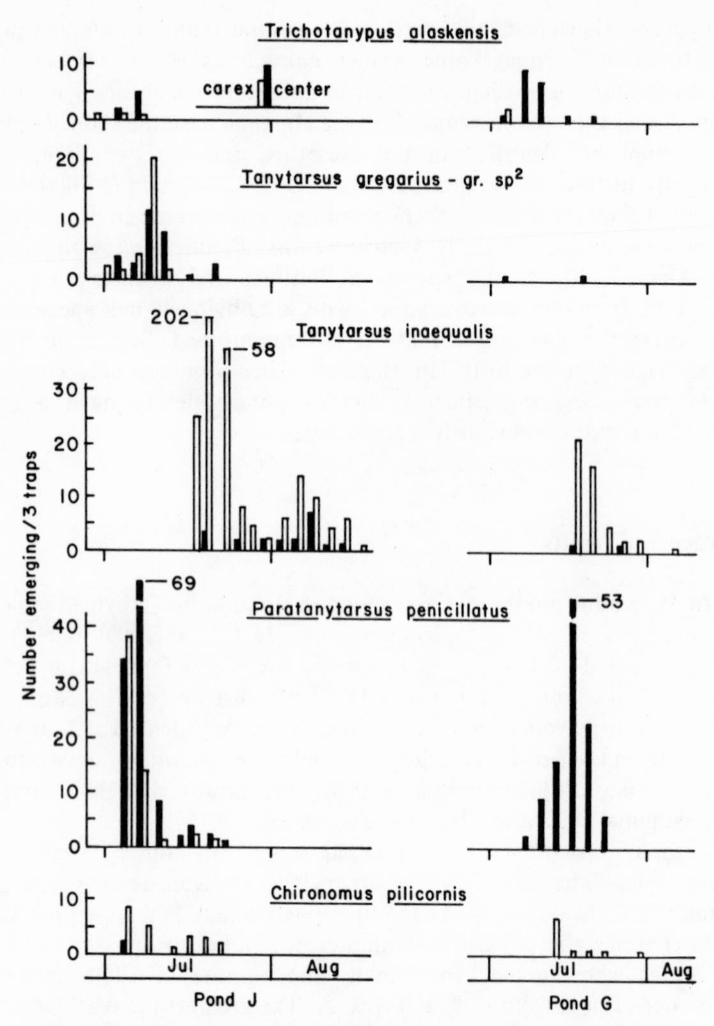

FIGURE 7-5. *Emergence phenologies of five chironomid species from both* Carex *(solid bars) and central sediment (open bars) habitats in Ponds G and J during 1976. Traps were cleared at 2- to 3-day intervals from late June through early August. Bar heights represent the number of midges emerging into three traps whose total area was 0.15 m².*

Tanytarsus inaequalis from Pond J, nearly all individuals emerging more than 3 days after the population maximum were infested with nematodes. In other ponds emergence of this species was more synchronous.

The sequence of emergence was consistent between all ponds and years, even when species composition varied. This again suggests the sort of species-specific thermal control of pupal development proposed by

Danks and Oliver (1972). Actual dates of first emergence of a species varied between ponds by as much as 1 week, but "early" ponds with respect to emergence were also the first to thaw in early June. Differences in thaw time appear to be related to degree of snow cover; ponds swept free of snow thawed earliest while those situated in areas of snow accumulation thawed later. Several weeks after the thaw when emergence began, there were no detectable differences in pond temperatures; this suggests that thermal cues at or soon after the time of thaw determine emergence phenology.

Emergence synchrony is an important requirement for mating success in species such as midges with short adult lives. This is especially true in the Arctic where there are highly unpredictable conditions for adult survival and mate location. Consistent synchrony of species' emergence periods may serve two functions. The potential for between-pond mating is increased, and the temporal isolation afforded these species may serve as a reproductive isolating mechanism. While two or more species were usually emerging simultaneously at any time during the season, the emergence periods of congeneric species overlapped little if at all.

Two genera of Orthocladiinae, *Psectrocladius* and *Corynoneura,* provided the only cases of prolonged emergence throughout several weeks. Since the taxonomy of these genera is not well established, it is possible that these collections represent a sequence of sympatric species with synchronously emerging populations.

Life Cycles

At first, cohort distinction and the determination of lifespans of chironomid species seemed straightforward, but it proved to be a complex task as more data accumulated. Our early definitions of cohorts of *Chironomus* indicated a 4-year cycle with 1 year spent in the egg and first 2 larval instars, 1 year as a third instar and 2 years as fourth instar, then a brief period as pupa and adult. Both Bierle's data and ours seemed to show two cohorts, a small and a large one, in the fourth instar. However, when examined in detail our later samples showed an almost continuous variation in size through the last larval instar, and moreover indicated that extremely high mortality ($\sim$83%) would have to occur in the older of the presumed fourth instar cohorts in the last weeks before they emerged as adults to account for the numerical discrepancy.

A reexamination of fourth instar larval growth at more frequent intervals during two summers showed that at least 7 years were required to complete growth in *Chironomus* of the pond centers. For distinction of cohorts, it was necessary not only to measure lengths of larvae precisely, but also to determine their relative maturity by use of the Wülker and Götz (1968) definitions of developmental phases (of adult primordia) in the larvae. When fourth instar animals were sorted into the nine phases of Wülker and Götz, three cohorts became apparent. This analysis was checked by following changes in phase and size class of stronger cohorts at 3- or 4-week intervals. It has not been possible to study each species with the same intensity, but we can say at least that *Tanytarsus* species have 2-year cycles and that no species of chironomid in the pond centers matures in less than 2 years. Analysis of this type is complicated by the presence of pairs of closely related species, so far indistinguishable as larvae, in important genera.

The 7-year generation time of *Chironomus pilicornis* in Pond J is portrayed in Figure 7-6. Larvae hatched from eggs within 2 weeks of oviposition. Egg masses appear to have been concentrated near the pond edges either by oviposition or wind, as we found few newly hatched first instars in the center. Early in the second summer the larvae molted to the second instar and apparently completed their move to the center of the pond. During the third summer, at about 2 years of age, the larvae molted to the third instar and in 2 more years, to the fourth. Growth and development in this last larval instar continued through the next 3 years. When the pond thaws in the eighth summer the larvae have reached a length of 16 to 18 mm (females are generally larger than males), and are in developmental phase 8 or 9. Pupation occurs soon after thawing, and it is

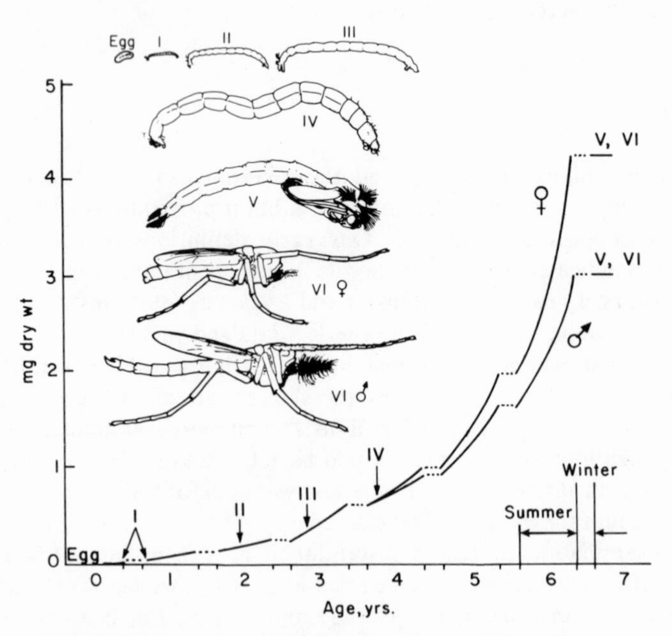

FIGURE 7-6. *Life cycle of* Chironomus pilicornis. *Arrows indicate time of molting to the next instar. Growth and molting schedules were determined from data taken in Pond J from 1975 to 1977.*

likely that no feeding or growth occurs during this final season. Instead, as suggested by Danks and Oliver (1972), the final winter diapause serves as a population synchronizing mechanism with prepupal and pupal development filling the time between thawing in mid- to late June and emergence in early to mid-July. Emerging adults flew away from the pond surface and awaited a period of calm or low wind speeds. Males form aerial swarms and females mate and oviposit in the manner described by other authors for most *Chironomus* species (e.g., Fischer 1969).

When temperatures begin to fall in late summer, the larvae spin cocoons and then spend the winter frozen solid in the mud (Scholander et al. 1953). The return to activity in June was observed in *Chironomus*, but this aspect of their biology has been described already by Danks (1971b). The ability to tolerate sub-freezing temperatures may be an important selective factor for the species composition of arctic chironomids (Danks and Oliver 1972).

Tanytarsus inaequalis, a much smaller species also abundant in the pond centers, has a 2-year life cycle. Emergence and oviposition do not take place until late July to early August, yet the newly recruited larvae reach the second instar before overwintering. The molt to the third instar occurs during the second summer, and soon after the second winter the fourth instar is reached. Since these insects will emerge later this same year, it is obvious that much feeding and growth must take place between thawing and emergence. This species, and perhaps others in the Barrow area, do not fit the "absolute spring species" designation of Danks and Oliver (1972) for the midge fauna of a higher arctic location. In such species, no feeding takes place during the summer of emergence.

The adult behavior of *Tanytarsus inaequalis* and other small species in these ponds differs from that described for *Chironomus pilicornis.* Instead of flying away from the pond immediately upon emergence, the adults rest sheltered from the wind among emergent *Carex* at the pond edges; during periods of calm the males skate over the surface to find mates. This behavior is perhaps best developed in an undescribed sibling species to *T. inaequalis, T. gregarius* gr. sp. 2. Here, the organs most involved in flight, swarm behavior, and aerial location of a mate are reduced (male antennae, wings) while the genitalia are relatively robust and the abdomen is shorter and wider than in *T. inaequalis*. Several other sibling species pairs from Barrow include one member with a reduced antennal plume, a phenomenon described as an adaptation to reduced aerial swarming in the windy arctic environment (Wülker 1959).

Respiration

Oxygen uptake rates of *Chironomus* were measured on groups of 5 to 20 larvae in a selected millimeter length class in 6-ml chambers containing Millipore (HA) filtered pond water. Chambers were placed in a water bath at 5°, 10°, or 15°C (±0.01°C) and the oxygen concentration measured over at least two 30-minute periods with a YSI meter.

The respiration of *Chironomus* was a function of temperature, larval size, and acclimation period (Figure 7-7, Table 7-2). Larvae that were moved to 15°C from pond temperatures of 7 to 10°C increased their respiration greatly (non-acclimated in Figure 7-7); after 2 days larval respiration was reduced by 40% (acclimated). The reverse occurred when animals were transferred to 5°C. First, the rate was lowered, then increased again by 25% after 2 days. This is Precht's type 3 compensation

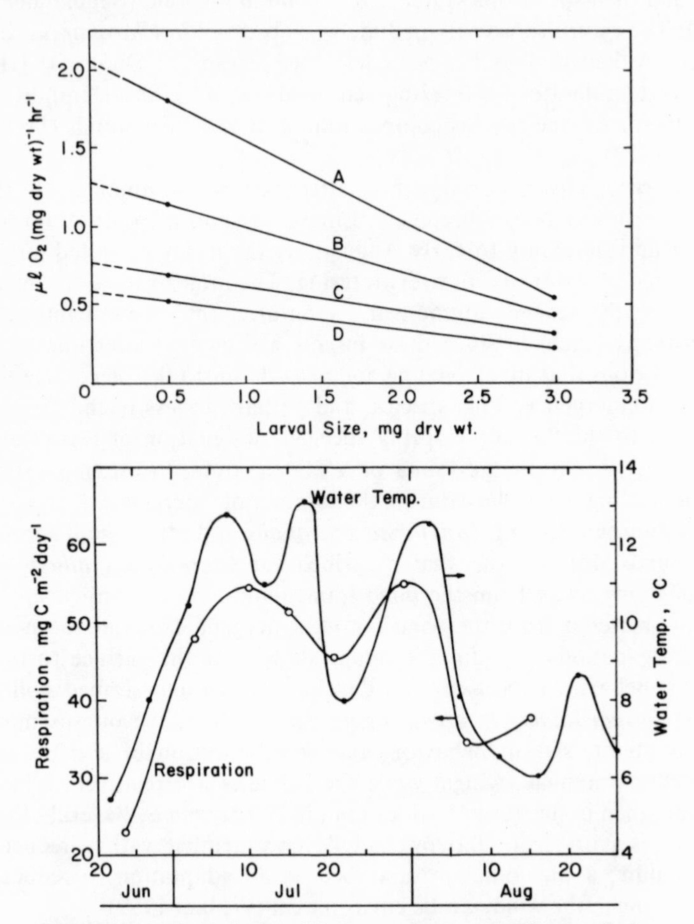

FIGURE 7-7. *Relationship between respiration and size for* Chironomus *larvae at 15°C (A and B) and 5°C (C and D) (top). Lines A and D represent nonacclimated larvae, while B and C represent acclimated larvae. These data were used to calculate the respiration rate of chironomids in Pond B for 1972 (bottom).*

TABLE 7-2 *Equations Describing the Relationship between Respiration and Size for* Chironomus *Larvae Represented in Figure 7-7*

Line designation	Temp (°C)		Equation of the line
A	15	Nonacclimated	-0.266 dry wt + 1.0968 = mg C (mg dry wt)$^{-1}$ hr^{-1}
			-0.496 dry wt + 2.048 = $\mu\ell$ O_2(mg dry wt)$^{-1}$ hr^{-1}
B	15	Acclimated	-0.163 dry wt + 7.24 = mg C (mg dry wt)$^{-1}$ hr^{-1}
			-0.304 dry wt + 1.352 = $\mu\ell$ O_2(mg dry wt)$^{-1}$ hr^{-1}
C	5	Acclimated	-0.0814 dry wt + 0.416 = mg C (mg dry wt)$^{-1}$ hr^{-1}
			-0.1520 dry wt + 0.776 = $\mu\ell$ O_2(mg dry wt)$^{-1}$ hr^{-1}
D	5	Nonacclimated	-0.064 dry wt + 0.311 = mg C (mg dry wt)$^{-1}$ hr^{-1}
			-0.120 dry wt + 0.580 = $\mu\ell$ O_2(mg dry wt)$^{-1}$ hr^{-1}

(Prosser 1973). Platzer (1967) has shown a similar response in tropical *Chironomus*. In other experiments the Q_{10} of unacclimated animals was 2.5 to 3.5 in the 5 to 15°C range and 2.0 in the 15 to 25°C range. A similar pattern of decreasing Q_{10} at higher temperatures has been observed for *Chironomus* sp. (Konstantinov 1971) and for stoneflies (Knight and Gaufin 1966, Prosser 1973). Thus, the chironomids appear to adapt to higher temperatures and keep their metabolic costs at a moderate level. Yet, they stay active at 5°C and respiration increases with acclimation to low temperatures. Respiration rates for fourth-instar *Chironomus* at 15°C ranged from 1.7 to 0.6 μg C (mg dry wt)$^{-1}$ hr^{-1}, with rates decreasing as larval size increased. Walsh-Maetz (1953) found a rate of 1.3 μg C (g dry wt)$^{-1}$ hr^{-1} (originally as 200 mm^3 (g fresh wt)$^{-1}$ hr^{-1} assuming 0.15 g dry weight per gram fresh weight) at 17°C for *Chironomus* larvae from France, so the respiration of unadapted thaw pond *Chironomus* was not notably different from other populations.

Chironomid respiration per unit area can be estimated for any of the ponds in which biomass and temperature are known by using the equations in Table 7-2. Non-acclimated rates were used because temperature fluctuated over a range of up to 10°C on many days. Mean daily respiration was 3.5 times higher in 1972 than in 1971 because both temperature and biomass were higher. Between 5 July and 16 August 1971 respiration was 630 mg C m^{-2}, or 15 mg C m^{-2} day^{-1}. Between 20 June and 17 August 1972 respiration was 2950 mg C m^{-2} or 52 mg C m^{-2} day^{-1} (Figure 7-7).

Temperature Experiments

In an experiment to determine the effect of temperature on growth rates, Bierle grew *Chironomus* from egg masses at different temperatures. At 5°, 10°, 15° and 25°C the eggs hatched in 14, 5, 3 and 2 days, respectively. At 5°C, larvae remained in the first instar throughout the 40-day experiment, an observation consistent with the evidence for threshold temperatures from Johnson and Brinkhurst (1971a) and Danks and Oliver (1972). At 10°C, all larvae reached the second instar. These responses were also reflected in field observations that larvae reach the second instar at about 50 to 60 thaw-days in their second summer. The next temperature level (15°C) allowed much more rapid growth with some larvae reaching the middle of the fourth instar in 40 days after hatching. Growth varied from chamber to chamber (Figure 7-8) but there was a strong negative correlation between size achieved and final larval density in the chamber. At 25°C larval growth (as length) was the same as at 15°C, but mean weight of a larva was less. The last temperature was much warmer than field populations would encounter; temperatures above 15°C were brief and infrequent in the ponds. This experiment illustrates the controlling effect which normal pond temperatures may have on larval growth and production.

Similar results were obtained in a pond experiment in 1972. Enclosures 1 m^2 in area containing sediments and water were established in one of the ponds and larval populations were counted (28 July). One

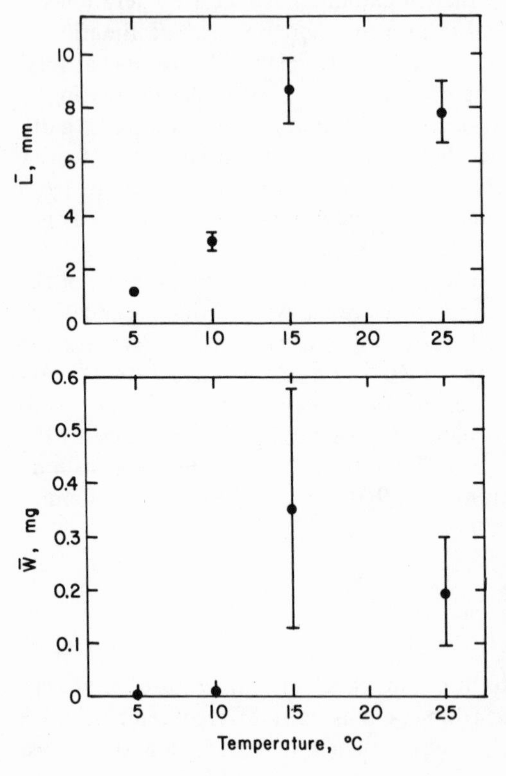

FIGURE 7-8. *Mean lengths and weights of* Chironomus *larvae grown in culture from single egg masses at four temperatures for 40 days. Bars indicate 95% confidence limits.*

enclosure was then heated about 5 to 6°C above the surrounding water with buried heat tapes and the larval populations were sampled again after 3 weeks. Fourth instar *Chironomus* in the heated enclosure grew 2 to 3 mm, while the same cohort in the control enclosure did not grow. Third instar *Chironomus* in the heated enclosure appeared to grow about 1 mm, but their density was low at the end of the experiment and the data are not definitive.

Vertical Distribution

Most chironomid larvae occupy the top 6 cm of sediments. In several series of analyses of cores at 2-cm intervals in 1971 and 1972, less than 10% of larvae occurred deeper than 6 cm (Figure 7-9). Between 55 and 90%, in fact, were in the top 2 cm. This pattern is consistent with those observed in many temperate lakes (Cole 1953, Milbrink 1973).

There is also a strong stratification of chironomid larvae by size. When larval size was plotted against depth in the sediments using data from August 1972 samples from Pond J, the relationship had a slope of 1.1 and an intercept of 5.4. Thus, the predicted size of an average larva at the mud-water interface was 5.4 mm, and for every centimeter of increasing distance into the sediment, average larval size increased by 1.1 mm. A few larvae in any large sample were found below 10 cm and these were usually in the fourth instar. This appears to represent superior ability to burrow and force water through a longer tube (Brundin 1951).

Trophic Structure

Most of the species of chironomids are detritivores and most of these are deposit feeders. *Chironomus*, the Tanytarsini, and probably most of the Orthocladiinae feed on material either gathered from near the surface of sediments or filtered from the water with nets; the gathering has been observed, but the filtration is presumptive and is based on previous studies of other *Chironomus* species. *Chironomus* has been found to feed on Oligochaeta opportunistically (Loden 1974), but the importance of this behavior is not known in arctic systems. *Procladius, Derotanypus* and the rare *Cryptochironomus* are primarily predators, although particularly in the first two instars both the tanypodines will ingest diatoms and detritus. *Trichotanypus, Corynoneura* and *Cricotopus* feed on fine detritus and algae adhering to the macrophytes.

The most common invertebrate predator in the pond centers was *Procladius* (900 to 1800 m^{-2}). Its stomach often contained head capsules and other sclerotized structures of chironomid prey; laboratory trials indicated a short-term predation rate (at 10°C) of about one third-instar

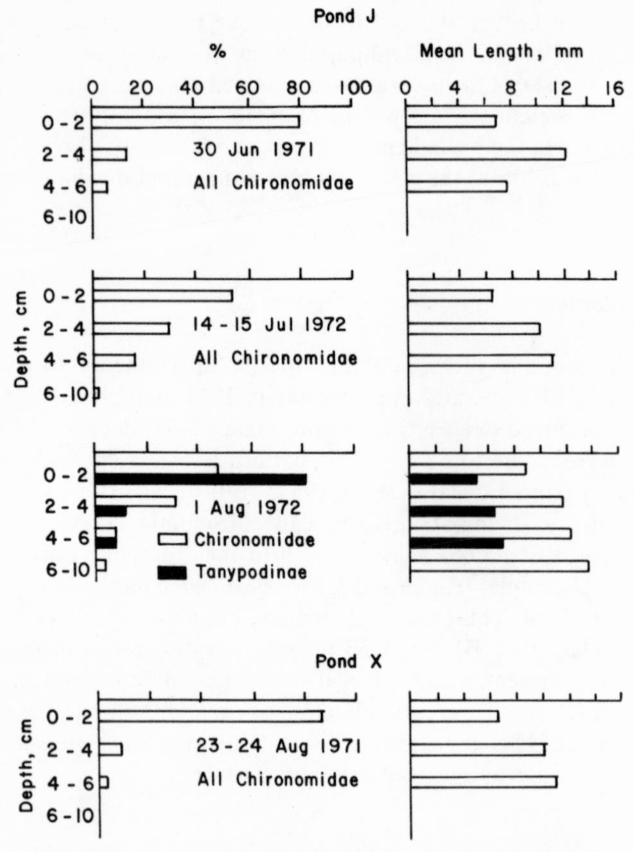

FIGURE 7-9. *Vertical distribution of Chironomidae larvae in the fine sediments of pond centers. Percent abundance and mean larval length in each depth interval below the sediment surface are shown from four dates in two different ponds.*

Tanytarsus every 2 days for a fourth-instar *Procladius*. *Procladius* was not strictly predaceous, for detritus and large diatoms were frequently observed in the guts of smaller instars and occasionally in larger ones. It is reasonable to assume that *Procladius* larvae would find it easy to locate the numerous, shallow-burrowing early instars and, indeed, most mortality occurred in the smallest instars of *Chironomus* (Figure 7-10). However, we have no direct evidence that *Procladius* is size-selective. Population density of *Procladius* and other predaceous chironomids was so high that each predator would have to ingest only one prey larva per week to account for all of the observed mortality. In addition to the mortality we could deduce from population measures (e.g., Figure 7-10),

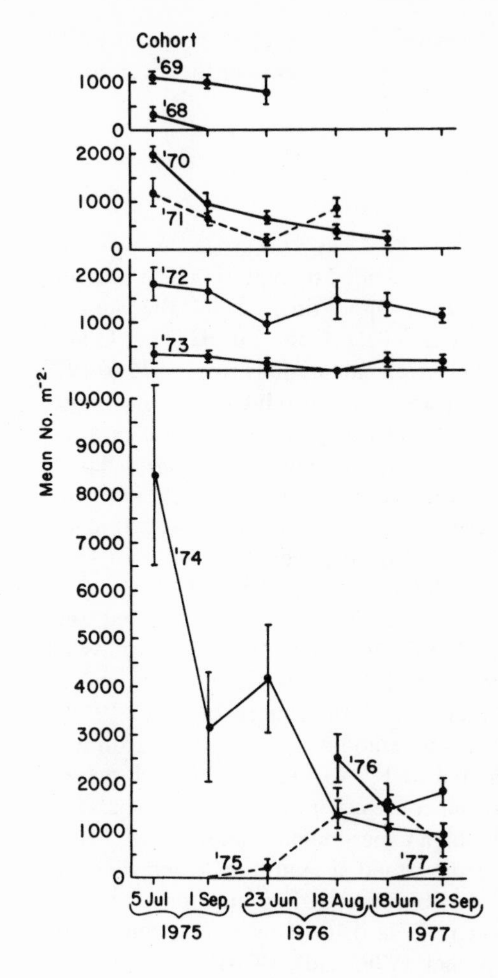

FIGURE 7-10. *Abundance of individual cohorts of* Chironomus pilicornis *in Pond J from 1975 through 1977. Mean number of larvae per* m^2 ± 1 *S.E. is shown for two dates in each year. Cohorts are named for the year in which they were recruited.*

there was undoubtedly a great deal of mortality in the very small first instars before they appeared in our pond-center samples.

Birds ate both larval and adult chironomids, as confirmed by the stomach contents of red phalarope (*Phalaropus fulicarius*), and by the study of Holmes and Pitelka (1968) in the pond area. Bierle examined a number of phalaropes in 1972, and found that 20% had eaten only immature stages. Another 30% had eaten adults as well as the pupae and larvae. Other bird species are also feeding their young during this period of most intense chironomid emergence in early July. It would therefore be reasonable to expect predation by birds to have some effect on chironomids but this does not seem to be the case. Quantification of bird-hours per pond in 1973 indicated that birds spent relatively little time feeding on any one pond (D. Schamel personal communication). Given that there are a hundred or more square meters in each pond and that each square meter contains several thousand larvae of appropriate size, the impact of bird predation must be small in comparison to the invertebrate predation.

Feeding Rates and Assimilation

Assimilation can be estimated and divided by ingestion as a measure of feeding efficiency. Assimilation is approximated by the sum of production and respiration, which was 3.01 g C m^{-2} or 32.7 mg C m^{-2} day^{-1}. This estimate combines production calculations from 1975 to 1977 and respiration estimates from 1971 and 1972. Production as dry weight was twice carbon (Waters 1977). The growing season was taken as 92 days. Ingestion was estimated by feeding individual *Chironomus* larvae with ^{32}P-labeled sediments. Natural sediments from the ponds were incubated in a solution of radioactive phosphate for 1 week with frequent shaking. Larvae were permitted to feed on the sediments at 12 to 15°C. Every 3 hours, the larvae were individually rinsed and transferred to scintillation vials. Cherenkov radiation in the water was measured, then larvae were returned to the labeled sediments for another 3 hours. When the rate of increase in ^{32}P uptake slowed, larvae were presumed to have filled their guts with labeled sediment. From the concentration of ^{32}P and organic carbon in the sediments, the amount of carbon in sediments ingested by the larvae was estimated as 120 to 150 mg C (g C larvae)$^{-1}$ day^{-1} or 12 to 15% of their carbon content per day. This is probably an overestimate because of relatively high experimental temperatures. Again using data on larval biomass from Pond J and the lower ingestion estimate, the community feeding rate becomes 325 mg m^{-2} day^{-1} in 1972 and the assimilation to ingestion ratio is 0.10. This is not unusual for deposit feeding invertebrates (Cammen 1978, Ladle 1974).

In Barrow ponds neither algae nor diatoms in whole sediments are abundant enough to supply the chironomid demands. Data from Chapter 4 for the carbon content and specific gravity of the upper 2 cm support a calculation of 7.5 cm^3 m^{-2} as the volume of whole sediment ingested by the chironomid larvae each day. According to studies of epipelic algae in the ponds there are only about 0.21 mg of algal organic carbon in the 7.5 cm^3. This is less than 1% of the organic carbon requirements of the larvae, so if epipelic algae are an important component of the diet, the larvae must feed on them selectively. Kajak and Warda (1968) indicate that *Chironomus* can achieve 5 times selectivity for diatoms (*Melosira* sp.) relative to detritus on a volume basis in an alga-rich sediment (2.2% by volume). If thaw pond larvae are selecting algae over detrital carbon with a factor twice that observed in Polish lakes, or 10 times, they would obtain just 2 mg C m^{-2} day^{-1}, less than 10% of their requirements, from this source.

The bacteria and fungi of the sediments are another source of food in the sediment. The carbon contained in each is about twice the algal carbon so it is difficult to believe that microbial biomass could be the sole food of the chironomids. There have been no data in other studies to suggest selective feeding for a relatively large, generalized detritivore such as these larvae although it is likely that smaller animals such as nematodes do select microbes. The only other study in which bacterial biomass was measured at the same time as feeding rates and energy requirements was that of Cammen (1978) for a marine worm, *Nereis*. The worm population needed to assimilate 8% of all the organic carbon in the sediment it ingested (per year) but the living portion of the organic carbon was only 3%. Thus, Cammen's results agree with ours. However, other studies where microbial biomass was not measured seemed to conclude that only living cells are used (Fenchel 1970, George 1964, Newell 1965, Hargrave 1970). It is also possible that microbes produce a great deal of extracellular polysaccharide and that it is this, rather than the microbial biomass, that is important to detritivores (Hobbie and Lee in press).

We conclude that the chironomid larvae must be consuming detritus as well as algae, fungi, and bacteria. This conclusion cannot be proven and the whole question has not been solved in any aquatic system.

Production of Chironomidae

Bierle estimated total production of the midge fauna in Ponds J, B, and D with larval and adult population data from 1972 (Table 7-3). The production was the sum of the increase in larval standing crop over the season, the biomass of emerging adults, the estimated biomass lost through mortality, and approximations of weight losses as exuviae at molts and as mucus secretions (drawn from the literature). Adult biomass was estimated as the number of emerging midges times the mean weight of a mixed assemblage. The data were adjusted for size by dividing adults into large (*Chironomus, Derotanypus,* and *Procladius*) and small (remaining species) size classes and using separate conversion factors.

TABLE 7-3 *Production of Chironomidae in 1972 in Three Ponds*

		Production (g m^{-2})				
Pond	Mean biomass g dry wt m^{-2}	Increase in standing crop	Mortality	Exuviae and excretions	Emergence of imagines	Total production
J	5.42	3.921	0.477	0.469	2.142	7.009
D	3.55	0.822	0.204	0.242	0.977	2.245
B	1.48	0.170	0.103	0.212	1.121	1.606

Mortality was estimated from the decrease in density of species-size classes (assumed to be cohorts) corrected for emergence losses. This density change was converted to biomass by a length-weight relationship.

In spite of some serious problems with Bierle's method of calculation (see below), these estimates probably reflect the actual relative secondary benthic production to be expected in these ponds. Production in Pond J exceeded that in Ponds D and B by factors greater than 3 and 4, respectively. The relatively large number and size of emerging adults and large weight gain of larvae during the summer contributed the most to the high value estimated for Pond J. Such differences may be primarily determined by between-pond variation in populations of large species. Emergence trapping and qualitative larval collections in a number of ponds in 1975-1977 support this explanation, as some ponds contain good populations of larger species like *Chironomus* while others are dominated by smaller species.

Production estimates such as these are subject to a number of errors. When a population takes several years to mature, as is true with all of these chironomids, it is not correct to attribute the total biomass emerging during a season to annual production for that year, for some of this biomass was produced during previous years. Some fraction of the biomass increment over a season will emerge in a future year and would thus be counted twice. Production lost to mortality before emergence can only be calculated when cohorts can be consistently recognized, clearly a difficult problem when up to seven cohorts of one species can coexist. Finally, Bierle's estimates of adult weights appears to have been too high, judging from later measurements in the same ponds.

Data collected in 1975-1977 on the population dynamics of *Chironomus pilicornis* larvae in the center of Pond J (data will be presented in M. Butler's Ph.D. thesis, University of Michigan) were used to calculate production of this species by the increment summation method. Here, the mean number of larvae in a given cohort during one summer is multiplied by the change in weight of an average individual over that summer. These cohort production values are summed for all coexisting cohorts (seven in this case) to provide a value for annual production by the entire species population. If only development, and not actual feeding and growth, occurs during the final summer (see Life Cycles above), emerging adult biomass during a season includes no actual production from that year.

Annual production estimates of 2.77, 1.42, and 1.80 g dry wt m^{-2} were determined for *Chironomus pilicornis* in Pond J during 1975, 1976, and 1977 respectively, based on samples from two dates each year. Since sampling intervals were different in different years, these values have been adjusted to a standard growing season of 92 days (approximately June 15 through September 15). The mean value is 2.00 g dry wt m^{-2} (about 1.0 g C m^{-2} yr^{-1}), with a range $\pm 38\%$ over the 3 years. Most of the production in any season results from growth by cohorts which are in the fourth instar. Consequently, variation in the strength of these cohorts in different years explains most of the variance in annual production. The cohort recruited in 1970 averaged 1475 fourth instar larvae m^{-2} in 1975, and contributed 61% of the total population production in that year. Similarly, the 1972 cohort was in the fourth instar in 1977, and its 1261 individuals m^{-2} produced 66% of the total. The corresponding cohort in 1976 (recruited in 1971) had a mean density of only 542 individuals m^{-2} and produced only 52% of the total. Production by cohorts in earlier instars is almost always lower due to lower individual growth (Figure 7-6) regardless of cohort size; hence the total 1976 *C. pilicornis* production was reduced relative to the other 2 years.

In reviewing single species estimates of annual production for chironomid larvae, Waters (1977) reports values ranging from as low as 0.024 g dry wt m^{-2} to as high as 161.6 g m^{-2} (for larvae in sewage treatment lagoons). Most values were in the range 0.5 to 3.5 g m^{-2}. The 2.3 g m^{-2} yr^{-1} estimated for *Chironomus pilicornis* in Pond J is certainly within the range of these values from temperate habitats. Although other species in the pond were not studied in sufficient detail to determine production, this species dominates the biomass of the pond centers due to its large maximum size. Other large species (mainly Tanypodinae) may contribute substantially, but sizes and densities suggest the total is less than double the production of *Chironomus*.

Controls over the production of this one species are probably multiple and interactive. The most striking aspect of its biology in this arctic environment is the seven years required for growth to maturity, but production is not limited by the long life cycle. Arctic winters essentially separate the usual two-year life span of cold water *Chironomus* (e.g., Jonasson 1972) into seven disjunct phases, each with its own cohort. The three cohorts of *C. pilicornis* which are in the fourth instar at any one time constitute most of the standing stock and contribute most to production.

Despite buffering factors within a species, life cycle production for different cohorts will depend primarily on their abundances. For the midge community as a whole, differences in species composition can be important. Factors influencing cohort sizes of large species through both recruitment and mortality will have the greatest effect on year-to-year and pond-to-pond variations in production.

Therefore, major control over production may occur during emergence and reproduction. As weak fliers and synchronous emergers, arctic chironomids are vulnerable to the high winds and cool temperatures which the adults may experience during their short life in the terrestrial environment. Larger species such as *Chironomus* and *Procladius* may experience reproductive failure if swarming is inhibited. Smaller species with surface mating habits may be less vulnerable, but also contribute less to benthic faunal production. Welch (1976) has suggested that weather conditions during emergence may have a major effect on the size of chironomid populations in Char Lake.

Once a cohort has been recruited, the pattern and intensity of mortality over its life span will influence production. Mortality appears to be high among small early instars of *C. pilicornis*, with a much lower death rate through the fourth instar. Individual cohorts need to be followed through most of their life cycle to confirm this apparent pattern, and to assess the variability of mortality in different ponds and years. The most obvious source of mortality is predation by other chironomid larvae in the family Tanypodinae, which includes four species in the genera *Procladius* and *Derotanypus* in these ponds.

Procladius is the most abundant predator in the community. Its potential importance is illustrated by data from Lake Sniardwy (Kajak and Dusoge 1970). In one experiment 534 *Procladius* m^{-2} were present with a prey population of 9,760 m^{-2}. In 5 to 9 days, *Procladius* consumed 530 mg of chironomids and 330 mg of benthic crustaceans or about 95 to 170 mg day^{-1}. Since prey were small, we estimate a weight of 0.3 mg wet weight per individual or 300 to 570 prey day^{-1} for this density of *Procladius*. *Procladius* population densities in Pond J in 1972 were similar to those in Lake Sniardwy. If *Procladius* populations were unusually abundant in a pond or year, their predation on early instars of *Chironomus* might limit cohort sizes and, consequently, benthic secondary production, for several years.

Ultimately, the general magnitude of chironomid production may be set by food availability resulting from primary production in the system. In fact, Welch (1973) has proposed that emerging insect biomass is a constant proportion of total primary production in lakes. The biomass emerging from a square meter of Pond J in 1976 averaged about 0.4 g and from Pond G about 0.2 g. Primary production in these ponds, including macrophytes, phytoplankton, and benthic algae, was about 50 g C m^{-2} yr^{-1}, or about 100 g of dry matter. The resulting ratio of emerging chironomid biomass to primary production was therefore about 0.2 to 0.4%, which agrees with the values cited by Welch from other systems.

The causality of this relationship has not been defined, and there are several variables which preclude its acceptance as anything more than a very general correlation. Emergence of *C. pilicornis* adults from Pond J in 1976 constituted only about 30% of the prepupal cohort predicted to emerge on the basis of larval sampling. This suggests high pupal mortality. The percentage of larvae emerging successfully is unlikely to be constant for different species or in different years. Hence emerging biomass may correlate poorly with secondary production. In addition, since most of the chironomids are detritivores, their food source is somewhat removed from the actual process of primary production. The variations in annual secondary production which could be introduced by differential species composition, by weather influences on recruitment, and by different rates of larval mortality preclude a very close coupling between primary production and secondary production of midge larvae.

TADPOLE SHRIMP*

Introduction and Life History

The tadpole shrimp, *Lepidurus arcticus* (Order Notostraca, Crustacea), only inhabits bodies of water with no fish. In temperate regions, it is limited to ephemeral ponds; in the Arctic it inhabits ponds as well as any lakes shallow enough to freeze to the bottom.

In the Barrow ponds, studied by Kallendorf (1974), the tadpole shrimp overwinter in the egg stage (33±0.5 μg C) and grow to maturity (10 to 12 mg C) in 60 days. Like the fairyshrimp, there is but one generation per year. Eggs hatched in the ponds sometime in the third week in June 1972, 17 days after the onset of runoff. The smallest naupliar instar contained 1.96±10.2 μg C. At this time, there were 2.6 animals m^{-2} in Pond C and 14.6 in Pond A. By 2 July, more eggs had hatched and Pond A contained 31 animals m^{-2}. These early stages, the first four instars, were planktonic but about the same time as the molt to the fifth instar (about 2 July) the animals moved into the upper layers of the sediment and became extremely difficult to sample (Table 7-4). When they reappeared at the surface of the sediments on 3 August, they had grown from 2.6 mm to 17 mm in length (Figure 7-11) and their density had dropped to 1.4 to 1.6 m^{-2}. Their density remained about the same during August and the animals reached 25.6 mm in length. Length was related to dry weight by a log-log regression where log dry weight (mg)=2.71 (log length in mm) −2.593. For this regression, r=0.91, n=47, and the regression was highly significant.

In the laboratory, growth was nearly continuous, with a molt every 48 to 60 hours at a constant temperature. Over 60 days, the animals will undergo 24 to 30 molts.

Egg production began in mid-August when organisms longer than 17.3 mm deposited eggs in their ovisacs. At this stage, animals are hermaphrodites and possess self-fertilizing ovotestes (Arnold 1967, Longhurst 1955). Adult animals carried 29 to 78 developing eggs beneath the carapace on the dorsal side (this is 13 to 35% of the total body weight). One to three eggs at a time are carried externally in each of the paired ovisacs on the 11th postcephalic appendage. These eggs (final diameter 0.71 mm) harden and turn orange in the 3 days they are held before release. The modal number of eggs in each ovisac was one; thus each animal could deposit about 20 eggs in the 30 days after sexual maturity.

* M.C. Miller.

TABLE 7-4 *Mean Length*[1] *and Mean Density*[2] *of* Lepidurus *in Pond A (PA), North Meadow Lake (NML), and North Meadow Pond (NMP) in 1973*

	PA		NML		NMP	
Date	Length	Density	Length	Density	Length	Density
June 27	1.96 ± 0.04	14.6				
June 29	1.93 ± 0.03	14.6				
July 2	2.44 ± 0.07	30.6				
July 4	2.45 ± 0.19	30.6				
July 5	2.62 ± 0.05	2.0				
July 30			6.79 ± 0.45	7.0		
Aug. 3	17.16 ± 0.72	0.3				
Aug. 6			10.75 ± 3.26	N.D.		
Aug. 8					27.78 ± 1.16	0.12
Aug. 11	18.61 ± 0.59	0.5				
Aug. 13			15.89 ± 1.74	7.3	28.94 ± 3.73	0.15
Aug. 16	21.45 ± 1.59	1.4	21.07 ± 1.04	N.D.		
Aug. 20	24.56 ± 1.66	1.6	21.96 ±3.06	7.3	31.30 ± 3.51	0.07
Aug. 24	25.10 ± 1.79	1.3				
Aug. 25					33.27 ± 1.67	0.06
Aug. 26	25.62 ± 2.64	1.3				

[1] mm
[2] ind m^{-2}

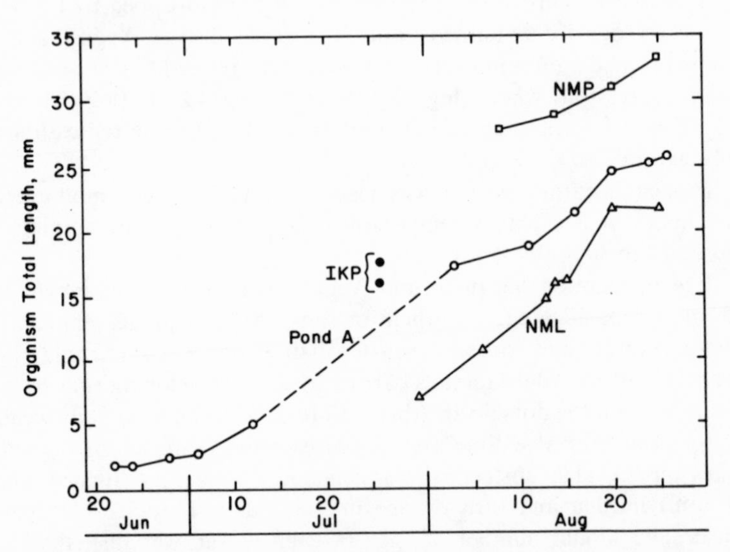

FIGURE 7-11. *Total length of* Lepidurus *during the 1973 growing season in Barrow ponds and lakes. For Pond A, length (mm) = 0.4262 (days after hatching) + 0.4126.*

Despite the fact that the total number of developing eggs carried internally is linearly related to total length of the organism (Figure 7-12), even the smallest animal has enough developing eggs to continue production at the maximum rate until the freeze in mid-September.

Accurate counting of the *Lepidurus* population was difficult because of their low numbers and because of the benthic stage of their life cycle. The planktonic larvae, in contrast, were quite easy to collect with a large net (10×50 cm). Both core and Ekman dredge samples were examined for the benthic forms but only a few were found. In contrast, when the adults moved into the surficial sediments in mid-August, the active animals could be easily seen and picked out with a small scoop-net. However, on cloudy and cool days they became inactive and could not be collected. To test the estimates, three 2.25-m^2 circular chambers were constructed and their edges pressed into the sediments on 3 July when the larvae were still planktonic. In August, an average of 1.33 adults m^{-2} were collected from these enclosures, which is close to the 1.3 given in Table 7-4.

Lepidurus were sampled in three other bodies of water (Table 7-4) and in the largest, North Meadow Lake, the animals were the most abundant but smallest in size. The largest animals were found in a pond even shallower and smaller than Pond A. It appeared that the animals in North Meadow Lake were concentrated along the lee shore in small, protected embayments.

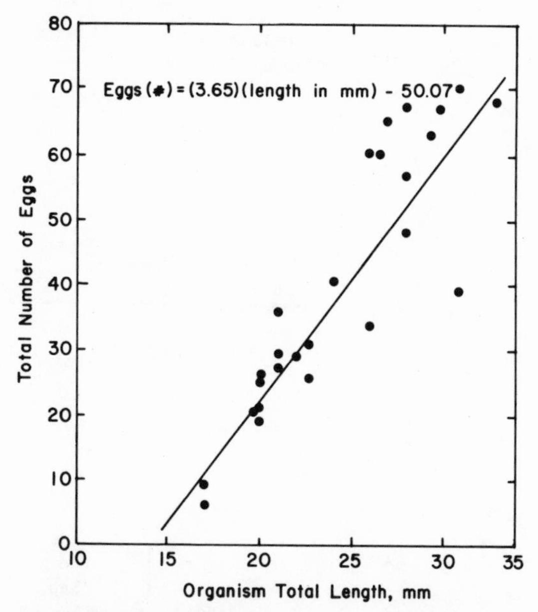

FIGURE 7-12. *Number of eggs carried internally by* Lepidurus *of various lengths.*

In summary, *Lepidurus arcticus* has a high reproductive rate, short generation times, and a resistant resting egg. Thus, its life history characteristics in tundra ponds are not significantly different from those of its temperate relatives (*Triops*) which exploit ephemeral ponds. Its strategy is to release as many eggs as it is able, one to three at a time, and to begin releasing eggs as early in the summer as possible in order to minimize the probability of the loss of a generation caused by an early freeze or by pond desiccation.

Effects of Temperature

The tadpole shrimp's maximum size is a function of the time of hatching in spring, the length of the growing season, the average temperatures, and the availability of food from year to year or pond to pond (Longhurst 1955). In East Greenland, Poulsen (1940) found the maximum average size of *Lepidurus arcticus* to be 24 mm. Adults in North Meadow Lake reached 21.9 mm, in Pond A they were 25.6 mm, and in North Meadow Pond 34.4 mm (Figure 7-11). The differences in maximal length between the habitats corresponded to the differences in size when the animals returned to the surface of the sediments in late July

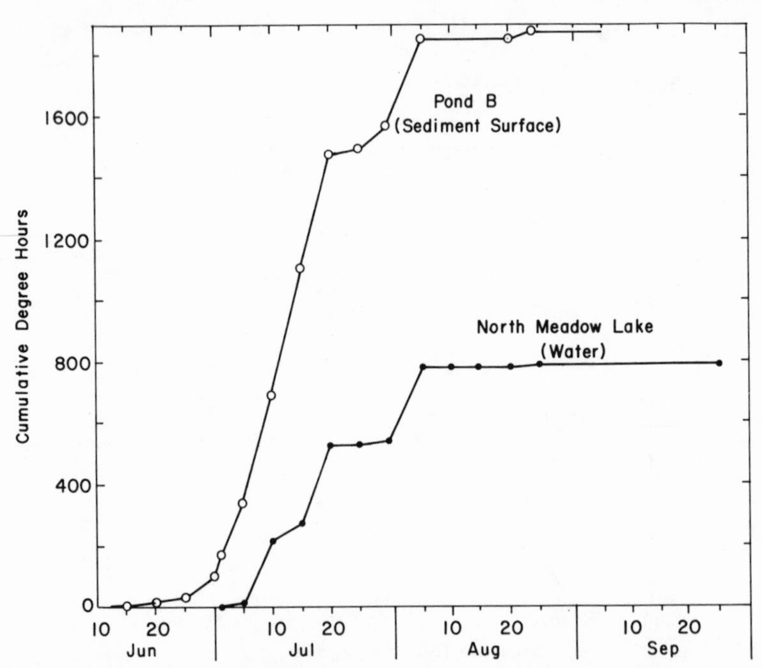

FIGURE 7-13. *Cumulative degree-hours greater than 10°C in Pond B and North Meadow Lake (1973).*

1973. Thus, individuals in North Meadow Lake were the smallest at 7 mm, those in North Meadow Pond the largest at an estimated 20 mm. Animals from several small, shallow inland ponds near Ikroavik Lake on 26 July, 1973, were very large for that date, 16 to 18 mm.

The size of the animals when they first appear on the surface of the sediments and the maximum size attained during the summer vary inversely with the depth and volume of the pond or lake. Accordingly, North Meadow Lake was 100+ cm deep; Pond A, 25 cm; North Meadow Pond, 10 cm; and Ikroavik Lake Ponds, 6 to 10 cm. The reason for this relationship is that the temperature on warm sunny days is higher in shallow ponds than in deeper ponds and lakes. For example, the cumulative degree hours above 10°C in 1972 in Pond C, which is equivalent to Pond A, was 2.4 times greater than the degree hours in North Meadow Lake by September (Figure 7-13). Similarly, Stross and Kangas (1969) measured twice as many degree days in a trough pond (0.5 m maximum depth) as in Imikpuk Lake (maximum depth 2.8 m).

The greatest differences in temperature in various habitats occurred in early summer when insolation was highest. At this time, most of the light energy received at the lake surface is reflected or goes to melting ice (Figure 2-4). The ponds, in contrast to the lakes, thaw early because of their thin ice cover and become warmer much sooner than the lakes. Thus, *Lepidurus* in shallow ponds have an earlier hatching date and, depending upon June temperatures, a faster initial growth rate than in lakes.

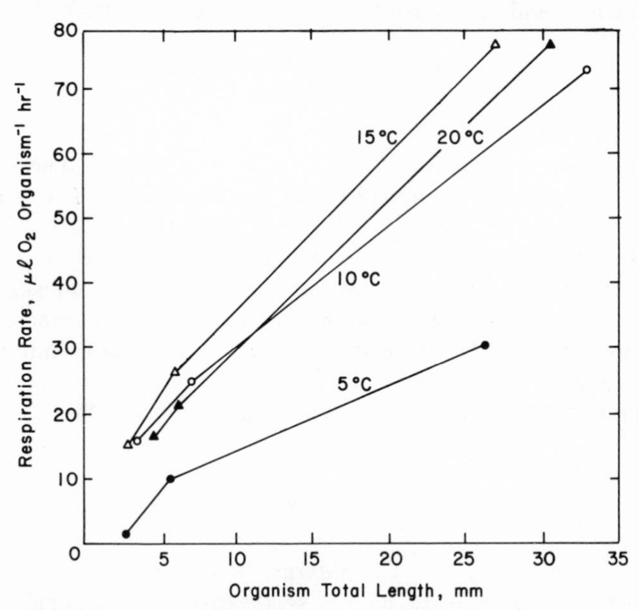

FIGURE 7-14. *Respiration of* Lepidurus arcticus *at four temperatures.*

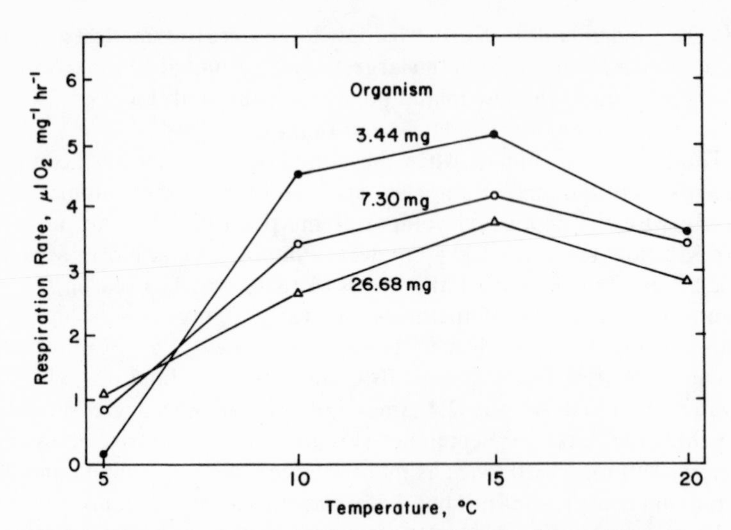

FIGURE 7-15. *Respiration per milligram dry weight of three different sizes of* Lepidurus *at four temperatures.*

However, in 1973 the mean daily water temperatures in the pond did not reach 10°C until 30 June; this was not too different from the temperature pattern in the lake. The difference in the size of organisms in these two habitats was caused not so much by differences in rates of growth but by the time of the melt and the time of hatching of the eggs (Figure 7-11).

Respiration

Oxygen uptake was measured in a Gilson respirometer. The small size of the flask restrained the animals so that the measured rates are likely minimal.

The respiration rates were highest at 15°C and decreased at higher and lower temperatures (Figure 7-14); inhibition at 20°C is unusual for animals but this is a higher temperature than they ever encounter in the ponds. The smaller animals have higher respiration rates per milligram of dry weight (Figure 7-15). In general, the rates per milligram agree with those found for zooplankton (D. Kangas personal communication).

Production

The annual net production of *Lepidurus*, 11.5 mg *Lepidurus* C m^{-2}, was calculated from the increase in biomass, the loss from the population, the mass of exuviae, and the amount of eggs produced (Table 7-5). This estimate covered only 27 June to 27 August, however, and some additional

TABLE 7-5 *Production Budget for* Lepidurus arcticus *in Pond A, June 27 to August 27, 1973*

Date	Length (mm)	Dry wt (mg)	Number m^{-2}	A Biomass increase per period ($mg C m^{-2}$)	B Egg prod. per period	C Exuviae ($mg C m^{-2}$)	D Mortality ($mg C m^{-2}$)	E Respiration
27 June	1.96	0.063	30.6	0.27	0.0	0.02	0.24	0.709
5 July	2.19	0.084	2	3.41	0.0	0.75	0.68	0.956
3 Aug	17.2	4.35	1.6	1.73	0.0	0.21	0.0	0.945
11 Aug	18.61	7.05	1.6	2.12	0.18	0.19	0.0	0.507
16 Aug	21.45	10.36	1.6	2.94	0.14	0.22	0.0	0.788
20 Aug	24.56	14.95	1.60	0.58	0.28	0.22	0.1	0.993
24 Aug	25.10	15.85	1.33	0.48	0.12	0.11	0.0	0.397
26 Aug	25.62	16.76	1.33	—	—	—	—	0.431
				8.07	0.72	1.72	1.02	
						Total respiration		5.73

Net production = A + B + C + D = 11.53 mg C m^{-2} released .
Assimilation = net production + E = 17.27 mg C m^{-2}

production undoubtedly occurred in early September. In some years, snowstorms or very low temperatures in August can kill most of the adults.

Egg mass produced was considerable and made up as much as 35% of the total weight of a small animal. Most of the eggs were not released and the increase in eggs beneath the carapace merely increased the biomass (Column A in Table 7-5). The mass of eggs actually released was only 6.2% of the total net production.

Another 24% of total production was calculated from the exuviae given off and from the mortality that occurred before the end of the season. Because of the 24 to 30 molts that take place during growth, the exuviae represent a major loss of carbon (15%) while the mortality is only 9% of total production. Most of this mortality occurs in the younger stages when the density of animals falls from 30 to 2 m^{-2}.

During the cold summer of 1973, *Lepidurus* appeared to respire only 33% of assimilated carbon, an extremely high yield indeed. Assimilation was the sum of net production plus respiration (Table 7-5). Respiration was calculated from mean daily temperatures, biomass estimates and data (Figures 7-14 and 7-15) on respiration rates for different temperatures and body weights (see Kallendorf 1974 for details). Of course, this yield of 67% is possible only because the temperature during 1973 was mostly below 10°C. Also, as noted, the Gilson measurements may be too low as this is normally an active animal.

Even if there is some error in the respiration measurements, the growth efficiency of the tadpole shrimp must have been high in order to produce such rapid growth. In 60 days, animals grew from a small egg (33 μg C) to a 25-mm adult (6800 μg C). The efficiency of close to 70% is very high compared with other aquatic crustaceans (Table 7-6) such as *Hyalella* (15 to 22%) and *Leptodora* (8%). The laboratory population of *Simocephalus* and the *Chironomus* in the ponds at Barrow did approach this efficiency.

The total carbon flux through this animal is extremely small (17 mg $m^{-2} yr^{-1}$) and would be small even if any respiration error were corrected. This may be due in part to the very low density of *Lepidurus* during 1973. On the other hand, the importance of a top carnivore such as this to the ecosystem is usually not reflected in the rate that it processes food.

Food Sources and Ingestion

In the planktonic form the nauplii of *Lepidurus* are only a little larger than *Daphnia*; yet they are not effective filter feeders. When the clearance rate was measured with $H^{14}CO_3$-labeled algae and glucose-^{14}C-labeled bacteria, the results of 0.16 ml hr^{-1} and 0.10 ml hr^{-1}, respectively, were very low for a 1.96 mm *Lepidurus*. Large *Daphnia* (2.0 mm) clear up to 3.4

TABLE 7-6 *Energy Budgets of Arctic* Lepidurus arcticus, Chironomus *sp. and Several Temperate Crustacea*

Parameter		*Lepidurus*[1]	*Chironomus*[2]	*Hyalella*[3]	*Leptodora*[4]	*Asellus*[5]	*Simocephalus*[6]
Assimilation efficiency	A/I	10% (of detritus)	10%	15%	40%	30%	31-72%
Ecological efficiency	P/A	70%	54-76%	15-22%	8%	18%	65-73%
	R/B day^{-1}	3.8%	1.4-4.1%	3.6%	-	4.5%	-
	P/R	231%	118-321%	31%	8%	23%	272%
	R/A	30%	24-146%	49%	92-94%	118%	27-35%
Conditions		field lab 60 days	field lab 60 days	field lab 1 yr	field lab 1 yr	Lab 21-23°C Alder leaves 117 days	22°C 27 days

A = assimilation I = ingestion P = net production

R = respiration B = biomass (ave.)

[1] Kallendorf 1974.
[2] Bierle personal communication.
[3] Hargrave 1971. Mathias 1971.
[4] Moshiri et al. 1969. Cummins et al. 1969.
[5,6] Klekowski 1970.

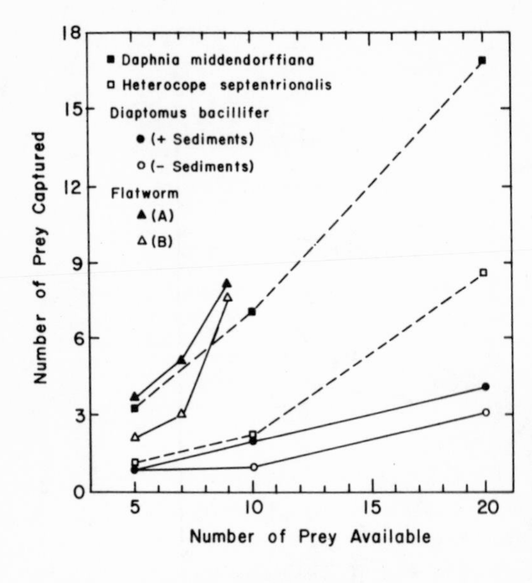

FIGURE 7-16. *Number of prey captured by* Lepidurus arcticus *adults at different densities of available prey.*

ml day^{-1} (Table 6-8). At this rate of clearing, this *Lepidurus* nauplius would ingest 0.043 μg algal C day^{-1} and 0.089 μg bacterial C day^{-1} at typical concentrations of phytoplanktonic algae (Figure 5-2) and bacteria (Hobbie and Rublee 1975). This total of 0.133 μg C day^{-1} is only 0.2% of the body weight but these animals respire 11% of their weight each day (Table 7-5). From this indirect evidence, we conclude that the nauplii are either predators or can use detritus.

Exactly the same results with pre-adult forms, 12.8 mm long, led to the conclusion that these too were possibly predators. In one experiment an animal ingested 0.17 mg C hr^{-1} of benthic detritus labeled with ^{32}P. This equalled 0.1% of the body weight day^{-1} and so indicates that they are predators.

Adult tadpole shrimp preyed on zooplankters but their preferred prey was flatworms. The observations were made in square containers, either 38.5 or 63.6 cm^2 in area. Each such experiment used one *Lepidurus*, one of three prey densities, and a 48-hour incubation at 10°C. Saturation of the feeding rate was seen only when *Daphnia* were added at a concentration of 40 liter^{-1}. Usually the feeding rate increased as the number of prey increased (Figure 7-16).

In another experiment, *Daphnia* labeled with ^{32}P were fed upon by *Lepidurus* in a 500-ml vessel. Three different food levels were used (5, 10, 20 *Daphnia*) and the experiments were run for 48 hours. There was a positive correlation between the *Daphnia* missing from the containers and the cpm in the *Lepidurus* (r=0.754, n=6, NS). Assimilation was low, with only 12.9% of the cpm in the missing *Daphnia* found in the *Lepidurus*.

We conclude that *Lepidurus* at all life stages is a rather non-selective predator that eats everything that moves. Its prey will vary from pond to

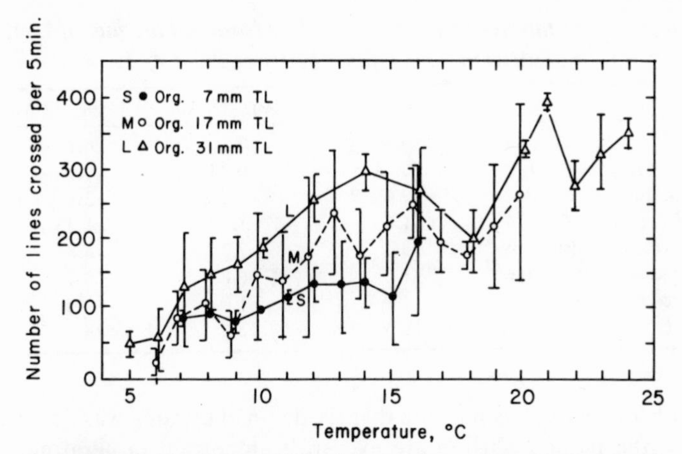

FIGURE 7-17. *Movement of three sizes of* Lepidurus arcticus *(in order of total length) at different temperatures. All animals were acclimated at 10°C. The units of movement refer to the lines on a 1-cm grid and mean and ranges are given for each observation.*

pond due to the considerable variation in prey density (Kangas 1972) and to the various prey's differing abilities to escape this predator. Some detritus is ingested by *Lepidurus* but assimilation is likely to be low. For example, small *Lepidurus* kept in the laboratory, with only surface sediments for food, did not grow nearly as fast as animals in the field.

Activity

The movement of *Lepidurus* was tested in a 14.5-cm-diameter shallow container with a gridded bottom. An observer counted the number of lines that each animal crossed in 5 minutes. Actual measurements of the path showed that each crossing equals 0.73 cm of path.

Although the ranges frequently overlapped, the general picture was of increasing activity for larger animals and increasing activity at higher temperatures (Figure 7-17). Over the 5° to 15°C temperature range, the means increased 2.7 times for the 7-mm animals, 5.3 times for the 17-mm animals and 5.8 times for the 30-mm animals. Field observations, using a gridded Plexiglas sheet placed above the sediments, showed a movement of 9.6 cm min^{-1}. Similar animals in the lab moved 7.3 cm min^{-1} at 5°C and 26 cm min^{-1} at 10°C so that the laboratory data appear to be reasonable. Therefore, we can next calculate the amount of sediment surface area that can be disturbed each day by the animals in each square meter (Table 7-7) from the movement per day and the width of the carapace. The disturbed area in each square meter was 40.9 m^2 for the 33-day interval or 1.2 m^{-2}

TABLE 7-7 *Computation of the Area Mixed by Normal Foraging of* Lepidurus arcticus *in Each Square Meter of Pond A, 1973*

	26 July - 5 Aug	5 Aug.-18 Aug	18 Aug -27 Aug
Mean size organism (mg dry wt)	3.4	7.3	26.7
Mean width carapace (cm)	0.61	0.74	0.88
Mean temp. during interval	9.4 ± 9°C	5.7 ± 1.1°C	3.1 ± 2.4°C
Density of organism (# m^{-2})	1.6	1.33	1.33
Movement (cm organism min^{-1}	13.2	7.3	5.8
Area covered (m^2 org^{-1} day^{-1}	1.2	0.8	0.7
Area covered (m^2 day^{-1})	1.9	1.0	1.0
Area covered (m^2)	18.5	13.5	8.9

day^{-1}. Other experiments indicate that the depth of plowing was about 0.5 cm. Thus, the tadpole shrimp are extremely important in aerating and mixing the sediment.

Control of Density

Lepidurus is an "r" selected species which maximizes its reproductive output by reaching reproductive age quickly (within 30 days). In addition, the reproductive output is maximized by having all individuals in the population reproductive and self-fertilizing (hermaphrodites). Also, the population produces its eggs during mid-summer while food is readily available. However, there is a lessened dependence upon food availability during late summer because the eggs have been previously formed (unlike *Daphnia*'s parthenogenic broods) and stored under the carapace.

There is good survival of eggs over the winter but the exact number produced is unknown. If the maximal density of the first naupliar stage was 30.6 m^{-2} and the population was in a steady state from year to year, then an adult survival of 1.5 animals m^{-2} would mean that each animal had to deposit 20 eggs if they all were to hatch. We observed that by the end of August the average animal carried 55 eggs. If the animals deposited eggs from mid-August through to the end of September at 2 to 4 eggs once every 3 to 4 days, then egg deposition would have been between 45 and 67 eggs m^{-2}. Thus, were the population in steady state, then half the egg production could have been lost, assuming all surviving eggs hatched. Unfortunately, our observations always end at the end of August, so we can not say how long egg production continued into September.

The greatest mortality we observed occurred between the time the planktonic nauplii disappeared (5 July) and the young adults reappeared on the surface of the sediments (3 August). In the laboratory all but two animals raised for several molts died on the fifth molt for an unknown reason. However, death in the laboratory incubators may not be analogous to what occurred in the pond. It should be noted that by the time of the fifth molt *Lepidurus* nauplii are probably predaceous and are larger (2.6 mm) than any other benthic organism except the chironomids.

Intraspecific competition may be one mechanism operating to determine density of *Lepidurus* adults. We noted that the ponds with the largest individuals had the lowest density and vice versa (Table 7-4). This could be a density-dependent effect such as competition for food or cannibalism. On one occasion in the laboratory cannibalism was observed (Kallendorf 1974). This interpretation of competition is confounded by the positive relationship between cumulative degree hours above 10°C and the maximum size of adults. Still we have observed that where the highest total biomass of *Lepidurus* was found the individuals were the smallest (on 20 August) and where the total biomass was smallest the individuals were the largest.

The length of time before the freeze probably exerts the dominant control over egg production and the next year's initial density. Yet many animals died before the freeze and in 1972 many *Lepidurus* were found dead in the pond in August before the ponds had frozen or reached a temperature of close to 0°C. In 1971, animals died during a mid-August snowstorm (19 August). It would appear that the probability of dying increases with normal stresses in the environment in the late summer. Animals taken into the laboratory in late summer frequently died within hours, whereas they would live for days under the same conditions in mid-summer. Hence, variations in density of *Lepidurus* from year to year are caused, at least in part, by random weather events.

SUMMARY

Macrobenthos

The larger benthic organisms, mostly insect larvae, dominate animal biomass and productivity in these ponds. They affect the pond system by mixing the surface sediments and promoting exchanges of dissolved materials between sediments and water. They also dominate major trophic pathways as detritivores, grazers, predators, and prey.

Larvae of the dipteran family Chironomidae constitute most of the macrobenthic fauna, although oligochaete worms, snails, mites, and turbellarians are also important. Non-dipteran insects and some larger crustaceans are poorly represented in the Barrow ponds relative to some other arctic freshwater systems. Many other invertebrate classes and insect orders common in temperate ponds, such as mayflies, odonates, leeches, and clams, are totally missing.

The soft sediments of the pond centers and the borders of emergent *Carex* and *Arctophila* represent two distinct habitats for the macrobenthos. The *Carex* habitat is less stable, as the low water levels at the end of each summer may expose the sedge beds. The non-dipteran insects, the snail *Physa*, and the water mite *Lebertia* are rarely found in the pond centers. Oligochaetes and midges occupy both habitats, but various species show a great deal of site specificity. Chironomids in the subfamily Orthocladiinae, the principal midge group reported from most arctic habitats, are most often found among the vegetation. The large midge *Chironomus* dominates the pond centers along with some of the smaller Tanytarsini.

Over 36 species of chironomids have been collected from these ponds; many of these were either first described from Barrow or were new to science. Several species have adapted to the windy arctic environment by dispensing with the normal adult swarming behavior. Instead, mating takes place on the pond surface and males show morphological modifications such as a reduced antennal plume. Many genera include two or more species, and congeneric larvae can not always be distinguished.

In most cases emergence is very synchronous within a species and sibling species emerge at different times; this makes species identifications easier. *Chironomus pilicornis*, the most abundant species in the pond centers (up to 15,000 m^{-2}), had two distinct emergence pulses. The two pulses may represent two distinct sibling species as the animals had two slightly different adult morphologies.

The timing of the emergences is apparently controlled by temperature. The same sequence of emergence was observed in all ponds every year, though certain species were sometimes missing or actual dates of emergence were sometimes shifted by early or late thaws. Synchrony of emergence was highest in species emerging early in the season and in at least one case was altered by nematode parasitism. Such synchrony maximizes mating success in these insects with short-lived adult forms and permits timing of emergence to act as a reproductive isolating mechanism.

Total chironomid emergence over the entire season varied from 500 to over 5000 adults m^{-2}. Pond to pond differences were considerable, but between-habitat variability within a pond was greater; the highest numbers emerged from the *Carex* beds.

The life history was studied in detail only for *Chironomus pilicornis*. Coexistence of seven distinct cohorts implied a 7-year life cycle for this species, with the larvae diapausing for seven winters in the frozen sediments. *Tanytarsus inaequalis* showed a 2-year life cycle. The life cycles of other species are presumed to fall between these two extremes, as no species appeared to reach maturity in less than 2 years.

The midge fauna appears to include many so-called "absolute spring species" which stop feeding and growth prior to their last winter, then pupate and emerge soon after thaw. There are also some species which feed, grow, and emerge during their final summer.

Temperature plays an important role in larval physiology and development. Respiration rates of *Chironomus* were similar to those of temperate species and decreased as the larval size increased. When larvae were placed in laboratory vessels which were 5°C above or below the *in situ* temperatures, there was some acclimation after two days. However, in the ponds the temperature changes this much each day so there was likely no acclimation. At 5°C, egg development times are long and growth of first instars does not occur. At a constant 15°C, growth requiring three to four summers in the field will take place in 40 days.

Between 55 and 90% of all chironomid larvae are found in the top 2 cm of sediment, and less than 10% occur deeper than 6 cm. The deeper larvae are usually fourth instar *Chironomus* so there is a strong vertical stratification of larvae by size.

Species living within the sediments are primarily deposit feeding detritivores or predators, while those found on the plants, including the snail, the stonefly nymph, the caddisfly larvae, and some midge larvae graze on epiphytic algae. *Chironomus*, the Tanytarsini, and probably most of the Orthocladiinae are deposit feeding detritivores. *Trichotanypus* and the orthoclads *Corynoneura* and *Cricotopus* were observed grazing epiphytic material from plant stems. The Tanypodinae *Procladius* and *Derotanypus* are at least facultative predators, as many head capsules of other midge larvae were observed in their guts; thus, *Procladius* predation may account for much of the mortality observed for early instars of *Chironomus*. In addition, chironomid larvae, pupae, and adults were eaten by the red phalarope (*Phalaropus fulicarius*). While the insects may be an important food for the birds, the impact of this feeding on the insects is small.

Estimates of assimilation and ingestion by *Chironomus* larvae lead to a feeding efficiency of 0.10, which is normal for a deposit feeding invertebrate. Yet, the source of the food is not clear; we have calculated that feeding on algal, bacterial, and fungal cells in the sediments at selectivities higher than those reported in the literature appears not to provide enough food. Digestion of organic detritus is a possible additional source of carbon.

Estimates of total benthic production during 1972 showed greater than a 4-fold variation among three ponds. One reason may be that late instars of large midge species such as *Chironomus* contribute most to total benthic production; therefore, the highly variable productivities may result from differences in species composition. On the other hand, because the population is made up of many cohorts, the effect of poor survival of a single cohort is reduced. For example, from 1975 to 1977 *Chironomus pilicornis* production was within 38% of the 3-year mean value of 2.0 g dry weight m^{-2} yr^{-1}. This production rate, which was equivalent to 1.0 g C, was well within the range of values for temperate zone midge populations.

As long as a new cohort is recruited every year, slow larval growth imposed by the arctic environment does not reduce potential productivity below levels expected for univoltine populations. Overall, the abundance of larvae is the determinant of productivity for a given species. The size of a cohort at recruitment can be influenced by weather conditions, and predation by invertebrates (especially predaceous midge species) can further reduce numbers and affect production.

The biomass of emerging adult chironomids, about 0.2 to 0.4% of total primary production, is similar to values reported from other aquatic systems. This may reflect a general dependence of benthic secondary production on the rate of carbon fixation by plants.

Tadpole Shrimp

The tadpole shrimp (*Lepidurus arcticus*) is a large (up to 26 mm) crustacean found in small numbers (about 1 m^{-2}) in or on the sediments of the ponds. This animal overwinters as a resistant egg and hatches in late June (about 2 mm long). The early instars are planktonic but in the first week in July the animals (2.6 mm in length) move into the upper layers of the sediment. When they reappear at the surface in early August, they are 17 mm in length. When the animals become adult, in mid-August, they begin to lay 1 to 2 eggs every 3 days. These eggs then overwinter.

Temperature is important to the growth rate of *Lepidurus*. For example, in a very shallow, warm pond the adults reached 34 mm in length while in a deeper, cold lake they were 22 mm long. Their respiration rates were highest at 15° and lower at 5, 10, and 20°C.

Production of 11.5 mg C m^{-2} yr^{-1} was calculated from the increase in biomass, the mortality, and the eggs. Some 69% of the total was increase in biomass while 9% was mortality and 15% was exuviae given off. In the cold summer of 1973, 67% of the carbon assimilated (production plus respiration) went to production, which is a very high percentage indeed. In part, this could reflect an error of the respiration measurement which had to be carried out on restrained animals.

The nauplii, pre-adult, and adult forms are likely all predators. In experimental vessels, adult *Lepidurus* preferred flatworms but also captured large *Daphnia* quite effectively. Given the low density of the tadpole shrimp, the predation rate in the pond should only be about 1 *Daphnia* m^{-2} day^{-1} out of a population of 800.

Perhaps the most important effect of *Lepidurus* was in disturbing and mixing the top 0.5 cm of sediment. The adults moved about 10 cm min^{-1} in the pond, which is equivalent to mixing 0.7 to 1.2 m^2 day^{-1} for each animal during the month of August.

The tadpole shrimp maximizes its reproductive output by having all the individuals reproducing and self-fertilizing. The actual number of eggs deposited in the pond, however, depends to some extent on the weather during late summer. At the beginning of September, the average animal still carried 55 eggs and had already deposited about ten eggs. At the rate of laying of a pair of eggs every 3 days, it is likely that not all eggs were laid before the pond froze unless the weather was unusually warm.

Production is likely limited by the mortality that occurred during the entire month of July when the pre-adults disappeared into the sediment. The cause for the mortality, a decrease from 30 to 1 animal m^{-2}, is not known. There is, however, the possibility of cannibalism, as the largest adult *Lepidurus* were always found at the same time as the smallest total number of animals. An alternate hypothesis is that there was a food limitation.

8

Decomposers, Bacteria, and Microbenthos

J. E. Hobbie, T. Traaen, P. Rublee, J. P. Reed, M. C. Miller, and T. Fenchel

BACTERIA*

Cell Numbers and Biomass

Direct counts of bacteria in the ponds were made during the ice-free periods in 1971 to 1973 using a modification of the method described by Francisco et al. (1973). Briefly, this involved adding a solution of acridine orange to a water sample, incubating for about 1 minute, and filtering 1 or 2 ml of the mixture through a black membrane filter. The filter was immediately examined under epi-illumination with blue light and fluorescent bacteria on the surface of the filter were counted. A more detailed description of this method may be found in Daley and Hobbie (1975) and an improved version in Hobbie et al. (1977). Sediment samples were diluted at least 100:1 before mixing with a high-speed Waring Blendor. Several milliliters were then treated as above.

Direct counts from three ponds in 1971 (weekly samples), one pond in 1972 (eight samples), and two ponds in 1973 (two samples per week) (see Hobbie and Rublee 1975) showed that total bacterial cell numbers in the water column generally ranged from 0.1 to 6.0×10^6 cells ml^{-1}. In the surface sediments the range was 0.1 to 55.0×10^9 cells (g dry wt)$^{-1}$ of sediment; cell numbers decreased with sediment depth (Table 8-1). When the sediments and water column are compared, a square meter of the sediments (5 cm depth) had a population of bacteria 3 to 4 orders of magnitude greater than the water column (20-cm depth). Direct counts made in 1973 (Figure 8-1) are representative of both the numbers and the seasonal pattern of bacteria in these ponds. In the plankton, almost all of the bacteria are free-living; that is, they were not attached to particles. In contrast, most of the sediment bacteria are attached.

Unfortunately, we did not find out until 1973 that when the planktonic system was examined in detail, there was a great deal of

*P. Rublee and J. E. Hobbie

TABLE 8-1 *Numbers of Bacteria at Different Depths in the Sediment of Pond B, Subpond 6, on 7 August 1972*

Depth	10^{10} cells (g dry wt)$^{-1}$
1 cm	5.8 ± 0.7
2 cm	3.3 ± 0.7
3 cm	2.5 ± 0.2
4 cm	1.4 ± 0.2
5 cm	0.8 ± 0.3

variability from day to day, and even from hour to hour, in the numbers of bacteria. This was caused by wind-driven currents that resuspended the bacteria from the sediment up into the water column of these shallow ponds. For example, on 24 July 1973 the bacteria in Pond A were sampled over a 20-hr period (Figure 8-2) during which the wind increased until early afternoon and then decreased through the rest of the sampling period. Wind mixing of the water column and sediment resuspension did appear to cause vertical movement of the bacteria (Figure 8-2). A similar experiment on 12 to 13 August 1974, when the wind was steady, showed little vertical movement over a 28-hr period (Hobbie and Rublee 1974).

Carbon in the bacteria was calculated from cell counts using a conversion factor of 0.87×10^{-8} μg C per cell (Ferguson and Rublee 1976). Peak biomass in the water column was generally 5 mg C m^{-2}. As with cell counts, sediment bacterial biomass was significantly higher than plankton

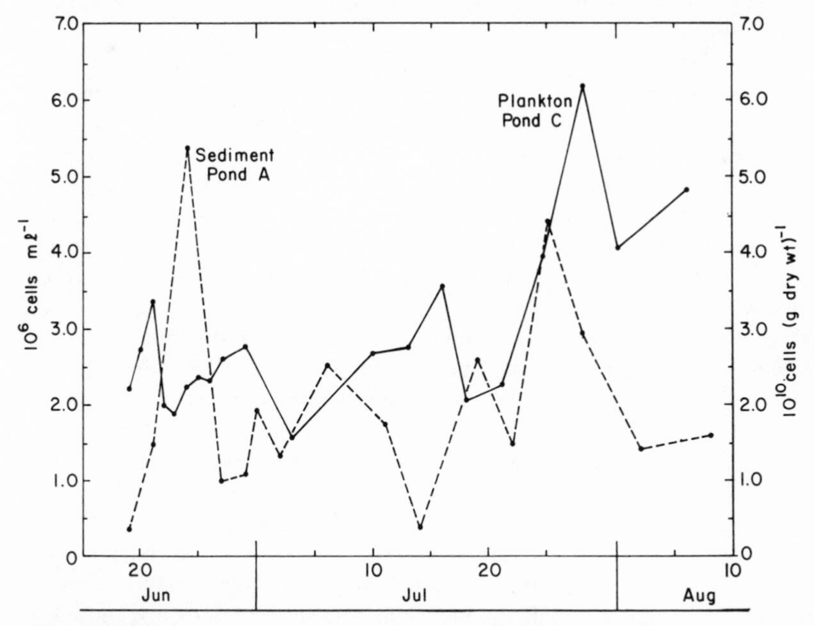

FIGURE 8-1. *Numbers of bacteria in the sediment of Pond A and the plankton of Pond C, 1973.*

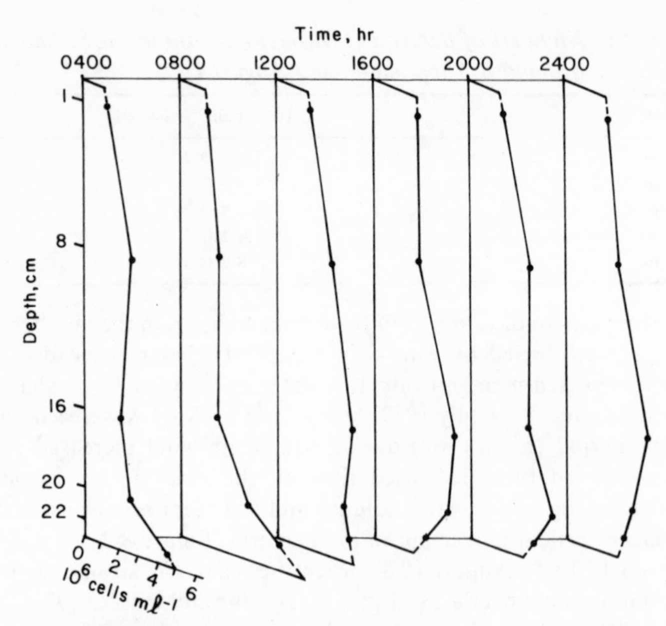

FIGURE 8-2. *Numbers of bacteria at various depths over 24 hours in the plankton of Pond A on 24 July 1973.*

biomass; the maximum quantity was 2.6 g C m^{-2} and the average was half that (assuming 5-cm depth of sediment).

Comparison of these values for cell density and biomass with other studies is difficult because of the variety of methods that have been used. The most common methods currently in use are plate count procedures and direct counts utilizing one of a variety of stains. Plate count techniques do not work for the determination of numbers and biomass; for example, Francisco (1970) reported a ratio of direct to plate counts ranging from 21 to 8900 with no apparent explanation for the differences. There are also smaller differences among the various direct count techniques but recent improvements in the microscope systems and in the membrane filter have led to agreement among a number of techniques with fluorescent dyes (acridine orange, FITC) and methods based on the scanning electron microscope (Bowden 1977) and lipopolysaccharide content of the bacterial cell wall (Watson et al. 1977).

The numbers of bacteria found in the tundra ponds agree with other measurements made in temperate waters with acridine orange (Table 8-2) but are higher than numbers collected with other methods, such as the phase contrast plus erythrosin dye. The difference is caused by better visibility with acridine orange or FITC that allows very small bacteria to be seen. As already noted, the high numbers in the ponds are caused, in part, by resuspension from the sediments. Even so, there are always high numbers of bacteria in both eutrophic and oligotrophic waters. The range

TABLE 8-2 *Bacterial Cell Counts and Biomass in the Plankton of Selected Water Bodies*

Location	Cell counts (ml^{-1})	Method*	Biomass ($mg\ C\ m^{-3}$)	Reference
Imikpuk Lake Alaska	$0.0025 - 3.4 \times 10^4$ $0.0025 - 7.4 \times 10^4$	PC, 22°C PC, 2°C	- -	Boyd and Boyd (1963)
Meretta Lake Canada	$2.1 - 82.7 \times 10^5$	EBDC	-	Morgan and Kalff (1972)
Char Lake Canada	$1.0 - 2.3 \times 10^4$	EBDC	-	Morgan and Kalff (1972)
Vorderer Finstertaler See, Austria	-	-	70 - 115	Tilzer (1972)
Castle Lake California	$1.1 - 19.0 \times 10^4$	EBDC	-	Jassby (1973)
Rybinsk Reservoir USSR	$0.1 - 3.9 \times 10^6$	EBDC	-	Romanenko (1971)
University Lake North Carolina	$1.0 - 8.0 \times 10^6$ $0.05 - 10.0 \times 10^4$	AODC PC, 25°C	- -	Francisco (1970)
Tundra Ponds Alaska	$0.09 - 6.3 \times 10^6$	AODC	1.6 - 114	this study
Ikroavik Lake Alaska	$0.05 - 5.8 \times 10^6$	AODC	0.9 - 104	this study
Loon Lake Canada	$0.5 - 0.9 \times 10^6$	AODC	-	Hobbie et al. (1977)

* PC = Plate count; EBDC = Direct count with erythrosin B stain; AODC = Direct count using acridine orange stain and epifluorescent illumination.

in lakes may be from 0.5×10^6 ml^{-1} in oligotrophic water to 8×10^6 ml^{-1} in eutrophic waters. The reason is that the bacteria spend most of their life in an inactive or only slightly active state. They will be very active and reproduce only when an algal bloom occurs or when they are in contact with a detritus particle. Thus, their presence does not reflect activity and their numbers do not reflect trophic states. As discussed later, bacteria are also grazed upon by zooplankton which would tend to reduce the numbers in rich bodies of water where animals are abundant.

Seasonal Pattern and Growth Rates

A characteristic seasonal pattern in populations was found in this study. Three phases may be distinguished: runoff phase, growth phase, and winter phase.

There is a peak in bacterial numbers and biomass in the water during the runoff phase in mid-June, a period characterized by flow of meltwater from snow through the ponds. This peak is primarily a result of the input of soil bacteria in the runoff water where concentrations may reach as high as 4×10^6 cells ml^{-1} (Boyd and Boyd 1963, Morgan and Kalff 1972). Secondarily, a nutrient input from the runoff water may stimulate growth of bacteria. Cell numbers in both runoff water and ponds correlate significantly with volume of runoff flow during this period. As flow decreases during the latter stages of this phase (late June), bacterial numbers also decrease. There are two possible causes of the decrease: first, the dilute pond water may be such a change for the soil bacteria that many die; or second, the bacteria may be eaten by zooplankton (see later discussion). Bacteria in the sediment also have an early peak which is likely caused by wintertime release of nutrients.

The growth phase encompasses the majority of the summer season. During this period, beginning in late June, there is a logistic growth of the population leading to a second peak in numbers in early August. An average net growth constant of 4.8% day^{-1} was calculated from standard microbiological equations. This average constant does not differ statistically from those of the individual planktonic populations (Table 8-3). Population constants of the sediment bacteria can not be compared statistically to those of the plankton but are similar (Table 8-3). Comparable values recalculated from other studies include approximately 2.31% day^{-1} for the bacteria at 10-m depth in Castle Lake, California (Jassby 1973), and approximately 1.5, 1.7, and 1.3% day^{-1} (spring minimum to fall maximum) for the 0.5-, 4.0-, and 7.0-m depths in a temperate reservoir (Francisco 1970). Because all these values are *in situ* ones, they are minimal values that do not include the effect of predation.

For approximately nine months of the year the ponds are totally frozen and there is little bacterial activity. Total numbers decrease by an order of magnitude between mid- and late August and the time of snow-

TABLE 8-3 *Net Growth Constants and Apparent Doubling Times for Bacterial Populations of Tundra Ponds*

	Pond	Year	Net growth constant K_n (day^{-1})	Apparent doubling time (days)
Plankton				
	B	1971	0.0600	11.6
	C	1971	0.0507	13.7
	D	1971	0.0426	16.3
	B	1972	0.0467	14.8
	A	1973	0.0446	15.5
	C	1973	0.0343	20.2
Sediment				
	J	1971	0.0514	13.5
	A	1973	0.0529	13.1
Average			0.0478	14.8

melt the following June. The death of these bacteria plus the mechanical damage to cells during freezing and thawing may release organic compounds that then allow the bacterial pulse in the early spring.

There is a close similarity of this seasonal pattern with earlier arctic studies (Boyd and Boyd 1963, Morgan and Kalff 1972) and with studies in alpine and temperate locations (Tilzer 1972, Francisco 1970, Romanenko 1971).

Heterotrophic Activity

For the other organisms in the ponds, the best measure of their effect and importance is a production rate or perhaps a feeding rate. For bacteria, however, this is not easy to measure as there are no good ways to measure bacterial growth and they feed on a great number of substrates. Heterotrophic bacteria do take up the simple compounds from solution and the process occurs by means of specific transport systems. They appear to be taking up a number of substrates simultaneously so that the rate of uptake of a single compound (or a group of compounds) will be an indicator of the activity of this general group of microbes. This is an imperfect measure but the data from a variety of lakes, estuaries, and oceans show that heterotrophic activity changes in step with primary productivity over a range of 4 or 5 orders of magnitude.

Heterotrophic activity was quantified by measuring the uptake kinetics of ^{14}C-labeled organic compounds. The basic method described by Hobbie and Crawford (1969) was to incubate subsamples with four different low concentrations of the substrate. After 1 to 2 hours the samples were killed with H_2SO_4, the $^{14}CO_2$ collected onto phenethylamine-saturated filter paper, and the subsample filtered through a membrane filter. Both filter paper and membrane filters were counted with liquid scintillation. The curve of uptake plotted against the substrate concentration resembles an enzyme saturation curve and is analyzed according to Michaelis-Menten kinetics. If the substrate concentration, S_n, could be measured, then a velocity of uptake could be determined and would be the best activity measurement. However, S_n is only a few parts per billion and is difficult to measure. Instead, the useful parameters are the maximum uptake velocity (V_{max}), turnover time (T_t), and the half-saturation constant plus substrate concentration ($K_t + S_n$).

Kinetic parameters were determined in Pond B in 1971 using ^{14}C-labeled glucose, proline, and aspartic acid (Table 8-4). The V_{max} and turnover times calculated for glucose in these experiments indicate a moderately high bacterial activity in the plankton and a relatively low bacterial activity in the sediments compared with other studies (Table 8-5).

Diurnal variation of V_{max} may be significant; it ranged from 0.122 to 0.761 μg C liter^{-1} hr^{-1} in an experiment conducted on 19 to 20 August 1972 in Pond B (Figure 8-3a). This was likely due to temperature alone as regression of temperature vs. V_{max} for this experiment shows a correlation coefficient of 0.71 (10 *df*)(Figure 8-3b).

In ponds B and C the V_{max} for uptake of acetate in the water column correlated positively (r=0.62), at the 0.05 level of significance, with primary productivity (Figure 8-4). These data exclude the period of spring runoff when much of the DOC in the ponds likely comes from outside the

TABLE 8-4 *Kinetic Parameters for the Uptake of Organic Compounds from Water of Pond B, 1971*

	Glucose			Proline			Aspartic acid		
	V^*_{max}	$K + S$†	T**	V_{max}	$K + S$	T	V_{max}	$K + S$	T
14 June	1.69	17.7	11	0.42	19.8	46	6.67	90.5	14
21	0.47	29.2	62	0.11	13.7	124	1.79	21.6	12
23	0.92	41.9	46						
28	0.88	36.0	41	0.11	9.3	84	2.17	38.9	18
5 July	0.86	10.8	13	0.12	2.6	21	2.26	26.6	11
11	0.67	11.1	25	0.16	10.6	66	2.08	40.1	19
19	0.51	17.8	35				2.27	61.4	27
26	0.41	44.4	122				1.05	47.5	45
2 August	0.67	75.1	115				1.05	35.0	33
9	0.58	22.1	38				2.86	36.7	13
16	1.25	131	105	0.07	5.4	81	2.56	21.6	8

* V_{max} in μg liter^{-1} hr^{-1}

† ($K + S$) in μg liter^{-1}

** T in hours

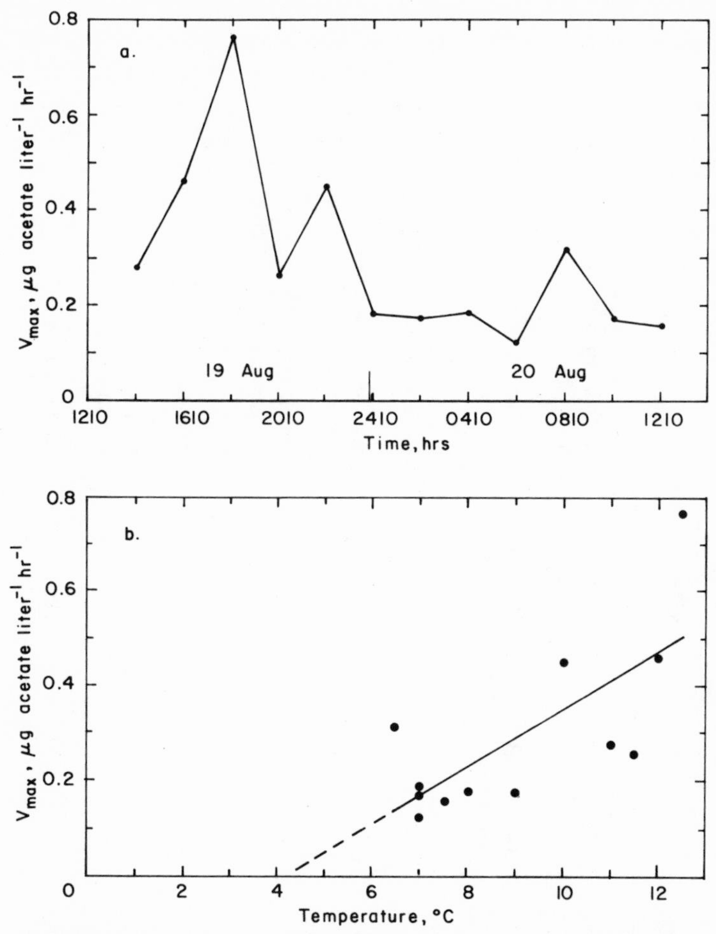

FIGURE 8-3. *a)* V_{max} *for acetate uptake, 19 to 20 August 1972, in the water of Pond B-6. b) Relationships between* V_{max} *and temperatures (same dates as above).*

ponds. Correction of the V_{max} values to a standard temperature did not improve the correlation.

Variation in the heterotrophic activity of plankton from pond to pond was also considered in an experiment on 9 August 1973. This comparison utilized the uptake of labeled acetate and glucose at only one concentration (192 and 2.1 μg C liter^{-1}, respectively). This method (Parsons and Strickland 1962) allows calculation of velocity of uptake, *v*, at that substrate concentration and of turnover time, T_t, but not V_{max} or (K_t+S_n). A transect of 25 ponds yielded glucose uptake velocities of 0.07 to 3.10 μg C liter^{-1} hr^{-1}, with a mean of 0.514 μg C liter^{-1} hr^{-1}. Results for acetate were similar, with a range of 0.04 to 1.51 μg C liter^{-1} hr^{-1} and

TABLE 8-5 *Kinetic Parameters for Planktonic and Sediment Bacteria from Selected Water Bodies*

Water Body	Glucose V_{max}[1]	Glucose T_t[2]	Acetate V_{max}	Acetate T_t	Remarks and Reference
Plankton					
Pond B	0.41 - 1.69 (0.81)	11 - 122 (55.7)	0.071 - 0.238[3] (0.123)	34.3 - 771.7 (253.4)	Ice-free period 1971 this study
Ikroavik Lake	-	-	0.043 - 1.115[4] (0.203)	161.8 - 947.9 (411.9)	Ice-free period 1971 this study
Char Lake	0.001 -- 0.008	100 - 1000+	-	-	Feb. to Dec. 1970 Morgan and Kalff (1972)
Lake Erken	0.004 - 0.18[3]	10 - 1000	0.006 - 0.142[3]	10 - 1000	Sept. 1964 - Aug. 1965 Hobbie (1966)
Meretta Lake	0.01 - 0.75	5 - 100+	-	-	Feb. to Dec. 1970 Morgan and Kalff (1972)
University Lake	0.2 - 13	0.5 - 80	1 - 43	0.5 - 34	Feb. 1969 - Jan. 1970 Francisco (1970)
Sediment					
Pond B	1.3 - 27.9	-	-	-	Ice-free period 1971 this study
Pond B	-	-	0.15 - 7.03[3] (2.16)	0.18 - 36.1 (16.02)	Ice-free period 1972 this study

Marion Lake	4 – 38 (9.4)	0.061 – 0.4 (0.23)	7.1 – 70.5 (20.4)	0.04 – 2.98 (0.128)	June 1971 - May 1972 Hall et al. (1972)
Pamlico Sound	(299.4)	(0.06)	-	-	Wood, L. W. (1970)

[1] V_{max} in μg C liter^{-1} hr^{-1} (average)
[2] T_t in hours (average).
[3] not respiration corrected.
[4] not respiration corrected; analyses at 5°C.

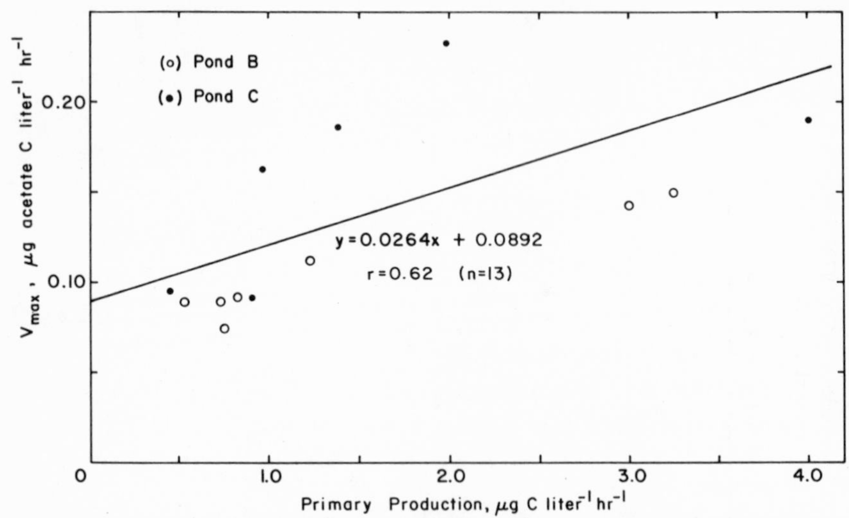

FIGURE 8-4. V_{max} *for acetate uptake by planktonic bacteria at different levels of primary production of phytoplankton algae. Data are from Ponds B and C in 1971. June data omitted.*

a mean of 0.635 µg C liter^{-1} hr^{-1}. It was further noted that velocity of uptake of both glucose and acetate in these ponds correlated significantly with primary production at the 99% level (Figure 8-5). For glucose r equaled 0.84 (n = 25) and for acetate r equaled 0.76 (n = 25).

The data on heterotrophic activity suggest that the interaction between bacteria and an active fraction of the DOC is the same in the ponds as in other aquatic systems that have been studied. That is, bacteria take up small organic molecules (sugars, amino acids, fatty acids) from the water even though the concentrations are only a few parts per billion. Presumably each bacterium takes up a number of substrates simultaneously in order to obtain enough energy. Over a certain range, the systems for transporting the substrates through the cell walls are undersaturated so any increase in concentration of the substrate is met by an automatic increase in the transport rate. It appears that there is a nearly constant level of substrate; this level is controlled by the bacteria. Therefore, the changes in uptake rates and turnover times are a reflection of changes in the rates of supply of the substrate.

In the ponds, much of the substrate comes directly (excretion) or indirectly (decomposition) from algae. A primary production rate of 10 to 15 mg C m^{-2} day^{-1} for the plankton and 100 to 150 mg C m^{-2} day^{-1} for benthic algae could come close to supplying the 6 mg C and 30 mg C m^{-2} day^{-1} that is the calculated uptake for the planktonic and benthic bacteria, respectively. Even larger amounts of organic carbon are photosynthetically fixed by the grasses and sedges. Some will be excreted

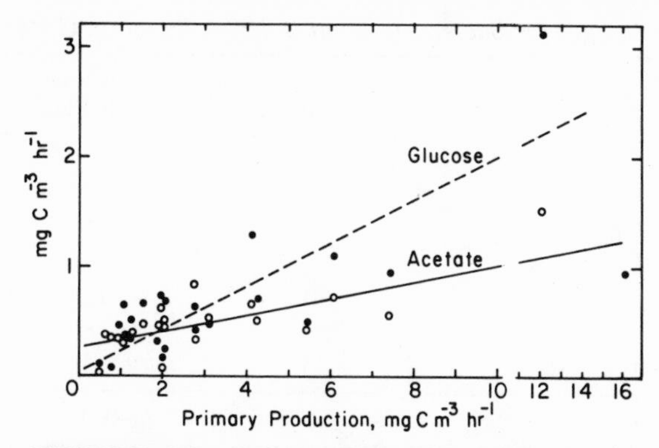

FIGURE 8-5. *Relationship of velocity of uptake of glucose and acetate to primary productivity of phytoplankton in a transect of 25 ponds.*

or lost during photosynthesis and even more will become available during decomposition.

Attached Bacteria

The discussion so far has dealt mostly with the free-living bacteria of the plankton or with those bacteria whose activity may be measured by gently mixing radioactive substrate into the sediment. However, a large number of bacteria are attached to the detrital particles of the sediment and many of these may be imbedded in the detritus or covered with masses of adhesive material (Costerton et al. 1978). The result is that many of these bacteria do not come into contact with the radioisotopes.

Thus, while the attached bacteria were counted by our methods their activity was not measured with the isotope uptake method. The whole-system respiration measurements would measure their activity.

Production

The production of bacteria was calculated from the changes in bacterial biomass, from respiration, from production losses, and from methane production (Hobbie and Rublee 1975). Their sum, gross production, was then apportioned to the two input parameters of total uptake of DOC and hydrolysis (breakdown) of detritus (Table 8-6).

Respiration in the ponds could not be measured *in situ* for the bacteria alone. In the water column the respiration rate was too low to be measured but there was measurable respiration in the sediments (see Sediment Respiration). The total quantity of about 13.7 g CO_2-C m^{-2}

TABLE 8-6 *Carbon Budget for Bacteria in Pond B, 1971, Based on Outputs*

	$g\ C\ m^{-2}\ yr^{-1}$	% of total production
Outputs		
1. Respiration	9.4	37
2. Biomass change	9.0	35
3. Loss to predation		
a. Chironomids[1]	1.0	15
b. Microbenthos	2.0	8
4. Methane production	1.1	4
Gross Production	22.5	100
Inputs		
1. Total uptake of DOC[2]	3.6	16
2. Hydrolysis[3]	18.9	4

[1] Estimated from total chironomid assimilation and assuming selective ingestion of algae over bacteria by a ratio of 2:1.
[2] Estimated from ^{14}C kinetic data.
[3] Estimated by subtraction of total DOC uptake from total production.

yr^{-1} was produced by bacteria, benthic algae, and micro- and macrobenthic animals. We have made measurements of the respiration of the macrobenthic animals, which make up most of the biomass, and made estimates of the respiration of the algae and of the microbenthos. Overall, the estimate is 4.3 g CO_2-C m^{-2} yr^{-1} for the respiration of the benthic animals plus the algae. The macrobenthos respired 1.5 g C, benthic algae about 1.5 g C (about 10% of primary production), and microbenthos about 1.4 [this is 14% of their biomass of around 100 mg C m^{-2} and is half of their feeding rate of 28 mg C m^{-2} day^{-1} (see Microbenthos)]. By subtraction, therefore, the bacteria respired 9.4 g CO_2-C m^{-2} yr^{-1}. This appears to be a large quantity but the amount of CO_2-C that left the entire pond was estimated to be about 50 g C m^{-2} yr^{-1} (Coyne and Kelley 1974). The remainder is likely root respiration (the sediment respiration measurements were made in the pond center where there were no macrophytes).

The change in biomass of bacteria over the summer season was 9 g C m^{-2} yr^{-1}. Over 99% of this value was attributable to the benthic bacteria.

Bacteria in the sediments are consumed by microbenthos and chironomids. Fenchel (1975) has estimated the microbenthos (protozoan and micrometazoan) predation in the tundra ponds as 20 mg C m^{-2} day^{-1} or 2.0 g C m^{-2} yr^{-1} on a seasonal (100 days) basis. Chironomid assimilation of all food was previously noted as 325 mg C m^{-2} day^{-1} (Chapter 7). Yet the measured feeding rates were only 7.5 cc sediment m^{-2} day^{-1}. This 7.5 cc contains only about 1.5 mg bacterial C and 0.75

mg algal C so it cannot support the chironomids unless detritus carbon is consumed also. However, some selectivity has been found in the feeding of chironomids (Kajak and Warda 1968) so they could do better than this for microbial grazing. Even a 6-fold selectivity, however, would give only about 1 g C m^{-2} yr^{-1}. This value has been used in Table 8-6, but it is the weakest section of our argument.

Zooplankton grazing of total planktonic POC has been estimated at approximately 1 g C m^{-2} yr^{-1} (Chapter 6). Particulate organic carbon includes detritus, algae, and bacteria; since bacterial biomass is usually significantly smaller than the algal and detrital components, zooplankton predation of planktonic bacteria is insignificant relative to predation in the sediments. However the absolute amount of zooplankton predation is important for the planktonic bacteria, as Peterson et al. (1978) have shown that *D. middendorffiana* will filter bacteria but at an efficiency only one-third that for algae. In the ponds, all the bacteria are, therefore, removed every 6 days or so. Predation by protozoans and micrometazoans in the water column is insignificant compared to total predation since the biomass of these organisms in the plankton is small.

Evolved gas from the sediments was collected ten times during 1973 for chemical analysis. From these collections, methane production from anaerobic metabolism was determined to be 11 mg CH_4-C m^{-2} day^{-1}. If we assume that the 1971 production was the same as the 1973 production, then methane production was 1.1 g CH_4-C m^{-2} yr^{-1}.

The sum of these values for respiration, biomass change, predation, and methane production yielded 22.5 g C m^{-2} yr^{-1} gross bacterial production for Pond B in 1971.

Apportionment of this total to the two input parameters, total uptake of DOC and hydrolysis of detritus, is difficult since neither of these inputs was measured. The first input, total DOC uptake, was estimated from kinetic data to be about 3.6 g C m^{-2} yr^{-1}. There may be two sources of this labile DOC. The first is carbohydrates leached from macrophytes and animal and algal secretion, estimated to be 9.5 g C m^{-2} yr^{-1} (Sediment Respiration section). All of this is not available to the bacteria, however, since some DOC is lost during meltwater runoff. A second source is labile DOC formed as a result of hydrolytic decomposition of detritus. No measurement or estimate was made of this source; by subtraction it may be around 19 g C m^{-2} yr^{-1}.

The value for gross bacterial production in Pond B in 1971 (22.5 g C m^{-2} yr^{-1}) is not necessarily representative of other ponds, or even the same pond in different years, since populations, macrophyte input, temperature, and other factors may vary. Further, this carbon budget is based on bacterial production near the center of the pond, which may be significantly lower than bacterial production in the littoral areas and macrophyte beds where considerable decomposition of the first three age-classes of detritus occurs (Decomposition section).

Controls of Bacterial Biomass

The sediments of the tundra ponds are highly organic and, indeed, the tundra is covered with plants and their remains. Why then are not the bacteria, and other microbes, even more abundant than they are? In fact, the abundance of bacteria in the water and the sediments is entirely typical for temperate ponds in general. Unfortunately, we know little about production of bacteria in any environment so comparisons are extremely difficult.

The biomass of bacteria in these ponds is likely controlled by the complete freeze each year and predation by animals. As noted, bacteria can become inactive and remain inactive for long periods of time. Thus, the actual amount of biomass can be independent of the rate of production. Despite this ability to survive, the bacteria of the sediments were reduced in number over the winter (Figure 8-1). This was caused presumably by freezing and by mechanical destruction by ice crystals.

Three types of animals graze on bacteria: the specialized bacterivores, the generalized detritus feeders, and the filter feeders. In the plankton, the filter feeding zooplankton, especially *Daphnia*, filter large amounts of water each day; in fact, they filter the entire volume of the pond every two days. Recent experiments (Peterson et al. 1978) have shown that *D. middendorffiana* is about 30% efficient when it feeds on bacteria compared with its feeding rate on larger particles. Thus, about 15% of the bacteria could be removed each day. This is clearly enough predation to hold the numbers fairly constant in the plankton and thus it appears that zooplankton grazing controls the bacteria.

The specialized bacterivores of the sediment are mostly flagellate and ciliate protozoans. According to Fenchel (1975), the protozoans in the ponds remove up to 20 mg C m^{-2} of bacterial carbon each day (this is about 1% of the total amount present). This overall rate of predation is low and does not appear to constitute any sort of control on the bacterial biomass. However, the potential for control is there as the protozoans can grow rapidly in response to increased numbers of bacteria. There is also a good possibility that our scale of sampling is wrong for the heterogeneous sediment and that micropockets of high numbers of protozoans are present. If this were true, then grazing by protozoans would control the bacterial numbers because of their selectivity and ability to move to localized pockets of bacterial abundance.

The chironomids, oligochaetes, and other detrital feeders of the sediment appear to be somewhat selective but it is not known how well they can select for bacteria. Assuming a 6-fold selectivity factor, the calculations then give 1.0 g C m^{-2} yr^{-1}, or a rate about half that of the protozoans. This additional sediment grazing does not change the conclusions about the lack of control by sediment grazers.

Controls of Bacterial Productivity

The primary control on bacterial productivity is the rate of supply of low molecular weight organic compounds within the pond. In the water column, these compounds come mainly from excretion or leakage from algae during photosynthesis (Figures 8-4 and 8-5). Some of the excreted material may be used directly while some must be hydrolyzed or otherwise slightly broken down before it can be used. If our assumptions about the constancy of the concentration of the substrate (S_n) are correct, then the actual flux of material into the water is about 1 to 5% of the most rapid rates known for polluted waters (Allen 1969). There is also the complication that significant amounts of DOC might be transferred from the sediments and from the rooted aquatic plants to the water.

Despite these other sources of DOC in the water column, there is a clear and positive relationship between the potential uptake rates and the rates of primary production in the various ponds. This has been observed previously; for example, for a series of lakes and ponds in the temperate zone (Morgan and Kalff 1972). Within the water column of a lake, Wright (1970) found that the rates of uptake of glycolic acid exactly paralleled photosynthesis rate.

In the sediments, the primary source of the low molecular weight compounds is the decomposition of the detrital material, mainly *Carex* stems and leaves. As described in the DOC section (Chapter 4) and in the next section (Decomposition of Macrophytes), about 20% of the organic matter leaches from dead plant material within a day or two of death. The remaining organic matter decomposes at about 20 to 70% per year; most of this is likely due to enzymes secreted by bacteria or fungi. In contrast, some plant material, such as the red marine alga *Gracilaria*, decomposes completely within a few days (Tenore 1977). Thus, the slow rate of supply of the DOC from the particulate matter is controlled in part by the chemical makeup and resistance to breakdown of the *Carex* leaves.

The low temperatures are also a control, but we judge temperature to be relatively unimportant. Certainly microbial processes, and likely decomposition too, would proceed faster at higher temperatures. For example, a Q_{10} for bacterial growth of 1.7 was found (4° and 10°C) and Figure 8-3 shows the tight coupling between V_{max} and temperature (the Q_{10} was about 7). However, if the process did go faster, would there be enough carbon to sustain the microbes at the faster rates? In the tundra near Barrow an interesting experiment was conducted in which heating pipes were buried in the tundra soil. Eventually, the increased rate of decomposition caused the loss of so much organic matter that the volume of the soil decreased, the level of the ground fell, and a small pond formed. Thus, it appears that the organic matter in the soil was in a long-term steady state before the disturbance. By analogy, the input and decomposition of organic matter in the ponds must also be in a rough steady state at the present time. An increase in temperature would increase the decomposition rate and upset this balance so that eventually the process would run out of substrates. Again, the low temperatures are a control but the rate of supply is the most important control. Another bit of evidence for this comes from the study of Morgan and Kalff (1972) on Char Lake, an unpolluted arctic lake, and on nearby Meretta Lake, which receives sewage. Despite the similar water temperatures, the bacterial activity was much higher in Meretta than in Char Lake.

Another control that has been suggested is animal grazing. Hargrave (1970) reported that bacterial activity in grazed systems was generally higher than activity in ungrazed systems. Some experiments carried out at Barrow with *Carex* for a substrate indicated that phosphorus cycled much more rapidly in microcosms where protozoans were present than in microcosms without protozoans (Barsdate et al. 1974). In a different study, Fenchel (1977) found that barley straw decomposed more completely (80%) when protozoa were present than when bacteria alone were present (20%). The phosphorus experiments showed that the enhancement of activity did not come from release of PO_4 by the animals. Although other nutrients could have affected the microbes, it is more likely that the grazing removed inactive cells and somehow kept the population active. Thus, the "grazing effect" does occur in the Barrow ponds but to an unknown degree.

FUNGI

Hyphal Length and Biomass

The species and biomass of the fungi in the terrestrial soils at Barrow were intensively studied during the IBP project (Bunnell et al. 1980). Most of the fungal biomass was hyphae; these were measured by direct counts with the light microscope. The average amounts of hyphae ranged from 496 to 1445 meters per gram dry weight of soil. Although this appears to be a large amount, woodland soils usually contain about 7000 m $(gdw)^{-1}$. Most of the hyphae in the Barrow soils are in the top 7 cm and their total mass is around 4 g m^{-2}. Bunnell et al. (1980) point out that 75% of the microbial biomass in this habitat is bacterial.

The amount of hyphae in the pond sediments was measured on 29 July 1978 by Gary A. Laursen (personal communication) with the same techniques used in the terrestrial study. He found that sediments of Pond Omega (Miller et al. 1978) contained 1662 m $(gdw)^{-1}$ at 1-2 cm and 296 m $(gdw)^{-1}$ at 6-7 cm. Sediments in Pond C contained 998 m and 140 m $(gdw)^{-1}$ at the same depths.

These length measures in Pond C can be converted into biomass by multiplication with 0.16 (gdw) cm $^{-3}$ (the bulk density of surface sediments in Pond C, Table 4-1) and with 1.0 x 10 $^{-3}$ mg m $^{-1}$ (hyphal dry weight per meter, from Bunnell et al. 1980). After further conversions to a square meter basis, the data become 1600 mg m $^{-2}$ at 1-2 cm and 220 mg at 6-7 cm depth. The assumption of a linear decrease with depth gives 7070 mg m $^{-2}$ in the top 8 cm or 3500 mg C m $^{-2}$.

Based on this single measurement of 3500 mg C m $^{-2}$, it would appear that fungal hyphae make up 70% of the microbial biomass (in Figure 1-5 there are 1500 mg C m $^{-2}$ of bacteria). Given the changes expected in both the bacteria (Figure 8-1) and the fungi (Figure 8-6 in Brown et al. 1980) this fungal portion may vary greatly. For example, the bacterial biomass data is based upon a volume of 0.1 μm^3 per cell. Rublee et al. (1977) found that the mean cell size in sediments was 0.2 μm^3 (this would double our biomass). Yet, the difference between these two volumes is equivalent to deciding that a bacterial cell has a diameter of 0.75 μm instead of 0.60 μm. In the absence of any other fungal data, the arguments about the importance of microbes in this book have used only the bacterial data. However, even if fungi were to be included, the conclusions would not change.

DECOMPOSITION OF MACROPHYTES*

Methods

On 18 June 1972, leaves of *Carex* were collected for litterbag experiments and separated into green, yellow, and brown fractions. The green fraction consisted of live leaves that were produced late in the previous season (leaf numbers 5, 6, and 7) but that retained all or a part of their chlorophyll. The yellow fraction was leaves which died at the end of the previous season, while the brown leaves were dead for the entire previous season or perhaps for several seasons. After separation, the leaves were air dried for 6 days, and then weighed amounts (88 to 600 mg) were placed in 7.5×15.0-cm litterbags made of 20-μm Nitex netting. We also tested oven drying at 105°C; it gave an additional loss of 4.4% from green leaves, 5.0% from yellow leaves, and 5.5% from brown leaves (Hobbie 1973). These 20-μm mesh bags were submerged in Pond B during the first week of July 1972; three litterbags of each fraction were removed on each subsequent sampling date which occurred after 1, 2, 4, 8, and 58 weeks.

In a similar fashion, yellow leaves were also placed in 1-mm mesh bags. These mesh bags were incubated along with 20-μm mesh bags in the

*T. Traaen

TABLE 8-7 *Dry Weight Disappearance from Litter Bags, 1972 and 1973**

Type	Mesh	Location	1 Week	2 Weeks	4 Weeks	8 Weeks	58 Weeks
Green	20μm	submerged	20.8 ± 0.5	23.4 ± 0.5	28.3 ± 0.3	37.5 ± 3.7	33.3 ± 1.7**
Yellow	20μm	submerged	12.1 ± 1.0	16.0 ± 1.0	20.6 ± 0.6	25.5 ± 2.4	35.3 ± 2.5
Brown	20μm	submerged	3.4 ± 0.1	3.8 ± 0.8**	2.8 ± 1.0**	4.9 ± 0.5	59.6 ± 1.2
Yellow	1mm	submerged	12.0 ± 0.4	-	29.4 ± 5.0	37.4 ± 4.6	29.8 ± 5.2**
Yellow	20μm	on sediment	-	14.2 ± 1.0	18.0 ± 1.5	24.2 ± 2.2	34.2 ± 2.2
Yellow	1 mm	on sediment	-	12.3 ± 1.2	18.8 ± 1.5	31.5 ± 3.8	66.7 ± 0.4
Yellow	20μm	in sediment	-	-	16.1 ± 0.9	19.1 ± 1.4	45.4 ± 3.0
Yellow	20μm	in air	-	-	0.2 ± 0.3	3.0 ± 0.3	-
Yellow	1 mm	in air	-	-	0.8 ± 0.4	4.0 ± 0.9	-

* Values are given as % loss from original ± standard deviation (n = 3).
**Not significantly different from the previous measurement.

water column, in the air, on the sediment surface, and in the sediment to test the effect of mesh size and location.

Loss of Weight from Macrophyte Detritus

The green litter incubated under water lost the greatest amount of weight, 37.5%, over the 8-week summer (Table 8-7). Over half of this loss, 20.8%, occurred during the first week. This early loss is similar in every way to the initial leaching that takes place in tree leaves placed in water. For example, Kaushik and Hynes (1971) measured some 15% loss in *Ulnus* leaves after 4 days at 10°C. Much of this leached material is carbohydrate and Chapin et al. (1975) have shown that much of this soluble carbohydrate is retranslocated below ground when the *Carex* begins to senesce. Thus the yellow leaves should have had less soluble carbohydrate and, indeed, lost only 12.1% in the first week. The summer weight loss from the yellow leaves was 25.5% or 13.4%, after the initial loss; the post-initial loss was similar, therefore, to that from the green and yellow leaves (13.4 vs. 16.7%). In contrast to the green and yellow leaves, the brown leaves lost only 4.9% of their weight over the entire summer.

From 8 to 58 weeks, which included the winter frozen period, the green litter gained weight (Table 8-7). This gain (4.2% of original weight) is not statistically significant, but the increase was likely due to microbial and microfaunal colonization of the detritus and of the mesh. The yellow fraction lost an additional 9.8% over this 58-week period, and the brown litter lost a highly significant 54.7%. The large loss is presumably a result of the winter freeze-thaw which physically broke up (triturated) the weakened structural material.

The yellow leaf detritus decomposed slower when buried in the sediments than when placed on the sediments or submerged in the water (Table 8-7); litterbags placed in air lost the least weight of all types (3 to 4% over 8 weeks). The leaves placed in 1-mm mesh bags had greater weight loss than those in the 20-μm mesh bags, indicating that trituration and subsequent loss may be a significant component of detrital breakdown. Such a breakdown of litter is probably a result both of animal foraging and of breakdown of microfibers by physical forces; the effect would be greatest on the older material which was already weakened.

Partitioning of Carbon Loss

The decomposition processes of macrophytes may be separated into leaching, hydrolysis, and trituration. While all of these are operating simultaneously, different stages of decomposition may be characterized by the dominance of a single process. Leaching, as discussed above, is

TABLE 8-8 *Decomposition Coefficients,* K_1* *and* K_2***, from Various Studies*

Location	K_1	K_2	Remarks
Tundra Ponds	99.99	69.89	Green *C. aquatilis*, submerged
(this study)	98.94	21.53	Yellow *C. aquatilis*, submerged
	63.54	9.50	Brown *C. aquatilis*, submerged
	98.15	13.90	Yellow *C. aquatilis*, on sediment
	-	12.60	Yellow *C. aquatilis*, in sediment
Frains Lake (Saunders 1972)	-	1.0–65.6	Radioactively labeled phytoplankton (pre-leached)
Marion Lake (Hall 1972)	99	98.50	Radioactively labeled leaf material
Laboratory Study (Otsuki and Hanya	99.99	98.64	Freeze-dried *Scenedesmus* sp. under aerobic conditions
1972a, 1972b)	99.99	94.81	Freeze-dried *Scenedesmus* sp. under anaerobic conditions

*K_1 = coefficient of loss of total leachable fraction (% yr^{-1}).
**K_2 = coefficient of loss due to hydrolysis (% yr^{-1}).

dominant in the initial stages of decomposition when soluble, low molecular weight organic compounds are lost. Hydrolysis, the long-term enzymatic breakdown of structural plant matter, dominates next. Trituration, mechanical breakdown by animals or physical forces (water movement, freeze-thaw), dominates the final stages of decomposition.

The combination of leaching and hydrolysis can be represented as the sum of two first-order equations of the form $W(t) = We^{-Kt}$ where $W(t)$ is the litter remaining at time t, W is the litter originally present, and K is a coefficient of decrease (Bunnell and Tait 1974). From this a K_1 may be calculated for leaching and a K_2 for hydrolysis (in percent weight loss per year of total substrate available to that process). These values are approximations only and have been calculated on the basis of an entire year so that they may be compared with other measurements (Table 8-8). While the K_1 rates are comparable with the other values, they only obtain for a 3-month decomposition season, so presumably the loss would be 33% per season in the arctic ponds. The K_2 rates are noticeably lower in the arctic leaf material but a comparison of tree leaves and sedge leaves may not be valid. In fact, the data summarized by Saunders (1976) show that the rate of decomposition is species-dependent and can range from 0.2 to 1.75% day^{-1} for tree leaves. Perhaps the best analogy in Table 8-8 would be between the yellow *Carex* leaves and leaf material reported by Hall (1972), as both underwent the natural process of retranslocation before being subject to decomposition. However, the low temperatures at which the arctic decomposition took place may well be even more important than the differences between types of plant material.

The comparisons above are with decomposition rates of litter incubated in water. Many studies of decomposition on land have shown that the amount of moisture is a primary controlling factor. Indeed, at Barrow *Carex aquatilis* litter (yellow leaves) lost around 15% of their weight per year when incubated in the air (Flanagan and Bunnell 1980).

Other experiments carried out in 1970 confirmed that about half of the first year decomposition of green leaves was abiotic leaching. These experiments were made with fresh green *Carex*, dried at 65°C and placed both in nylon mesh bags (1.5-mm openings) and in plastic Whirl-Pak bags with pond water plus 100 ppm $HgCl_2$. After 33 days the leaves in the six mesh bags had lost 46% of their original weight while those in the plastic bags had lost only 24.1%. Since the $HgCl_2$ treatment prevents microbial growth, the plastic bag treatment is taken as the abiotic loss. Therefore, we can say that 24% of the total weight loss in the mesh bags was caused by abiotic factors such as leaching and mechanical breakdown and 22% was caused by biological activity. Similar results were obtained in 1972.

Timecourse of Decomposition

A decomposition budget may be calculated in terms of loss over time from a single year's production of *Carex* litter in Pond J (Table 8-9). It was assumed that the detritus that entered the water at the beginning of the

TABLE 8-9 *Annual Input, Decomposition, Loss by Leaching, Hydrolysis, and Trituration of* Carex aquatilis *Detritus in Pond J*[1]

Age-class	Composition	Weight[2]	Total wt loss[3]	Loss by leaching[4] Usable DOC[5]	Loss by leaching[4] Refractory DOC[6]	Loss by Hydrolysis[7]	Loss by trituration[8]
0	Green	32.00 (100)	0	0	0	0	0
1	Green	10.67	4.00 (12.4)	2.50 (7.7)	0	1.50 (4.7)	0
	Yellow	21.33	5.44 (17.1)	2.35 (7.4)	1.79 (3.4)	1.31 (6.4)	0
2	Brown	22.60	3.68 (16.3)	0.75 (3.3)	0.55 (0.5)	2.38 (12.5)	0
3	Brown	18.90	11.24 (59.5)		0.11[9] (1.0)		11.09
4	Brown	7.70	7.70 (100)				7.70
TOTALS			32.09	5.60	2.45	5.19	18.79

[1]Values in g C $m^{-2}yr^{-1}$, % of total for age-class in parenthesis.
[2]Estimated macrophyte detrital input (based on 1971 macrophyte input in Pond J).
[3]20 μm mesh litterbag dry weight loss during ice-free period.
[4]Whirl-Pak dry weight loss.
[5]Whirl-Pak dry weight loss not recovered as DOC 2 weeks after thaw.
[6]Whirl-Pak dry weight loss recovered as DOC 2 weeks after thaw.
[7]20 μm mesh litterbag dry weight minus total DOC loss.
[8]Loss due to trituration from 20 μm mesh litterbag over the winter period.
[9]Estimated from sediment leach in center of pond.

first year was composed of 33% green and 67% yellow material. The total weight loss of each fraction was calculated from the loss from litterbags. This total loss was partitioned into loss by leaching and loss by hydrolysis based on experiments described above with plastic bags and $HgCl_2$. The leaching loss was further partitioned by experiments in which litter was incubated in plastic bags with pond water but without a killing agent. After two weeks, some of the organic matter lost from the leaves was still present as DOC; this was called refractory DOC. The organic matter lost from the leaves and oxidized by the microbes was called usable DOC. After the second year, the loss from the brown material can not be partitioned and almost all loss is attributed to trituration. By the fourth year, the remaining material was less than 20 μm in diameter and was easily moved around the pond. It is treated in the budget as if it were all completely decomposed by the end of this period but we cannot be certain of this.

Over the 4 years, most of the plant litter was lost by trituration (18.7 g C) while only 8.05 g C was lost by leaching and 5.19 g C by hydrolysis (Table 8-9). These totals can also be used for the total amount of decomposition occurring in a single year in this pond. Obviously the data are not accurate enough to determine whether some plant material remains undecomposed and is eventually added to the sediments. If only 5 to 10% (or 1.6 to 3.2 g C m^{-2}) of the total produced was added each year, it could account for the organic accumulation in the sediments.

Another source of organic matter to the pond is the roots of aquatic plants. These add up to 100 g C m^{-2} yr^{-1} to the sediments in the plant bed or 33 g C m^{-2} to the whole pond (see Chapter 5). An unknown amount of this production remains undecomposed.

Changes in Inorganic Composition during Decomposition

The inorganic composition of the *Carex* samples removed after 1, 2, 4, and 8 weeks of decomposition was analyzed in 1972. The elements measured were P, K, N, Ca, Mg, Fe, Mn, and Zn. In addition, laboratory measurements of loss from fresh lemming feces were made.

The fresh green leaves of *Carex* lost almost all of their P and K within one week. The yellow and brown litter had very low concentrations of P and K at the start of the experiment while the P and K of lemming feces was mostly lost during the first 4 days. The elements tied to the structural material of the plants, Mg, Ca, and Fe, increased relatively over 8 weeks as the proportion of structural material in the total remaining material increased. All other elements remained at essentially the same relative concentrations in the detritus and in the lemming feces throughout the experimental period.

SEDIMENT RESPIRATION *

Decomposition can also be examined at the level of a whole system. In this section, the respiration rate of the sediments is described as a measure of the microbial processes occurring.

Methods

Sediment respiration was determined from the accumulation of inorganic carbon in Plexiglas core tubes containing undisturbed sediment and overlying water. After the sediments were collected, the tubes were closed at the top with a rubber stopper and incubated *in situ* for 4 to 6 hr. An infra-red gas analyzer was used to measure the amount of carbon dioxide produced during an incubation (see Reed (1974) and Miller and Reed (1975) for details).

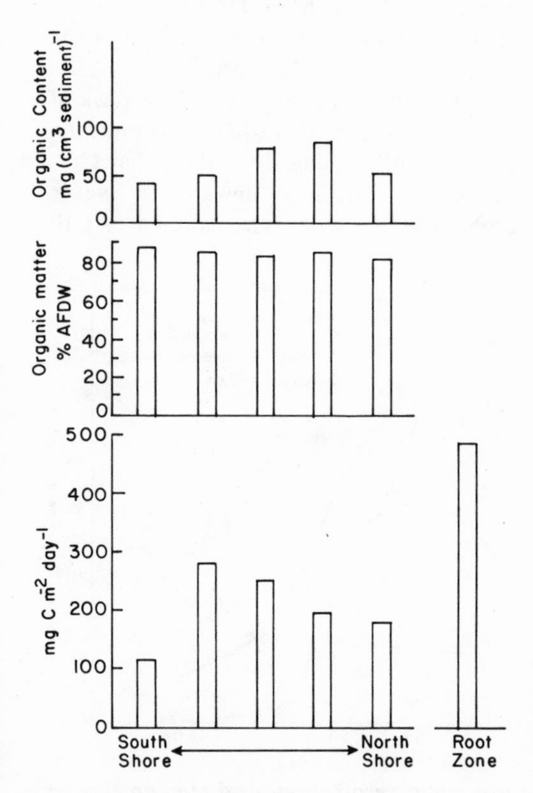

FIGURE 8-6. *Sediment respiration, organic matter, and organic content from a transect in Pond A, 16 August 1973.*

*J.P. Reed and M.C. Miller.

Results

The respiration of the sediment varied greatly from sample to sample. Replicates had a coefficient of variation of 50% (range was 4 to 390%). Because of this variation, the respiration rates of a series of samples collected along a transect across pond A (Figure 8-6) were not statistically different. The increased respiration in the zone of rooted plants is obvious, however.

Most of the respiration in the plant-free areas of the pond takes place in the top few centimeters of sediment. For example, when the top 4 cm of sediment were removed there was an average reduction of 77% in the respiration rates (four experiments). As noted earlier (Table 8-1), the bacteria also decreased greatly in the top 4 cm and it is likely that the fungi are confined entirely to the top few centimeters because of their need for oxygen. In addition, 99.6% of the algae and 89% of the chironomids were found in the top 4 cm.

During a single day the respiration rate varied with the temperature at the sediment surface; however, the wave of diffusion of CO_2 from below the surface layer lags behind and correlates with the temperature wave within the sediment. As measured in the water above the surface of the sediment (Figure 8-7), the maximum respiration occurred between 1300 and 1900 and coincided with the temperature peak but not with the peak of solar radiation. However, the respiration increased at 0400 in the morning at a time when the surface layers were still cooling. It is likely that CO_2 from below the sediment surface is diffusing upward in waves corresponding to the temperature oscillations within the sediment. At 10

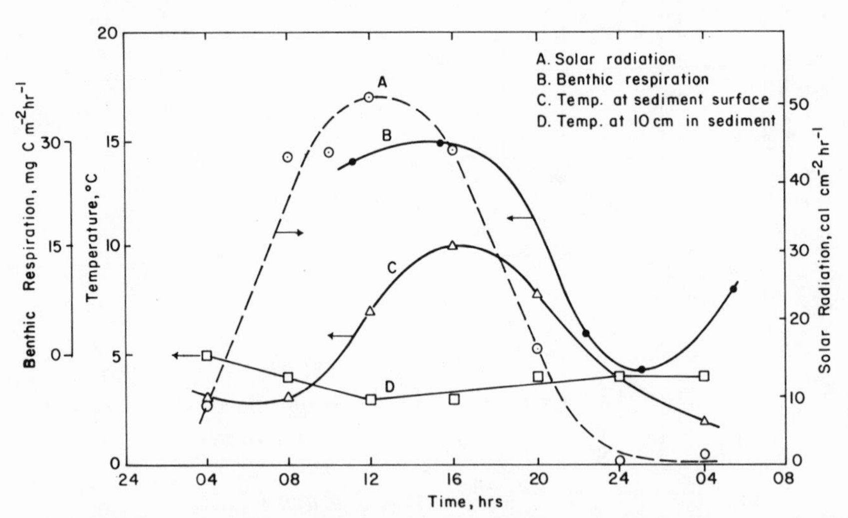

FIGURE 8-7. *Benthic respiration, solar respiration, and temperature in Pond B on 24 and 25 July 1972.*

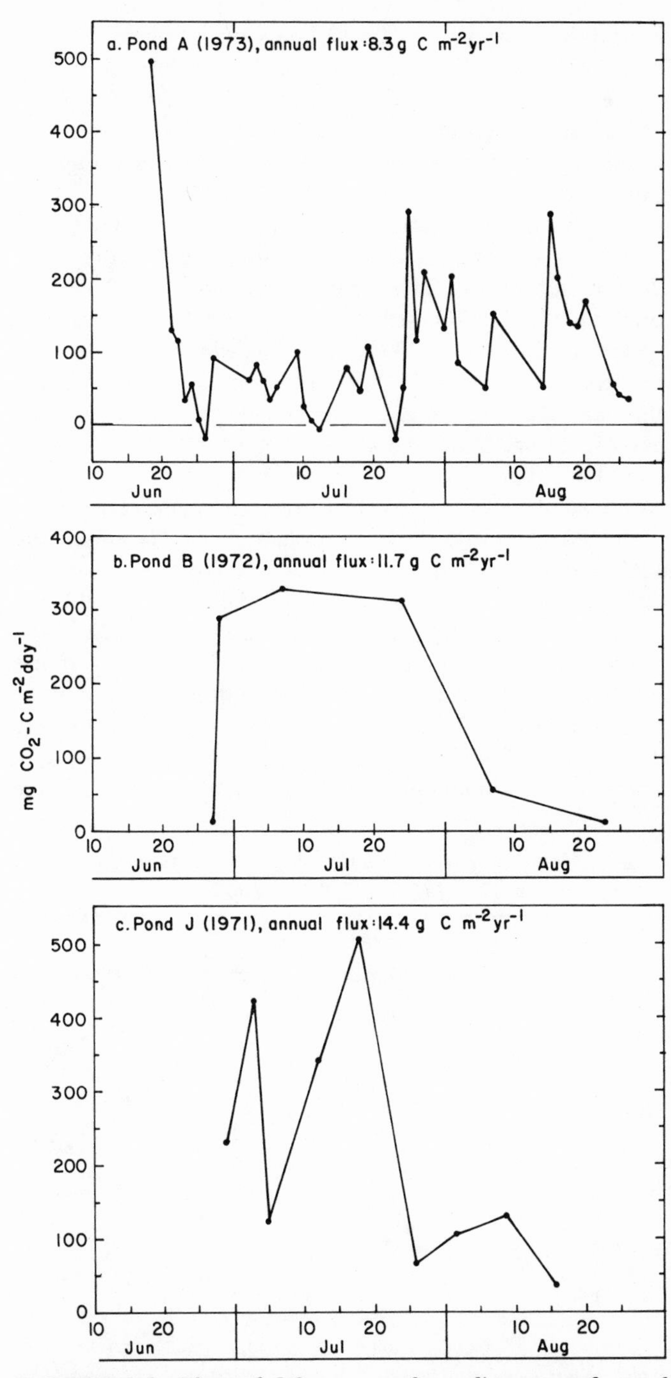

FIGURE 8-8. *Flux of CO_2 across the sediment surface of Pond A in 1973, Pond B in 1972, and Pond J in 1971.*

cm depth, this oscillation appeared to be 12 hours out of phase with the surface temperatures (Figure 8-7, Figure 3-5). The diffusion rate within the sediment was tested by injecting $^{14}CO_2$ and proved to be 0.167 cm hr^{-1} at 7°C. This rate will vary, of course, as a gradient of the CO_2 concentrations and of the temperature change but it is rapid enough to account for the discrepancies.

From day to day, the respiration rates within a pond fluctuated greatly (Figure 8-8). The highest rates were measured in June 1973 immediately after the ice melted. This peak corresponds to early-season peaks in bacterial numbers in the sediment (Figure 8-1) and peaks in activity in the water (Table 8-4); it is likely caused by wintertime release of organic nutrients from microbes and other organic matter. This high rate of respiration undoubtedly occurred each year but was missed in 1971 and 1972. Similar early season respiration maxima have been found in the soils at Barrow (Brown et al., in press).

The total amount of CO_2 released from the sediments varied from year to year but did correlate with the total amount of solar radiation and with benthic algal production (Figure 8-8, Table 8-10). When the amount released is corrected for the additional input from the rooted plant zone, where rates are 2.5 times higher, the whole-pond average is 24.9, 22.9, and 15.1 g C m^{-2} yr^{-1} for 1971, 1972, and 1973, respectively. These values are 47 to 78% of the macrophyte decomposition rate of 32 g C m^{-2} yr^{-1} (Table 8-9) but the variability of the respiration data do not allow too fine a comparison to be made. Core respiration rate is also much lower than the total pond respiration estimate of 50 g C m^{-2} made in 1971 (Coyne

TABLE 8-10 *Annual Measurements of Solar Radiation, Temperature, Algal Production and CO_2 Flux for 1971 to 1973*

	1971	1972	1973
Solar radiation (cal cm^{-2} summer^{-1}) (June-August)	50,048	39,573	32,225
Water temperature (°C average daily)	5.7	8.5	5.8
Benthic algal production (g C m^{-2} summer^{-1})	10.1	8.9	4.1
Annual CO_2 flux pond center (g C m^{-2} summer^{-1})	13.7	11.7	8.3
Annual CO_2 flux pond center and macrophyte zone (g C m^{-2} summer^{-1})	24.9	22.9	15.1

TABLE 8-11 *Comparison of Algal Production and Sediment Respiration*

Location	Temp (°C)	Primary production $g\,C\,m^{-2}yr^{-1}$	Sediment Respiration $g\,C\,m^{-2}yr^{-1}$	$mg\,C\,m^{-2}\,day^{-1}$	Reference
Arctic ponds	9.6	10.0	22.9	254.5	Present study 1972
Char Lake		6.2	15.0		Welch 1974
Temperate Spring	25	78.8	38.8	106.4	Teal 1957
Grane Lang Lake	11	153	43.4	118.9	Hargrave 1973
Marion Lake	15	56.8	49.9	136.8	Hargrave 1969
Lake Esrom	6	191	93.8	256.8	Hargrave 1973
Fure Lake	7	310	117.3	321.4	Hargrave 1973
Lawrence Lake	5	169	137.7	377.2	Hargrave 1973
Castle Lake	10	580	201.9	553.3	Hargrave 1973
Tuel Lake	8	595	306.3	565.1	Hargrave 1973
Bagsuaerd Lake	10	522	229.8	629.6	Hargrave 1973
Arctic wet meadow					
Soil	5		159	1800	Flanagan and Bunnell (in press)

and Kelley 1974). As explained in Chapter 4, this rate was calculated from measurements of the partial pressure of CO_2 in the water. The discrepancy cannot be explained although the whole pond measurement was made in pond C while the core measurements were made in Pond B.

The daily rates of sediment respiration are about the same as those from other oligotrophic to mesotrophic waters (Table 8-11). The short ice-free season, however, results in a very low annual rate for the ponds. The terrestrial soils at Barrow have respiration rates that are seven times higher than those of the pond.

Effect of Temperature

The effect of increases of temperature on the respiration rate of sediment cores was measured in 1972 on cores transported to the laboratory. The resultant Q_{10} averaged 3.1, which is an impossibly high value for a biologically mediated process (Table 8-12). The measurements in 1973, when the incubations were made in the field with a minimum of disturbance, resulted in an average value of 1.6. While the average is believable, the amount of variability in the measurements (Table 8-12) is large and unexplained. As noted earlier, the Q_{10} estimated in this way measured not only the respiration but also any increases due to changes in the diffusion of CO_2 from below the surface layers and from the sediment to the water.

In the pond, as well as in the experiments, the respiration rate was correlated with temperature. However, over a whole season the effective temperature was altered by effects of different amount of photosynthesis, etc. As a result, the correlation of respiration rate with average daily water temperatures was only significant at the 5% level ($r = 0.368$, $n = 40$).

Other workers have also noted a correlation between temperature and sediment respiration. This correlation was noted over a 24-hour cycle by

TABLE 8-12 *Experimental* Q_{10}*'s for Sediment* CO_2*-C Flux 1972, 1973*

Date	Temperature range (°C)	Q_{10}
29 - 30 June 1972	5-15	3.23[1]
21 - 22 August 1972	5-15	3.07[1]
25 June 1973	14-20	1.24
26 June 1973	8-20	1.85
2 July 1973	14-20	1.64
3 July 1973	14-20	2.50
20 August 1973	2-8	1.7
	2-14	0.8
	2-20	2.1
	14-20	1.5
26 August 1973	14-20	1.0
Average 1972	5-15	3.1
Average 1973	2-up	1.6
	14-20	1.6
	all	1.6

[1] Average of four different determinations made over a 24-hour period.

TABLE 8-13 *Percentage Change in Dark CO_2-C Flux in Sediment Cores from Pond A–B Enriched with Inorganic Phosphorus, Acetate, or Glucose During 1972, 1973*

Date	Substrate added[1]	Concentration	Average percent change over controls
1972	Phosphorus (3)[2]	310 μg P liter^{-1}	+124%
19 July 1973	Phosphorus (3)	270 μg P liter^{-1}	+ 56%
24 July 1973	Phosphorus (3)	72 μg P liter^{-1}	+141%
25 July 1973	Phosphorus (3)	74 μg P liter^{-1}	+ 22%
26 July 1973	Phosphorus (3)	126 μg P liter^{-1}	− 59%
1972	Acetate (3)[2]	10 mg C liter^{-1}	+162%
1972	Glucose (3)[2]	10 mg C liter^{-1}	+ 35%

[1]The number in parenthesis is the number of replicates.
[2]Sediment core mixed; therefore, substrate uniformly distributed.

Wetzel et al. (1972) by Rich (1970), and by Hargrave (1969). In this same work on Marion Lake, Hargrave found that temperature accounted for 71% of the annual variation in oxygen uptake.

Nutrients

In these ponds the availability of phosphorus and of labile organic compounds appears to restrict microbial decomposition. This was determined by adding nutrients to the water above the sediments in core tubes; the respiration rates were measured after 4-hr incubations. Most of the experiments involved the addition of phosphate (Table 8-13) which usually increased respiration. Acetate also increased respiration significantly but the variance was high. The results were surprising in view of the large quantities of phosphorus and organic carbon that are present in sediments. However, mere presence does not mean that the nutrients are available. Other workers have also concluded that low levels of nutrients limit decomposition (Waksman et al. 1933, Waksman and Renn 1936, Ryhänen 1968).

Effect of Animals

In a number of experiments in other systems, the presence of animals appeared to stimulate benthic respiration. To test if this occurred in the ponds, an 18-day experiment was set up in which the equivalent of 7600

TABLE 8-14 *Effect of the Addition of Chironomid Larvae (fourth instar) and* Lepidurus *Adults to the Measured Sediment Respiration in Pond A, 6-24 August 1973*

	Average temperature (°C)	AVERAGE RESPIRATION (mg C m^{-2} day^{-1})				Percent change over controls
		Control[a] cores	Experimental[a] cores	Animal[b] cores	Difference between experimental and animal	
Chironomid addition						
6 August	5.6	103 ± 10	146 ± 2	69.7	76.8	−26%
17 August	7.6	172 ± 43	98 ± 5	55.7	42.3	−75%
19 August	5.7	95 ± 16	98 ± 5	46.9	50.6	−47%
24 August	1.9	152 ± 8	179 ± 44	?	176	+15%
Lepidurus arcticus						
7 August	5.4	92	166 ± 27	12	154	+67%
24 August	1.9	152 ± 8	251 ± 35	6	245	+61%

[a]Values are expressed as the ± 1 standard error (n = 2).
[b]Respiration of chironomids = 0.326 μg C (mg dry wt)$^{-1}$ hr^{-1} for 1.1 mg animals.

fourth instar chironomid larvae or 700 *Lepidurus* (13-mm) per square meter were added to cores.

Our results (Table 8-14) show that the respiration of the microbes actually decreased in four of five cores when chironomids were added. To calculate this, the respiration of the animals has to be subtracted; as noted earlier, these respiration data were measured on confined animals and may not be correct. Also, the experiment run at 1.9°C (Table 8-14) assumes that the animal respiration was effectively zero.

Lepidurus appears to stimulate respiration of the sediment and this species does not eat detritus and microbes. However, the calculation does depend upon a separate measure of the animals' respiration. It is possible that the chironomids eat enough of the microbes to affect the results but this is doubtful, as the concentrations of animals added approximated the natural levels. In an additional test, *Lepidurus* also stimulated the release of $^{14}CO_2$ from labeled barley straw which had been boiled to remove the soluble compounds (Figure 8-9). These experiments indicate also that the number of chironomid larvae is important as ten animals were twice as effective as five animals. We have already described how the protozoans increased microbial activity (Microbenthos section).

The increase in microbial activity may be due to an increase in the flux of oxygen into the sediment. A similar effect was found in Marion Lake where Hargrave (1969) found that respiration decreased when the oxygen fell below 8.6 mg O_2 liter^{-1} in cores. In contrast, Pamatmat (1965) and Edwards and Rolley (1965) found that at low temperatures a moderate stirring had no effect on the respiration rates. Here, it is possible the activity of the *Lepidurus* in the top 0.5 cm of sediment might affect oxygen diffusion; on a microscale the protozoans might have the same effect on diffusion around particles.

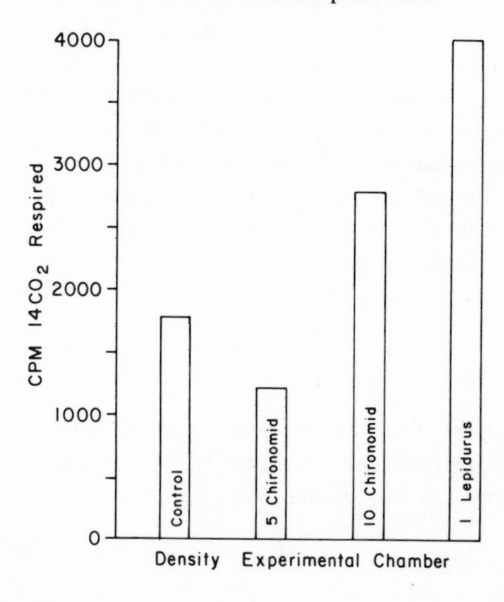

FIGURE 8-9. *Effect of invertebrates on the decomposition of barley as measured by $^{14}CO_2$ release. Experiments were run at 10°C for 24 hours in 500-ml flasks with 60 μm fragments of barley straw.*

Chironomid burrows may also allow oxygen to penetrate into the anaerobic zone of the sediments. Some species actually pump water through their burrow; other workers have seen a distinct effect of the burrows on sediment pH and Eh (Edwards 1958, Davis 1974).

The activities of the macrobenthic organisms may also influence sediment respiration rates by decreasing the size of the detritus particles. Small particles have a much higher bacterial biomass and respiration rate than do large particles (Fenchel 1969, Hargrave 1972). At our research site the highest rates of respiration occurred in ponds with the finest of sediments (X, J, D).

MICROBENTHOS*

Introduction

The general objectives of the pond research, to investigate nutrient cycling and energy flow, led us to approaches that precluded analyses of components of the ecosystem which may well be controlling the microbenthos. Thus, scant attention was paid to predation and competition, to name only two mechanisms. In addition, the microbenthos is a complex but non-functional grouping that in itself includes a large number of species and several complicated food webs. One example is the ciliates, which were studied in some detail but not necessarily because of their importance. It is also true that no one has ever before attempted to analyze the controls of any aquatic microbenthic community so we cannot even make comparisons or gain insights from other studies.

The sediments of aquatic environments usually sustain a varied and rich microfauna. These animals include protozoans as well as such micrometazoans as turbellarians, nematodes, and crustaceans. Few aquatic studies examine the microbenthos and little is known of its role in the ecosystem. Because of the importance of sediment processes in these tundra ponds, we have attempted to quantify the numbers and production of the microbenthos. A full report on this work is given in Fenchel (1975); data for 1971 and 1972 come from Dillon and Hobbs (1973).

Methods

During 1971 and 1972, samples from the sediment of Pond B were fixed in Schaudinn's fluid and diluted, and then subsamples were counted. In 1973, four cores (0.65-cm diameter) were collected to a depth of 6 cm

*T. Fenchel

on each sampling date from the central part of Pond B (no plants). On any date, all the samples were taken within an area of 0.5 m^2. In the laboratory, the cores were sectioned into 2-cm layers. Large organisms were microscopically enumerated and identified by placing each section into a petri dish, adding filtered pond water, and then removing the animals one by one. Small zooflagellates, microalgae, and bacteria were counted on the surface of sediment particles with a fluorescent microscope (Fenchel 1970). Most ciliates were identified to species; some small ciliates, zooflagellates, rotifers, and gastrotrichs were identified only to genus; no attempts were made to identify nematodes, ostracods, or harpacticoids. Counts were converted to biomass using volume estimates of Fenchel (1967) or from linear dimensions and reasonable geometric shapes.

In order to estimate feeding rates of protozoans in nature, the egestion rates of two species of bacterivorous and two species of algivorous protozoans were determined in laboratory cultures at different temperatures (Fenchel 1975). By assuming that the feeding rate in the field is a continuous process, the actual feeding rate can then be estimated by counting the food organisms in the feeding vacuoles in protozoa immediately after collection from the sediments; the feeding rate is the egestion rate times the actual counts.

Vertical Distribution and Composition of the Microfauna

All of the groups were most abundant in the surface sediments and their numbers decreased with depth. One reason for this decrease is the anaerobic conditions below about 2 cm (Chapter 4, Figure 4-11). Only nematodes, which are among the few metazoans that live in anaerobic sediments (Fenchel 1969), were regularly found below this depth. However, some ciliates were found in the 2- to 4-cm sections; these (*Caenomorpha medusula, Saprodinium* sp. and *Metopuses*) are known as anaerobic forms from other studies (Bick 1958, Fenchel 1969).

In these ponds, however, the mechanical properties of the sediment are probably the most important factor which limits the ciliates to the top 3 or 4 cm. The upper layers of the sediment consist of large particles (a mean particle diameter of 500 μm (Figure 4-1)) that are loosely packed and contain large interstitial spaces. In contrast, the deeper sediments are made up of finer particles that clog these spaces and prevent penetration by the ciliates.

Organisms may also be moved within the sediment by the actual mixing of the sediment itself. This occurs by the movements of chironomid larvae and oligochaetes, by wind action, and by deposition of suspended material (see also discussion in Chapter 5).

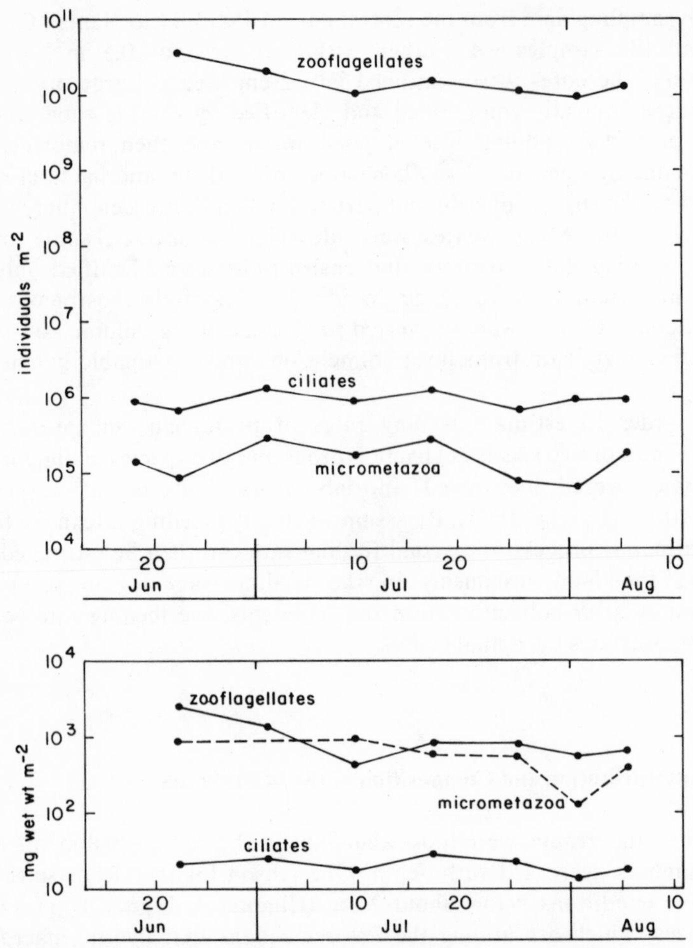

FIGURE 8-10. *Numbers and biomass of zooflagellates, ciliates, and micrometazoa in the sediment of the center of Pond B, 1973.*

In pond B, the small zooflagellates are numerically the most important element of the microfauna; their wet weight per square meter is equalled by the micrometazoans (Figure 8-10). These zooflagellates have only rarely been quantitatively studied in nature (see Fenchel 1970) but they are important bacterial grazers here and undoubtedly in many other aquatic systems as well. Common genera were *Monas, Oikomonas,* and *Bodo.* All measure 5 μm or less in diameter and all graze on bacteria.

In these detritus-rich sediments, the robust burrowing forms of the micrometazoa, such as nematodes, ostracods, harpacticoids, and turbellarians, are more successful than the ciliates (Figure 8-10), which are more abundant in sandy capillary sediments. Both the numbers and

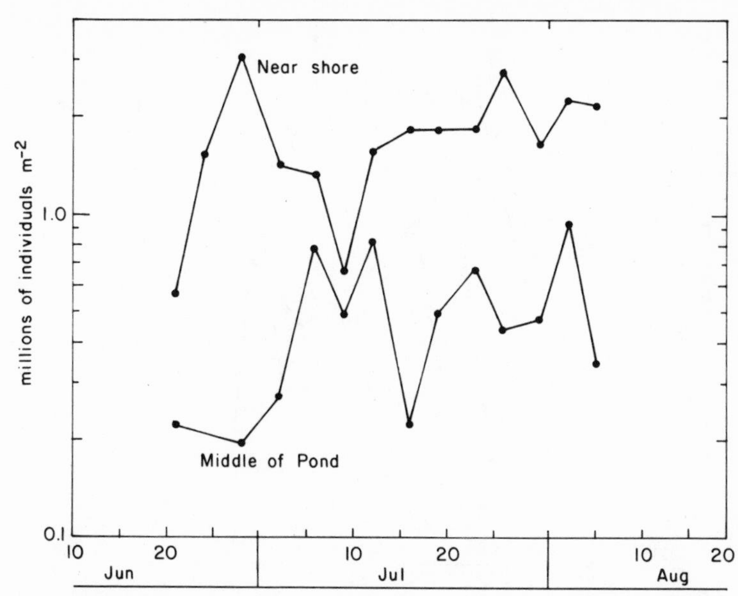

FIGURE 8-11. *Numbers of ciliates in Pond B sediments, 1971.*

biomass found here of ciliates (10^6 m^{-2}, 20 to 40 mg m^{-2}) and of micrometazoans (10^5 m^{-2}, 1000 mg m^{-2}) are typical of detrital and muddy sediments from temperate freshwater and marine environments. In fine sandy marine sediments, Fenchel (1967, 1969) found up to 20×10^6 m^{-2} ciliates with a weight of 1000 mg m^{-2}

Protozoans and micrometazoans were more abundant in the littoral areas of the pond among the emergent vegetation than in the central area (Figure 8-11). This is illustrated by data for ciliates but nematodes and rotifers followed the same pattern (2 to 5 times higher). This distribution reflects the higher numbers of bacteria produced by the decomposition of leaves and excretion of organic matter in the littoral zone as well as the greater depth of the aerobic layer of sediments (Chapter 4, Figure 4-11).

Ciliates in the center area of the pond appeared to be most abundant in 1972 and least abundant in 1971 (Figure 8-12). However, slightly different parts of the pond were sampled each year and a different counting technique was used in 1973. Therefore, the differences between years may not be significant. Rotifer counts were also highest in 1972 (3.3×10^5 m^{-2}) but lowest in 1973 (0.47) and intermediate (0.83) in 1971 (as averages for July of each year). Figure 8-12 does show two peaks in ciliate abundance, one around the beginning and one around the end of July. This is likely in response to the bacterial peaks in the beginning and end of the summer (Figure 8-1) and to the continued increase in numbers of epipelic algae (Figure 5-10) over the summer. Actually, most of the ciliates at the beginning and end of the summer were bacteria feeders

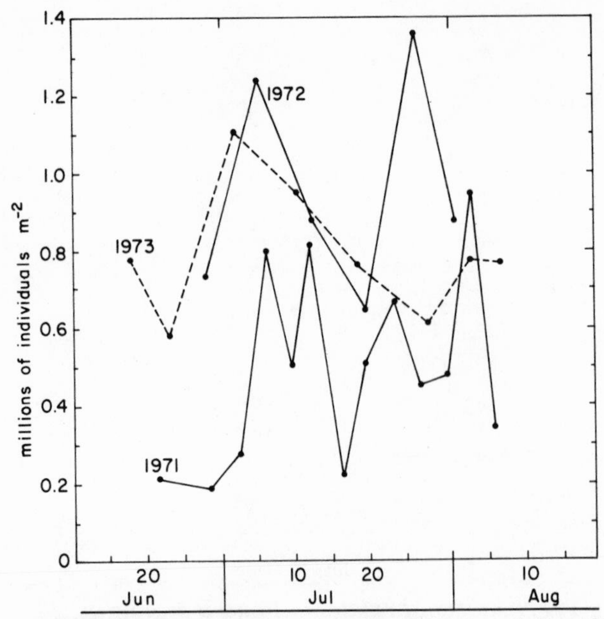

FIGURE 8-12. *Numbers of ciliates in Pond B sediments during the summers of 1971, 1972 and 1973.*

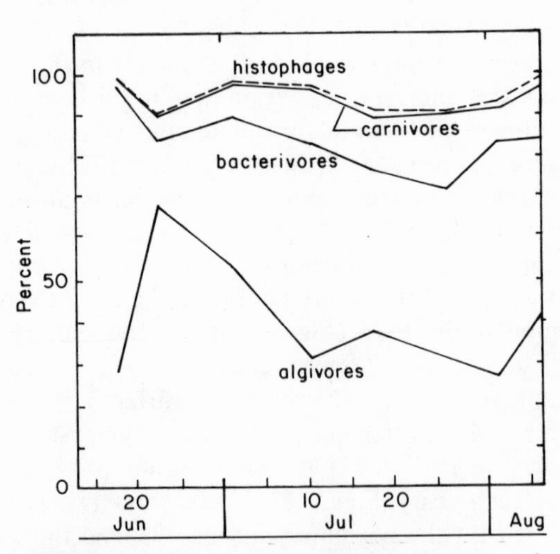

FIGURE 8-13. *Percentage distribution of feeding types among the total ciliate population of the pond sediments, based on numbers.*

(Figure 8-13) while the majority of the ciliates in late June were algae feeders.

Among the important algae feeders were *Nassula* and *Cryptogramma* sp. (both feed mainly on blue-green algae), *Frotonia acuminata, F. leucas, Prorodon teres, Strombidium* sp. and a number of oxytrichids. Species of *Stentor, Stylonychia mytilus*, and *Euplotes patella* grazed on algae but also ate other ciliates and zooflagellates. The bacterivorous ciliates were mostly small holotrichs such as *Colpidium, Tetrahymena,* and *Trimyema*, but two large forms, *Spirostomum teres* and *Blepharisma lateritium*, were also common while *Vorticella* regularly occurred on particles of detritus. The most important carnivorous ciliates were *Lacrymaria* sp., *Dileptus anser, Spathidioum* sp., and a number of holophryids (e.g., *Pseudoprorodon* sp.) which fed on zooflagellates and other ciliates. Finally, the category "histophagous" in Figure 8-13 refers to ciliates that specialize in attacking and eating damaged but living metazoans such as oligochaetes and chironomids (Mugard 1949). This quantitatively unimportant group was represented by *Ophryoglena* sp. and *Coleps hirtus*. In general, the ciliate assemblage in this arctic pond is strikingly similar to that of temperate freshwater ponds of similar sediment type (Bick 1958, Picken 1937, Noland and Gojdics 1967).

Rhizopods such as amoebae and testaceans are also present in the sediments of the pond, but it is likely that the testaceans were washed in

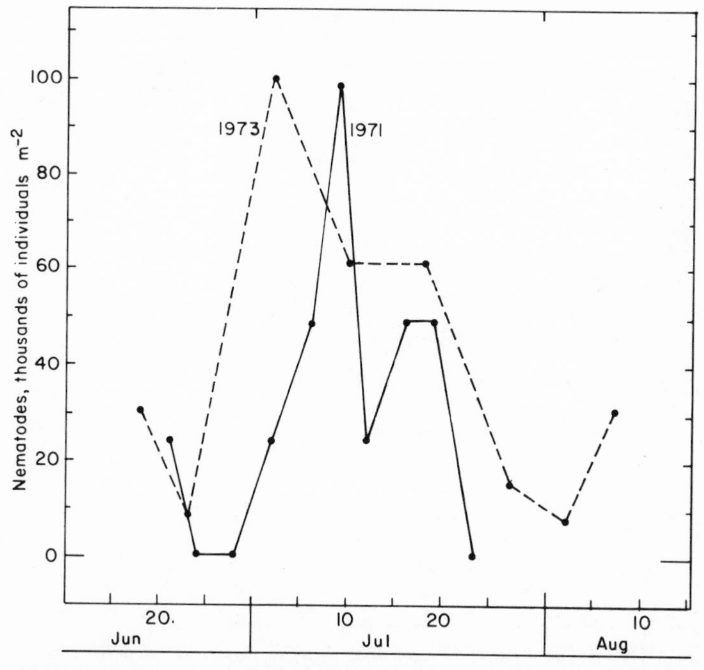

FIGURE 8-14. *Nematodes in the center of Pond B, 1971 and 1973.*

from the terrestrial moss community. Although tests of *Arcella* and *Diflugia* were sometimes seen, they were rare and were not counted as it requires observation of each individual to tell if the tests are occupied. Small "limax-type" amoebae were occasionally observed with the fluorescence microscope but could not be counted.

Nematodes were the most abundant metazoan (Figure 8-14); their numbers peaked at the beginning of July (10^5 m^{-2}) but most of these were juveniles. By August, the number had declined to 1 to 3×10^4 m^{-2}. This pattern suggests that nematodes have only one generation per year in these ponds. A similar pattern was noted for the harpacticoid copepods (Figure 8-15). During June and July nauplii were frequently found in the samples but the first adults did not appear until the beginning of August. Rotifers occurred throughout the summer at 3 to 8×10^4 m^{-2}. *Philodina* was the most abundant of the six genera seen. Other metazoans found included chaetonotid gastrotrichs, turbellarians, tardigrades, and ostracods.

Carbon Flow

From the biomass and feeding rates, it is possible to calculate the carbon flow through the protozoans of the sediment (Figure 8-16). This figure is based on the quantitative sample taken on 10 July 1973 and

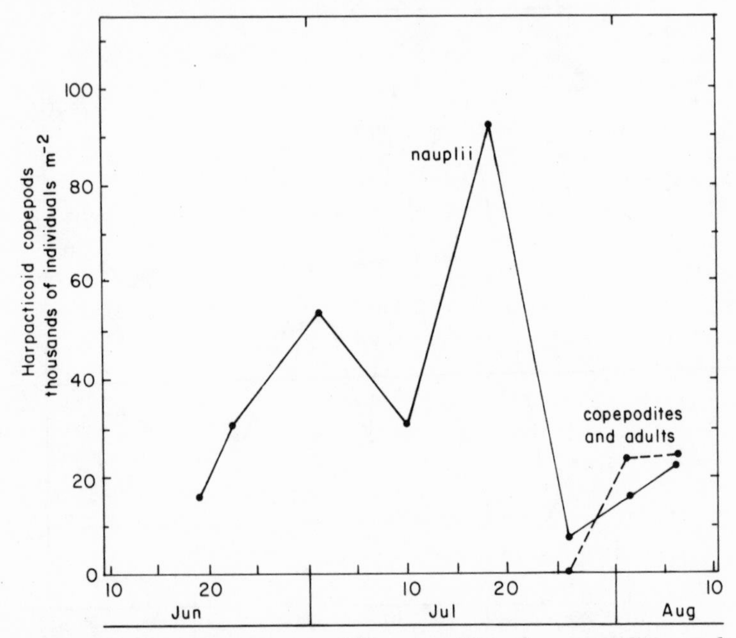

FIGURE 8-15. *Numbers of the nauplii and copepodites and adults of harpacticoid copepods in the center of Pond A–B, 1973.*

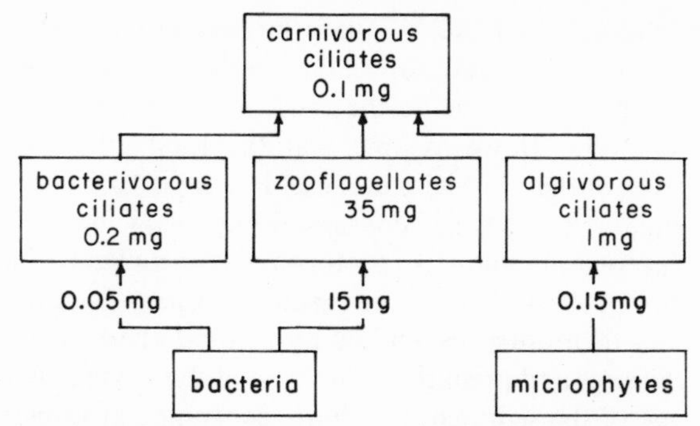

FIGURE 8-16. *Carbon flow through the most important groups of the protozoans based on samples from 10 July 1973 and a temperature of 12°C. Units are mg C m^{-2} for the standing stock (inside rectangles) and mg C m^{-2} day^{-1} for the rates (arrows). (After Fenchel 1975.)*

assumes a temperature of 12°C at the sediment surface (during the period 10 to 14 July, the temperature varied from 6 to 14°C). The rates would change with temperature, of course, but the general picture will be unchanged.

The values given in Figure 8-16 are probably underestimates both of the biomass and of the rates. We feel, however, that better estimates would not change the relative rates or biomasses. The zooflagellates consume about half their body weight (expressed here as carbon) of bacteria every 24 hours, whereas the ciliates feeding on bacteria take up only one-quarter of their weight. Yields (grams of protozoa produced per gram of bacteria) of protozoans feeding on bacteria are quite high and values of 0.58 and 0.78 have been reported for *Tetrahymena* and *Colpoda*, respectively (Curds and Cockburn 1968, Proper and Garver 1966). Thus, the pond data agree with previous estimates of a 24-hour generation time for other ciliates under optimum conditions (12°C) calculated by Fenchel (1968). Ciliates that eat algae consume an amount equal to 15% of their biomass every 24 hours, but this is likely an underestimate because many of the algae are larger than the 100 μg C assumed here. Also, many of the ciliates grouped as algivores are also predators but the predation rate has not been quantified. Despite this lack of precision, we conclude that ciliates are responsible for only a modest amount of the grazing on algae. Overall, the calculated feeding rates of protozoa on bacteria and algae are probably correct within a factor of 10.

Estimates of micrometazoan grazing rates are much less certain due to a lack of knowledge of even the qualitative aspects of their feeding biology. It is known that some rotifers, nematodes, chaetonotids, and microcrustaceans feed on bacteria and that some rotifers, nematodes, and turbellarians are predators (see Fenchel (1969) for references). In tundra ponds, we conclude that the most important single food item of the micrometazoans is epipelic algae, while the most important food of the protozoans is the bacteria. If we assume that the food intake of the micrometazoans is 20% of their biomass every 24 hours (at 12°C) and that half of their food is algae, then they will consume 7 mg algal C m^{-2} hr^{-1}.

Overall, the microbenthic animals (protozoa and metazoans) have a standing stock of about 100 mg C m^{-2} during the summer. The different elements of this fauna form intricate and complex food chains complete with specialized herbivores and predators. The base of these food chains is the bacteria and algae of the sediment,which are consumed at rates of 20 mg bacterial C m^{-2} $(24\ hr)^{-1}$ and 8 mg algal C m^{-2} $(24\ hr)^{-1}$. If the standing stock of bacteria is 1500 mg C m^{-2} and that of algae is 700 mg C m^{-2}, these grazer food chains would appear to have a very broad base and to only graze a small amount (1 to 2%) each day. However, both the algae and bacteria have a large standing stock and a proportionately small daily production of about 100 mg bacterial C m^{-2} and 150 mg algal C m^{-2}. The microfauna thus consume 20% of the bacterial production and 5% of the algal production each day. This may well be enough grazing to exert some degree of control.

Comparisons with Temperate Communities

One characteristic of arctic freshwater communities is a low diversity and the absence of whole groups of organisms found in temperate freshwaters. This is certainly true of the macrofauna of these tundra ponds that lack Hemiptera and have but a single beetle species. While the microfauna has only been incompletely investigated, the evidence from the relatively detailed study of the ciliates is that the diversity of the ciliate fauna is in no way lower than in similar temperate ponds. Even the species, or at least the morphologically defined species, are exactly the same as in temperate, oligotrophic ponds.

In addition, there are no apparent differences between the physiological rates of the pond ciliates and those of temperate ones. *Tetrahymena pyriformis* from the ponds had growth rates and Q_{10}'s for generation times quite comparable, between 5° and 25°C, to temperate forms (Barsdate et al. 1974). At 25°C, the generation time was around 3 hours. In the same way, the ingestion rates of two species of ciliates from these arctic ponds had a Q_{10} similar to that for temperate species in the 5° to 25°C range (Fenchel 1975). These data imply that there is no temperature compensation in the tundra pond species and, indeed, a similar result was found in a marine ciliate isolated from antarctic pack ice (Fenchel and Lee 1972). Its growth optimum was between 5° and 10°C and it did not survive above 17°C. In contrast, the pond ciliates from the Arctic survive to at least 25°C.

The protozoans are versatile animals broadly adapted to a variety of environmental conditions. For example, by the middle of June when only a few centimeters of sediment have thawed and temperatures are at 2°C, the protozoa fauna is fully developed. Some forms are known to have cysts but it is not known how *Paramecium* can survive as it does not form cysts (Kudo 1966). In any case, overwinter survival of the protozoans is high. The low temperatures throughout the summer do slow their activities, but even at the average summer temperatures at Barrow the protozoans will have generation times between 15 and 200 hours (Fenchel 1968). Over the summer, the smallest forms will have at most about 80 generations while the largest forms will have at most only 6. By extrapolation of this relationship, the larger animals will have a generation time of over a year; this is the case for macrobenthic animals.

The small metazoans of the microfauna, i.e., rotifers, tardigrades, turbellarians, and gastrotrichs, are similar to the protozoans in that numbers and species resemble those in temperate ponds. These metazoans also have mechanisms for winter survival such as resting eggs or anabiotic stages.

Slightly larger organisms, which may have several generations a year in temperate ponds (Gerlach and Schrage 1972, Muus 1967) have only one generation in the tundra ponds. In the case of nematodes, some adults apparently survive the winter and reproduce the following summer. Harpacticoids likely overwinter as juveniles, as a mature or near-mature form was seen in mid-June. Because only one developmental stage at a time is present, these arctic animals exploit much less of their total range of foods than temperate populations do. Unfortunately, we have not identified many of the larger microfauna so can not tell to what extent this limitation affects the diversity and community structure.

The larger animals of the macrofauna are strongly affected by the stresses of the 9-month freeze and the short summers. Their diversity is drastically reduced and they often have special adaptations in their life cycles or physiology. Some groups are completely eliminated and the absence of competition and predation (e.g., no fish or amphibians) alters the community structure. Thus, in both the flora and the fauna, the smallest organisms are least adapted and affected by the arctic environment while the largest organisms are the most adapted and show the greatest community disruption.

Control of Microfaunal Populations

The number of protozoans in the ponds appears to be a function of food availability, especially the bacteria and the benthic microalgae. Some of the food chains have already been described. One piece of evidence is that the populations of ciliates and zooflagellates have peaks close to the two peaks in bacterial numbers (Figures 8-1, 8-10, 8-11). Also, the higher numbers of ciliates along the shore than in the center of the pond correspond to the higher rate of production of bacteria in this zone where the fresh leaf detritus is abundant. Other evidence for the close link between population size and food availability comes from the changes in the food preference of the ciliate community over the summer (Figure 8-13). Bacteria feeders dominate early and late in the summer in response to the two peaks of bacteria. Algal feeders make up more than 50% of the ciliates in late June and early July, a period coinciding with the maximum epipelic algal productivity.

The significance of predation in controlling the microfauna is difficult to assess. Undoubtedly most predation on this group is by other microfauna; the best data are about the ciliates. In the ponds at Barrow, the carnivorous ciliates mostly preyed on other ciliates, although ciliates are also known to feed on zooflagellates and even small metazoans. As we might expect with predators, the carnivorous forms constitute only a few percent of the total ciliate fauna in early summer, but they are most abundant in late July when they make up 15 to 20% of the total (Figure 8-13). In mid-July their biomass is about 0.1 mg C m^{-2}. If their food needs are similar to those of the bacterivorous ciliates, then they will daily ingest about 30% of their body weight or about 3% of the total fauna. They could, therefore, be responsible for a part of the decline in ciliate numbers in the middle of the summer.

Metazoans also consume microfauna but we could not quantify this predation at all. Certainly rotifers, nematodes, and turbellarians do feed on protozoa and other micrometazoans. There are only slightly more data on the feeding of macrofauna. In total, the chironomids and oligochaetes consume about 100 mg C $m^{-2}day^{-1}$. The amount of food that is made up of detritus, of algae, and of bacteria is not known. Assuming that the detritus is not eaten and that the bacteria, algae, and microfauna are eaten in the same percentage as their biomass (15:7:1), then some 2% of the microfauna would be consumed each day. If the macrofauna are able to select their food at all, then this percentage may be even higher.

Feedback of Grazing on Bacterial and Algal Production

It is now well known that microbial activity is higher when grazers are present than when they are absent. For example, Fenchel (1970) and Hargrave (1970) showed that detritivores increased the productivity of sediment algae and bacteria. Fenchel (1977) also found that the decomposition of plant material by bacteria was much faster when a bacterivorous protozoan was present. This was true even though the protozoan reduced the bacterial numbers by half.

This phenomenon was studied in microcosms at Barrow during 1973 using *Carex* and organisms from the ponds. Details are given in Barsdate et al. (1974). Briefly, the experimental systems were 1-liter flasks containing 500 ml of autoclaved pond water, sterilized *Carex*, and an inoculum of organisms. Three inocula were used: pond bacteria obtained by filtering sediment suspensions through a 3-μm pore size Nuclepore Filter; bacteria plus the bacteria-feeding ciliate *Tetrahymena pyriformis* isolated from Pond B; or detritus direct from the pond and containing all the living organisms. The organisms in these microcosms attained a plateau after 50 to 100 hours. In grazed systems, the bacterial density was always 2 to 5 times lower than in the ungrazed systems. One theory to explain the increased activity and productivity was that the grazers were recycling nutrients. Accordingly, measurements were made of the movements of phosphorus such as the transfers of ^{32}P from the water to the bacteria, from bacteria to the water, from bacteria to ciliates, and from ciliates back to the water. Also, measurements were made of the actual mineralization rate of PO_4 from the *Carex*. Finally, a conceptual model of the phosphorus cycle of the microcosm was constructed.

The turnover rate of the phosphorus was always higher in the grazed than in the ungrazed systems over a wide range of phosphate concentrations in the water. This was mainly caused by a 4-fold increase in the rate of uptake and release of phosphorus by the grazed bacteria; the ciliates only contributed 4 to 5% of the total phosphorus excretion of the system. Therefore, we could not confirm the experiments from which Johannes (1965) concluded that protozoa excreted a significant amount of phosphorus that otherwise is tied up in the bacterial biomass. Rather, these results indicate that bacterial activity is stimulated by some mechanism other than the protozoan excretion of phosphorus. This increased activity then results in the increased rate of phosphorus cycling. Finally, the actual mineralization of phosphorus from the *Carex* was increased 40 times by the presence of the grazer.

The mechanism or process responsible for this stimulation of bacterial activity is not known. Some possibilities are the creation of micro-turbulence, the excretion of growth-promoting substances other than phosphorus, the stopping of competition among the bacteria for some factor that limits growth or density, and the selection for rapidly growing bacteria.

Although these microcosms did mimic the natural system to some degree by using *Carex* and natural flora and fauna, it is clear that only the qualitative results can be applied to our conception of the natural system. Thus, we can say that phosphorus does not appear to be limiting the bacterial growth and that protozoan grazing somehow acts to increase bacterial decomposition rates. In this way, protozoa and other microfauna may well be exerting a control on the rate of carbon flow within the entire sediment ecosystem that is far greater than their biomass or predation rates would indicate.

SUMMARY

Bacteria

Bacteria are abundant in the water (10^6 ml^{-1}) and in the sediment (10^9 ml^{-1}) of the ponds. In a square meter of pond (20 cm deep, 5 cm of sediment) there are 2 to 4 mg C in the plankton bacteria and an average of 1000 to 3000 mg C in the sediment bacteria. The numbers peak as soon as the ponds melt; in the case of the plankton this is likely caused by the influx of soil bacteria from the surrounding tundra, while in the case of the sediment bacteria this peak may be growth caused by the overwinter release of nutrients from dead microbes and other organic matter. A rapid decrease in numbers and then slow buildup takes place, with another peak in late summer.

The small molecules of the dissolved organic carbon are turned over every 2 days in the water and every 5 hours in the sediment. If the total amount of this DOC fraction is 120 μg C liter^{-1}, the total turnover is 3.6 g C m^{-2} yr^{-1}, of which only 20% is planktonic. In the plankton, the heterotrophic activity was positively correlated with the primary productivity rate so it is likely that much of the DOC used by the planktonic bacteria comes directly or indirectly from the algae. The sediment uptake was too high to come from algal loss so most likely comes from decomposition of particulate matter.

Bacteria in the plankton were grazed by the zooplankton at a daily rate of 15% while the sediment bacteria were grazed at a daily rate of about 1.5% (1% by microbenthos, 0.5% by macrobenthos). Total biomass removed was around 1 g C m^{-2} yr^{-1} at a maximum. The respiration of bacteria in the plankton was too low to measure but in the sediment the bacteria respiration was 9.4 g C m^{-2} yr^{-1} (all other biota was 4.3 g C). The increase in biomass of the sediment bacteria was 9 g C m^{-2} yr^{-1}. Methane production accounted for another 1.1 g C m^{-2}. The sum of these changes and losses was a bacterial production rate of 22.5 g C m^{-2} yr^{-1}.

Bacteria die in great quantities during the winter freeze and this may be the most important control on the biomass. Animals of the sediment are unimportant controls of biomass but grazing by the zooplankton may control the number of planktonic forms.

Bacterial productivity is controlled by the rate of supply of small molecular weight molecules. In the plankton, most of this usable material comes from algal excretion while in the sediment it comes mostly from decomposition of detritus. Warmer temperatures would increase bacterial growth but it is doubtful if the high rates could continue for very long before the substrate (detritus) was used up. Therefore, temperature is a control but the low temperatures help to keep the close balance between input and decomposition. One stimulus to bacterial production is grazing by animals. The mechanism for this effect is unknown but may be the result of a constant removal, keeping the bacteria in an active growth stage.

Bacteria in these ponds are likely inactive for most of their existence. This is an excellent life-strategy as it allows a large number of bacteria to exist and to be ready to quickly take advantage of good growth conditions.

Fungi

A single measurement in 1978 showed 998 m of hyphae (gram dry weight)$^{-1}$ at 1 to 2 cm of depth in the sediment of Pond C and 140 m at 6 to 7 cm. This is slightly lower than amounts in terrestrial soils at Barrow and much less than amounts in forest soils. The total quantity of hyphae in Pond C is 3500 mg C m^{-2}, which is 70% of the total microbial carbon (taking bacterial biomass as 1500 mg C m^{-2}).

Decomposition and Respiration

Leaves of *Carex* were incubated in nylon mesh litterbags in the ponds. Fresh green leaves lost nearly 40% of their weight during the 8-week summer; half the loss occurred in the first week so was likely due to leaching. Yellow leaves had died at the end of the previous summer and so were already leached; they lost 25% of their weight over the summer. Finally, brown leaves had been dead for at least one entire season; they lost only 5% of their weight over the entire summer. Over the winter, the green and yellow leaves did not change appreciably but the weakened brown leaves broke up so that 60% was lost.

Based on the litterbag measurements, a description of the decomposition can be constructed. If 32 g C of *Carex* leaves enters the pond at the beginning of the first summer, it will take 4 years to decompose. The leaves start as 33% green and 67% yellow. By the end of the first summer, 9.4 g C is lost (green loses 4 g and yellow loses 5.4 g). At the start of the second summer, there is 22.7 g C present, all brown. At the start of the third summer, there is 18.9 g C, and at the start of the fourth, 7.7 g C. Over the 4 years, most of the leaves are lost by mechanical breakdown (18.8 g C), while 8.1 g C is lost by leaching as DOC and 5.2 g C lost by hydrolysis (enzymatic breakdown). Our measurements were not accurate enough to tell whether or not a small amount (5 or 10%) was being preserved in the sediments. In addition to the 32 g C of *Carex* litter that is added and decomposed in the pond each year, a large amount of roots of *Carex* and grass, perhaps as much as 33 g C m^{-2}, is also produced. Little is known of its decomposition, but again only about 10% could be accumulating each year so most of this must be decomposed.

After 8 weeks of decomposition, both green *Carex* leaves and lemming feces lost almost all of their P and K while N, Mn, and Zn remained in the same proportion to the total weight of the material. The elements tied to the structural material, Mg, Ca, and Fe, increased.

Another way of looking at microbial activity and decomposition is by measuring the respiration of an entire system. This was investigated in cores using 4 to 6 hour incubations. Most of the activity occurred in the top 4 cm (77%) and a great deal (2.5 times higher than in the pond center) in the macrophyte beds (presumably due to root respiration). Over a single day, the peak of respiration lagged the peak of sediment temperature by a few hours. Total annual respiration over the entire pond was 15 to 25 g C m^{-2} over 3 years. Additional CO_2 could have been lost by transfer from the sediment into (and through) the macrophyte roots. However, the rate of CO_2 loss from the sediments measured in these core experiments was only about 50% of the rate measured by the CO_2 evasion technique.

The respiration rate of the sediments was directly related to temperature, with an average Q_{10} of 1.6. Other factors affected the respiration too. For example, added phosphorus increased respiration by 70% even in these short-term incubations. Acetate additions also increased respiration. When chironomids were added to the cores, the respiration decreased slightly. *Lepidurus* appeared to stimulate respiration. However, the errors in the respiration estimates for the animals themselves may have affected these results.

Microbenthos

The sediments of the Barrow ponds contain a variety of protozoans and metazoans (turbellarians, nematodes, ostracods, and harpacticoid copepods). The species composition, relative abundance, and even absolute numbers of this diverse community are very similar to those in temperate ponds. Most of the organisms in these Alaska ponds are found only in the top 2 cm because of anaerobic conditions deeper in the sediment.

In one pond, the small zooflagellates were numerically the most important part of the microfauna (10^{10} to 10^{11} m^{-2}). Micrometazoa were present at 10^5 to 10^6 ml^{-1} but the biomass of both these groups was equal (about 1 g wet weight m^{-2}). Ciliates were also abundant (about 10^6 ml^{-1}) but their biomass was low (20 to 40 mg wet weight m^{-2}). These numbers are also typical of detrital and muddy sediments in temperate marine and freshwater systems. There are two peaks of abundance of the ciliates, one after each peak of bacterial abundance. One group of ciliates fed on bacteria and another group fed upon benthic algae.

Nematodes were the most abundant metazoan at 1 to 3×10^4 m^{-2}. Rotifers were equally abundant but harpacticoid copepods reached 9×10^4 m^{-2}. All of these animals had but one generation per year and all except the copepods overwinter as eggs (copepods overwintered either as eggs or as juveniles).

Most of the carbon flow in the microbenthos is through the zooflagellates; ciliates and metazoans provide less than 10% of the total. All of the microbenthic grazers remove about 28 μg C m^{-2} day^{-1} or about 1 to 2% of the bacteria and algae. This is, however, about 20% of the bacterial and 5% of the algal production each day.

The microfauna are likely controlled in part by predation; most of this is by other microfauna. This predation is difficult to quantify, however. Predators do make up about 20% of the total ciliate fauna at their peak abundance. Predation by macrofauna is likely only about 2% day^{-1}.

Experiments with microcosms containing *Carex*, bacteria, and a protozoan (*Tetrahymena*) showed both a decrease in bacterial numbers (2 to 3-fold) and an increase in bacterial activity when the protozoan was present. This increase was not caused by any release of phosphate by the protozoan. A direct measurement revealed that the ciliates contributed only 4 to 5% of the total phosphorus excretion in the system. Actual mineralization of phosphorus from the *Carex* was increased 40 times by the presence of the grazer.

Temperature seemed to have only a slight effect on the microfauna. For example, in the middle of June only a few centimeters of sediment had thawed out and the water was still at 2°C. Yet the protozoan fauna was already fully developed. Later, the animals were able to reproduce and the smallest forms had up to 80 generations during the summer.

9

Oil Spill Effects

R. J. Barsdate, M. C. Miller, V. Alexander, J. R. Vestal, and J. E. Hobbie

INTRODUCTION

Oil production at Prudhoe Bay stimulated interest in research on effects of a possible oil spill on ponds. In fact, there are natural oil seeps at Cape Simpson, not very far from Barrow, and a producing gas well is in sight of the ponds. For this reason, NSF funded pond research in 1970 and an experimental oil spill was carried out on Pond E. This pond was sampled during the years of intensive research and then the Department of Energy provided funds for an additional oil spill (Pond Omega in 1975) and follow-up studies from 1975 through 1979. This research allowed additional studies on chironomids, zooplankton, and algae; much of this basic research is included in this book. In addition to the basic research, the oil spills were experimental treatments that produced information on such things as the control of phytoplankton algae by zooplankton and recolonization of ponds by insects.

Most of the research on effects of oil has been carried out on marine species; there are, however, several studies of effects on arctic freshwaters. In Canada, both lake and stream spills have been studied in the Mackenzie Delta (Brunskill et. al. 1973, Snow and Rosenberg 1975a, 1975b). In Alaska, a spill was studied in an arctic lake (Jordan et al. 1978, Miller et al. 1978a) and the natural seeps at Cape Simpson have been investigated (Barsdate et al. 1973).

Spill on Pond E, 1970

On 16 July, 1970, 760 liters (4 barrels) of crude oil from Prudhoe Bay (ARCO) were applied to the surface of Pond E (Figure 9-1). Pond E has a surface area of 300 m^2, or 490 m^2 if the surrounding marsh is included, and an approximate volume of 800 m^3. This amount of oil is about 16,000 liters ha^{-1} or 25 times the dose used for the Mackenzie Delta study.

Several hours after the spill on Pond E the oil covered the entire pond; 24 hours later the wind had moved the oil to the west side of the pond where it accumulated in a band 3 to 5 m wide. Throughout August and early September, about half of the applied oil moved back and forth in the

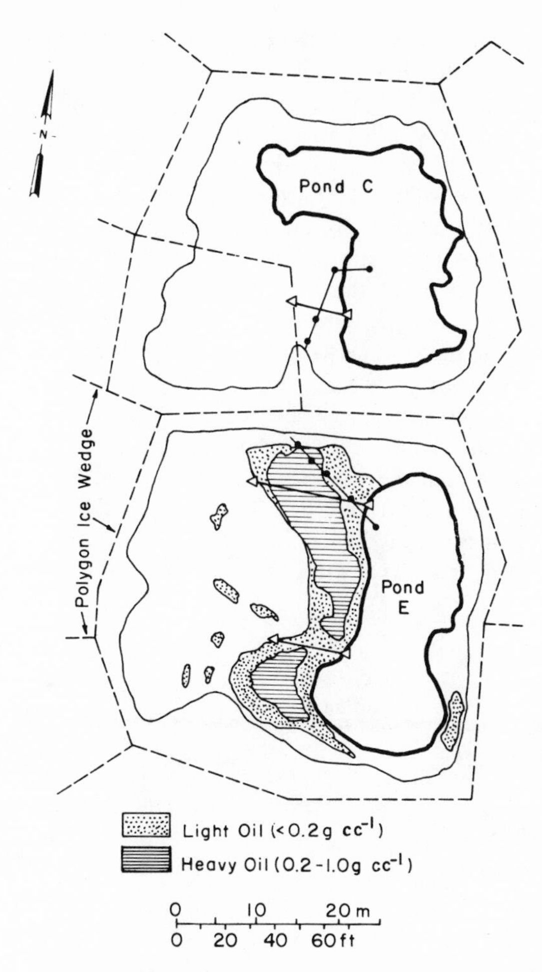

FIGURE 9-1. *Ponds C and E showing location of sampling points for temperature and oxygen in 1970 (solid points), the distribution of oil in July 1971, and the location of the 1971 and 1972 vegetation transect (triangles).*

pond as the wind changed while the rest remained in the emergent-vegetation zone. Some of the lighter fractions of oil evaporated and as the remainder became more viscous, it began to adhere to the stems of the emergent plants (first noticed in mid-August). Only a small amount of oil penetrated into the sediment or into the wet litter. A lighter (rainbow-colored) scum surrounded the margin of the ponds and extended over much of the surface whenever the wind fell. Later, the lighter scum folded into brown floating fans whenever it struck a blade of grass. These fans were easily broken up by wave action and sunk. In September, the floating oil retarded freezing so that the water beneath the oil accumulations did not freeze until the ice was 2 cm thick on the rest of the pond.

The following spring (1971), these same areas of accumulation thawed more rapidly than the rest of the pond. Pond E did overflow during the runoff but little oil moved out of the polygon due to filtering by the snowpack and by the emergent plants. A few small clumps of plant litter soaked with oil were found some 20 m from the polygon. Through late June and early July, 1971, some floating oil was still visible (Figure 9-1) and the oil odor was evident. By mid-July, the thin film of oil surrounding the heavy accumulation along the west side had changed from a typical oil slick to a thin brown scum. At the same time, the thick accumulations began to sink, and became attached to the plant litter in the *Carex* bed.

A rough determination of the quantity of oil remaining in July 1971 revealed that at least half of the oil was still present. This was measured from aerial and ground observations of the area covered by the heavy accumulation of oil, the oil slick, and the light scum. Samples were then taken 0.5 m to either side of the division between the light and the heavy accumulation to obtain an average weight of oil for each area. Approximately 150 m^2 of the total area was covered with oil weighing greater than 2 kg m^{-2} and about 190 m^2 with oil weighing less than this. These values included the oil attached to the vascular plants and litter. This attached oil made collection and extraction of the oil very difficult so there may well be a large error associated with this estimate. Certainly a lot of the oil was still present nearly 1 year after the spill, for a minimum

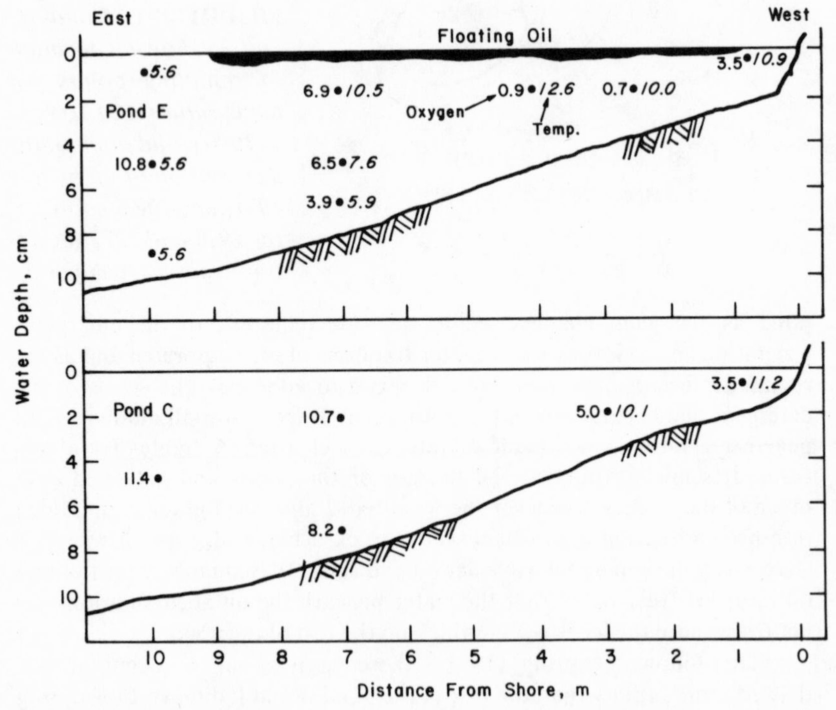

FIGURE 9-2. *Temperature (°C) and oxygen concentrations (mg O_2 liter^{-1}) in Ponds E and C, 23 July 1970.*

estimate of the quantity is 395 kg remaining of the 680 kg added (this is 150 x 2 + 190 x 0.5).

During the remainder of 1971 and in later years, the location and general appearance of the oil did not change and there was no further effect on the timing of the freeze and melt. Occasional oil slicks and small patches of floating oil were seen in 1972 and 1973; and in June, 1975, oil was visible in the *Carex* beds on the west side of the pond as a low ridge several centimeters high in the sediments. Although the surface of the ridge was covered with a brown scum and organic detritus, any mechanical disturbance caused a renewed oil slick to well up from the ridge.

Spill on Pond Omega, 1975

The experimental spill in 1975 was made on Pond Omega, a 260 m^2-pond located several hundred meters from Pond B (see Miller et al. 1978a for exact location). The 0.24 liter m^{-2} dose on 10 July was about one-tenth the amount added to Pond E. The initial movement of the oil slick was controlled by the direction of the wind and within 24 hours after the spill, the oil had collected in the *Carex* beds around the pond. This was similar to the 1970 oil spill.

PHYSICAL AND CHEMICAL MEASUREMENTS

Temperature

There was practically no change in the physical regime of the ponds as a result of the oil spill. The water temperature of Pond E did increase by about 4°C for 3 days after the spill but returned to normal after the oil moved into the pond margin and after this was exactly the same as the temperature in Pond C. As long as the floating oil covered a large area, it is likely that there was a microstratification of temperature both horizontally and vertically. For example, 7 days after the spill the range of temperature beneath the oil was slightly higher than in a similar location in Pond C (Figure 9-2). Evaporation in Pond E was about the same as in the control Pond C as shown by their similar water levels (Figure 9-3). Finally, it was possible that the oil could change the heat budget of the sediment through a decrease in albedo or an increase in heat conduction. Again the similarity of the depth of thaw (Figure 9-3) to that in the control pond indicates that any changes were too slight to measure.

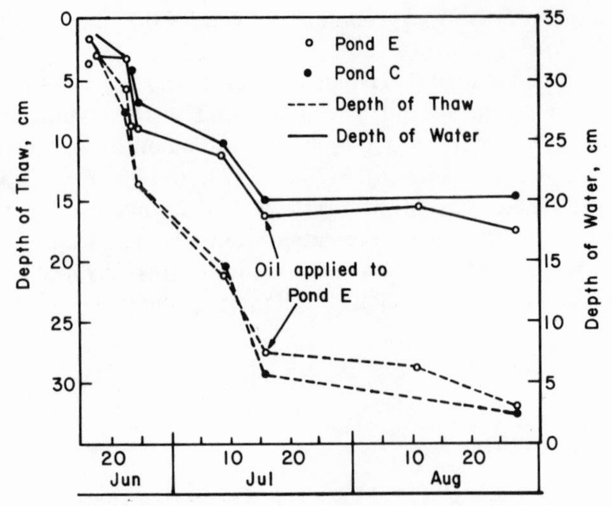

FIGURE 9-3. *Depth of thaw and of the water level in Ponds C and E, 1970.*

Oxygen

The only measured chemical change was in the oxygen concentration. There is normally a decrease in oxygen near the pond margins caused by the high rate of sediment respiration (Chapter 8), by the shallow water containing a small absolute amount of oxygen, and by the restricted circulation and mixing in the shelter of the emergent plants (Pond C in Figure 9-2). This decrease is intensified by the floating oil which not only drastically restricts water movement but also reduces the diffusion of oxygen through the water surface. These data are for a single date and there is every possibility that oxygen could be lowered even further on warm days or at night. If so, then the benthic animals may be affected.

Nutrients and Water Chemistry

There was no detectable change in the inorganic chemistry of the water after the oil spills in 1970 and 1975. We measured pH, alkalinity, Ca, Ha, K, Mg, Fe, and conductivity as well as nutrients (Table 9-1). In fact, the oil itself contains by weight 0.23% N and 0.82% S (Thompson et al. 1971) so the amounts added are large (20 mg N and 70 mg S $liter^{-1}$ of Pond E) . Also, the organic matter added can change the systems as Brunskill et al. (1973) found anaerobic conditions and H_2S beneath the ice of Lake 4 (Mackenzie Delta). However, in Ponds E and Omega the strong reducing conditions did not develop and evidently the N remains tied up in the heavier fractions (Ball 1962).

TABLE 9-1 *Nutrient Concentrations in Control Pond C and Before and After Whole Pond Spills on Pond E and Pond Omega*

Parameter-	Pond/Year	1970 prespill	1970 postspill	1971	1972	1975 prespill		1975 postspill
Dissolved reactive phosphate (μg at. PO_4—P liter^{-1})	Pond C	0.066[2] (.042)	0.082[2] (.029)	0.071[2] (.04)	0.064[2] (.01)	0.24[3] (.13)		0.38[3] (.58)
	Pond E	0.052 (0.03)	0.076 (.048)	0.059 (.05)	0.082 (.04)		0.28 (.21)	
	Pond Omega					0.43 (.30)		0.43 (.62)
Ammonium (μg at. NH_3—N liter^{-1})	Pond C	2.9 (1.6)	1.4 (0.4)	2.6 (0.8)		2.4 (2.3)		2.7 (2.0)
	Pond E	2.8 (1.2)	1.2 (0.3)	2.1 (1.9)			1.6 (0.8)	
	Pond Omega					2.8 (1.2)		2.2 (1.5)
Nitrate (μg at. NO_3—N liter^{-1})	Pond C	0.64 (.47)	0.03 (.05)	1.30 (1.1)		1.0 (1.1)		0.88 (.74)
	Pond E	0.57 (.58)	0.27 (0.8)	2.9 (2.6)			2.8 (3.8)	
	Pond Omega					2.6 (1.5)		1.0 (0.7)
Silicate (μg at. SiO_3—Si liter^{-1})	Pond C	4.2 (5.7)	9.4 (7.6)	2.3 (0.9)		3.4 (2.6)		1.5 (1.5)
	Pond E	3.5 (2.4)	8.6 (4.3)	2.9 (1.8)			8.9 (10.3)	
	Pond Omega					1.8 (1.0)		1.6 (0.7)

[1] Data expressed as yearly average (± standard deviation).
[2] 1970-72 DRP was extracted into isobutanol and read in 10 cm cells.
[3] Autoanalyzer, unextracted 1975.
Source: Miller et al (1978a).

The only chemical change noted was a detectable amount of nitrogen fixation in Pond E 3 years after the spill (Miller et al. 1978a). This fixation occurred in only one of three samples; the only other N fixation ever found in the ponds was in an experimental subpond of Pond B which was heavily fertilized with phosphorus. This may be bacterial fixation similar to that found in oiled experimental ponds along the Ottawa River (Shindler et al., 1975).

Some of the oil is lost by volatization and some lost by degradation; the exact amount lost is difficult to measure in the field but at least half of the oil disappeared in the first several months after a spill (details given in Miller et al. 1978b). For example, the initial loss rate of the volatile components of Prudhoe Bay crude oil was 10.3% day^{-1} at 5°C and 12.5%

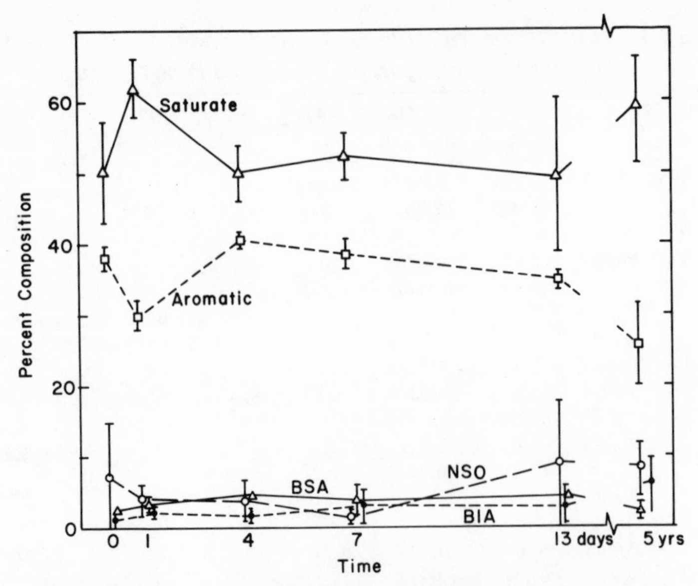

FIGURE 9-4. *Composition of Prudhoe Bay crude oil as a percentage of the total. The classes are benzene-soluble asphaltenes (BSA), benzene-insoluble asphaltenes (BIA), the saturate fraction, the aromatic fraction, and the nitrogen-, sulfur- or oxygen-containing fraction (NSO). (After Miller et al. 1978b.)*

day^{-1} at 25°C in a laboratory hood with moving air. The loss of all fractions of the oil was 18 to 19% in 36 days when the oil was placed in darkened Petri dishes in the field. When clear dishes were placed in the sun, 24% of the oil was lost, so this photo-decomposition could be important in the arctic. The loss was even higher when the oil was placed in plastic tubes containing natural water and sediment and incubated in the pond. After 45 days, 75% of the oil disappeared, which indicates that biotic processing was not too important. Somewhat slower rates were reported by Federle et al. (1979); after 1 year 58% of the oil remained in core tubes.

Despite the rapid loss of oil, the overall composition of the remaining oil does not change appreciably over time (Figure 9-4). The different classes were separated on solid-liquid chromatographic columns as described by Jobson et al. (1972) after separation from the water and sediment by Freon extraction. Even after 5 years in the pond, the composition of the oil is about the same. However, the saturate fraction (pentane soluble) did show some biological degradation, as Miller et al. (1978b) found that in the saturate fraction all of the hydrocarbons with fewer than 13 carbon atoms were lost within 13 days (measured by gas-liquid chromatography).

TABLE 9-2 *Average Bacterial Numbers and Turnover Times for Acetate in the Plankton and Benthic Respiration in Ponds B, C and E*

Pond	Treatment	Bacteria * (10^6 ml^{-1})	Benthic** respiration (dark) (mg C m^{-2} day^{-1})	Turnover* of acetate (hr)
B	Control	1.02	165	224
C	Control	1.90	113	129
E	Oil	1.18	51	110

* 1971
**1972

The loss rate of Prudhoe crude (10% in the first day) is much less than that measured by Snow and Rosenberg (1975a) who found that 34 to 43% of Canadian crude oil was lost in the first day. Our results are similar to those of Atlas (1975) who also worked with Prudhoe crude. Thus it is important to realize that oil from different sources will weather in different ways. Also, the amount of oil that will actually dissolve in water will differ for different sources. Federle et al. (1979) reported that Barrow pond water took up 15 mg oil liter^{-1} after 2 hours irrespective of the amounts of added oil. At 8°C, vigorous shaking resulted in 90 to 125 mg oil liter^{-1} in solution. The water soluble fraction is the most lethal part of the oil.

BIOLOGICAL MEASUREMENTS

Bacteria

One year after the oil spill in Pond E, the bacterial activity in the plankton was higher than that in control ponds but total numbers of bacteria were unchanged. Activity, as measured by the turnover time of ^{14}C-acetate (Wright and Hobbie 1966), was most rapid in Pond E (Table 9-2). Bacterial numbers were within the range of the controls. However, the direct count method enumerates all bacteria and we know that only a small fraction of these are active. These active forms could have increased in Pond E without any measurable change in direct counts.

Two years after the oil spill, the sediment respiration (one date) was less than half that of the controls (Table 9-2).

Bergstein and Vestal (1978) used plate-count techniques to enumerate the types of bacteria in the water and sediment in oiled and unoiled portions of Pond Omega, 2 years after the spill. There were no differences in the numbers of bacteria that could grow on crude oil, nutrient agar, mineral salts agar, or hexadecane between oiled and unoiled areas of the pond. If there was a toxic or stimulatory effect on the microflora after the spill, the microflora were back to the control-pond levels within 2 years.

TABLE 9-3 *Rates of Photosynthesis of the Benthic Algae on 16 August 1971 in Ponds E, C and J*

Pond	Treatment	mg C m^{-2} hr^{-1}	mg C m^{-2} day^{-1}
E	Oil	5	60
C	Control	13	160
J	Control	11	130

Source: D. Stanley (personal communication).

The general results from spills in lakes in arctic Alaska are similar to the pond results. For example, Jordan et al. (1978) found no significant increases in numbers (direct counts) in the water or sediment of a lake one year after the addition of oil (0.25 liters m^{-2} or 0.18 ml $liter^{-1}$). They also found that glucose uptake rates were unaffected by the oil but that the uptake of hexadecane and napthalene increased drastically over controls (but only after 110 hours of incubation). Another study by Horowitz et al. (1978) measured high numbers of hydrocarbon-utilizing organisms in the sediments of a lake near Barrow 7 years after a large gasoline spill. From these and other studies it appears that large numbers of hydrocarbon-utilizing microbes are present only when there are continual additions of oil or gasoline.

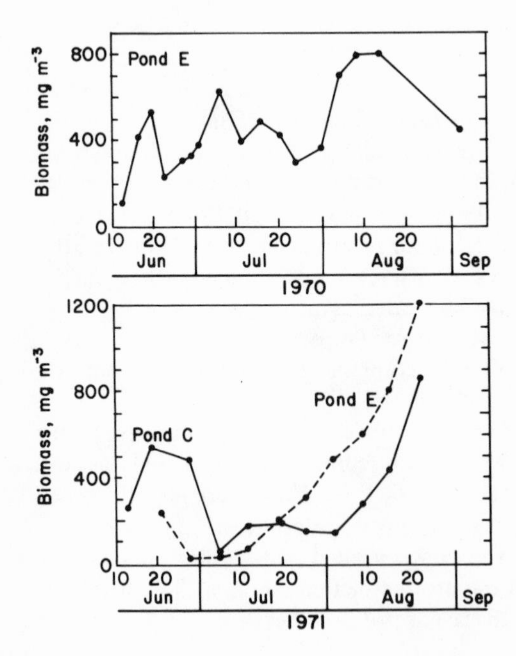

FIGURE 9-5. *Biomass of phytoplankton algae (mg wet weight m^{-3}) in Ponds E and C.*

Algae

One year after the spill on Pond E, the photosynthesis of benthic algae was reduced by more than 50% over the controls (Table 9-3). Only a single measurement was made, however.

The phytoplankton algae were studied in great detail in the oil experiment ponds. The results are clear; there was some increase in primary production and a change in species. However the causes are difficult to separate into direct toxicity effects and indirect effects due to the killing of zooplankton. This death of zooplankton will be documented later; they are extremely sensitive to oil and were eliminated from Pond E.

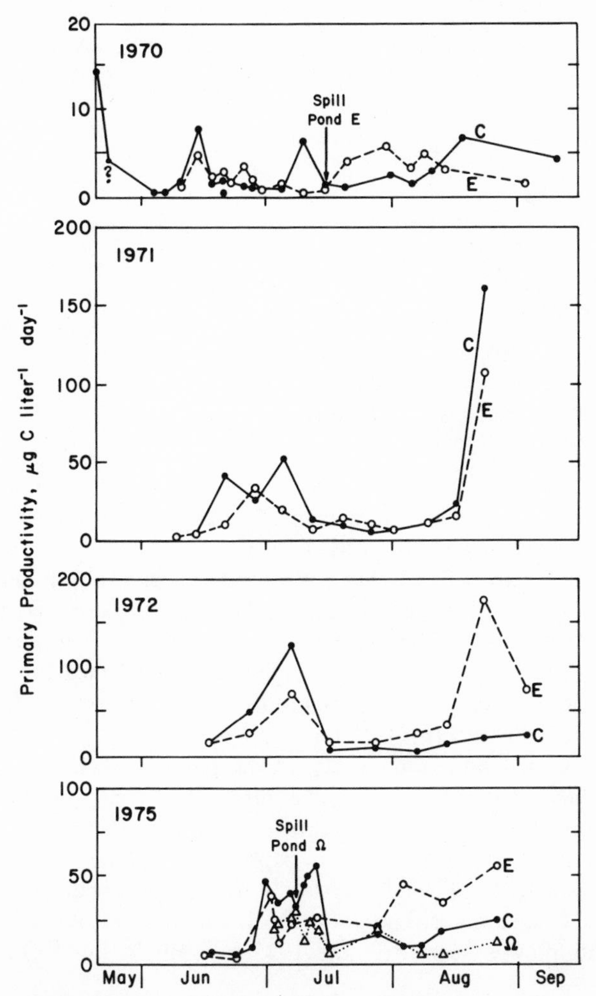

FIGURE 9-6. *Primary productivity in Ponds E, Ω and C. (After Miller et al. 1978a.)*

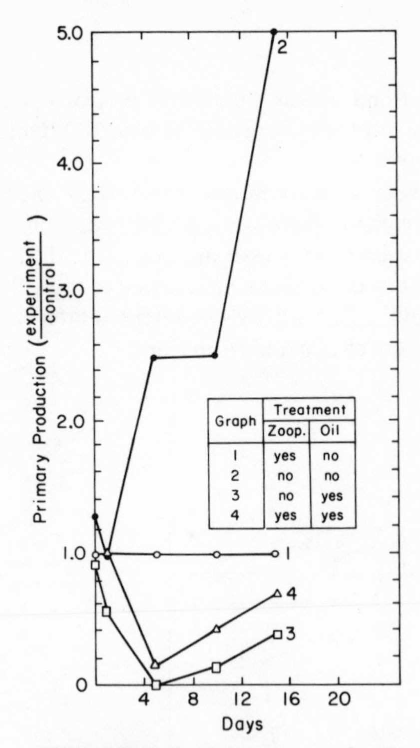

FIGURE 9-7. *Primary production in subponds after various treatments. (After Federle et al. 1979.)*

The quantity of phytoplankton, measured by direct counting with an inverted microscope, changed only a little as a result of the oil spill in Pond E on 16 July 1970 (Figure 9-5). Within 3 weeks of the spill, the amount of algae did double but a similar pattern was seen the following summer. This may be an advancement of the late summer bloom that had been released by removal of the zooplankton. A similar increase, measured as particulate carbon, occurred after the oil addition in the Mackenzie Delta lakes (Snow and Rosenberg 1975b).

The long-term primary productivity of the phytoplankton was not changed appreciably by the oil spills (Figure 9-6). There is enough variability from pond to pond and year to year that the differences between Pond E and Pond C (the control) cannot be attributed to the oil. However, the productivity of Pond E certainly was not lessened by the oil except during the first summer. The same effect, a temporary reduction in primary production, was seen after the 1975 oil spill in Pond Omega (Figure 9-6).

The reduction in primary productivity is directly proportional to the amount of oil added. Federle et al. (1979) found that when various quantities of oil were shaken with Pond C water, the short-term productivity was reduced by 50% (by 15 μl oil liter^{-1}) and stopped completely by 30 μl. Oil layered on top of experimental vessels and not shaken was only half as toxic. Thus, the water-soluble fraction of the oil is inhibitory to the algae.

The inhibition lasts at least several weeks (Figure 9-7) but the algae begin to recover after one week. This inhibition and recovery has also been

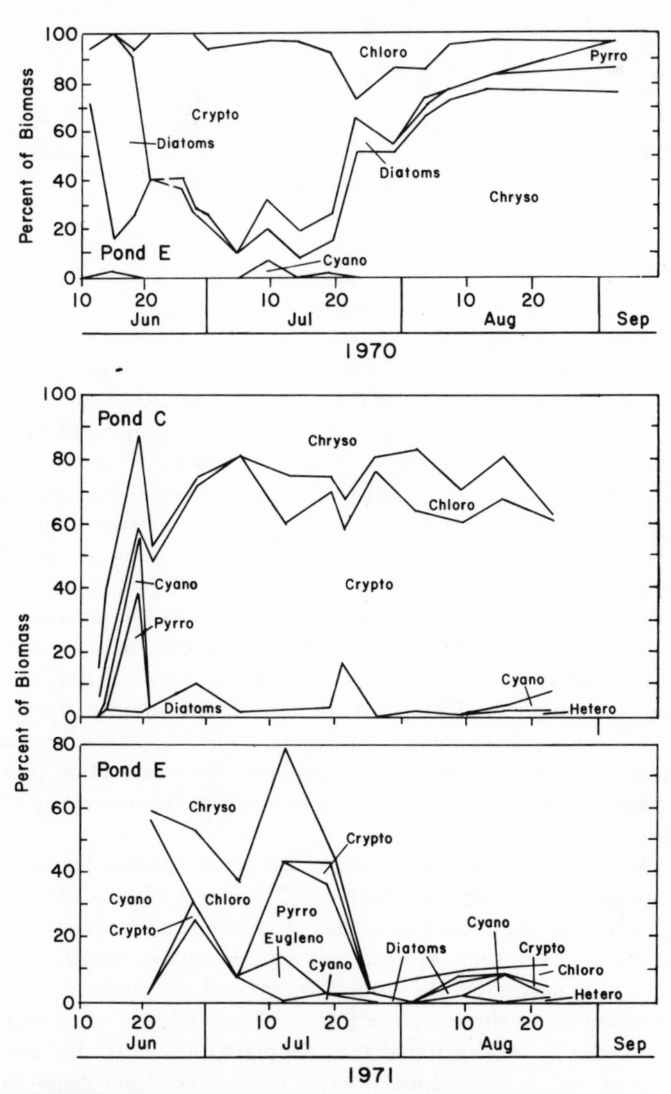

FIGURE 9-8. *Groups of phytoplankton as a percentage of the total wet weight in Ponds C and E.*

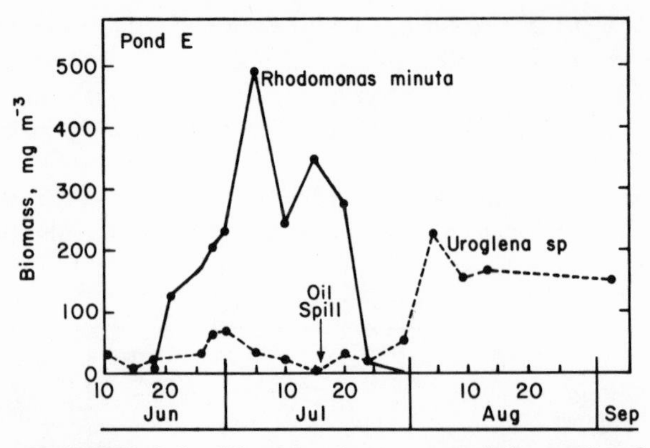

FIGURE 9-9. *Biomass (mg wet weight m^{-3}) of* Uroglena *sp. and* Rhodomonas minuta *in Pond E, 1970.*

found in *Chlamydomonas* cultures (Soto et al. 1975) and was caused by volatile fractions of the oil.

The most dramatic effect of the oil experiment in the subponds (Figure 9-7) was the replacement of the cryptomonad species of algae by chrysophytes (Federle et al. 1979). The same replacement of species occurred in Pond E in 1970 and 1971 (Figure 9-8) and in Pond Omega in 1975 (Miller et al. 1978a). In all the unaltered ponds we studied, the usual seasonal progression of forms was an early chrysophyte domination followed by a cryptophyte peak; a typical pattern is given for Pond C, 1971, in Figure 9-8. One of the obvious changes is a complete replacement of the small flagellate *Rhodomonas minuta* by *Uroglena* (Figure 9-9). This replacement took place in Pond E in 1970 and the *Rhodomonas* did not return to this pond until 1976, the same year that *Daphnia* and fairyshrimp returned in any numbers (Federle et al. 1979). The same replacement occurred in Pond Omega in 1975 (Miller et al. 1978b) and in the oiled experimental subponds (Figure 9-7). Similarly, Barsdate et al. (1973) noted decreased densities of cryptophytes in ponds at the natural oil seeps at Cape Simpson.

It is likely that the *Rhodomonas minuta* are eliminated because the zooplankton are killed rather than because of special sensitivity to oil. This species, a worldwide planktonic form, has never been cultured so we could not test it in the laboratory. We did run experiments in which the only treatment was removal of the zooplankton and concluded that this removal was enough to eliminate the *Rhodomonas* (Figure 9-7, treatment 2). Other workers have also found that zooplankton control the species composition of the phytoplankton (Porter 1973, Weers and Zaret 1975) but the mechanism of this interaction in the arctic ponds remains unknown. It is possible that the grazing pressure and the zooplankton's enhancement of the nutrient cycling rate control the competition among the various algal species.

In other experiments with oil in the Arctic, green and blue-green algae become abundant. For example, Hanna et al. (1975) and Snow and Scott (1975) report that *Oscillatoria* increased after oil spills and some of our experimental subponds followed the same pattern (Federle et al. 1979). This change did not occur in the whole-pond experiments we performed but did occur when experiments were run in subponds without sediment. It is likely that the phosphorus in the natural ponds is kept low and relatively unavailable because of adsorption by the sediment; in lakes and in experimental vessels without sediment the phosphorus becomes more available because of the elimination of zooplankton.

Rooted Plants

No damage to the vascular plants, especially the dominant *Carex aquatilis*, was observed in the ponds during 1970, but growth was affected during following years. Immediately after the spill much of the oil became attached to the stems of this sedge along the shore of the pond. As long as the oil touched only the stem there was no damage; some damage did occur when the oil contacted leaves. In the same manner the new leaves of *Carex* that appeared the next spring (1971) were killed in the area of the pond where they had to actually push through the floating layer of oil. This mechanical effect, therefore, could have been the cause of the low biomass of *Carex* in the areas of heavy oil accumulation (Table 9-4). The area covered with a light accumulation of oil had a lowered *Carex* biomass in 1971 but was back to normal in 1972. There was no evident effect on the rooted plants of the 10-fold smaller spill in 1975 in Pond Omega.

TABLE 9-4 *The Above-water Live Biomass of* Carex aquatilis *in Ponds E and C in 1971 and 1972**

Treatment	8 July 1971	17 July 1972
Heavy oil		
A (north transect, Pond E)**	3 (15)	6 (8)
Light oil		
B (south transect, Pond E)**	18 (9)	34 (8)
Control		
(Pond C transect)**	30 (8)	27 (5)

* Data are expressed in g dry wt m^{-2}; (n) = number of samples.
**Location of transects given in Figure 9-1.

TABLE 9-5 *Sequence of Disappearance of Zooplankton Species from Pond Omega Following the Experimental Oil Spill on 9 July 1975**

Species Observed	Observed presence after spill								
	7/10	7/11	7/12	7/14	7/16	7/18	7/21	7/23	7/29
Fairyshrimp (both species)	X	O	O	O	O	O	O	O	O
Daphnia middendorffiana	X	X	X	O	O	O	O	O	O
Heterocope septentrionalis	X	X	X	X	X	O	O	O	O
Cyclops spp.	X	X	X	X	X	X	X	X	X

*An "X" indicates that a representative of that species or group of species was found alive on that day. An "O" indicates that no individual of that species or group was found alive on that day.
Source: O'Brien 1978.

TABLE 9-6 *Numbers per Liter of Four Groups of Crustaceans in Ponds C and E June through August 1971*

Pond	Jun 14	Jun 21	Jun 28	Jul 5	Jul 12	Jul 19	Jul 26	Aug 2	Aug 16
Cyclops vernalis									
E	2.3	1.5	1.3	1.6	0.0	0.5	0.2	0.1	0.2
C	0.8	6.7	4.6	4.0	2.3	2.9	1.6	0.6	1.0
Daphnia middendorffiana									
E	0.0	0.1	0.0	0.0	0.0	0.0	0.0	0.0	0.0
C	0.0	0.1	3.0	1.8	3.6	1.2	3.9	1.4	1.7
Fairyshrimps									
E	0.4	0.7	0.0	0.0	0.0	0.0	0.0	0.0	0.0
C	0.1	4.9	3.0	0.4	0.2	0.2	0.1	0.1	0.1
Calanoid Copepods									
E	0.6	3.4	0.2	0.2	0.1	0.0	0.0	0.0	0.1
C	0.0	93.5	32.4	10.0	10.1	5.4	4.4	2.7	2.7

Source: R. Stross (personal communication).

TABLE 9-7 *Zooplankton Production in Ponds B, C and E in 1971, 1972 and 1973*

Pond	Year			
	1971	1972	1973	$\overline{X}$
E (Oil)	0.05	0.08	0.04	0.06
B (Control)	0.53	1.07	0.44	0.68
C (Control)	1.40	1.19	0.82	1.14

Source: R. Stross (personal communication).

Zooplankton

The major effect of the oil spills on the ponds was the rapid kill of the zooplankton. O'Brien (1978) studied the animals during the Pond Omega experiment and showed the great sensitivity of the fairyshrimp and *Daphnia* and the lesser sensitivity of the copepods (Table 9-5). In pond E, which received 10 times more oil than Pond Omega, the copepods were eliminated in 1970 but a few were found in 1971 (Table 9-6). Some animals of each species were present at the beginning of every year, presumably due to transfer of animals during the spring flooding of the tundra. These were killed each year by the small amounts of the water-soluble fraction (WSF) released each year from the oil in and near the pond (Figure 9-1). *Daphnia* did not return to Pond E until 1977 (Butler and Keljo personal communication); therefore, production of the zooplankton community was extremely low even though some copepods were present (Table 9-7).

The differential sensitivity of the zooplankton to oil was confirmed in aquarium studies of O'Brien (1978). *Daphnia* were killed at all levels of added oil including 0.2 ml oil liter^{-1} or about 15% of the Pond Omega treatment level. Fairyshrimp were most sensitive, *Daphnia* next, *Heterocope* next, and *Cyclops* least, exactly duplicating the field results of Table 9-5. The toxicity could be eliminated by vigorous aeration.

The results of the pond studies are similar to those from other marine and freshwater studies. For example, Busdosh and Atlas (1977) found that 3.0 ml liter^{-1} of Prudhoe crude oil killed marine amphipods and that it was the WSF that was toxic. A tidepool copepod was killed by 1 ml diesel oil liter^{-1} within 3 days (Barnett and Kontogiannis 1975). Aeration of oil and water dispersions resulted in a loss of 80 to 90% of the WSF within 24 hours (Anderson et al. 1974). Because the aeration did eliminate the toxic effects of small amounts of oil, O'Brien (1978) suggests that aeration might be used in an actual spill.

Aquatic Insects

Insects from the ponds are not killed by oil in aquarium studies. Mozley (1978) added up to 8.4 ml liter^{-1} in 13-liter aquaria and found no change in survival of several kinds of chironomid larvae and eggs, of trichopteran larvae, and of plecopteran nymphs. Unfortunately, up to 50% of the animals in the controls of these experiments died during the 12-day test so that a low level of toxicity would have been missed.

Despite the lack of toxicity of oil in the laboratory, some insects and other invertebrates were eliminated or drastically reduced in numbers in the oiled ponds at Barrow (Mozley and Butler 1978). The *Agabus* (beetles), *Asynarchus* and *Micrasema* (caddisflies), *Nemoura* (stoneflies) and *Physa* (snails) were especially affected while *Libertia* (mites) remained present in all ponds. Most of these animals live only in the plant beds and

TABLE 9-8 *Two-year Mean Densities and Taxonomic Composition of Benthic Macroinvertebrates from Hand-core Samples in Barrow Thaw Ponds**

Measure	Pond J	Pond G	Pond Ω	Pond E
Mean density m^{-2}	26,800	5,380	7,810	14,700
Standard Error	4,800	570	800	3,800
%Chironomini	35.9	63.9	18.3	28.4
%Tanytarsini	51.1	30.0	64.0	9.9
%Tanypodinae	5.8	4.4	6.7	15.0
%Orthocladiinae	0.5	1.8	2.4	20.6
%Oligochaeta	6.8	0	7.9	25.9

*Standard errors are based on variation between mean densities for each sampling date.

Source: Mozley, 1978.

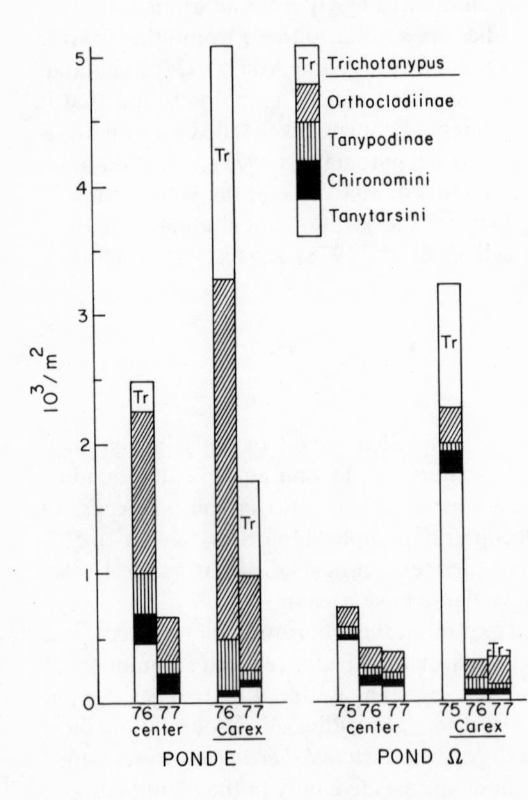

FIGURE 9-10. *Number of emerging chironomid adults per square meter by subfamily in the two ponds treated with oil. Trapping was continuous through the emergence season. (After Mozley and Butler 1978.)*

may have become trapped in the oil on the plant stems and in the floating oil. Snow and Rosenberg (1975a) reported similar entrapment of insects in the surface film of a Mackenzie Delta lake. At Barrow, recovery must take more than 6 years as many of the insects were still absent in Pond E in 1977.

The chironomid larvae were virtually unaffected by the oil spills. The numbers present in the sediments did vary a great deal from pond to pond but the highest and lowest numbers occurred in the control ponds J and G (Table 9-8). Emergence was affected by the oil, however, and the metamorphosis of *Tanytarsus* was strongly reduced after the spill in Pond Omega (Figure 9-10). This genus of filter feeders did not recover in Pond E either and was also strongly affected in Mackenzie Delta Lake 4 (Snow and Rosenberg 1975a).

These observations on the aquatic insects indicate that the oil-induced changes are on the species level and that such measures as secondary production and carbon flux are virtually unchanged. For example, the emerging cohort of *Chironomus* in 1977 presumably hatched in the year of the spill in Pond E and yet its numbers in Pond E were the same as in the control Pond J (Mozley and Butler 1978). These authors conclude that a light spill such as these reported here might best be treated by merely attempting to absorb floating oil onto inert materials and possibly by flooding the ponds to float oil away from the littoral plants.

SUMMARY

Crude oil from Prudhoe Bay was added to Pond E in 1970 (1.6 liter m^{-2}) and to Pond Omega in 1975 (0.24 liter m^{-2}). The wind moved the floating oil to the edge of the ponds and some oil floated for about a month. By the end of the summer all the oil was trapped along the pond edge and much had sunk. No oil left the pond during runoff the next spring but oil was still visible at the edge. After several years, at least half the oil was still present and was covered by debris and organic matter; it still welled up and created a scum when disturbed.

The only physical-chemical change caused by the oil was a slight decrease in the oxygen concentration of the shallow pond margins. This was likely the result of reduced diffusion and water movement. There was no change in the pH, alkalinity, or nutrient concentrations.

At least half of the oil was lost during the first year after the spill, mostly by volatilization and chemical degradation; there was also a small effect of biological degradation. In Pond E, for example, the oil remaining after five years had virtually the same chemical composition, but there was some loss of those hydrocarbon compounds with fewer than 13 carbon atoms (presumably from biological degradation).

The number of bacteria in the plankton was unaffected by the addition of oil but their activity increased during the first year. Two years after the spill in Pond Omega, there were no differences in the types and numbers of bacteria in the oiled and unoiled parts of the pond (plate count technique, nutrient and oil agar). Thus it appears that a single addition of crude oil briefly stimulated microbial activity but the microflora were back to control-pond levels within 2 years.

The effect of oil on benthic algae was only briefly studied; 1 year after the spill the photosynthesis in Pond E was 50% that of a control pond. In contrast, the phytoplankton algae were intensively studied and we found that the water-soluble fraction of the oil strongly reduced photosynthesis for several days. The amount of algae, however, did not change as a result of the spill and productivity reached normal levels within several months.

The added oil drastically changed the species composition of the planktonic algae in both the ponds and in experimental chambers. This change, a rapid replacement of the cryptophyte *Rhodomonas* by the chrysophyte *Uroglena,* continued for 6 years. It is likely that the *Rhodomonas* are eliminated because the zooplankton are killed; experimental removal of the zooplankton caused the same elimination. It is not known whether the algae responded to a release of grazing pressure or to a cessation of the zooplankton's recycling of nutrients.

No damage to the vascular plants was observed in the ponds during the first year, but growth of *Carex aquatilis* was reduced in later years. Much of this reduction was caused when new leaves encountered a barrier of floating oil.

In experiments in the laboratory, the fairyshrimps were most sensitive to oil, *Daphnia* were next, *Heterocope* next, and the *Cyclops* were least sensitive. This sequence duplicates the field results; all the *Daphnia* and fairyshrimps were killed immediately by the whole-pond treatments and did not return for 7 years. The less-sensitive copepods returned within a year.

In laboratory tests, aquatic insects and other invertebrates were not sensitive to oil. In the field, the spill had no major effect on the numbers and production of chironomids, but there were some minor effects on their emergence. One genus, *Tanytarsus,* was nearly eliminated from the ponds. Beetles, caddisflies, stoneflies, and snails were also drastically affected; most of these animals live only in the plant beds and may have become trapped in the oil on plant stems and in the floating oil. These insects were still absent in Pond E 6 years after the spill.

When the oil spills are relatively light, as in these experiments, then the best treatment would be to absorb the floating oil and perhaps to flood the marsh to float oil away from the littoral plants. The biota of ponds will recover within a few years with this simple treatment. More drastic cleanup measures will induce greater changes into the ecology of ponds.

10

Modeling

J. L. Tiwari, R. J. Daley, J. E. Hobbie, M. C. Miller, D. W. Stanley and J. P. Reed

MODELING IN THE AQUATIC PROGRAM OF THE TUNDRA BIOME

The mathematical modeling of whole ecosystems was one of the scientific tools that all parts of the U.S. IBP studies were to use. It is obvious that a model of an ecosystem that was biologically correct would be of great aid in predicting the effect of changes, such as a temperature decrease or an increased rate of nutrient input. Yet no large-scale model had been constructed when IBP began; the personnel were inexperienced and it was not entirely clear how to proceed. To some scientists, the modeling was a tool to be learned about; to others, it was already suspect because of past failures or abuses. Some scientists looked at the whole U.S. IBP as an experiment in the use of models; they asked the questions "Can a large-scale mathematical model be constructed?" and "Is this approach a useful one for a detailed study of an ecosystem?"

The experiences of the aquatic program were by no means unique; other groups that used modeling reached the same conclusions. Yet many of these conclusions have not been published and there are programs beginning each year that appear to be making some of the same mistakes that we did. In the hope of at least making people aware of the problems, we present here some of our experiences and general conclusions.

The first important step is to assemble a group of ecologists who are used to thinking at the whole-system level. This is relatively easy in limnology, for studies of lakes or watersheds lend themselves to simple input-output models and the studies of cycling of nutrients are further advanced in lakes than in soils, for example. A majority of the scientists on the project must be able to think like this. Some specialists, such as taxonomists and chemists, are necessary but every attempt should be made to attract scientists who understand more of the ecosystem than their own specialty. One reason for the need for this type of scientist is that there are many questions of judgment that arise during modeling, such as deciding what are the important processes and organisms that have to be in the model. Another reason for having this type of scientist is that modeling often requires different types of data than might be collected usually. For example, zooplankton counts may have to be transformed into carbon amounts, yet the factors for the transformation are not constants and must be separately measured.

The next problem is to choose a modeler. The number of ecologists who are really good modelers is quite small so we chose instead a geneticist with a good mathematical background. It is important that a modeler be familiar with the uncertainty and variability that are characteristic of biological systems. Too many modelers with engineering backgrounds appear to believe that a constant is inviolate and that the literature can be trusted. Our choice of modeler meant that all of the biological insight had to be provided by the ecologists and that modeling could not go on without them.

The model has to be detailed enough that it is biologically sophisticated and satisfying to the ecologists. Thus, photosynthesis could not be put into the model as a daily sine wave but, instead, had to be a function of light, of nutrients, of temperature, etc. The actual functional relationships, for example those between light and photosynthesis at different temperatures, had to be incorporated into the equations. Other details of the models are given in the next section.

As a general philosophy, we tried to measure every parameter, constant, and initial value that went into the model. This was about 90% successful but some processes that we knew were important could not be measured yet had to be incorporated into the model. One example is the natural, or non-predation, death rate of microbes such as algae. Most models of lakes or reservoirs contain this process but no one has measured it directly. We asked one of the foremost engineering modelers about this process. He replied that it had often been measured and that all one had to do was follow the populations of plankton algae after enclosing them in a bottle. This is absurd and naive. The situation becomes even more upsetting when coefficients like that for algal death rate are used to "tune" the model by adjusting them so that the output of the model agrees with nature.

A great deal of time, effort, and money was spent in developing a computer data bank for the Tundra Biome. Most of this was wasted as far as the aquatic research was concerned. Only the most routine type of data can be easily put into such a data bank and we stopped taking very much of this type of data after the first year. From then on, most of the data came from experiments which were run under conditions that changed each time. Pages of explanation were necessary to put this type of experimental data into the data bank. One extreme view of the situation was that "if the result goes easily into the data bank it probably isn't worth getting." This is really a reflection of the fact that we had lots of routine data but needed more information about the relationships and controls.

Once the modeling began, it became important to have some interchange between the modelers and experimental scientists so that the experiments could measure at least some of the things that were necessary for the modeling. In our case, it did not prove profitable to have the modeler in the field and so most of the interchange took place during the

9-month winter season. The interchange was also important for deciding whether or not the output from the models was reasonable. We believe strongly that in our present state of understanding the model output must not be believed unless it is biologically reasonable. The biological insight of the scientist must prevail and models must be judged incorrect if the output is biologically unrealistic. This may seem trivial but it is evidently all too easy to begin believing the model output and to use this as the basis for rejecting contradictory data.

It is obvious that the ecologists need to be intimately involved at all stages in the modeling. The judgments on the processes to be included, on which papers in the literature are to be believed, and on the reality of the output all call for knowledgeable, experienced scientists who are critical thinkers and completely up-to-date with the latest developments in the field.

WHOLE SYSTEMS MODELS

Introduction

It was originally hoped that the relatively simple aquatic ecosystems of a tundra pond would allow whole system models to be constructed and that these models would be good enough for testing hypotheses or predicting the effects of perturbations. It is now clear that even an ecosystem so simple and diligently studied as a tundra pond is not understood well enough for predictive modeling. We believe that the fault lies with our understanding of the biological processes and not with the modeling techniques.

As will be demonstrated, we have been able to simulate the complete pond ecosystem using the initial conditions of biomass and concentration actually observed as well as the measured parameter values such as half saturation constants for photosynthesis, etc. The deterministic model, however, proved very sensitive to small changes in certain parameters—too sensitive, in fact, to be satisfying to biologists. Obviously some ecological feedbacks and changes in parameters over the season are not known; yet, these may be controlling the rates of predation or adjusting the respiration rates.

Some subsections of the model did contribute to our understanding of the limnology of the pond by showing how various components could be interacting. An example is the effect of light and temperature on algal photosynthesis (Stanley and Daley 1976). This same study also used the model to determine the importance and necessary values of downward mixing of epipelic algae in the surface sediments.

Finally, the model and its construction were excellent management tools. The modeling procedure forced us to examine certain interactions, such as the interaction of phosphate with the sediments, that proved to be important keys to our eventual understanding of the controls of the ecosystem. In addition the construction of the model over three years kept us on a relatively narrow track and helped us avoid the innumerable branch lines we could have taken. These, while interesting, could not all be followed with the limits of time and manpower placed on this project.

Overall, we conclude that the modeling effort was well worthwhile. Yet, the value lies mainly in the construction and not in the output from the model; this value is difficult to document here. Also, it is difficult to document the interaction of the modeling with the design of experiments. This difficulty is added to by the organization of this book; logical development required presentation of driving variables (such as temperature and light) initial conditions (such as concentrations of organisms at the beginning of the year) and parameters and constants (such as respiration coefficients and Q_{10}'s) before the presentation of the model. For these reasons, the modeling is presented last in this report and may appear somewhat separate from the rest of the synthesis. As noted, the modeling was an important part of the whole project.

General Description

The study of ecological systems using the formalisms of system science requires an understanding of the whole system, of its living and non-living component entities, and of their various relationships in abstract form. Thus the structure of the system under consideration can be conceptualized as a collection of "objects" coupled together by some form of interactions. Each object is characterized by a finite number of "attributes" which can be measured and assigned some value. In general, these attributes are time-dependent quantities. Various interactions between the objects and between the attributes of each object can be described by mathematical formulae. (For details and rigorous definitions and formulations of these concepts see Zadeh and Desoer 1963, Zadeh and Polak 1969, Caswell et al. 1972, Klir 1969, Rosen 1970.)

Thus an object is essentially a set of variables together with a set of relationships between them; a time sequence of instantaneous values of these attributes describes the behavior of the system. The dynamics of the system are characterized by the interactions between the elements of the system, for these interactions impose a set of constraints on the output variables of the system.

The objects and interactions between them provide a basis for a "state-space" description of the system. Each basic entity of the system is a state variable which is chosen by the investigator for the particular system. Over some range, the values attained by a state variable are the state space of that variable. The state of the total system at time t can be defined as the values of the state variables. This state can be represented as a vector in n-dimensional phase space which will indicate the movement of the system. The time sequence of these vectors will form trajectories which show the long term behavior of the system.

The state of the system at any time t can be determined by solving the equations describing the interactions. Various types of equations can be used to describe these interactions, depending upon the nature of the problem and the objectives of the study. Thus,differential equations may be used for a continuous system and difference equations for a discrete system. For a certain class of problems, differential-difference equations might be more appropriate. To include the space effects in the model (for "distributed" systems), the system can be formulated by partial differential equations. In certain situations integral equations might be more appropriate and useful.

Once the system is defined and described by a set of mathematical equations we can proceed to analyze its dynamic behavior. The equations describing the biological systems are generally of the non-linear type and are analytically intractable; often they can only be analyzed by computer simulation techniques. This is essentially a numerical approximation of the results for a particular set of parameter values and initial conditions. It is not a general solution to the differential equations describing the dynamics of the system.

Once the model is successfully duplicating the observed behavior of the system under consideration, the effects of various perturbations can be investigated. We can ask the questions: what is the effect of changing an interaction or a set of interactions between the state variables, of changing parameter values, of adding or deleting some state variables and interactions, etc.

The basic questions and problems which emerge in the analysis of all systems (Zadeh and Desoer 1963) are essentially the specification of the objects and their attributes to be considered in the study. Furthermore, the interactions between the attributes of each object and the interactions between the attributes of different objects must be characterized in appropriate mathematical forms. The relationships between system inputs and outputs, and also a system of relationships between the attributes of the system as a whole, should be defined.

The complexity of the internal structure of the system is dependent upon the complexity of the patterns of interactions between the variables of the system. The "connectance"—the probability that any pair of variables will interact—of a large and complex ecosystem may be an important factor affecting the stability of the system (Gardener and Ashby 1970, Somorjai and Goswami 1972, May 1971, 1972, 1973, Siljak 1974, Maynard Smith 1974). These analytical and computer simulation results show that in large systems there is a very sharp transition from stable to unstable behavior as the connectance or the strength of interaction in the system exceeds a critical value. Thus both the complexity of the ecosystems and the magnitude of the interactions between its components must be taken into consideration and analyzed, for they will affect the behavior of the system.

The first step in the modeling process was for the biologists to identify and define the appropriate variables and the interactions between these variables to be included in the models. Experiments were designed to quantify the relevant interactions and necessary data were collected to measure and estimate variables, parameters, and constants. The models evolved through a series of workshop meetings where interactions were identified and their mathematical forms discussed and evaluated. These ideas were incorporated into the model and computer simulations were made to evaluate these concepts and results. The mathematical model of the aquatic system presented here represents our present "best" understanding of the structure and function of these systems. It is appropriate to mention here that the model does not include every existing species and interaction. Only those variables and processes are included which were measured experimentally and whose parameters could be estimated from the existing data. The exceptions to this were processes that we knew were very important yet could not measure directly. In part because of these unknowns, we do not present our models as finished products, but instead believe they are mere steps towards eventual understanding of aquatic ecosystems. They should provide a basis for further experimentation and more refined mathematical analyses of such systems.

General Formulation and Notational Convention

Let the state of the system under consideration be defined at any instant of time t by parameters x_i, $i+1, 2, \ldots$, n, which we shall denote for convenience by the vector **X**:

$$\mathbf{X} = \begin{bmatrix} x_1 \\ x_2 \\ \bullet \\ \bullet \\ \bullet \\ x_n \end{bmatrix} \tag{1}$$

The rate of change of the system with respect to time will be expressed by the derivation relation

$$\dot{x} = \mathrm{d}x_i/\mathrm{d}t \qquad i = 1, 2, \ldots, n$$

which may be written in vector form as

$$\dot{\mathbf{X}} = d\mathbf{X}/dt = \begin{bmatrix} dx_1/dt \\ dx_2/dt \\ \cdot \\ \cdot \\ \cdot \\ dx_n/dt \end{bmatrix} \tag{2}$$

These rates will be assumed to be functions of the instantaneous values of the states, the time t, known inputs u_i (t), $i=1, 2, \ldots, n$, and random inputs w_i (t) $i=1, 2, \ldots, J$, such that we have a system of differential equations

$$\dot{x}_i = f_i \left[x_1(t), x_2(t), \ldots, x_n(t), u_1(t), u_2(t), \ldots, u_n(t), w_1(t), w_2(t), \ldots, w_J(t) \right] \tag{3}$$

with initial conditions that may also be random:

$$x_i(t) = x_{i_0} \qquad i = 1, 2, \ldots, n. \tag{4}$$

These can be expressed conveniently in the form of a vector differential equation and initial condition vector

$$\dot{\mathbf{X}} = f\left[\mathbf{X}(t), \mathbf{U}(t), \mathbf{W}(t), t\right], \quad \mathbf{X}(t_0) = \mathbf{X}_0 \tag{5}$$

The behavior of the system can be represented by a set of trajectories in n-dimensional space. The state of the system at any instant of time is determined by the values of the variables, $x_1, x_2, \ldots, x_n$. The variables will usually be the biomasses of the species and the concentrations of dissolved organic carbon and nutrients constituting the system. The state of the system can then be represented as a point in n-dimensional phase space. To each point in this space we can attach a vector indicating the movement of the system. These vectors can be joined to form trajectories which show the long-term dynamic behavior of the system.

For convenience, the mathematical model of the total aquatic system has been divided into two submodels, benthic (Tiwari et al. 1978) and planktonic. The state variables of the models are types of organisms (algae, bacteria, *Daphnia*, chironomids), nutrients, detritus and dissolved organic carbon (see Tables 10-1 to 10-8). These variables are the objects of the system and the processes associated with each of the variables are the attributes. The dynamic behavior of the system is characterized and constrained by the set of interactions between these objects and attributes.

A diagrammatic representation of the system structure involving state variables and their interactions can be illustrated in the form of a block diagram (Figures 10-1 and 10-2). The square boxes in the diagrams represent the state variables and the arrows joining these boxes depict the processes of the system. The incoming arrows express the positive processes, which are increasing the rate of change of the biomass, and the outgoing arrows describe the negative processes, which are contributing towards the loss of carbon from a state variable. The RB and RP numbers associated with each of the arrows are the equation numbers describing the mathematical form of the processes. Thus, the net rate of change of a particular state variable at time t can easily be obtained by taking the difference between the magnitudes of the positive and negative processes.

The dynamics of the system can be visualized in terms of loss and gain of carbon ascribable to its component biological processes. The effects of these processes are to either increase or decrease the magnitude of carbon from a state variable. Thus the time-dependent net rate of change of a particular state variable at any instant of time t is equal to the difference between the sum of all the processes adding to the amount of carbon and the sum of all the processes subtracting from the amount of carbon.

Let the loss of variable z_i to z_j and that of z_j to z_i (i.e. the gain of z_i due to z_j) attributable to some process be denoted by $z_{i_{z_j}}$ and $z_{j_{z_i}}$, respectively. The quantity z_i at any time t can be computed by summing all $z_{i_{z_j}}$ and all $z_{j_{z_i}}$ and taking the difference. Thus

$$\dot{z}_i = \sum_j z_{j_{z_i}} - \sum_j z_{i_{z_j}} \qquad i \neq j \tag{6}$$

The values of z_i can easily be obtained by integrating $\dot{z}_i$ using any one of the standard numerical methods of integration.

To illustrate this let us consider the dynamics of the benthic algal biomass. Photosynthesis is the only process contributing towards the growth of this population. On the other hand, there are several negative processes (respiration, excretion of usable DOC (U-DOC), death and burial) whose net effect is to reduce the algal biomass. Therefore, the net rate of change in algal biomass (Table 10-1) is given by

$$\dot{x}_1 = x_{602_{x_1}} - x_{1_{x_2}} - x_{1_{x_{201}}} - x_{1_{x_{202}}} - x_{1_{x_{401}}} - x_{1_{x_{501}}} - x_{1_{x_{602}}} - x_{1_{x_{801}}} - x_{1_{x_{802}}} - x_{1_{x_{803}}} - x_{1_{x_{804}}} \tag{7}$$

where

$\dot{x}_1$ = time- dependent rate of change of algal biomass,
$x_{602}{}_{x_1}$ = rate of photosynthesis,
$x_1{}_{x_2}$ = rate of burial of surface algal cells,
$x_1{}_{x_{201}}$ = rate of loss of algal cells due to ciliates,
$x_1{}_{x_{202}}$ = rate of loss of algal cells due to micrometazoans,
$x_1{}_{x_{401}}$ = rate of loss of algal cells to detritus due to death,
$x_1{}_{x_{501}}$ = rate of loss of algal cells to U-DOC due to excretion,
$x_1{}_{x_{602}}$ = rate of algal respiration, and
$x_1{}_{x_{801}}$ to $x_1{}_{x_{804}}$ = rates of loss of algal cells due to chironomids (four cohorts).

The state variables, rate processes, and constants of the two submodels are identified by two types of algebraic symbols. This notation avoids confusion and ensures easy identification of algae, bacteria, detritus, U-DOC and R-DOC (refractory DOC) which are all state variables of both benthic and planktonic systems. This dual notation system was also helpful in developing these two identical but separate computer programs for simulation studies.

The state variables and constants of benthic submodels are denoted by subscripted x and k, respectively (Tables 10-1 and 10-4). The mathematical equations for the processes of this system are numbered with the prefix RB (Table 10-3). The state variables of the planktonic system are indicated by subscripted y and the constants by c. The process equations of this submodel are numbered with the prefix RP (Tables 10-5 to 10-8).

Abiotic Input Variables

Light and temperature are the two abiotic input variables which are driving the aquatic system. Based on the data of the past 4 years an empirical periodic function of the following form was developed to simulate the approximate light available for epipelic algal photosynthesis:

$$I_z = I_0 \exp(-Ez) \qquad (8)$$

where

I_z = irradiance at the depth z (cal cm^{-2} hr^{-1}),
I_0 = water surface-incident irradiance (cal cm^{-2} hr^{-1}),
E = vertical extinction coefficient as a function of organic color in the water
= $2.4 \times 10^{-6} y_{502}$

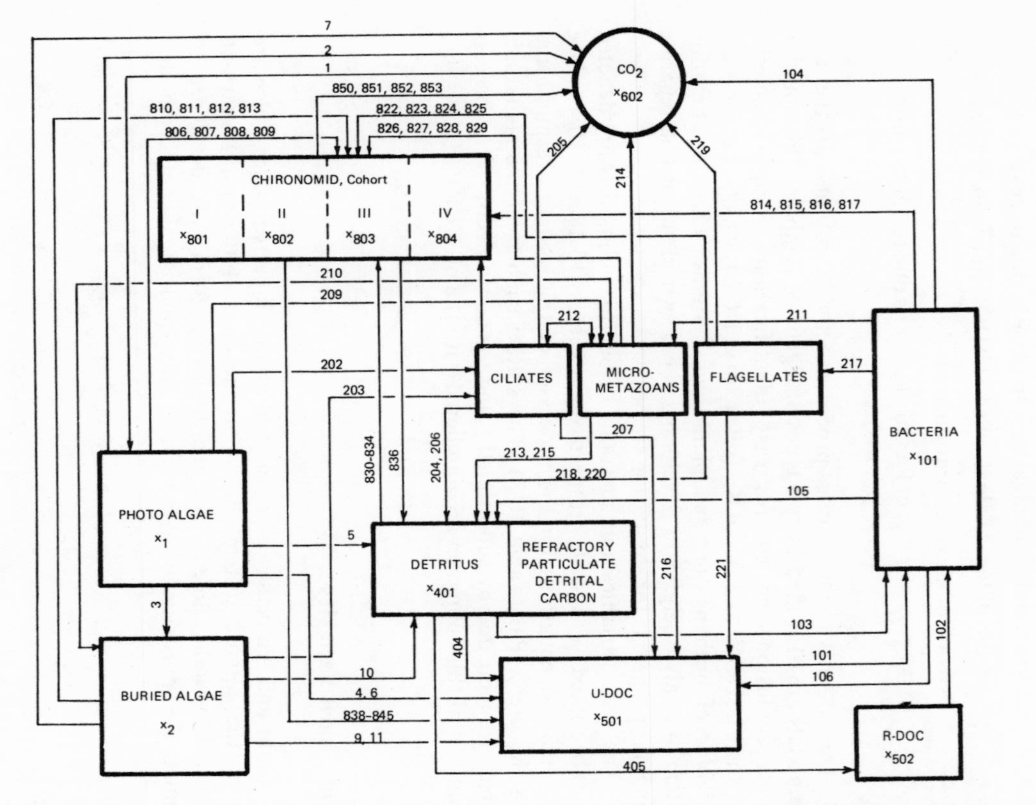

FIGURE 10-1. *Interrelationships contained in the benthic carbon model of a tundra pond.*

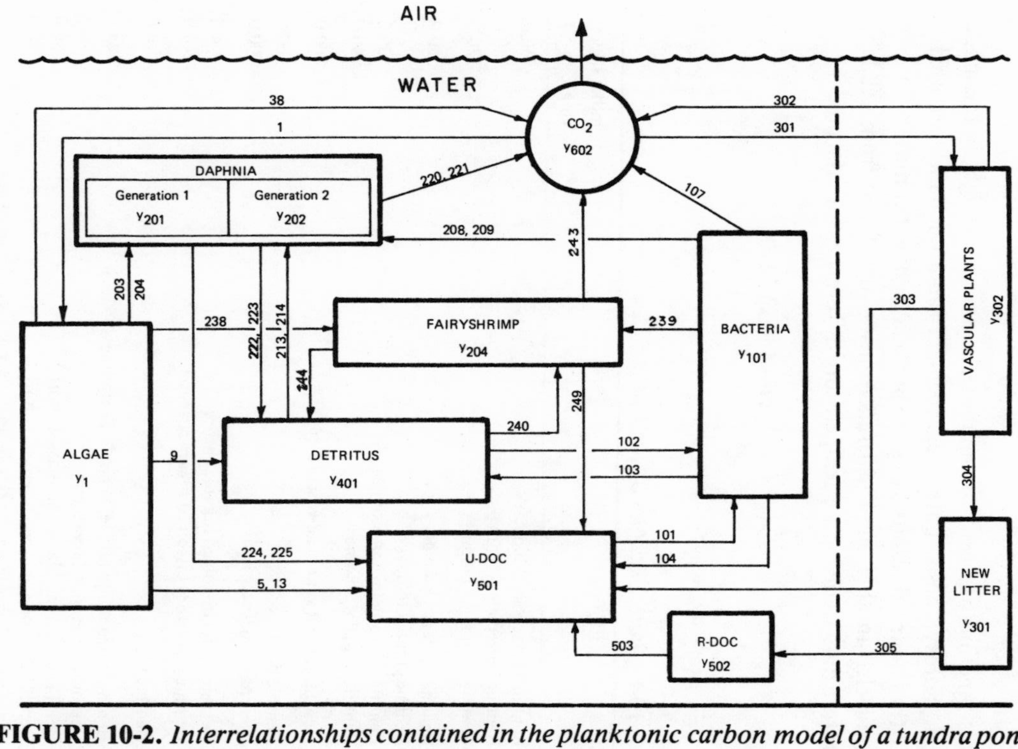

FIGURE 10-2. *Interrelationships contained in the planktonic carbon model of a tundra pond.*

where

y_{502} = concentration of R-DOC in the water column, and
z = depth of pond as a function of time, mimics the average pond water level and changes seasonally.
$= 16+4 \sin \left[(2 \pi t .000349) + 1.5708\right]$.

This function was used to simulate the amount of light in the model for algal photosynthesis. Later, real incident radiation data were used to validate the empirical results using the periodic light generator.

Similarly, another empirical function was developed from available temperature data to introduce temperature differences into the model:

$$\overline{T} = 12 \sin \left[(\pi t - 480\pi)/2400\right] \qquad (9)$$

TABLE 10-1 *State Variables of the Benthic Model*

Symbol	Description	Initial values ($mg\ C\ m^{-2}$)
x_1	Benthic surface algal biomass	40.0
x_2	Benthic buried algal biomass	80.0
x_3	Benthic total algal biomass $= x_1 + x_2$	120.0
x_{101}	Benthic bacterial biomass	350.0
x_{201}	Benthic ciliate biomass	1.0
x_{202}	Benthic micrometazoan biomass	30.0
x_{203}	Benthic flagellate biomass	40.0
x_{301}	New litter (first year dead vascular plants)	80000.0
x_{401}	Benthic detritus	500.0
x_{402}	Inert particulate carbon	1700000.0
x_{501}	Benthic useful dissolved organic carbon	4.0
x_{502}	Benthic refractory dissolved organic carbon	1240.0
x_{602}	Benthic CO_2	0.0
x_{801}	Benthic chironomid larvae, cohort 1 beginning at 530 hr	144.0
x_{802}	Benthic chironomid larvae, cohort 2	70.0
x_{803}	Benthic chironomid larval biomass, cohort 3	56.0
x_{804}	Benthic chironomid larval biomass, cohort 4	70.0
x_{805}	Total benthic chironomid larvae biomass $= x_{801}+x_{802}+x_{803}+x_{804}$	
x_{806}	$x_1+x_2+x_{101}+x_{201}+x_{202}+x_{203}+x_{401}$	
x_{807}	$(x_3 k_{804})+(x_{101} k_{805})+(x_{201} k_{806})+(x_{202} k_{807})+(x_{203} k_{808})+(x_{401} k_{809})$	
x_{1401}	Benthic detrital phosphorus ($mg\ P\ m^{-2}$)	0.0
x_{1501}	Benthic colloidal phosphorus ($mg\ P\ m^{-2}$)	0.0
x_{1601}	Benthic phosphate phosphorus ($mg\ P\ m^{-2}$)	0.04

TABLE 10-2 *Equations for the Dynamic State Variables of the Benthic System*

$$x_1 = x_{602x_1} - x_{1x_{602}} - x_{1x_2} - x_{1x_{501}} - x_{1x_{401}} - x_{1x_{501}} - x_{1x_{201}} - x_{1x_{202}} - x_{1x_{801}} - x_{1x_{802}} - x_{1x_{803}} - x_{1x_{804}}$$

$$x_2 = x_{1x_2} - x_{2x_{602}} - x_{2x_{501}} - x_{2x_{401}} - x_{2x_{501}} - x_{2x_{201}} - x_{2x_{202}} - x_{2x_{801}} - x_{2x_{802}} - x_{2x_{803}} - x_{2x_{804}}$$

$$x_{101} = x_{501x_{101}} + x_{502x_{101}} + x_{401x_{101}} + x_{402x_{101}} - x_{101x_{602}} - x_{101x_{401}} - x_{101x_{501}} - x_{101x_{202}} - x_{101x_{203}} - x_{101x_{801}} - x_{101x_{802}} - x_{101x_{804}} - x_{101x_{803}}$$

$$x_{201} = x_{1x_{201}} + x_{2x_{201}} - x_{201x_{401}} - x_{201x_{602}} - x_{201x_{401}} - x_{201x_{501}} - x_{201x_{202}} - x_{201x_{801}} - x_{201x_{802}} - x_{201x_{803}} - x_{201x_{804}}$$

$$x_{202} = x_{1x_{202}} + x_{2x_{202}} + x_{201x_{202}} - x_{202x_{401}} - x_{202x_{602}} - x_{202x_{401}} - x_{202x_{501}} - x_{202x_{801}} - x_{202x_{802}} - x_{202x_{803}} - x_{202x_{804}}$$

$$x_{203} = x_{101x_{203}} - x_{203x_{401}} - x_{203x_{602}} - x_{203x_{401}} - x_{203x_{501}} - x_{203x_{801}} - x_{203x_{802}} - x_{203x_{803}} - x_{203x_{804}}$$

$$x_{401} = x_{1x_{401}} + x_{2x_{401}} + x_{101x_{401}} + x_{201x_{401}} + x_{201x_{401}} + x_{202x_{401}} + x_{202x_{401}} + x_{203x_{401}} + x_{203x_{401}} + x_{801x_{401}} + x_{801x_{401}} + x_{802x_{401}} + x_{802x_{401}} + x_{803x_{401}} + x_{803x_{401}} + x_{804x_{401}} + x_{804x_{401}} - x_{401x_{101}} - x_{401x_{801}} - x_{401x_{802}} - x_{401x_{803}} - x_{401x_{804}} - x_{401x_{402}} - x_{401x_{501}} - x_{401x_{502}}$$

$$x_{402} = x_{401x_{402}} - x_{402x_{101}}$$

(continued)

TABLE 10-2 (Continued)

$$x_{501} = x_{1}x_{501} + x_{1}x_{501} + x_{2}x_{501} + x_{2}x_{501} + x_{101}x_{501} + x_{201}x_{501} + x_{202}x_{501} + x_{203}x_{501} + x_{401}x_{501} + x_{801}x_{501}$$
$$+ x_{801}x_{501} + x_{802}x_{501} + x_{802}x_{501} + x_{803}x_{501} + x_{803}x_{501} + x_{804}x_{501} + x_{804}x_{501} + x_{501}x_{501} - x_{501}x_{101}$$

$$x_{502} = x_{401}x_{502} + x_{502}x_{502} + x_{502}x_{101}$$

$$x_{602} = x_{1}x_{602} + x_{2}x_{602} + x_{101}x_{602} + x_{201}x_{602} + x_{202}x_{602} + x_{203}x_{602} + x_{801}x_{602} + x_{802}x_{602} + x_{803}x_{602} + x_{804}x_{602}$$

$$x_{801} = x_{1}x_{801} + x_{2}x_{801} + x_{101}x_{801} + x_{201}x_{801} + x_{202}x_{801} + x_{203}x_{801} + x_{401}x_{801} - x_{801}x_{401} - x_{801}x_{401} - x_{801}x_{501}$$
$$- x_{801}x_{501} - x_{801}x_{602}$$

$$x_{802} = x_{1}x_{802} + x_{2}x_{802} + x_{101}x_{802} + x_{201}x_{802} + x_{202}x_{802} + x_{203}x_{802} + x_{401}x_{802} - x_{802}x_{401} - x_{802}x_{401} - x_{802}x_{501}$$
$$- x_{802}x_{501} - x_{802}x_{602}$$

$$x_{803} = x_{1}x_{803} + x_{2}x_{803} + x_{101}x_{803} + x_{201}x_{803} + x_{202}x_{803} + x_{203}x_{803} + x_{401}x_{803} - x_{803}x_{401} - x_{803}x_{401} - x_{803}x_{501}$$
$$- x_{803}x_{501} - x_{803}x_{602}$$

$$x_{804} = x_{1}x_{804} + x_{2}x_{804} + x_{101}x_{804} + x_{201}x_{804} + x_{202}x_{804} + x_{203}x_{804} + x_{401}x_{804} - x_{804}x_{401} - x_{804}x_{401} - x_{804}x_{501}$$
$$- x_{804}x_{501} - x_{804}x_{602}$$

TABLE 10-3 *Processes and Their Equations for the Benthic System*

Symbol	Process	Equation
${x_{602}}_{x_1}$	RB1 Photosynthesis	$x_1 k_1 k_3^{\mathrm{TEMP}}(\mathrm{Light}/\mathrm{Light} + k_2 k_{15}^{\mathrm{TEMP}})$ $(x_{1601}/x_{1601} + k_4)$
${x_1}_{x_{602}}$	RB2 Respiration	$x_1 k_5 k_6^{\mathrm{TEMP}}$
${x_1}_{x_2}$	RB3 Burial	$x_1 k_7 k_{803}^{\mathrm{TEMP}}$
${x_1}_{x_{501}}$	RB4 Excretion	$(x_1 k_8) + ({x_{602}}_{x_1} k_9)$
${x_1}_{x_{401}}$	RB5 Death	$x_1 k_{11} k_{14}$
${x_1}_{x_{501}}$	RB6 Death	$x_1 k_{11}(1 - k_{14})$
${x_2}_{x_{602}}$	RB7 Respiration	$x_2 k_{114} k_6^{\mathrm{TEMP}}$
${x_2}_{x_{501}}$	RB9 Excretion	$x_2 k_8$
${x_2}_{x_{401}}$	RB10 Death	$x_2 k_{11} k_{14}$
${x_2}_{x_{501}}$	RB11 Death	$x_2 k_{11}(1 - k_{14})$
${x_{501}}_{x_{501}}$	RB101 Uptake	$x_{101} k_{101} \left(\frac{x_{501}}{x_{501} + k_{102}}\right) k_{103}^{\mathrm{TEMP}}$
${x_{502}}_{x_{101}}$	RB102 Uptake	$x_{101} k_{104} \left(\frac{x_{502}}{x_{502} + k_{105}}\right) k_{106}^{\mathrm{TEMP}}$
${x_{401}}_{x_{101}}$	RB103 Hydrolysis	$x_{101} x_{401} k_{107}$
${x_{402}}_{x_{101}}$	RB107 Cometabolism	$({x_{501}}_{x_{101}} + {x_{401}}_{x_{101}}) k_{113}$
${x_{101}}_{x_{602}}$	RB104 Respiration	$({x_{501}}_{x_{101}} + {x_{502}}_{x_{101}} + {x_{401}}_{x_{101}} +$ $({x_{501}}_{x_{101}} + {x_{401}}_{x_{101}}) k_{113}) k_{110}$
${x_{101}}_{x_{401}}$	RB105 Death	$x_{101} k_{108} k_{109}$
${x_{101}}_{x_{501}}$	RB106 Death	$x_{101} k_{108}(1 - k_{109})$

(continued)

TABLE 10-3 (Continued)

Symbol	Process	Equation
$x_1 x_{201}$	RB202 Ingestion	$x_{201} k_{201} (\frac{x_3}{x_3 + k_{203}}) \frac{x_2}{x_3} k_{202}{}^{\text{TEMP}}$
$x_2 x_{202}$	RB203 Ingestion	$x_{201} k_{201} (\frac{x_3}{x_3 + k_{203}}) \frac{x_2}{x_3} k_{202}{}^{\text{TEMP}}$
$x_{201} x_{401}$	RB204 Egestion	$x_{201} k_{201} (\frac{x_3}{x_3 + k_{203}}) k_{202}{}^{\text{TEMP}} (1 - k_{211})$
$x_{201} x_{602}$	RB205 Respiration	$x_{201} k_{212} k_{202}{}^{\text{TEMP}}$
$x_{201} x_{401}$	RB206 Death	$x_{201} k_{214} k_{226}$
$x_{201} x_{501}$	RB207 Death	$x_{201} k_{214} (1 - k_{226})$
$x_1 x_{202}$	RB209 Ingestion	$[x_{202} k_{204} (\frac{x_3 + x_{101} + x_{201}}{x_3 + x_{101} + x_{201} + k_{206}}) k_{205}{}^{\text{TEMP}}] (\frac{x_1 k_{219}}{(x_3 k_{219}) + (x_{101} k_{220}) + (x_{201} k_{221})})$
$x_2 x_{202}$	RB210 Ingestion	$[x_{202} k_{204} (\frac{x_3 + x_{101} + x_{201}}{x_3 + x_{101} + x_{201} + k_{206}}) k_{205}{}^{\text{TEMP}}] (\frac{x_2 k_{219}}{(x_3 k_{219}) + (x_{101} k_{220}) + (x_{201} k_{221})})$
$x_{101} x_{202}$	RB211 Ingestion	$[x_{202} k_{204} (\frac{x_3 + x_{101} + x_{201}}{x_3 + x_{101} + x_{201} + k_{206}}) k_{205}]^{\text{TEMP}} (\frac{x_{201} k_{220}}{(x_3 k_{219}) + (x_{101} k_{220}) + (x_{201} k_{221})})$
$x_{201} x_{202}$	RB212 Ingestion	$[x_{202} k_{204} (\frac{x_3 + x_{101} + x_{201}}{x_3 + x_{101} + x_{201} + k_{206}}) k_{205}{}^{\text{TEMP}}] (\frac{x_{201} k_{221}}{(x_3 k_{219}) + (x_{61} k_{220}) + (x_{201} k_{221})})$
$x_{202} x_{401}$	RB213 Egestion	$[x_{202} k_{204} (\frac{x_3 + x_{101} + x_{201}}{x_3 + x_{101} + x_{201} + k_{206}}) k_{205}{}^{\text{TEMP}}] (1 - k_{223})$
$x_{202} x_{602}$	RB214 Respiration	$x_{202} k_{224} k_{205}{}^{\text{TEMP}}$
$x_{202} x_{401}$	RB215 Death	$x_{202} k_{225} k_{228}$
$x_{202} x_{501}$	RB216 Death	$x_{202} k_{225} (1 - k_{228})$
$x_{101} x_{203}$	RB217 Ingestion	$x_{203} k_{207} (\frac{x_{101}}{x_{101} + k_{209}}) k_{208}{}^{\text{TEMP}}$
$x_{203} x_{401}$	RB218 Egestion	$[x_{203} k_{207} (\frac{x_{101}}{x_{101} + k_{209}}) k_{208}{}^{\text{TEMP}}] (1 - k_{216})$

TABLE 10-3 (Continued)

Symbol	Process	Equation
$x_{203}x_{602}$	RB219 Respiration	$x_{203}k_{217}k_{208}^{\text{TEMP}}$
$x_{203}x_{602}$	RB220 Death	$x_{203}k_{218}k_{227}$
$x_{203}x_{501}$	RB221 Death	$x_{203}k_{218}(1-k_{227})$
$x_{401}x_{402}$	RB403 Excretion	$(x_{401}x_{101})k_{112}$
$x_{401}x_{501}$	RB404 Leaching	$x_{401}x_{404}2.56 \times 10^{-4}$
$x_{401}x_{502}$	RB405 Leaching	$x_{401}(1-k_{404})\ 2.56 \times 10^{-4}$
$x_{501}x_{501}$	RB501 Exchange	$[x_{501}-(x_{501}k_{901})]k_{501}(-1)$
$x_{502}x_{502}$	RB502 Exchange	$[x_{502}-(x_{502}k_{901})]k_{502}(-1)$
x_1x_{801}	RB806 Ingestion	$x_{801}k_{801}(\frac{x_{806}}{x_{806}+x_{802}})k_{803}^{\text{TEMP}}(\frac{x_1k_{804}}{x_{807}})$
x_2x_{801}	RB810 Ingestion	$x_{801}k_{801}(\frac{x_{806}}{x_{806}+x_{802}})k_{803}^{\text{TEMP}}(\frac{x_2k_{804}}{x_{807}})$
$x_{101}x_{801}$	RB814 Ingestion	$x_{801}k_{801}(\frac{x_{806}}{x_{806}+x_{802}})k_{803}^{\text{TEMP}}(\frac{x_{101}k_{805}}{x_{807}})$
$x_{201}x_{801}$	RB818 Ingestion	$x_{801}k_{801}(\frac{x_{806}}{x_{806}+x_{802}})k_{803}^{\text{TEMP}}(\frac{x_{201}k_{805}}{x_{807}})$
$x_{202}x_{801}$	RB826 Ingestion	$x_{801}k_{801}(\frac{x_{806}}{x_{806}+x_{802}})k_{803}^{\text{TEMP}}(\frac{x_{202}k_{807}}{x_{807}})$
$x_{203}x_{801}$	RB822 Ingestion	$x_{801}k_{801}(\frac{x_{806}}{x_{806}+x_{802}})k_{803}^{\text{TEMP}}(\frac{x_{203}k_{808}}{x_{807}})$
$x_{401}x_{801}$	RB830 Ingestion	$x_{801}k_{801}(\frac{x_{806}}{x_{806}+x_{802}})k_{803}^{\text{TEMP}}(\frac{x_{401}k_{809}}{x_{807}})$
$x_{801}x_{401}$	RB834 Death	$x_{801}k_{816}k_{817}$
$x_{801}x_{501}$	RB838 Secretion	$(x_1x_{801}+x_2x_{801}+x_{101}x_{801}+x_{201}x_{801}+$ $x_{202}x_{801}+x_{203}x_{801}+x_{401}x_{801})k_{821}$
$x_{801}x_{501}$	RB842 Death	$x_{801}(1-k_{816})k_{817}$

(continued)

TABLE 10-3 (Continued)

Symbol	Process	Equation
$x_{801}{}_{x_{401}}$	RB846 Egestion	$x_1{}_{x_{801}}(1-k_{810}) + x_2{}_{x_{801}}(1-k_{810}) +$ $x_{101}{}_{x_{801}}(1-k_{811}) + x_{201}{}_{x_{801}}(1-k_{812}) +$ $x_{202}{}_{x_{801}}(1-k_{813}) + x_{203}{}_{x_{801}}(1-k_{814}) +$ $x_{401}{}_{x_{801}}(1-k_{815})$
$x_{801}{}_{x_{602}}$	RB850 Respiration	$x_{801}k_{822}k_{823}{}^{\mathrm{TEMP}}$
$x_1{}_{x_{802}}$	RB807 Ingestion	$x_{802}k_{828}(\frac{x_{806}}{x_{806}+k_{802}})\,k_{803}{}^{\mathrm{TEMP}}(\frac{x_1k_{804}}{x_{807}})$
$x_2{}_{x_{802}}$	RB811 Ingestion	$x_{802}k_{828}(\frac{x_{806}}{x_{806}+k_{802}})\,k_{803}{}^{\mathrm{TEMP}}(\frac{x_2k_{804}}{x_{807}})$
$x_{101}{}_{x_{802}}$	RB815 Ingestion	$x_{802}k_{828}(\frac{x_{806}}{x_{806}+k_{802}})\,k_{803}{}^{\mathrm{TEMP}}(\frac{x_{101}k_{805}}{x_{807}})$
$x_{201}{}_{x_{802}}$	RB819 Ingestion	$x_{802}k_{828}(\frac{x_{806}}{x_{806}+k_{802}})\,k_{803}{}^{\mathrm{TEMP}}(\frac{x_{201}k_{806}}{x_{807}})$
$x_{202}{}_{x_{802}}$	RB827 Ingestion	$x_{802}k_{828}(\frac{x_{806}}{x_{806}+k_{802}})\,k_{803}{}^{\mathrm{TEMP}}(\frac{x_{202}k_{807}}{x_{807}})$
$x_{203}{}_{x_{802}}$	RB823 Ingestion	$x_{802}k_{828}(\frac{x_{806}}{x_{806}+k_{802}})\,k_{803}{}^{\mathrm{TEMP}}(\frac{x_{203}k_{808}}{x_{807}})$
$x_{401}{}_{x_{802}}$	RB831 Ingestion	$x_{802}k_{828}(\frac{x_{806}}{x_{806}+k_{802}})\,k_{803}{}^{\mathrm{TEMP}}(\frac{x_{401}k_{809}}{x_{807}})$
$\overline{x}_{802}{}_{x_{401}}$	RB835 Death	$x_{802}k_{816}k_{818}$
$x_{802}{}_{x_{501}}$	RB839 Secretion	$(x_1{}_{x_{802}} + x_2{}_{x_{802}} + x_{101}{}_{x_{802}} + x_{201}{}_{x_{802}} +$ $x_{202}{}_{x_{802}} + x_{203}{}_{x_{802}} + x_{401}{}_{x_{802}})\,k_{821}$
$x_{802}{}_{x_{401}}$	RB847 Egestion	$x_1{}_{x_{802}}(1-k_{810}) + x_2{}_{x_{802}}(1-k_{810}) +$ $x_{101}{}_{x_{802}}(1-k_{811}) + x_{201}{}_{x_{802}}(1-k_{815})$
$x_{802}{}_{x_{501}}$	RB843 Death	$x_{802}(1-k_{816})\,k_{818}$
$x_{802}{}_{x_{602}}$	RB851 Respiration	$x_{802}k_{822}k_{823}{}^{\mathrm{TEMP}}$
$x_1{}_{x_{803}}$	RB808 Ingestion	$x_{803}k_{829}(\frac{x_{806}}{x_{806}+k_{802}})\,k_{803}{}^{\mathrm{TEMP}}\ (\frac{x_1k_{804}}{x_{807}})$
$x_2{}_{x_{803}}$	RB812 Ingestion	$x_{803}k_{829}(\frac{x_{806}}{x_{806}+k_{802}})\,k_{803}{}^{\mathrm{TEMP}}\ (\frac{x_2k_{804}}{x_{807}})$
$x_{101}{}_{x_{803}}$	RB816 Ingestion	$x_{803}k_{829}(\frac{x_{806}}{x_{806}+k_{802}})\,k_{803}{}^{\mathrm{TEMP}}\ (\frac{x_{101}k_{805}}{x_{807}})$
$x_{201}{}_{x_{803}}$	RB820 Ingestion	$x_{803}k_{829}(\frac{x_{806}}{x_{806}+k_{802}})\,k_{803}{}^{\mathrm{TEMP}}\ (\frac{x_{201}k_{806}}{x_{807}})$

TABLE 10-3 (Continued)

Symbol	Process	Equation
$x_{202x_{803}}$	RB828 Ingestion	$x_{803}k_{829}(\frac{x_{806}}{x_{806}+k_{802}})k_{803}^{\mathrm{TEMP}}(\frac{x_{202}k_{807}}{x_{807}})$
$x_{203x_{803}}$	RB824 Ingestion	$x_{803}k_{829}(\frac{x_{806}}{x_{806}+k_{802}})k_{803}^{\mathrm{TEMP}}(\frac{x_{203}k_{808}}{x_{807}})$
$x_{401x_{803}}$	RB832 Ingestion	$x_{803}k_{829}(\frac{x_{806}}{x_{806}+k_{802}})k_{803}^{\mathrm{TEMP}}(\frac{x_{401}k_{809}}{x_{807}})$
$x_{803x_{401}}$	RB836 Death	$x_{803}k_{816}k_{819}$
$x_{803x_{501}}$	RB840 Secretion	$(x_{1x_{803}}+x_{2x_{803}}+x_{101x_{803}}+x_{201x_{803}}+$ $x_{202x_{803}}+x_{203x_{803}}+x_{401x_{803}})\,k_{821}$
$x_{803x_{501}}$	RB844 Death	$x_{803}(1-k_{816})\,k_{819}$
$x_{803x_{401}}$	RB848 Egestion	$x_{1x_{803}}(1-k_{810})+x_{2x_{803}}(1-k_{810})+$ $x_{101x_{803}}(1-k_{811})+x_{201x_{803}}(1-k_{812})+$ $x_{202x_{803}}(1-k_{813})+x_{203x_{803}}(1-k_{814})+$ $x_{401x_{803}}(1-k_{815})$
$x_{803x_{602}}$	RB852 Respiration	$x_{803}k_{822}k_{823}^{\mathrm{TEMP}}$
$x_{1x_{804}}$	RB809 Ingestion	$x_{804}k_{830}(\frac{x_{806}}{x_{806}+k_{802}})k_{803}^{\mathrm{TEMP}}(\frac{x_{1}k_{804}}{x_{807}})$
$x_{2x_{804}}$	RB813 Ingestion	$x_{804}k_{830}(\frac{x_{806}}{x_{806}+k_{802}})k_{803}^{\mathrm{TEMP}}(\frac{x_{2}k_{804}}{x_{807}})$
$x_{101x_{804}}$	RB817 Ingestion	$x_{804}k_{830}(\frac{x_{806}}{x_{806}+k_{802}})k_{803}^{\mathrm{TEMP}}(\frac{x_{101}k_{805}}{x_{807}})$
$x_{210x_{804}}$	RB821 Ingestion	$x_{804}k_{830}(\frac{x_{806}}{x_{806}+k_{802}})k_{803}^{\mathrm{TEMP}}(\frac{x_{201}k_{806}}{x_{807}})$
$x_{202x_{804}}$	RB829 Ingestion	$x_{804}k_{830}(\frac{x_{806}}{x_{806}+k_{802}})k_{803}^{\mathrm{TEMP}}(\frac{x_{202}k_{807}}{x_{807}})$
$x_{203x_{804}}$	RB825 Ingestion	$x_{804}k_{830}(\frac{x_{806}}{x_{806}+k_{802}})k_{803}^{\mathrm{TEMP}}(\frac{x_{203}k_{808}}{x_{807}})$
$x_{401x_{804}}$	RB833 Ingestion	$x_{804}k_{830}(\frac{x_{806}}{x_{806}+k_{802}})k_{803}^{\mathrm{TEMP}}(\frac{x_{401}k_{809}}{x_{807}})$
$x_{804x_{401}}$	RB837 Death	$x_{804}k_{816}k_{820}$
$x_{804x_{501}}$	RB841 Secretion	$(x_{1x_{804}}+x_{2x_{804}}+x_{101x_{804}}+x_{201x_{804}}+$

(continued)

TABLE 10-3 (Continued)

Symbol	Process	Equation
		$x_{202_{x_{804}}} + x_{203_{x_{804}}} + x_{401_{x_{804}}}) k_{821}$
$x_{804_{x_{501}}}$	RB845 Death	$x_{804}(1-k_{816}) k_{820}$
$x_{804_{x_{401}}}$	RB849 Egestion	$x_{1_{x_{804}}}(1-k_{810}) + x_{2_{x_{804}}}(1-k_{810}) +$
		$x_{101_{x_{804}}}(1-k_{811}) + x_{201_{x_{804}}}(1-k_{812}) +$
		$x_{202_{x_{804}}}(1-k_{813}) + x_{203_{x_{803}}}(1-k_{814}) +$
		$x_{401_{x_{804}}}(1-k_{815})$
$x_{804_{x_{602}}}$	RB853 Respiration	$x_{804}k_{822}k_{823}{}^{\mathrm{TEMP}}$

TABLE 10-4 *Parameters of the Benthic Model*

Symbol	Description	Value
k_1	Specific (per mg) V_{max} for photosynthesis of benthic surface algae at reference temperature (mg C (mg C hr)$^{-1}$)	0.05
k_2	Half-saturation constant for light for benthic surface algal	10.0
	photosynthesis (cal cm^{-2}hr^{-1})	2.5
k_3	Q_{10} for benthic surface algal photosynthesis	
k_4	Half-saturation constant for benthic surface algal photosynthesis	0.01
	as a function of size of interstitial PO_4-P pool (mg P m^{-2})	0.01
k_5	Specific (per mg) respiration rate for surface benthic algae at	
	reference temperature (mg C (mg C hr)$^{-1}$)	2.0
k_6	Q_{10} for surface benthic algal respiration	
k_7	Specific (per mg) burial rate of surface benthic algae (mg C (mg C hr)$^{-1}$)	0.15
k_8	Specific (per mg) dark excretion rate of benthic surface algae (mg C(mg C hr)$^{-1}$)	0.001
k_9	Fraction of gross benthic surface algal photosynthesis excreted to U-DOC	0.2
k_{11}	Specific (per mg) death rate of benthic surface algae (mg C (mg C hr)$^{-1}$)	0.015
k_{12}	Mean sediment depth for benthic surface algae (cm)	0.3 cm
k_{13}	Mean sediment depth for benthic buried algae (cm)	2.7 cm
k_{14}	Fraction of algae, both surface and buried, transferred to detritus as a result of death	0.60
k_{15}	Q_{10} for photosynthetic half-saturation constant	3.0
k_{101}	Specific (per mg) V_{max} for uptake of U-DOC by benthic bacteria at reference temperature (mg C (mg C hr)$^{-1}$)	0.05
k_{102}	Half-saturation constant for U-DOC uptake by benthic bacteria (mg C m^{-2})	0.05
k_{103}	Q_{10} for U-DOC uptake by benthic bacteria	2.0
k_{104}	Specific (per mg) V_{max} for uptake of R-DOC by benthic bacteria at reference temperature (mg C (mg C hr)$^{-1}$)	0.005
k_{105}	Half-saturation constant for R-DOC uptake by benthic bacteria (mg C m^{-2})	500.0
k_{106}	Q_{10} for R-DOC uptake by bacteria	2.0
k_{107}	Rate of bacterial hydrolysis of benthic detritus per mg of bacteria and per mg of detritus mg C (mg C)$^{-1}$ (mg C hr)$^{-1}$	3.7×10^{-8}

TABLE 10-4 (Continued)

Symbol	Description	Value
k_{108}	Specific (per mg) death rate of benthic bacteria (mg C (mg C hr)$^{-1}$)	0.01
k_{109}	Fraction of dead bacteria transferred to benthic detritus	0.55
k_{110}	Benthic bacterial respiration rate as a fraction of total carbon uptake by bacteria	0.4
k_{111}	Mean sediment depth for bacteria (cm)	2.5
k_{112}	Proportion of detritus attached but not taken up by bacteria	0.1
k_{113}	Ratio of particulate carbon uptake by bacteria to ratio of bacterial uptake of U-DOC, R-DOC and detritus	
k_{117}	Specific (per mg) V_{max} respiration rate for buried benthic algae at reference temperature (mg C (mg C hr)$^{-1}$)	
k_{201}	Specific (per mg) V_{max} for ingestion of surface and buried algae by ciliates at reference temperature (mg C (mg C hr)$^{-1}$)	0.0025
k_{202}	Q_{10} for ciliate ingestion and respiration	2.0
k_{203}	Half-saturation constant for ingestion of surface and buried algae by ciliates (mg C m^{-2})	50.0
k_{204}	Specific (per mg) V_{max} for total ingestion of all food sources by micrometazoans at reference temperature (mg C (mg C hr)$^{-1}$)	0.003
k_{205}	Q_{10} for micrometazoan feeding and respiration	2.0
k_{206}	Half-saturation constant for total ingestion of all food sources by micrometazoan (mg C m^{-2})	200.0
k_{207}	Specific (per mg) V_{max} for ingestion of bacteria by flagellates at reference temperature (mg C (mg C hr)$^{-1}$)	0.007
k_{208}	Q_{10} for flagellate ingestion and respiration	2.0
k_{209}	Half-saturation constant for ingestion of bacteria by flagellates (mg C m^{-2})	150.0
k_{210}	Ciliate assimilation efficiency as a fraction of the specific (per mg) ciliate ingestion rate (mg C (mg C hr)$^{-1}$)$^{-1}$	24.0
k_{211}	Maximum ciliate assimilation efficiency	0.70
k_{212}	Specific (per mg) ciliate respiration rate (mg C (mg C hr)$^{-1}$)	0.0015
k_{214}	Specific (per mg) death rate of ciliates (mg C (mg C hr)$^{-1}$)	0.01
k_{215}	Flagellate assimilation efficiency as a fraction of the specific (per mg) flagellate ingestion rate (mg C (mg C hr)$^{-1}$)$^{-1}$	11.0
k_{216}	Maximum flagellate assimilation efficiency	0.90
k_{217}	Specific (per mg) flagellate respiration rate (mg C (mg C hr)$^{-1}$)	0.00005
k_{218}	Specific (per mg) death rate of flagellates (mg C (mg C hr)$^{-1}$)	0.01
k_{219}	Selectivity coefficient for micrometazoan ingestion of algae, both surface and buried	4.0
k_{220}	Selectivity coefficient for micrometazoan ingestion of bacteria	1.0
k_{221}	Selectivity coefficient for micrometazoan ingestion of ciliates	20.0
k_{222}	Micrometazoan assimilation efficiency as a fraction of specific (per mg) micrometazoan ingestion rate (mg C (mg C hr)$^{-1}$)$^{-1}$	19.0
k_{223}	Maximum micrometazoan assimilation efficiency	0.70
k_{224}	Specific (per mg) micrometazoan respiration rate (mg C (mg C hr)$^{-1}$)	0.002
k_{225}	Specific (per mg) death rate of micrometazoans (mg C (mg C hr)$^{-1}$)	0.01
k_{226}	Fraction of dead ciliates transferred to detritus	0.6
k_{227}	Fraction of dead flagellates transferred to detritus	0.6
k_{228}	Fraction of dead micrometazoans transferred to detritus	0.6
k_{229}	Mean sediment depth of ciliates (cm)	2.0

(continued)

TABLE 10-4 (Continued)

Symbol	Description	Value
k_{230}	Mean sediment depth of flagellates (cm)	2.0
k_{231}	Mean sediment depth of micrometazoans (cm)	2.0
k_{501}	Fractional exchange constant for the equilibrium between water column U-DOC and benthic U-DOC (mg C (mg C hr)$^{-1}$)	0.0
k_{502}	Fractional exchange constant for the equilibrium between water column R-DOC and benthic R-DOC (mg C (mg C hr)$^{-1}$)	0.0
k_{801}	Specific (per mg) V_{max} for total ingestion of all food sources by all chironomid larvae at reference temperature (mg C (mg C hr)$^{-1}$)	0.02
k_{802}	Half-saturation constant for ingestion of all food sources by all chironomid larvae (mg C m^{-2})	250.0
k_{803}	Q_{10} for chironomid larval ingestion and activity	3.0
k_{804}	Chironomid selectivity coefficient for surface and buried benthic algae	3.0
k_{805}	Chironomid selectivity coefficient for bacteria	1.0
k_{806}	Chironomid selectivity coefficient for ciliates	5.0
k_{807}	Chironomid selectivity coefficient for micrometazoans	5.0
k_{808}	Chironomid selectivity coefficient for flagellates	2.0
k_{809}	Chironomid selectivity coefficient for benthic detritus	0.01
k_{810}	Chironomid assimilation efficiency, all cohorts, for benthic algae, both surface and buried	0.4
k_{811}	Chironomid assimilation efficiency, all cohorts, for bacteria	0.4
k_{812}	Chironomid assimilation efficiency, all cohorts, for ciliates	0.4
k_{813}	Chironomid assimilation efficiency, all cohorts, for flagellates	0.4
k_{814}	Chironomid assimilation efficiency, all cohorts, for micrometazoans	0.4
k_{815}	Chironomid assimilation efficiency, all cohorts, for detritus	0.01
k_{816}	Fraction of dead chironomids (all cohorts) transferred to detritus	0.8
k_{817}	Specific (per mg) death rate of chironomid larvae, cohort 1 (mg C (mg C hr)$^{-1}$)	0.05
k_{818}	Specific (per mg) death rate of chironomid larvae, cohort 2 (mg C (mg C hr)$^{-1}$)	0.02
k_{819}	Specific (per mg) death rate of chironomid larvae, cohort 3 (mg C (mg C hr)$^{-1}$)	0.001
k_{820}	Specific (per mg) death rate of chironomid larvae, cohort 4 (mg C (mg C hr)$^{-1}$)	0.001
k_{821}	Fraction of chironomid ingestion rate secreted as mucus	0.1
k_{822}	Specific (per mg) respiration rate for chironomids, all cohorts (mg C (mg C hr)$^{-1}$)	0.0002
k_{823}	Q_{10} for chironomid respiration, all cohorts	2.0
k_{824}	Mean sediment depth for chironomid larvae, cohort 1 (cm)	0.3
k_{825}	Mean sediment depth for chironomid larvae, cohort 2 (cm)	1.0
k_{826}	Mean sediment depth for chironomid larvae, cohort 3 (cm)	1.0
k_{827}	Mean sediment depth for chironomid larvae, cohort 4 (cm)	3.0
k_{828}	Specific (per mg) V_{max} for total ingestion of all food sources by chironomid larvae (cohort 2) (mg C(mg C hr)$^{-1}$)	0.006
k_{829}	Specific (per mg)) V_{max} for total ingestion of all food sources by chironimid larvae (cohort 3) (mg C(mg C hr)$^{-1}$)	0.008
k_{830}	Specific (per mg) V_{max} for total ingestion of all food sources by chironomid larvae (cohort 4) (mg C(mg C hr)$^{-1}$)	0.006
k_{901}	Volume/area conversion factor for benthic system (m^2 m^{-3})	20.0
k_{902}	Drag coefficient of sediment	0

where

$\overline{T}$ = mean water temperature in °C.

The effect of temperature on physiological processes of the system was expressed in the form of a Q_{10} function:

$$T_e = Q_{10}{}^{(T - T_r)/10} \tag{10}$$

where

T_e = temperature effect on the process, and
T_r = reference temperature.

It has been recognized that most organisms do not show clear Q_{10} responses as the rates do not increase exponentially over the entire range of temperatures suitable for life. The models are built to cope only within a narrow range, 0 to 15°C. Temperature responses are frequently curvilinear or linear for *Lepidurus* respiration and activity rates, for zooplankton respiration and ingestion rates, and for algal P_{max}, etc. The present simplification in the models limits their accuracy at temperature extremes. The simulation results were validated using real temperature data as input.

Formulation of Process Rates

One of our basic assumptions is that processes such as uptake of nutrients and algal photosynthesis, uptake of carbon and nutrients by bacteria, and ingestion of food by consumers can be characterized approximately by Michaelis-Menten type functions. Thus,

$$v = V_{max} x_i / (x_i + k_s) \tag{11}$$

where

v = rate of photosynthesis, uptake of nutrients, ingestion, etc.,
V_{max} = maximum uptake velocity,
x_i = concentration of food, nutrient, etc., and
k_s = half saturation constant.

This is based on our data on uptake of nutrients and primary productivity from arctic ponds (Stanley 1974) and several published reports (O'Brien 1974, Toerien et al. 1971, Golterman et al. 1969, Eppley and Thomas 1969, Caperon 1968) which indicate this type of relationship in the growth rate of phytoplankton. Experimental measurements from these ponds and from lakes and estuaries also suggest that bacterial uptake of nutrients follows Michaelis-Menten types of curves (Hobbie and Crawford 1969, Wright and Hobbie 1966, Crawford et al. 1974). The data of Chisholm et al. (1975) on the feeding rates of arctic *Daphnia* also imply that the ingestion rates can be approximated by Michaelis-Menten type equations.

It is further assumed that excretion, secretion, and death rates can be represented by the following simple functional relationship:

$$x_{i_{x_j}} = x_i k_i \tag{12}$$

where

$x_{i_{x_j}}$ = process representing excretion, secretion or death,
x_i = biomass, and
k_i = a constant.

Wherever it is appropriate this relationship was multiplied by a Q_{10} type of function to include the effect of temperature.

Computer Simulations

The sets of differential equations describing the benthic and planktonic models were simulated on an IBM computer using CSMP III (Continuous System Modeling Program). CSMP allows us to simulate the dynamic behavior of a continuous system expressed as a set of ordinary differential equations or a set of partial differential equations. It provides considerable flexibility in the choice of integration methods and has powerful capabilities for handling input and output and their specifications. The program automatically sorts user-supplied structure statements to establish a correct execution sequence. This sorting of structure statements of the program is very important, for an incorrect sequence would introduce a phase lag that could seriously affect the accuracy and stability of the solution. Complex problems involving non-linear and time-varying elements can be easily handled, for it is possible to incorporate FORTRAN statements into CSMP III programs. The fourth-order Runge-Kutta method was used for numerical integration. (For a detailed description of CSMP III see IBM manual SH19-7001-2.)

All the rates in the equations are expressed in units of hours. The models were simulated for a period of 1440 hours to cover our experimental observations recorded from 15 June to 15 August.

BENTHIC CARBON FLOW MODEL

The state variables of the benthic system and their initial conditions are given in Table 10-1. The block diagram of Figure 10-1 depicts the flow of carbon to and from these variables. The square boxes in the diagram represent the state variables and the arrows describe the dynamic processes which are adding and removing carbon from the variables. A list of all the equations describing the system is given in Table 10-3. The *k*'s in these equations are the parameters which are defined in Table 10-4.

Primary Producers

Although the benthic population of the sediment is distributed over 5 cm of depth, photosynthesis is limited to the top 3 mm. The algal cells of the bottom 4.7 cm are not exposed to the light and hence are not photosynthesizing. For this reason we have separated the total algal biomass into two groups: surface algae (which are photosynthesizing) and buried algae (which are not photosynthesizing). These two types of algal populations are treated as two separate state variables. The loosely packed sediment of these ponds is continuously disturbed by the movements and the activities of chironomids and tadpole shrimp (*Lepidurus arcticus*). This continual mixing contributes towards the burial (RB3) of algal cells. This exchange mechanism transfers the algal biomass between the two compartments and is a very important control on the exponential growth of the surface algal population.

Algal photosynthesis is modeled as a function of its biomass, of light intensity, of temperature, and of the concentration of phosphorus in the sediment. Experimental evidence from these ponds suggests that phosphorus is in limited supply and acts as a limiting factor (Chapter 5, Prentki 1976). As outlined earlier, the relationships between both light intensity and phosphorus and the rate of photosynthesis are of the Michaelis-Menten type. The loss terms in the differential equations describing the dynamics of the algal populations are respiration (RB2 and RB7), death (RB5 and RB10), and excretion of U-DOC (RB4 and RB11). It is assumed that the rate of excretion of U-DOC is proportional to the algal biomass; a constant fraction of the primary productivity is also excreted as U-DOC.

Decomposers

Bacteria are the only microbial decomposers included in this model. Their growth rate is proportional to their biomass and to the concentrations of U-DOC and R-DOC in the sediment (RB101 and RB102). The rate of uptake of U-DOC and R-DOC is assumed to follow a Michaelis-Menten type of kinetics. In addition to the uptake of U-DOC and R-DOC, these organisms are also obtaining some food by hydrolyzing detrital particles (RB103) and by utilizing particulate carbon by cometabolism (RB107) with usable DOC. The loss of bacterial biomass is a result of respiration (RB104), death (RB105), and predation.

Consumers

Chironomids, ciliates, flagellates, and micrometazoans are the predominant consumer species in the pond sediment and play a major role in the dynamics of the benthic carbon. Among these, chironomids are the dominant species and constitute about 80% of the total weight. The total chironomid population consists of four cohorts which differ in their feeding rate, mortality, and respiration; hence these cohorts are treated in the model as four separate state variables. These organisms feed on algae, bacteria, detritus, and even other consumer species. Ciliates and flagellates live mostly on algae and bacteria, respectively. Microbenthos consume algae, bacteria and ciliates. (For biological details see Chapter 8 and Fenchel 1975.)

The rate of ingestion of a particular food type by a consumer species is a function of its biomass, the density of food, and temperature. As discussed earlier, the effect of food density on the rate of ingestion follows a Michaelis-Menten type response curve. To take a specific example, let us consider the rate of ingestion of bacteria by flagellates. The functional form of this process is

$$\mathrm{RB217} = x_{203} k_{207} \left[x_{101} / (x_{101} + k_{209}) \right] T_e \qquad (13)$$

where

RB217 = rate of ingestion of benthic bacteria by benthic flagellates,
x_{101} = bacterial biomass,
x_{203} = flagellate biomass,
k_{207} = specific V_{max} for ingestion of bacteria by flagellates at the reference temperature,
k_{209} = half-saturation constant for ingestion of bacteria by flagellates, and
T_e = temperature effect.

If a consumer species feeds on more than one food source, then it is assumed that each such consumer has a total rate of ingestion that is a function of the biomass of the consumer, of the total biomass of all food sources, and of temperature. The expression for the food density is of the Michaelis-Menten type. The actual rate of ingestion of a food source, S_i, is then the product of the total rate of ingestion and S_i divided by the sum of all food sources available to the consumer. In the second expression each food source S_i is multiplied by its selectivity coefficient. For example, let us consider the rate of ingestion of bacteria by micrometazoans. Since these organisms also feed on algae and ciliates in addition to bacteria, the total rate of ingestion is given by the following functional relationship:

$$RB208 = x_{202}k_{204}\left[\frac{(x_3+x_{101}+x_{201})}{(x_3+x_{101}+x_{201}+k_{206})}\right]T_e \qquad (14)$$

where

RB208 = total rate of ingestion by benthic micrometazoans of all food sources,
x_3 = total algal biomass (surface and buried algal populations),
x_{101} = bacterial biomass,
x_{201} = ciliate biomass,
x_{202} = micrometazoan biomass,
k_{204} = specific V_{max} for total ingestion of all food sources by micrometazoans at the reference temperature, and
k_{206} = half-saturation constant for total ingestion of all food sources by micrometazoans.

Then the rate of ingestion of bacteria by micrometazoans is

$$RB211 = RB208(x_{101}k_{220})/(x_3k_{219}+x_{101}k_{220}+x_{201}k_{221}) \qquad (15)$$

where

RB211 = rate of ingestion of bacteria by micrometazoans,
k_{219} = selectivity coefficient for micrometazoan ingestion of algae (both surface and buried),
k_{220} = selectivity coefficient for micrometazoan ingestion of bacteria,
k_{221} = selectivity coefficient for micrometazoan ingestion of ciliates, and
x_3, x_{101} and x_{201} are defined as above.

The loss of carbon from these species is a result of respiration, egestion, and death (see Tables 10-2 and 10-3).

Detritus and Dissolved Organic Carbon

The two major inputs to the benthic detrital pool are death of living organisms of the system and egestion from chironomids, ciliates, flagellates, and micrometazoans. In addition, sedimentation of planktonic detrital material contributes towards this pool. Detritus is ingested by chironomids (RB830-RB833) and is also hydrolyzed by bacteria (RB103). Some detrital carbon is lost by leaching of U-DOC and R-DOC (RB404 and RB405). A fraction of benthic detritus is also lost to refractory particulate detrital carbon (RB403), which is one of the food sources for bacteria (RB107).

U-DOC leaks from algal cells and is secreted by chironomids in the form of mucus. However, not all of the biomass removed by the death of living organisms is transferred to the detrital pool. A constant fraction of this dead material, the value depending upon the species, is also leached into the U-DOC pool. This U-DOC is the major food source for the bacterial population of the sediment.

R-DOC is the second major food source for the bacteria. Leaching from the detrital material and exchange from the planktonic system are two inputs to this pool.

Respiration of all living organisms contributes towards the total CO_2 concentration. (For further details of the processes and their equations see Tables 10-2 and 10-3.)

PLANKTONIC CARBON FLOW MODEL

One of the important features of the Barrow ponds is the existence of a shallow water column (average depth of 20 cm or less) that is well mixed by the wind. This characteristic allows us to simplify the mathematical models of the planktonic system. Thus, the vertical distribution of organisms and nutrients in the water column can be disregarded and the dynamics of the system can be adequately represented by ordinary differential equations. The low number of species in the ponds enables us to model the total ecosystem with fewer equations than a temperate pond model would require. The computer simulation of these equations and the estimation of their parameters are also made easier by the fact that there are fewer species.

A diagrammatic representation of the plankton system is given in Figure 10-2. A list of all the processes associated with the planktonic system is given in Table 10-7. The c's in the equations are the parameters of the rate processes and are defined in Table 10-8.

Primary Producers

Algae and two species of vascular plants are the primary producers in these ponds (see Chapter 5). Although there are a number of species in the algal population, all are grouped together and considered as a single state variable. Other models (e.g. Lehman et al. 1975) have broken up the algal

TABLE 10-5 *State Variables of the Plankton Model*

Symbol	Description	Initial Values
y_1	Biomass of algae species (mg C m^{-3})	1.0
y_{101}	Bacterial biomass (mg C m^{-3})	10.0
y_{201}	First generation *Daphnia* biomass (mg C m^{-3})	2.0
y_{202}	Second generation *Daphnia* biomass (mg C m^{-3}) beginning at 600 hr.	2.0
y_{204}	Fairyshrimp biomass (mg C m^{-3})	2.0
y_{301}	New litter carbon (mg C m^{-2})	80000.0
y_{302}	Above-ground vascular plant biomass (mg C m^{-2})	26000.0
y_{401}	Microsestonic detrital carbon (mg C m^{-3})	3000.0
y_{402}	Total sestonic detrital carbon (mg C m^{-3})	510.0
y_{501}	Usable DOC (mg C m^{-3})	100.0
y_{502}	Refractory DOC (mg C m^{-3})	5000.0
y_{601}	CO_2 (mg C m^{-3})	3500.0

TABLE 10-6 *Equations for the Dynamic State Variables of the Planktonic System*

$$y_1 = {y_{601}}_{y_1} - {y_1}_{y_{201}} - {y_1}_{y_{202}} - {y_1}_{y_{204}} - {y_1}_{y_{401}} - {y_1}_{y_{401}} - {y_1}_{y_{501}} - {y_1}_{y_{601}}$$

$$y_{101} = {y_{401}}_{y_{101}} + {y_{501}}_{y_{101}} - {y_{101}}_{y_{201}} - {y_{101}}_{y_{202}} - {y_{101}}_{y_{204}} - {y_{101}}_{y_{401}} - {y_{101}}_{y_{501}} - {y_{101}}_{y_{601}}$$

$$y_{201} = {y_1}_{y_{201}} + {y_{101}}_{y_{201}} + {y_{401}}_{y_{201}} - {y_{201}}_{y_{401}} - {y_{201}}_{y_{401}} - {y_{201}}_{y_{501}} - {y_{201}}_{y_{601}}$$

$$y_{202} = {y_1}_{y_{202}} + {y_{101}}_{y_{202}} + {y_{401}}_{y_{202}} - {y_{202}}_{y_{401}} - {y_{202}}_{y_{401}} - {y_{202}}_{y_{501}} - {y_{202}}_{y_{601}}$$

$$y_{204} = {y_1}_{y_{204}} + {y_{101}}_{y_{204}} + {y_{401}}_{y_{204}} - {y_{204}}_{y_{401}} - {y_{204}}_{y_{401}} - {y_{204}}_{y_{501}} - {y_{204}}_{y_{601}}$$

$$y_{301} = {y_{302}}_{y_{301}} - {y_{301}}_{y_{501}} - {y_{301}}_{y_{502}}$$

$$y_{302} = {y_{60r}}_{y_{301}} - {y_{302}}_{y_{301}} - {y_{302}}_{y_{501}} - {y_{302}}_{y_{601}}$$

$$y_{401} = {y_1}_{y_{401}} + {y_{101}}_{y_{401}} + {y_{201}}_{y_{401}} + {y_{202}}_{y_{401}} + {y_{204}}_{y_{401}} - {y_{401}}_{y_{201}} - {y_{401}}_{y_{202}} - {y_{401}}_{y_{204}}$$

(continued)

TABLE 10-6 (Continued)

$$y_{402} = y_{401} + y_{201_{y_{402}}} + y_{202_{y_{402}}} + y_{204_{y_{402}}} + y_{401_{y_{402}}} - y_{402_{y_{101}}} - y_{402_{y_{401}}}$$

$$y_{501} = y_{1_{y_{501}}} + y_{1_{y_{501}}} + y_{101_{y_{501}}} + y_{201_{y_{501}}} + y_{202_{y_{501}}} + y_{204_{y_{501}}} + (y_{301_{y_{501}}} c_{901}) + (y_{302_{y_{501}}} c_{901}) + y_{502_{y_{501}}} - y_{501_{y_{101}}} - y_{501_{x_{501}}}$$

$$y_{502} = (y_{301_{y_{502}}} c_{901}) \pm y_{502_{x_{502}}} - y_{502_{y_{501}}}$$

TABLE 10-7 *Processes and Their Equations of the Planktonic System*

Symbol	Process	Equation*
$y_{602_{y_1}}$	RP 1 Photosynthesis	$y_1\ c_1 + c_{25} \sin\left(\frac{2\pi}{24} - \pi\right)\left(\frac{\text{Wlight}}{\text{Wlight} + c_9 c_{29}\ \text{Temp}}\right) \frac{y_{1005}}{y_{1005} + c_{21}\ y_1}\ c_5^{\text{Temp}}$
$y_{1_{y_{501}}}$	RP 5 Excretion	$y_1\ c_{16} + y_{602_{y_1}}\ c_{15}$
$y_{1_{y_{401}}}$	RP 9 Death	$y_1\ c_{17}\ c_{28}$
$y_{1_{y_{501}}}$	RP 13 Death	$y_1\ c_{17}\ (1 - c_{28})$
$y_{1_{x_{401}}}$	RP 17 Sedimentation	$y_1\ c_{34}\ c_5^{\text{Temp}}$
$y_{1_{y_{601}}}$	RP 38 Respiration	$y_1\ c_{13}\ c_{30}^{\text{Temp}}$
$y_{501_{y_{101}}}$	RP 101 Uptake	$y_{101}\ c_{102}\left(\frac{y_{501}}{y_{501} + c_{103}}\right) c_{104}^{\text{Temp}}$

* Wcur	Water current speed	Wlight	Light intensity in water
Alight	Light intensity in air	Temp	Q_{10} exponent $(T - T_r)/10$

TABLE 10-7 (Continued)

Symbol	Process	Equation
$y_{401}y_{101}$	RP 102 Hydrolysis	$y_{101}\,y_{401}\,c_{109}$
$y_{101}y_{401}$	RP 103 Death	$y_{101}\,c_{107}\ \ c_{110}$
$y_{101}y_{501}$	RP 104 Death	$y_{101}\,c_{107}\,(1-c_{110})$
$y_{101}y_{601}$	RP 107 Respiration	$(x_{501}x_{101} + x_{401}x_{101})\ c_{108}$
$y_{1}y_{202}$	RP204 Ingestion	$\text{RP 202}\left[\dfrac{\dfrac{\text{RP 202}}{c_{201}}}{\dfrac{\text{RP 202}}{c_{201}} + \dfrac{\text{RP 207}}{c_{207}} + \dfrac{\text{RP 212}}{c_{210}}}\right]$ where $\text{RP 202} = y_{202}\left(\dfrac{y_1}{y_1 + c_{202}}\right)c_{201} + c_{204}$ $\sin\left(\dfrac{2\pi}{12} + 1.047\right)\,c_{203}{}^{\text{Temp}}$ $\text{RP 207} = y_{202}\left(\dfrac{y_{101}}{y_{101} + c_{208}}\right)c_{207} + c_{209}$ $\sin\left(\dfrac{2\pi}{12} + 1.047\right)c_{203}{}^{\text{Temp}}$ $\text{RP 212} = y_{202}\left(\dfrac{y_{401}}{y_{401} + c_{211}}\right)\,c_{210} + c_{212}$ $\sin\left(\dfrac{2\pi}{12} + 1.047\right)c_{203}{}^{\text{Temp}}$
$y_{101}y_{202}$	RP 209 Ingestion	$\text{RP 207}\quad\left[\dfrac{\dfrac{\text{RP 207}}{c_{207}}}{\dfrac{\text{RP 202}}{c_{201}} + \dfrac{\text{RP 207}}{c_{207}} + \dfrac{\text{RP 212}}{c_{210}}}\right]$

(continued)

TABLE 10-7 (Continued)

Symbol	Process	Equation
$y_{401}y_{202}$	RP 214 Ingestion	RP 212 $\left[\dfrac{\frac{RP\ 212}{c_{210}}}{\frac{RP\ 202}{c_{201}} + \frac{RP\ 207}{c_{207}} + \frac{RP\ 212}{c_{210}}}\right]$
$y_{202}y_{401}$	RP 219 Egestion	$y_1y_{202}\ (1-c_{213}) + y_{101}c_{202}\ (1-c_{214}) + y_{401}y_{202}$ $(1-c_{219})$
$y_{202}y_{601}$	RP 221 Respiration	$y_{202}\ c_{215}c_{216}{}^{\text{Temp}}$
$y_{202}y_{401}$	RP 223 Death	$y_{202}\,c_{234}\,c_{217}$
$y_{202}y_{501}$	RP 225 Death	$y_{202}\,c_{234}\,(1-c_{217})$
y_1y_{204}	RP 238 Ingestion	RP 235 $\left[\dfrac{\frac{RP\ 235}{c_{220}}}{\frac{RP\ 235}{c_{220}} + \frac{RP\ 236}{c_{221}} + \frac{RP\ 237}{c_{222}}}\right]$ where $RP\ 235 = y_{204}\,c_{220}\left(\frac{y_1}{y_1 + c_{223}}\right) c_{224}{}^{\text{Temp}}$ $RP\ 236 = y_{204}\,c_{221}\left(\frac{y_{101}}{y_{101} + c_{225}}\right)c_{224}{}^{\text{Temp}}$ $RP\ 237 = y_{204}\,c_{222}\left(\frac{y_{401}}{y_{401} + c_{226}}\right)c_{224}{}^{\text{Temp}}$
$y_{101}y_{204}$	RP 239 Ingestion	RP 236 $\left[\dfrac{\frac{RP\ 236}{c_{221}}}{\frac{RP\ 235}{c_{220}} + \frac{RP\ 236}{c_{221}} + \frac{RP\ 237}{c_{222}}}\right]$

TABLE 10-7 (Continued)

Symbol	Process	Equation
$y_{401}y_{204}$	RP 240 Ingestion	$\text{RP 237}\left[\dfrac{\frac{\text{RP 237}}{c_{220}}}{\frac{\text{RP 235}}{c_{220}}+\frac{\text{RP 236}}{c_{221}}+\frac{\text{RP 237}}{c_{222}}}\right]$
$y_{204}y_{401}$	RP 242 Egestion	$y_1y_{204}(1-c_{227})+y_{101}y_{204}(1-c_{228})+y_{401}y_{204}(1-c_{233})$
$y_{204}y_{601}$	RP 243 Respiration	$y_{204}\,c_{229}\,c_{230}^{\text{Temp}}$
$y_{204}y_{401}$	RP 244 Death	$y_{204}\,c_{231}\,c_{232}$
$y_{204}y_{501}$	RP 249 Death	$y_{204}c_{231}(1-c_{232})$
y_1y_{201}	RP 203 Ingestion	$\text{RP 201}\left[\dfrac{\frac{\text{RP 201}}{c_{201}}}{\frac{\text{RP 201}}{c_{201}}+\frac{\text{RP 206}}{c_{207}}+\frac{\text{RP 211}}{c_{210}}}\right]$ where $\text{RP 201}=y_{201}\left(\frac{y_1}{y_1+c_{202}}\right)\left[c_{201}+c_{204}\sin\left(\frac{2\pi}{12}+1.047\right)\right]c_{203}^{\text{Temp}}$ $\text{RP 206}=y_{201}\left(\frac{y_{101}}{y_{101}+c_{208}}\right)\left[c_{207}+c_{209}\sin\left(\frac{2\pi}{12}+1.047\right)\right]c_{203}^{\text{Temp}}$ $\text{RP 211}=y_{201}\left(\frac{y_{401}}{y_{401}+c_{211}}\right)\left[c_{210}+c_{212}\sin\left(\frac{2\pi}{12}+1.047\right)\right]c_{203}^{\text{Temp}}$
$y_{101}y_{201}$	RP 208 Ingestion	$\text{RP 206}\left[\dfrac{\frac{\text{RP 206}}{c_{207}}}{\frac{\text{RP 201}}{c_{201}}+\frac{\text{RP 206}}{c_{207}}+\frac{\text{RP 211}}{c_{210}}}\right]$
$y_{401}y_{201}$	RP 213 Ingestion	$\text{RP 211}\left[\dfrac{\frac{\text{RP 211}}{c_{210}}}{\frac{\text{RP 201}}{c_{201}}+\frac{\text{RP 206}}{c_{207}}+\frac{\text{RP 211}}{c_{210}}}\right]$

(continued)

TABLE 10-7 (Continued)

Symbol	Process	Equation
$y_{201_{y_{401}}}$	RP 218 Egestion	$y_{1_{y_{201}}}(1-c_{213}) + y_{101_{y_{201}}}(1-c_{214}) +$ $y_{401_{y_{201}}}(1-c_{219})$
$y_{201_{y_{601}}}$	RP 220 Respiration	$y_{201}\, c_{215}\, c_{216}{}^{\text{Temp}}$
$y_{201_{y_{401}}}$	RP 222 Death	$y_{201}\, c_{218}\, c_{217}$
$y_{201_{y_{501}}}$	RP 224 Death	$y_{201}\, c_{218}(1-c_{217})$
$y_{601_{y_{302}}}$	RP 301 Photosynthesis	$y_{302}\, c_{302} \dfrac{\text{Alight}}{\text{Alight} + c_{303}} c_{304}{}^{\text{Temp}}$
$y_{302_{y_{602}}}$	RP 302 Respiration	$y_{302}\, c_{301}\, c_{309}{}^{\text{Temp}}$
$y_{302_{y_{501}}}$	RP 303 Excretion	$y_{601_{y_{302}}}\, c_{305} + y_{302}\, c_{310}$
$y_{302_{y_{501}}}$	RP 303 Excretion	$y_{601_{y_{302}}}\, c_{305} + y_{302}\, c_{310}$
$y_{302_{y_{401}}}$	RP 304 Death	$y_{302}\, c_{308}$
$y_{301_{y_{502}}}$	RP 305 Leaching	$y_{301}\, c_{306}\, c_{307}$
$y_{301_{y_{501}}}$	RP 306 Leaching	$y_{301}\, c_{306}\,(1-c_{307})$
$y_{401_{x_{401}}}$	RP 401 Sedimentation	$y_{401}\, c_{401}$

TABLE 10-7 (Continued)

Symbol	Process	Equation
$x_{401}y_{401}$	RP 402 Wind Re-suspension	$\frac{x_{401}\,k_{902}}{c_{901}}$ Wcur^4
$y_{501}x_{501}$	RP 501 Exchange	$y_{501}\,c_{501}$
$y_{502}y_{501}$	RP 503 Photooxidation	$y_{502}\,(c_{503}$ Wlight $+ c_{504})$

TABLE 10-8 *Parameters of the Plankton Model*

Symbol	Description	Value
c_1	Mean, specific (per mg) V_{max} for photosynthesis at reference temperature for algal species 1 (mg C (mg C hr)$^{-1}$)	0.06
c_5	Q_{10} for photosynthesis and respiration for algal species 1	2.0
c_9	Half-saturation constant for light for algal photosynthesis (cal cm^{-2} hr^{-1})	1.5
c_{13}	Specific (per mg) respiration rate for algal species (mg C (mg C hr)$^{-1}$)	0.002
c_{15}	Fraction of algal photosynthetic rate excreted to U-DOC (mg C (mg C)$^{-1}$)	0.25
c_{16}	Specific (per mg) dark excretion rate for algal species (mg C (mg C hr)$^{-1}$)	0.002
c_{17}	Specific (per mg) death rate for algal species (mg C (mg C hr)$^{-1}$)	0.0025
c_{21}	Half-saturation constant for photosynthesis as a function of size of available phosphorus pool for algal species (mg P (mg C)$^{-1}$)	0.0001
c_{25}	Amplitude of variation in the V_{max} for photosynthesis, algal species at reference temperature	0.002
c_{26}	Period of rhythm in V_{max} for photosynthesis, algal species (hr)	24.0
c_{27}	Phase shift parameter of rhythm in V_{max} for photosynthesis, algal species (radians)	−3.1416
c_{28}	Fraction of dead algae transferred to sestonic detritus	0.6
c_{29}	Q_{10} for photosynthetic half-saturation constant	3.0
c_{102}	Specific (per mg) V_{max} for uptake of usable DOC by bacteria at reference temperature (mg C (mg C hr)$^{-1}$)	0.014
c_{103}	Half-saturation constant for usable DOC uptake (mg C m^{-3})	200.0
c_{104}	Q_{10} for bacterial heterotrophy and respiration	2.0
c_{107}	Specific (per mg) bacterial death rate (mg C (mg C hr)$^{-1}$)	0.00412
c_{108}	Bacterial respiration rate as a fraction of bacterial carbon uptake (mg C (mg C)$^{-1}$)	0.35
c_{109}	Rate of bacterial hydrolysis of sestonic detritus per mg bacteria and per mg sestonic detritus (mg C (mg C)$^{-1}$ (mg C hr)$^{-1}$)	4×10^{-8}
c_{110}	Fraction of dead bacteria transferred to sestonic detritus	0.6

(continued)

TABLE 10-8 (Continued)

Symbol	Description	Value
c_{201}	Specific (per mg) mean V_{max} for ingestion of total algae by *Daphnia*, both generations, at reference temperature	0.09
c_{202}	Half-saturation constant for ingestion of total algae by *Daphnia*, both generations (mg C m $^{-3}$)	250.0
c_{203}	Q_{10} for ingestion by *Daphnia*, both generations	3.0
c_{204}	Amplitude of variation in the V_{max} for ingestion of algae by *Daphnia* at reference temperature	0.006
c_{205}	Period of rhythm in V_{max} for ingestion of all prey types by both generations of *Daphnia* (hr)	12.0
c_{206}	Phase shift parameter for rhythm in V_{max} for ingestion of all prey types by both generations of *Daphnia* (radians)	1.047
c_{207}	Specific (per mg) mean V_{max} for ingestion of bacteria by *Daphnia*, both generations, at reference temperature (mg C (mg C hr) $^{-1}$)	0.007
c_{208}	Half-saturation constant for ingestion of bacteria by *Daphnia*, both generations (mg C m $^{-3}$)	25.0
c_{209}	Amplitude of fluctuation in the V_{max} for ingestion of bacteria by *Daphnia* at reference temperature (mg C (mg C hr) $^{-1}$)	0.001
c_{210}	Specific (per mg) mean V_{max} for ingestion of sestonic detritus by *Daphnia*, both generations, at reference temperature (mg C (mg C hr) ^{1})	0.2
c_{211}	Half-saturation constant for ingestion of sestonic detritus by *Daphnia*, both generations (mg C m $^{-3}$)	700.0
c_{212}	Amplitude of fluctuation of the V_{max} for ingestion of sestonic detritus by *Daphnia* at reference temperature (mg C (mg C hr) $^{-1}$)	0.0005
c_{213}	Assimilation efficiency for both *Daphnia* generations for algae	0.7
c_{214}	Maximum assimilation efficiency for both generations of *Daphnia* for bacteria	0.3
c_{215}	Specific (per mg) respiration rate for *Daphnia* (mg C (mg C hr) $^{-1}$)	0.0057
c_{216}	Q_{10} for *Daphnia* respiration	2.0
c_{217}	Fraction of dead *Daphnia*, both generations, transferred to sestonic detritus	0.6
c_{218}	Specific (per mg) death rate of *Daphnia*, both generations (mg C (mg C hr) $^{-1}$)	0.0009
c_{219}	Assimilation efficiency of first and second generation *Daphnia* for sestonic detritus	0.1
c_{220}	Specific (per mg) V_{max} for ingestion of total algae by fairyshrimp at reference temperature (mg C (mg C hr) $^{-1}$)	0.02
c_{221}	Specific (per mg) V_{max} for ingestion of bacteria by fairyshrimp at reference temperature (mg C (mg C hr) $^{-1}$)	0.005
c_{222}	Specific (per mg) V_{max} for ingestion of sestonic detritus by fairyshrimp at reference temperature (mg C (mg C hr) $^{-1}$)	0.015
c_{223}	Half-saturation constant for ingestion of total algae by fairyshrimp (mg c m $^{-3}$)	50.0
c_{224}	Q_{10} for ingestion of all food types by fairyshrimp	3.0
c_{225}	Half-saturation constant for ingestion of bacteria by fairyshrimp (mg C m $^{-3}$)	5.0
c_{226}	Half-saturation constant for ingestion of sestonic detritus by fairyshrimp (mg C m $^{-3}$)	100.0
c_{227}	Assimilation efficiency for fairyshrimp as a fraction of total ingestion rate	0.6

TABLE 10-8 (Continued)

Symbol	Description	Value
c_{228}	Maximum assimilation efficiency of fairyshrimp	0.25
c_{229}	Specific (per mg) respiration rate of fairyshrimp (mg C (mg C hr)$^{-1}$)	0.00015
c_{230}	Q_{10} for fairyshrimp respiration	2.0
c_{231}	Specific (per mg) death rate of fairyshrimp (mg C (mg C hr)$^{-1}$)	0.002
c_{232}	Fraction of dead fairyshrimp transferred to sestonic detritus	0.6
c_{301}	Specific (per mg) respiration rate for vascular plants (mg C (mg C hr)$^{-1}$)	0.0003
c_{302}	Specific (per mg) V_{max} for vascular plant photosynthesis at reference temperature (mg C (mg C hr)$^{-1}$)	0.0015
c_{303}	Half-saturation constant for light for vascular plant photosynthesis (cal cm^{-2} min^{-1})	5.0
c_{304}	Q_{10} for vascular plant photosynthesis	2.0
c_{305}	Fraction of net photosynthesis lost to U-DOC due to excretion (mg C (mg C)$^{-1}$)	0.03
c_{306}	Specific (per mg) leaching rate of new litter carbon (mg C (mg C hr)$^{-1}$)	8.75×10^{-5}
c_{307}	Fraction of new litter carbon leached to R-DOC (mg C (mg C)$^{-1}$)	0.04
c_{308}	Specific (per mg) death rate of vascular plant leaves transferred to new litter (mg C (mg C hr)$^{-1}$)	5×10^{-5}
c_{309}	Q_{10} for vascular plant respiration	2.5
c_{310}	Specific (per mg) dark excretion rate of vascular plants (mg C (mg C hr)$^{-1}$)	0.0001
c_{401}	Specific (per mg) rate of sedimentation of sestonic detritus to benthic detritus (mg C (mg C hr)$^{-1}$)	0.0
c_{501}	Fractional exchange constant for equilibrium between water column U-DOC and benthic U-DOC (mg C (mg C hr)$^{-1}$) = K501 in benthic model	0.02
c_{502}	Fractional exchange constant for the equilibrium between water column R-DOC and benthic R-DOC (mg C (mg C hr)$^{-1}$)	0.0
c_{503}	Specific (per mg) rate of photooxidation of R-DOC to U-DOC per unit light (mg C (mg C hr)$^{-1}$ (cal (cm^2 hr)$^{-1}$)$^{-1}$)	2.4×10^{-5}
c_{504}	Specific (per mg) rate of photooxidation of R-DOC to U-DOC at 0 light intensity (i.e., Y intercept in photooxidation vs light intensity function) (mg C (mg C hr)$^{-1}$)	-1.2×10^{-4}
c_{901}	Volume/area conversion factor for planktonic system (m^{-2} m^{-3})	1.66

populations into groups based upon phyla (green, blue-green, etc.). We would have liked to be able to have a number of physiological groups in this model and tried to obtain information from autoradiography on the half-saturation constants for nutrient uptake or light responses; these measurements were not completely successful (Chapter 5). The rate of photosynthesis is modeled as a function of algal biomass, light, temperature, and phosphorus concentration (RPI). As discussed earlier the functional form of the effects of light and phosphorus on the rate of photosynthesis is a Michaelis-Menten type equation. Experimental observations (Stanley 1974) indicate that the magnitude of the half saturation constant (C_{29}) in the light function is influenced by temperature. Thus, this quantity is multiplied by a Q_{10} function. Furthermore, the Q_{10} type of temperature function can independently influence the rate of photosynthesis. The experimental data of Stross indicate that the photosynthetic capability in these algae is rhythmic and hence V_{max} is represented as an oscillatory quantity with a mean of 0.06 mg C (mg algal C)$^{-1}$ hr^{-1} and an amplitude of 0.002.

The loss of carbon from algae is due to death (RP9) and excretion of U-DOC (RP5). It is assumed that a constant fraction of the biomass is lost due to these processes. Furthermore, the algae are also transferred to the bottom of the pond by burial, which is facilitated by the settling of large detrital particles (RP17) (Rublee 1974).

Carex aquatilis and *Arctophila fulva* grow in the shallow areas of the ponds. *Carex*, in particular, is amphibious and is a part of the carbon cycle of both the aquatic and terrestrial systems. It has been studied and modeled in detail by the investigators of the terrestrial section of the Tundra Biome (see Brown et al. in press). Therefore, no attempt has been made to duplicate that work. Nevertheless, for the sake of completeness we have included some equations to describe the input from these plants. One equation gives the growth rate in terms of a gross photosynthesis rate (RP301). The losses from this box are due to respiration (RP302), excretion of U-DOC (RP303), and death (RP304).

Consumers

Though there are several consumer species in the ponds (see Chapter 6, Zooplankton), *Daphnia* and fairyshrimps are dominant and account for more than 95% of the total consumer biomass. Because of the major role of these two species in the carbon dynamics of these ponds, we have included them in the plankton model.

In the early part of the season, *Daphnia* are inactive. After the first week these organisms hatch and begin feeding on algae, bacteria and detritus present in the water column. *Daphnia* are parthenogenetic and start producing young after 3 weeks. These second generation organisms start growing and feeding on algae, bacteria, and detritus after about 4 to 5 weeks, but towards the end of the season (after 6 weeks) about 90% of the population dies. Thus, after about 4 to 5 weeks of the season we have two generations of these organisms and these two types have different feeding rates. For this reason we have followed them as two separate variables.

In modeling the dynamics of the consumer species we have assumed that the feeding rate for a single food source is different from the feeding rate when multiple food sources exist in the same environment. Thus, for each food type, the consumer possesses a potential rate of ingestion. The actual rate of ingestion of a particular food, in the multiple food situation, is a function of this potential rate and of a Michaelis-Menten form of the sum of all the potential rates of all the food types each divided by its own V_{max}. For example, consider the rate of ingestion of algae by ***Daphnia***. Since bacteria and detritus are also available in the water, the actual rate of ingestion of algae (see Table 10-6) is

$$\mathrm{RB203} = \mathrm{RP201}(\mathrm{RP201}/c_{201}) / \left[(\mathrm{RP201}/c_{201}) + (\mathrm{RP206}/c_{207}) + (\mathrm{RP211}/c_{210})\right] \qquad (16)$$

where

RP203 = rate of ingestion of algae,
RP201 = potential rate of ingestion of algae in the absence of another food type,
RP206 = potential rate of ingestion of bacteria in the absence of another food type,
RP211 = potential rate of ingestion of detritus in the absence of another food type,
c_{201} = V_{max} for the ingestion of algae,
c_{207} = V_{max} for the ingestion of bacteria, and
c_{210} = V_{max} for the ingestion of detritus.

The potential rate of ingestion is a function of the biomass of algae and of *Daphnia* and the temperature. Experimental evidence from these ponds (Chisholm et al. 1975) suggests that there is an endogenous feeding rhythm in these *Daphnia* with a cycle length of 12 hours that varies by a factor of 2. Therefore, to include this rhythmic effect on the feeding rate we have added a sine function, which generates these oscillations, to the V_{max} for each food type. Thus the V_{max} for each food type is not a constant but varies as a periodic function with a cycle length of 12 hours varying by a factor of 2. As an example, the potential rate of ingestion of algae is

$$\mathrm{RP201} = \left[y_{201} y_1 / (y_1 + c_{202})\right] c_{20} \left[1.5 + 0.5 \sin\left(\frac{\pi t}{12} + 1.047\right)\right] T_e \qquad (17)$$

where

RP201 = potential rate of ingestion of algae in the absence of another food type,
y_{201} = biomass of *Daphnia*,
y_1 = biomass of algae,
c_{201} = V_{max} for ingestion of algae,
c_{202} = half saturation constant for ingestion of algae, and
T_e = temperature effect

The loss of biomass of the consumers is due to respiration and death (RP221, RP223). It is also assumed that the loss of carbon due to egestion is a constant fraction of the rates of ingestion of algae, bacteria and detritus (RP204, RP209, RP219). A constant fraction of the dead organisms is converted to U-DOC.

Decomposers

Bacteria are the only decomposers experimentally studied and are included in the plankton model. Their growth rate is dependent upon the concentration of U-DOC in the water. The rate of uptake of U-DOC by these organisms follows a Michaelis-Menten type of equation (RP101). They also grow on detrital particles and derive some U-DOC by hydrolyzing these particles. The rate of hydrolysis is assumed to be proportional to the biomasses of bacteria and detritus (RP102).

The loss of bacterial carbon is due to respiration (RP107) and death (RP103); a fraction of this loss is transferred to U-DOC.

Detritus and Dissolved Organic Carbon

Most of the detrital material in the water is planktonic in origin; however, a small portion of it also comes from the bottom due to wind resuspension. The death of algae, of bacteria, and of the consumers also contributes to this component. The U-DOC comes from excretion from both algae and vascular plants. A fraction of the planktonic detrital material is also converted into U-DOC.

RESULTS AND DISCUSSION

Deterministic Framework

The primary objective of the modeling effort was to provide a framework around which the ideas and experimental work of a group of investigators could be organized. Thus, the structure of the model reflected our current biological and ecological understanding of this particular aquatic system. These ideas and the results of a number of experiments carried out by the investigators were used to construct the forms of mathematical equations of rate processes and dynamics of biomasses. The results obtained from the computer simulation studies of these equations are indicative of general seasonal changes in the biological processes and species of organisms (measured as the amount of carbon per unit area or volume). We do not claim to have formulated a model capable of predicting everything about the system; it can only be construed as a first step towards a model capable of describing the structure and function of such an ecosystem. When that goal is achieved, perhaps we will have a model which can be used as a powerful tool in management policy decisions involving ecosystems.

The equations used to describe the component processes (e.g. photosynthesis, feeding rate of consumers, nutrient uptake, etc.) are not specific to this tundra aquatic system; their general validity is supported by published results from a wide range of aquatic systems. Wherever detailed knowledge of some process was lacking, the functional form for that process was assumed to be simply a constant multiplied by the biomass (e.g. death rate, excretion rate, etc.). This can be considered a first approximation.

The basic structure of the model is specified by a set of nonlinear differential equations, and a computer simulation analysis of the model involves the numerical integration of these equations. Thus in a simulation study it is necessary to specify the numerical value of all the parameters, initial conditions,and also appropriate functional forms or tabular forms of data for the effects of input variables, such as light and temperature. Once these quantities are specified in the computer program the model is run for a period of 60 days (the active life of tundra ponds). The model output consists of the time-dependent behavior of system variables and rate processes. Data collected over a 3-year period produced the values of the parameters and the initial conditions of the state variables; this single set of parameter values was used to obtain the simulation results. For light and temperature, tables of recorded data were supplied to the program. Thus the results of the computer simulation reflect the behavior of the model (variables) with fixed initial conditions and parameters and variable

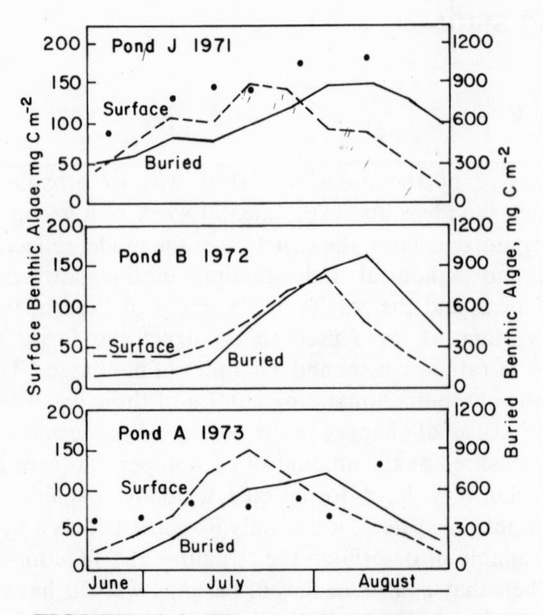

FIGURE 10-3. *The model simulation of surface and buried benthic algae biomass. The solid circles represent the measured algal biomass (0 to 5 cm).*

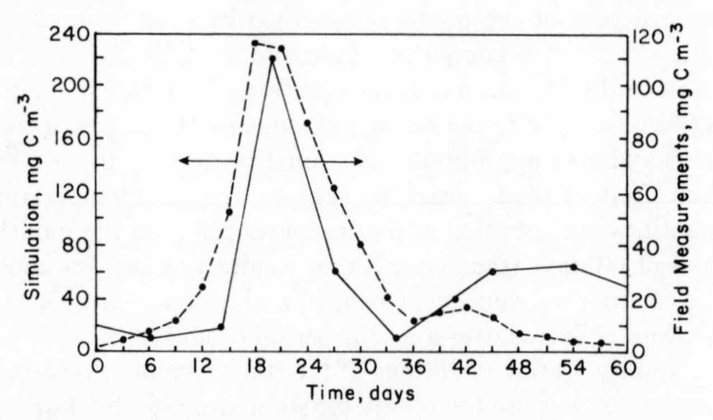

FIGURE 10-4. *Field measurements (solid line) and the model simulation (dashed line) of the biomass of planktonic algae in Pond B.*

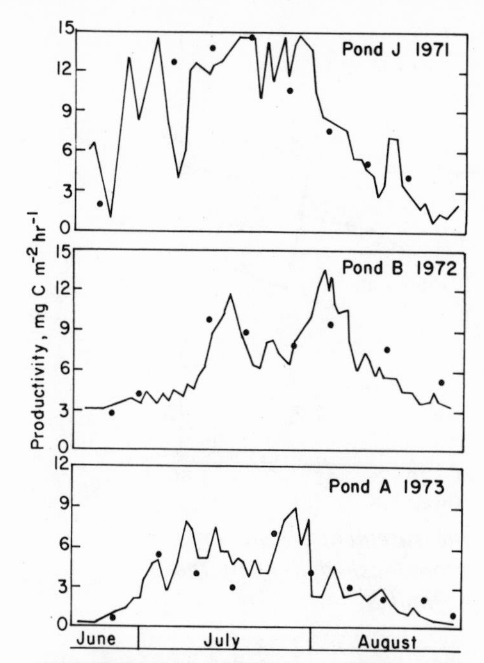

FIGURE 10-5. *Model simulation and measured estimates (black circles) of net photosynthesis of the benthic algae. The data are for rates at 1200. (After Stanley 1974.)*

(year to year) light and temperature. Although the behavior of all the state variables of the model have been simulated and the output data have been compared with the field observations, we present here the behavior of only one variable from each of the three major categories—primary producers, consumers and decomposers.

Figures 10-3 and 10-4 depict the seasonal trend in the biomasses of benthic and planktonic algae populations. In these figures experimental data are also included for a comparison. There is excellent agreement between the simulation and the field data. For the best studied case, the photosynthesis of the benthic algae (Stanley 1976a, 1976b), the parameters were measured in 1973 and tested against the photosynthesis measured in 1971, 1972, and 1973. Figure 10-5 shows the measured and simulated values of net photosynthesis for benthic algae.

Daphnia and chironomid biomasses are given in Figures 10-6 and 10-7. *Daphnia* in the water and chironomids in the sediment are the main consumers and they account for about 90% of the biomass of the consumer population of this system.

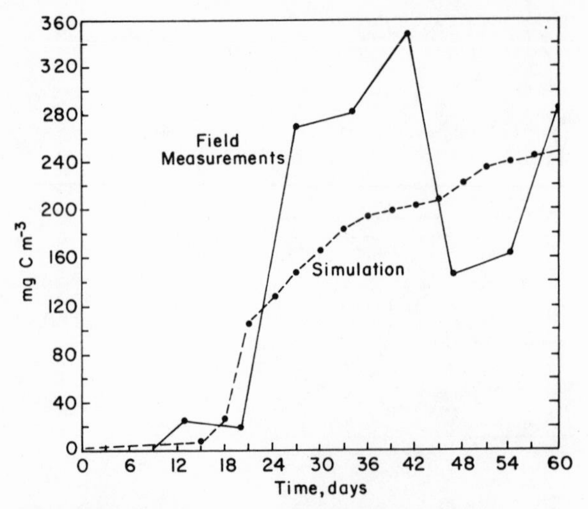

FIGURE 10-6. *Field measurements (solid line) and the model simulation (dashed line) of the biomass of* Daphnia *in Pond B.*

Bacteria are the only decomposers included in the model (both planktonic and benthic); the seasonal change in their biomass in the plankton is shown in Figure 10-8.

In general, the simulation shows the same seasonal pattern as the field measurements. The actual values of the simulations are less important, as we can raise or lower these by appropriate small changes in parameters or even in initial conditions. Thus, the algal biomass is too high in the simulation (Figure 10-4) but this can easily be corrected by some small changes in the parameters of the photosynthesis and respiration. It would

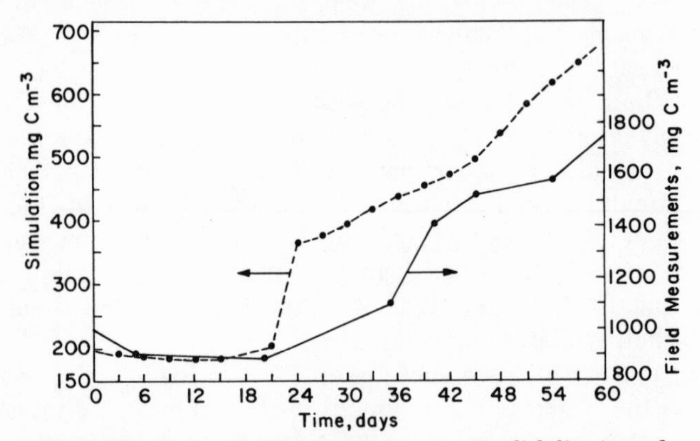

FIGURE 10-7. *Field measurements (solid line) and the model simulation (dashed line) of the biomass of chironomids in Pond B.*

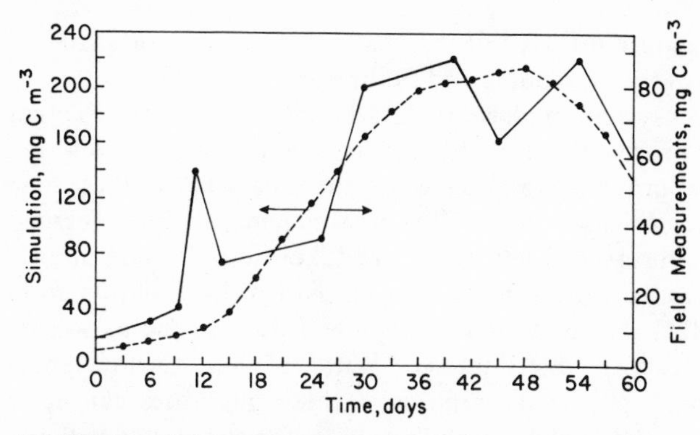

FIGURE 10-8. *Field measurements (solid line) and the model simulation (dashed line) of the biomass of bacteria in the water of Pond B.*

add nothing to our understanding, however, so we have not bothered to improve the simulation.

In a computer simulation type of modeling scheme for a complex ecosystem, the manipulation of the initial conditions and parameter values is known as tuning and calibration of the model. The objective of such procedures is to arrive at a set of curves almost identical and consistent with the observed data. This model with tuned parameters and initial conditions is used to predict or verify the results of a different year. The major difficulty with this procedure is that for a new set of input variables the "retuned" model does not produce a behavior of the variable similar to the one observed in the field or during experimental investigations. Thus a new cycle of tuning and calibration is required. This model is justified on the premise that experimental and field data for parameters and initial conditions do indeed show a range of values and so it is logical and acceptable to manipulate the parameters within the observed range to produce the desired behavior from the model. Even simple models of ecosystems include dozens of parameters, all of which can be manipulated within the given range. Thus, although it is relatively simple to produce a set of desired curves by tuning and calibration, the inherent danger in such a procedure is that one is liable to get the right results for the wrong reasons. This is a serious problem and it also imposes a severe restriction on the utility of these models for applied problems in ecology and resource management.

Stochastic Framework: Toward a More Realistic and General Model

One possible and potentially very powerful approach to the analysis of these complex ecological systems is to formulate the dynamics of the system in terms of stochastic differential equations. The computer simulation procedure involved in the analysis of such models can help us circumvent the process of tuning and calibration.

Since all ecosystem models are data-based and since the fluctuations of various rate processes are based on experimental observations, a more realistic and appropriate approach to the analysis of these specific ecological systems would be the one in which naturally occurring and experimentally observed fluctuations are taken into consideration. This will enable us to utilize the maximum amount of information from the experimental data and to assign a range of values to the parameters in terms of probability distributions. With Monte Carlo simulation techniques a probabilistic description of these quantities can easily be incorporated into the model and the results of the simulation will give the mean and variance of each of the variables of interest. This would be of practical value in any natural resource management scheme, for allowance can be made for the expected fluctuation from the average value. A probabilistic description of the system is also more realistic and desirable because of our ignorance of the exact values and because of uncertainties associated with these complex natural systems.

To formalize these concepts and to facilitate our discussion, we begin from our general equation (5) describing the dynamics of the system under consideration.

$$\dot{\mathbf{X}} = f\left[\mathbf{X}(t), \mathbf{U}(t), \mathbf{W}(t), t\right], \quad \mathbf{X}(t_0) = \mathbf{X}_0 \tag{5}$$

The introduction of random elements in this equation makes $\dot{\mathbf{X}}$ a vector of stochastic differential equations and the solution, $\mathbf{X}$, is now a vector of stochastic processes. There are three levels—initial conditions, input variables, and parameters—at which stochastic elements can be introduced. Results from a number of independent Monte Carlo realizations obtained through computer simulation procedures can then be used to estimate the mean and associated variance of the system variables or rate processes.

A simplified version of the plankton subsystem was analyzed by Tiwari and Hobbie (1976a) using random differential equation models. Initial values of the state variables, input variables (light and temperature), and parameters were defined as random variables having univariate Gaussian distributions. The mean and variances for these variables were estimated from the data. For each simulation run, numerical values of these variables were sampled from the prespecified distribution and this process was repeated nine times. From these realizations, mean and standard deviations were estimated for each of the state variables (for details see Tiwari and Hobbie 1976a).

The results of a simulation in which randomness was incorporated at all three levels are summarized in Figures 10-9 and 10-10. These two graphs show the temporal behavior of average biomasses together with the

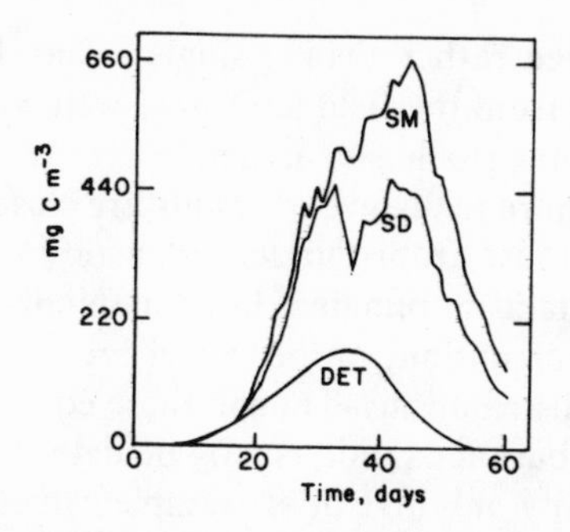

FIGURE 10-9. *Algae biomass in the plankton of a pond as simulated by a deterministic (DET) and a stochastic (SM) model. The stochastic model simulation is the mean of nine runs. The standard deviation (SD) is also given. (After Tiwari and Hobbie 1976a.)*

standard deviation of algae and *Daphnia*. Also included in these figures are the results of the corresponding deterministic model in which all the variances were set to zero. Some of the interesting properties of these models are worth mentioning here. In general the results of the deterministic models were different from that of the stochastic means. In fact, in most of the cases the differences were so great that the deterministic results were outside the 95% confidence limit. Although this is not very surprising, especially because of the nonlinear terms in the differential equations, it can be pointed out that one of the assumptions implicit in the rationale behind a deterministic formulation is that the deterministic model is a mean of a stochastic model. Some degree of increased stability was also observed in the stochastic models. The behavior of the deterministic model is very sensitive to the small changes in the parameter values; for some parameters a small change in the third decimal place can produce very different results in the resultant biomass. However, in our stochastic formulation where all the parameters were treated as random variables, this type of sensitivity was not observed. Thus it seems that simultaneous fluctuations in a number of parameters create in the system a buffering mechanism in which too great effects of some parameters are counterbalanced by small effects of others. But perhaps the most useful quantity, from an application point of view, that comes from these models is the standard deviation associated with the average value of the state variables. With the information from a standard deviation, a confidence interval around the mean can be constructed. This allows us to

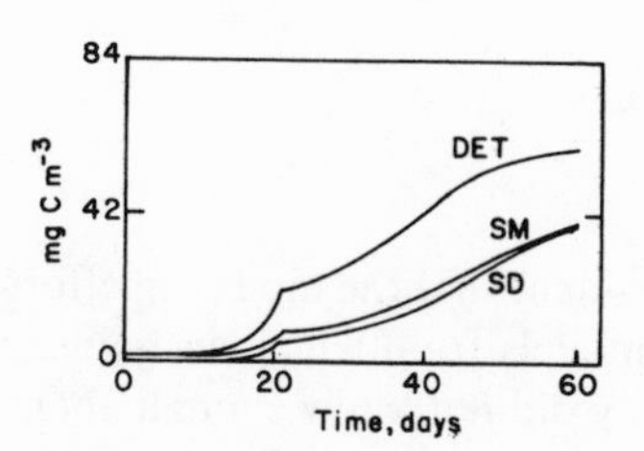

FIGURE 10-10. Daphnia *biomass in the plankton of a pond as simulated by a deterministic and a stochastic model. The stochastic model simulation is the mean of nine runs. The standard deviation is also given. (After Tiwari and Hobbie 1976a.)*

make predictions in terms of a range rather than a single value. It is interesting that all the sample values from the field data were well within the confidence interval obtained from the stochastic model.

Although a stochastic model is more realistic and results are closer to the experimental observations than those from the deterministic model, some difficulty in formulating such models remains. For this model, we assumed that parameters and initial conditions in the model are random variables with Gaussian probability distributions. Though these equations are variables, the nature of the distribution cannot usually be determined from the available data because usually only five or six sample values are available; this is not enough to decide the type of distributions. Thus it seems that we need a method by which we can make the best use of the available data without making explicit assumptions about the distribution. This can be accomplished by the formalisms of Information Theory (Tiwari and Hobbie 1976b). If and when more data are available, that information can be utilized to modify the shape of the distribution. This method can help us formulate a general realistic model based on the maximum amount of information that can be extracted from a given set of data.

In spite of our use of models, the discrepancies between model results and experimental and field observations on natural systems signify that there are some unresolved yet fundamental problems in biology. For example, what are the behavioral features of large, complex systems and what are the underlying causes responsible for these properties? Furthermore, the properties associated with the structures and functions of biological systems have been shaped by the forces of natural selection over thousands and millions of years. Some recent works in this area suggest that modes of behavior of large, complex systems can be influenced by the number of elements in the systems, the degree of connectance between the elements, the degree of non-linearity in the system, the hierarchical structure, and the magnitude of the entropy (May 1973, Gardner and Ashby 1970, Kauffman 1969, Siljak 1974, McMurtrie 1975). Perhaps the observed behavior of our models reflects the cumulative effect of all these and even more factors. During the last few years some progress has been made in this direction but much research is needed before we can understand the behavior of models; even more is needed before the models will effectively simulate real systems or ecosystems.

Conclusion

From the preceding sections, it is evident that the modeling effort did not produce predictive models or even models from which we gained new insights. It became obvious that we really did not know enough about the real controls operating within the ecosystem to construct a whole-system model that was predictive. Yet we were able to contruct a model that simulated the seasonal cycles of biomass and production quite well. If we had not known so much about this system, perhaps we would have believed that the model was entirely adequate. Some of the aspects of the model that made us distrust it were its sensitivity to extremely small changes in some parameters, its dependence upon tuning of some parameters or constants about which we really knew little, and our ability to make this model simulate any conceivable seasonal cycle by making slight changes (this is true of most nonlinear models). In addition, our development of stochastic versions of the deterministic models showed us how the slight variability found in values measured in nature often had a drastic effect on the output from the model. In contrast, our experience with this ecosystem indicated that this amount of variability had no profound effect on the behavior of the natural system. We concluded that perhaps a stochastic model is preferable to a deterministic one but that we could not construct a realistic stochastic model because of lack of data on the natural variability.

There were, however, several positive aspects of our experience with modeling. First, a number of scientists and students learned a great deal about the good and bad aspects of ecosystem modeling. Education in ecosystem research was one goal of IBP. Second, the modeling forced us to think critically about processes, interrelationships, and controls within our pond. We hope we would have done this anyway but it is all too easy to become caught up in the problems of rather routine data gathering and avoid the more difficult problems. Third, the modeling forced a whole-system approach and focused our attention upon some aspects of the ecosystem that had been little studied; for example, the feeding rates of protozoans on bacteria. Conversely, we became aware that some areas like primary productivity of planktonic algae had been over-studied and further intensive study was unnecessary. Fourth, the modeling effort quickly made us aware of how little we really knew about some of the processes and relationships in the ponds. This realization caused us to consolidate our efforts into one or two ponds. The natural tendency of ecologists seems to be to sample as many ponds as possible and to pick a different pond for the study of each process. Finally, a fifth positive aspect was the tests we were able to run with the small models or with small parts of the large models. These were tests of the importance of various factors or of the possible range of some rates that could not be measured. One example was the determination of the likely rates of mixing of benthic algae into the sediment.

Overall, the use of modeling as described here was probably worthwhile and was probably an aid to the project. The models did not turn out to be the predictive tools we had hoped for but the process of modeling was extremely valuable for a number of reasons.

From our experiences, we conclude that the construction of an ecological model should be a long-term effort measured in decades, not in months or a few years. It is too bad that the tundra pond models could not be used as such a long-term guide for research but the questions that were being asked during construction of the model were so basic that they could best be studied in temperate regions, not in the Arctic.

SUMMARY

Several small models of processes such as phosphorus cycling and primary production of benthic algae were constructed, and gave valuable information about the importance of such things as the mixing of algae into the sediments or the contribution of protozoans to the phosphorus flux. Two large models were also constructed. These models incorporated relationships of the physiology of populations to factors of the environment. Thus, algal photosynthesis was related to temperature, light, and algal biomass. Almost all of the parameters and constants used in the model were measured in the ponds; a few could not be measured but were incorporated because they are known to be important.

The output of the models successfully simulated the seasonal cycles of biomass and production in the ponds. However, some constants and parameters had to be tuned and some were very sensitive in that slight changes caused drastic changes in output. Because of this, and because we felt we did not know all of the controls operating in nature, we did not use the models for prediction of future states of the ecosystem.

The main value of the modeling came from the construction itself. We were obliged to attempt to answer the most difficult questions of controls and carbon flow. All the scientists had to look beyond the borders of their own specialties and to think about the whole system. This stimulated fruitful discussion and joint experimentation. It also caused us to realize how important it was to concentrate on one or two ponds rather than attempting to study a large number of ponds.

The large models were deterministic but we also constructed several stochastic models that incorporated variability into the initial conditions, the input variables (e.g., light and temperature), and various parameters. The introduction of the variability caused the models to become more stable and also allowed us to calculate a mean of output values along with a standard deviation. This approach appears promising because all of the information about the natural system may be incorporated and because all parts of the natural system are variable to some degree or other.

References

Alexander, M.R. (1971) *Microbial Ecology*. New York: John Wiley, 511 pp.

Alexander, V. and R.J. Barsdate (1971) Physical limnology, chemistry and plant productivity of a taiga lake. *Internationale Revue der Gesamten Hydrobiologie,* **56**: 825-872.

Alexander, V. and R.J. Barsdate (1975) Studies of nitrogen cycle processes in arctic tundra systems. In *Proceedings of the Circumpolar Conference on Northern Ecology, Ottawa, 1975*. National Research Council of Canada, Scientific Committee on Problems of the Environment, pp. III-53 to III-64.

Allen, H.L. (1969) Chemoorganotrophic utilization of dissolved organic compounds by planktonic bacteria and algae in a pond. *Internationale Revue der Gesamten Hydrobiologie,* **54**: 1-33.

Allen, P.W. and R.O. Weedfall (1966) Weather and climate. In *Environment of the Cape Thompson Region, Alaska* (N.J. Wilimovsky and J.W. Wolfe, Eds.). Oak Ridge, Tenn.: U.S. Atomic Energy Commission, PNE-481, pp. 9-44.

Allessio, M.L. and L.L. Tieszen (1973) Patterns of translocation and allocation of ^{14}C-photoassimilate *in situ* studies with *Dupontia fischeri* R.Br., Barrow, Alaska. In *Primary Production and Production Processes: Proceedings of the Conference, Dublin, Ireland, April 1973* (L.C. Bliss and F.E. Wielgolaski, Eds.). Stockholm: International Biological Programme, Tundra Biome Steering Committee, pp. 219-229.

American Public Health Association (1960) *Standard Methods for the Examination of Water and Wastewater Including Bottom Sediments and Sludges*. New York: American Public Health Association, Inc., 11th edition.

Amren, H. (1964) Ecological studies of zooplankton populations in some ponds on Spitsbergen. *Zoologiska Bidrag fran Uppsala,* **36**: 161-191.

Anderson, F.S. (1946) East Greenland lakes as habitats for chironomid larvae. *Meddeleser om Grønland,* **100**: 1-65.

Anderson, J.W., J.M. Neff, B.A. Cox, H.E. Tatem and G.M. Hightower (1974) Characteristics of dispersions and water-soluble extracts of crude and refined oils and their toxicity to estuarine crustaceans and fish. *Marine Biology,* **27**: 75-88.

Armstrong, W. (1964) Oxygen diffusion from the roots of some British bog plants. *Nature,* **204**: 801-802.

Arnborg, L., H.J. Walker and J. Peippo (1966) Water discharge in the Colville River, 1962. *Geografiska Annaler,* **48A**: 195-210.

Arnold, G.P. (1967) Observations on *Lepidurus arcticus* (Pallas) (Crustacea: Notostraca) in East Greenland. *Annals and Magazine of Natural History,* **13**(9): 599-617.

Atlas, R.M. (1975) Effects of temperature and crude oil composition on petroleum biodegradation. *Applied Microbiology,* **30**: 396-403.

Bache, B.W. and E.G. Williams (1971) Phosphate sorption index for soils. *Journal of Soil Science,* **22**: 289-301.

Ball, J.S. (1962) Nitrogen compounds in petroleum. *American Petroleum Institute Proceedings,* **42**: 27-30.

Barko, J.W. and R.M. Smart (1979) The nutritional ecology of *Cyperus esculentus,* an emergent aquatic plant, grown on different sediments. *Aquatic Botany,* **6**: 13-28.

Barko, J.W. and R.M. Smart (In press) Mobilization of sediment phosphorus by submersed freshwater macrophytes. *Freshwater Biology.*

Barlow, J.P. and J.W. Bishop (1965) Phosphate regeneration by zooplankton in Cayuga Lake. *Limnology and Oceanography,* **10** (Supplement): R15-R24.

Barnett, C.J. and J.E. Kontogiannis (1975) The effect of crude oil fractions on the survival of a tidepool copepod, *Tigriopus californicus. Environmental Pollution,* **8**: 45-54.

Barsdate, R.J. and V. Alexander (1971) Geochemistry and primary productivity of the Tangle Lakes system, an Alaskan alpine watershed. *Arctic and Alpine Research,* **3**: 27-41.

Barsdate, R.J. and V. Alexander (1975) The nitrogen balance of arctic tundra: pathways, rates and environmental implications. *Journal of Environmental Quality,* **4**: 111-117.

Barsdate, R.J., V. Alexander and R.E. Benoit (1973) Natural oil seeps at Cape Simpson, Alaska: Aquatic effects. In *Proceedings of the Symposium on the Impact of Oil Resource Development on Northern Plant Communities: Presented at the 23rd AAAS Alaska Science Conference, Fairbanks, Alaska, 17 August 1972* (B.H. McCown and D.R. Simpson, Coordinators). University of Alaska, Institute of Arctic Biology, Occasional Publications on Northern Life 1, pp. 91-95.

Barsdate, R.J. and W.R. Matson (1966) Trace metals in arctic and subarctic lakes with references to the organic complex of metals. In *Radioecological Concentration Processes: Proceedings of an International Symposium held in Stockholm, April 1966* (B. Åberg and F.P. Hungate, Eds.). New York: Pergamon Press, pp. 711-719.

Barsdate, R.J., R.T. Prentki and T. Fenchel (1974) Phosphorus cycle of model ecosystems: significance for decomposer food chains and the effect of bacterial grazers. *Oikos,* **25**: 239-251.

Bell, R.K. and F.J. Ward (1968) Use of liquid scintillation counting for measuring self-absorption of carbon-14 in *Daphnia pulex. Journal of the Fisheries Research Board of Canada,* **25**: 2505-2508.

Bennett, M.E. and J.E. Hobbie (1972) The uptake of glucose by *Chlamydomonas* sp. *Journal of Phycology,* **8**: 392-398.

Bergstein, P. and J.R. Vestal (1978) Crude oil biodegradation in arctic tundra ponds. *Arctic,* **31**: 158-169.

Berman, T. and U. Pollingher (1974) Annual and seasonal variations of phytoplankton, chlorophyll and photosynthesis in Lake Kinneret. *Limnology and Oceanography,* **19**: 31-54.

Bernard, J.M. (1974) Seasonal changes in the standing crop and primary production in a sedge wetland and in adjacent dry old-field in central Minnesota. *Ecology,* **55**: 350-359.

Bick, H. (1958) Ökologische Untersuchungen an Ciliaten fallaubreicher Kleingewässer. *Archiv für Hydrobiologie,* **54**: 506-542.

Billings, W.D., K.M. Peterson and G.R. Shaver (1978) Growth, turnover, and respiration rates of roots and tillers in tundra graminoids. In *Vegetation and Production Ecology of an Alaskan Arctic Tundra* (L.L. Tieszen, Ed.). New York: Springer-Verlag, Inc., pp. 415-434.

Birge, E.A. and C. Juday (1934) Particulate and dissolved organic matter in inland lakes. *Ecological Monographs,* **4**: 440-474.

Black, A.P. and R.F. Christman (1963) Chemical characteristics of fulvic acids. *Journal of the American Water Works Association,* **55**: 807-912.

Black, R.F. (1964) Gubik formation of Quaternary age in northern Alaska. *U.S. Geological Survey Professional Paper* 302-C, pp. 59-91.

Black, R.F. and W.L. Barksdale (1949) Oriented lakes of northern Alaska. *Journal of Geology,* **57**: 105-118.

Bowden, W.B. (1977) Comparison of two direct-count techniques for enumerating aquatic bacteria. *Applied and Environmental Microbiology,* **33**: 1229-1232.

Boyd, W.L. (1959) Limnology of selected arctic lakes in relation to water supply problems. *Ecology,* **40**: 49-54.

Boyd, W.L. and J.W. Boyd (1963) A bacteriological study of an arctic coastal lake. *Ecology,* **44**: 705-710.

Brewer, M.C. (1958) The thermal regime of an arctic lake. *Transactions of the American Geophysical Union,* **39**: 278-284.

Britton, M.E. (1957) Vegetation of the arctic tundra. In *Arctic Biology: Proceedings of the Annual Biology Colloquium* (H.P. Hansen, Ed.). 2nd edition. Corvallis: Oregon State University Press, pp. 67-113.

Broch, E.S. (1965) Mechanism of adaptation of the fairyshrimp *Chirocephalopsis bundyi* Forbes to the temporary pond. Ph.D. dissertation, Cornell University. Agricultural Experiment Station Memoir 392, 48 pp.

Brooks, J.L. (1957) The systematics of North American *Daphnia. Memoirs of the Connecticut Academy of Arts and Sciences,* **13**: 1-180.

Brooks, J.L. and S.I. Dodson (1965) Predation, body size, and composition of plankton. *Science,* **150**: 28-35.

Brown, J. (1965) Radiocarbon dating, Barrow, Alaska. *Arctic,* **18**: 36-48.

Brown, J., F.L. Bunnell, P.L. Miller and L.L. Tieszen (Eds.) (In press) *An Arctic Ecosystem: The Coastal Tundra at Barrow, Alaska.* Stroudsburg, Pa.: Dowden, Hutchinson and Ross.

Brown, J., S.L. Dingman and R.I. Lewellen (1968) Hydrology of a small drainage basin on the coastal plain of northern Alaska. *U.S. Army CRREL Research Report* 240, 18 pp.

Brown, J., C.L. Grant, F.C. Ugolini and J.C.F. Tedrow (1962) Mineral composition of some drainage waters from arctic Alaska. *Journal of Geophysical Research,* **67**: 2447-2453.

Brown, J. and P.V. Sellmann (1973) Permafrost and Coastal Plain history of arctic Alaska. In *Alaskan Arctic Tundra* (M.E. Britton, Ed.). Arctic Institute of North America Technical Paper 25, pp. 31-47.

Brown, J. and A.K. Veum (1974) Soil properties of the international Tundra Biome sites. In *Soil Organisms and Decomposition in Tundra: Proceedings of the Microbiology, Decomposition and Invertebrate Working Groups Meeting, Fairbanks, Alaska, August 1973* (A.J. Holding, O.W. Heal, S.F. MacLean, Jr. and P.W. Flanagan, Eds.). Stockholm: International Biological Programme, Tundra Biome Steering Committee, pp. 27-48.

Brundin, L. (1951) The relation of O_2-microstratification at the mud surface to the ecology of the profundal bottom fauna. *Institute of Freshwater Research Drottningholm Report,* **32**: 32-42.

Brunskill, G.J., D.M. Rosenberg, N.B. Snow, G.L. Vascotto and R. Wagemann (1973) Ecological studies of aquatic systems in the Mackenzie–Porcupine drainages in relation to proposed pipeline and highway development. Canadian Task Force on Northern Oil Development, Environmental–Social Committee Report 73-40 (1), 131 pp.; Report 73-41 (App.) (2), 345 pp.

Bunnell, F.L., S.F. MacLean, Jr. and J. Brown (1975) Barrow, Alaska, U.S.A. In *Structure and Function of Tundra Ecosystems: Papers Presented at the IBP Tundra Biome V International Meeting on Biological Productivity of Tundra, Abisko, Sweden, April 1974* (T.

Rosswall and O.W. Heal, Eds.). Ecological Bulletin 20. Stockholm: Swedish Natural Science Research Council, pp. 73-124.

Bunnell, F.L., O.K. Miller, P.W. Flanagan and R.E. Benoit (In press) The microflora: Composition, biomass and environmental relations. In *An Arctic Ecosystem: The Coastal Tundra at Barrow, Alaska* (J. Brown, F.L. Bunnell, P.C. Miller and L.L. Tieszen, Eds.). Stroudsburg, Pa.: Dowden, Hutchinson and Ross, Chap. 8.

Bunnell, F.L. and D.E.N. Tait (1974) Mathematical simulation models of decomposition processes. In *Soil Organisms and Decomposition in Tundra: Proceedings of the Microbiology, Decomposition and Invertebrate Working Groups Meeting, Fairbanks, Alaska, August 1973* (A.J. Holdingm O.W. Heal, S.F. MacLean, Jr. and P.W. Flanagan, Eds.). Stockholm: International Biological Programme, Tundra Biome Steering Committee, pp. 207-225.

Bunt, J.S. and C.C. Lee (1972) Data on the composition and dark survival of four sea-ice microalgae. *Limnology and Oceanography,* **17**: 458-461.

Burkholder, P.R., A. Repak and J. Sibert (1965) Studies on some Long Island Sound littoral communities of microorganisms and their primary productivity. *Bulletin of the Torrey Botanical Club,* **92**: 378-402.

Burns, C.W. (1969) Relation between filtering rate, temperature, and body size in four species of *Daphnia. Limnology and Oceanography,* **14**: 693-700.

Burns, C.W. and F.H. Rigler (1967) Comparison of filtering rates in *Daphnia rosea* in lake water and in suspensions of yeast. *Limnology and Oceanography,* **12**: 492-502.

Busdosh, M. and R.M. Atlas (1977) Toxicity of oil slicks to arctic amphipods. *Arctic,* **30**: 85-92.

Cady, F.B. and W.V. Bartholomew (1960) Sequential products of aerobic denitrification in Norfolk soil material. *Soil Science Society of America, Proceedings,* **24**: 477-482.

Caldwell, M.M., L.L. Tieszen and M. Fareed (1974) The canopy structure of tundra plant communities at Barrow, Alaska, and Niwot Ridge, Colorado. *Arctic and Alpine Research,* **6**: 151-159.

Cameron, J.N., J. Kostoris and P.A. Penhale (1973) Preliminary energy budget of the nine-spined stickleback (*Pungitius pungitius*) in an arctic lake. *Journal of the Fisheries Research Board of Canada,* **30**: 1179-1189.

Cammen, L.M. (1978) The significance of microbial carbon in the nutrition of the polychaete *Nereis succinea* and other aquatic deposit feeders. Ph.D. dissertation, North Carolina State University, Raleigh, 84 pp.

Caperon, J.W. (1968) Population growth response of *Isochrysis galbana* to nitrate variation at limiting concentrations. *Ecology,* **49**: 866-872.

Caperon, J.W. and J. Meyer (1972) Nitrogen-limited growth of marine phytoplankton. 2. Uptake kinetics and their role in nutrient-limited growth of phytoplankton. *Deep-Sea Research,* **19**: 619-632.

Carlson, R.F., W. Norton and J. McDougall (1974) *Modeling Snowmelt Runoff in an Arctic Coastal Plain.* University of Alaska Institute of Water Resources Report IWR-43, 72 pp.

Carpenter, E.J. and R.R.L. Guillard (1971) Intraspecific differences in nitrate half-saturation constants for three species of marine phytoplankton. *Ecology,* **52**: 183-188.

Carson, C.E. (1968) Radiocarbon dating of lacustrine strands in arctic Alaska. *Arctic,* **21**: 12-26.

Carson, C.E. and K.M. Hussey (1960) Hydrodynamics in three arctic lakes. *Journal of Geology,* **68**: 585-600.

Caswell, H., H.E. Koenig, J.A. Resh and Q.E. Ross (1972) An introduction to systems science for ecologists. In *Systems Analysis and Simulation in Ecology* (B.C. Patten, Ed.). Volume II. New York: Academic Press, pp. 3-78.

Chamberlain, W. and J. Shapiro (1969) On the biological significance of phosphate analysis; comparison of standard and new methods with a bioassay. *Limnology and Oceanography,* **14**: 921-927.

Chang, S.C. and M.L. Jackson (1957) Fractionation of soil phosphorus. *Soil Science,* **84**: 133-144.

Chapin, F.S., III (1974) Morphological and physiological mechanisms of temperature compensation in phosphate absorption along a latitudinal gradient. *Ecology,* **55**: 1180-1198.

Chapin, F.S., III, R.J. Barsdate and D. Barèl (1978) Phosphorus cycling in Alaskan coastal tundra: A hypothesis for the regulation of nutrient cycling. *Oikos,* **31**: 189-199.

Chapin, F.S., III, and A. Bloom (1976) Phosphate absorption: Adaptation of tundra graminoids to a low temperature, low phosphorus environment. *Oikos,* **26**: 111-121.

Chapin, F.S., III, K. Van Cleve and L.L. Tieszen (1975) Seasonal nutrient dynamics of tundra vegetation at Barrow, Alaska. *Arctic and Alpine Research,* **7**: 209-226.

Chisholm, S.W. (1974) Studies on daily rhythms of phosphate uptake in *Euglena* and their potential ecological significance. Ph.D. dissertation, State University of New York, Albany, 106 pp.

Chisholm, S.W., R.G. Stross and P.A. Nobbs (1975) Environmental and intrinsic control of filtering and feeding rates in arctic *Daphnia. Journal of the Fisheries Research Board of Canada,* **32**: 219-226.

Cogley, J.G. and S.B. McCann (1971) Information on a snowmelt runoff system obtained from covariance and spectral analysis. Paper presented at 1971 Annual Meetings of the American Geophysical Union, Washington, D.C.

Cole, G.A. (1953) Notes on the vertical distribution of organisms in the profundal sediments of Douglas Lake, Michigan. *American Midland Naturalist,* **49**: 252-256.

Comita, G.W. (1956) A study of a calanoid copepod population in an arctic lake. *Ecology,* **37**: 576-591.

Cook, F.A. (1960) Rainfall measurements at Resolute, N.W.T. *Révue Canadienne de Géographie,* **14**: 45-50.

Costerton, G.W., G.G. Geesey and K.-J. Cheng (1978) How bacteria stick. *Scientific American,* **238**: 86-95.

Coulon, C. and V. Alexander (1972) A sliding-chamber phytoplankton settling technique for making permanent quantitative slides with applications in fluorescent microscopy and autoradiography. *Limnology and Oceanography,* **17**: 149-152.

Coyne, P.I. and J.J. Kelley (1971) Release of carbon dioxide from frozen soil to the arctic atmosphere. *Nature,* **234**: 407-408.

Coyne, P.I. and J.J. Kelley (1974) Carbon dioxide partial pressures in arctic surface waters. *Limnology and Oceanography,* **19**: 928-938.

Coyne, P.I. and J.J. Kelley (1975) CO_2 exchange over the Alaskan arctic tundra: Meteorological assessment by an aerodynamic method. *Journal of Applied Ecology,* **12**: 587-611.

Crawford, C.C., J.E. Hobbie and K.L. Webb (1974) The utilization of dissolved free amino acids by estuarine microorganisms. *Ecology,* **55**: 551-563.

Cummins, K.W., R.R. Costa, R.E. Rowe, G.A. Moshiri, R.M. Scanlon and R.K. Zajdel (1969) Ecological energetics of a natural population of the predaceous zooplankton *Leptodora kindtii* (Focke) (Cladocera). *Oikos,* **20**: 189-223.

Curds, C.R. and A. Cockburn (1968) Studies on the growth and feeding of *Tetrahymena pyriformis* in axenic and monoxenic culture. *Journal of General Microbiology,* **54**: 343-358.

Daley, R.J. and J.E. Hobbie (1975) Direct counts of aquatic bacteria by a modified epifluorescent technique. *Limnology and Oceanography,* **20**: 875-882.

Danks, H.V. (1971a) Spring and early summer temperatures in a shallow arctic pond. *Arctic,* **24**: 113-123.

Danks, H.V. (1971b) Overwintering of some north temperate and arctic Chironomidae. II. Chironomid biology. *Canadian Entomologist,* **103**: 1875-1910.

Danks, H.V. and D.R. Oliver (1972) Seasonal emergence of some high arctic Chironomidae (Diptera). *Canadian Entomologist,* **104**: 661-686.

Davis, R.B. (1974) Tubificids alter profiles of redox potential and pH in profundal lake sediment. *Limnology and Oceanography,* **19**: 342-346.

de March, L. (1975) Nutrient budgets for a high arctic lake (Char Lake,

N.W.T.). *Verhandlungen Internationale Vereinigung für Theoretische und Angewandte Limnologie,* **19**: 496-503.

de March, L. (1978) Permanent sedimentation of nitrogen, phosphorus, and organic carbon in a high arctic lake. *Journal of the Fisheries Research Board of Canada,* **35**: 1089-1094.

Dennis, J.G. and P.L. Johnson (1970) Shoot and rhizome-root standing crops of tundra vegetation at Barrow, Alaska. *Arctic and Alpine Research,* **2**: 253-266.

Dennis, J.G., L.L. Tieszen and M.A. Vetter (1978) Seasonal dynamics of above- and belowground production of vascular plants at Barrow, Alaska. In *Vegetation and Production Ecology of an Alaskan Arctic Tundra* (L.L. Tieszen, Ed.). New York: Springer-Verlag, Inc., pp. 113-140.

Dillon, R.D. and J.T. Hobbs (1973) Estimating quantity and quality of the biomass of benthic protozoa. *Proceedings of the South Dakota Academy of Science,* **52**: 47-58.

Dingman, S.L. (1971) Hydrology of the Glenn Creek watershed, Tanana River basin, central Alaska. *U.S. Army CRREL Research Report* 297, 110 pp.

Dingman, S.L. (1973) The water balance in arctic and subarctic regions: Annotated bibliography and preliminary assessment. *U.S. Army CRREL Special Report* 187, 131 pp.

Dingman, S.L., R.G. Barry, G. Weller, C. Benson, E. LeDrew and C. Goodwin (In press) Climate, snow cover, microclimate and hydrology. In *An Arctic Ecosystem: The Coastal Tundra at Barrow, Alaska* (J. Brown, F.L. Bunnell, P.L. Miller and L.L. Tieszen, Eds.). Stroudsburg, Pa.: Dowden, Hutchinson and Ross, Chap. 2.

Dodson, S.I. (1972) Mortality in a population of *Daphnia rosea. Ecology,* **53**: 1011-1023.

Dodson, S.I. (1974) Zooplankton competition and predation: an experimental test of the size-efficiency hypothesis. *Ecology,* **55**: 605-613.

Dodson, S.I. (1975) Predation rates of zooplankton in arctic ponds. *Limnology and Oceanography,* **20**: 426-433.

Doty, M.S. (1959) Phytoplankton photosynthetic periodicity as a function of latitude. *Journal of the Marine Biological Association of India,* **1**: 66-68.

Downes, M.T. and H.W. Paerl (1978) Separation of two dissolved reactive phosphorus fractions in lakewater. *Journal of the Fisheries Research Board of Canada,* **35**: 1636-1639.

Dugdale, R.C. (1967) Nutrient limitation in the sea: dynamics, identification and significance. *Limnology and Oceanography,* **12**: 685-695.

Dugdale, R.C. and J.J. Goering (1967) Uptake of new and regenerated forms of nitrogen in primary productivity. *Limnology and Oceanography,* **12**: 196-206.

Dugdale, V.A. and R.C. Dugdale (1965) Nitrogen metabolism in lakes. III. Tracer studies of the assimilation of inorganic nitrogen sources. *Limnology and Oceanography,* **10**: 53-57.

Edmondson, W.T. (1955) The seasonal life history of *Daphnia* in an arctic lake. *Ecology,* **36**: 439-455.

Edwards, R.W. (1958) The effect of larvae of *Chironomus riparius* Meigen on the redox potentials of settled activated sludge. *Annals of Applied Biology,* **46**: 457-464.

Edwards, R.W. and H.L.J. Rolley (1965) Oxygen consumption of river muds. *Journal of Ecology,* **53**: 1-19.

Ekman, S. (1957) Die Gewässer des Abisko-Gebietes und ihre Bedingungen. *Kungliga Svenska Vetenskapsakademiens Handlingar* (Fjärde Serien), **6**(6): 3-68.

Elgmork, K. (1959) Seasonal occurrence of *Cyclops strenuus* in relation to environment in small water bodies in Southern Norway. *Folia Limnologica Scandinavica,* **11**: 1-196.

Eppley, R.W., J.N. Rogers and J.J. McCarthy (1969) Half-saturation constants for uptake of nitrate and ammonia by marine phytoplankton. *Limnology and Oceanography,* **14**: 912-920.

Eppley, R.W. and W.H. Thomas (1969) Comparison of half-saturation constants for growth and nitrate uptake of marine phytoplankton. *Journal of Phycology,* **5**: 375-379.

Federle, T.W., J.R. Vestal, G.R. Hater and M.C. Miller (1979) The effect of Prudhoe Bay crude oil on primary production and zooplankton in arctic tundra thaw ponds. *Marine Environmental Research,* **2**: 3-18.

Fenchel, T. (1967) The ecology of marine microbenthos. I. The quantitative importance of ciliates as compared with metazoans in various types of sediments. *Ophelia,* **4**: 121-137.

Fenchel, T. (1968) The ecology of marine microbenthos. III. The reproductive potential of ciliates. *Ophelia,* **5**: 123-136.

Fenchel, T. (1969) The ecology of marine microbenthos. IV. Structure and function of the benthic ecosystem, its chemical and physical factors and the microfauna communities with special reference to the ciliate protozoa. *Ophelia,* **2**: 1-182.

Fenchel, T. (1970) Studies on the decomposition of organic detritus derived from the turtle grass *Thalassia testudinum. Limnology and Oceanography,* **15**: 14-20.

Fenchel, T. (1975) The quantitative importance of the benthic microfauna of an arctic tundra pond. *Hydrobiologia,* **46**: 445-464.

Fenchel, T. (1977) The significance of bacterivorous protozoa in the microbial community of detrital particles. In *Aquatic Microbial Communities* (J. Cairns, Jr., Ed.). New York: Garland, pp. 529-544.

Fenchel, T. and C.C. Lee (1972) Studies on ciliates associated with sea ice from Antarctica. I: The nature of the fauna. II: Temperature responses and tolerances in ciliates from antarctic, temperate and tropical habitats. *Archiv für Protistenkunde,* **114**: 231-236 and 237-244.

Fenchel, T. and B.J. Straarup (1971) Vertical distribution of photosynthetic pigments and the penetration of light in marine sediments. *Oikos,* **22**: 172-182.

Ferguson, R.L. and P. Rublee (1976) Contribution of bacteria to standing crop of coastal plankton. *Limnology and Oceanography,* **21:** 141-145.

Fife, C.V. (1959) An evaluation of ammonium fluoride as a selective extractant for aluminum-bound soil phosphate. II. Preliminary studies on soils. *Soil Science,* **87**: 83-88.

Fischer, J. (1969) Zur Forlpflanzungs-biologie von *Chironomus nuditarsis* Str. *Revue Suisse de Zoologie,* **76**: 23-55.

Fisher, D.W., A.W. Gambell, G.E. Likens and F.E. Bormann (1968) Atmospheric contributions to water quality of streams in the Hubbard Brook Experimental Forest, New Hampshire. *Water Resources Research,* **4**: 1115-1126.

Fitzgerald, G.P. and T.C. Nelson (1966) Extractive and enzymatic analyses for limiting or surplus phosphorus in algae. *Journal of Physiology,* **2**: 32-37.

Flanagan, P.W. and F.L. Bunnell (In press) Microflora activities and decomposition. In *An Arctic Ecosystem: The Coastal Tundra at Barrow, Alaska* (J. Brown, F.L. Bunnell, P.L. Miller and L.L. Tieszen, Eds.). Stroudsburg, Pa.: Dowden, Hutchinson and Ross, Chap. 9.

Fogg, G.E. (1965) *Algal Cultures and Phytoplankton Ecology.* Madison: University of Wisconsin Press, 126 pp.

Francisco, D. (1970) Glucose and acetate utilization by the natural microbial community in a stratified reservoir. Ph.D. dissertation, University of North Carolina, Chapel Hill, 83 pp.

Francisco, D.E., R.A. Mah and A.C. Rabin (1973) Acridine orange-epifluorescence technique for counting bacteria in natural waters. *Transactions of the American Microscopical Society,* **92**: 416-421.

Frink, C.R. (1969) Chemical and mineralogical characteristics of eutrophic lake sediments. *Soil Science Society of America, Proceedings,* **33**: 369-372.

Fuhs, G.W., S.D. Demmerle, E. Canelli and M. Chen (1972) Characterization of phosphorus-limited plankton algae. In *Nutrients and Eutrophication: The Limiting-Nutrient Controversy* (G.E. Likens, Ed.). Special Symposia, Vol. 1. Lawrence, Kansas: American Society of Limnology and Oceanography and Allen Press, pp. 113-133.

Gardner, M.R. and W.R. Ashby (1970) Connectance of large dynamic (cybernetic) systems: Critical values for stability. *Nature,* **228**: 784.

Gardner, W.S. and G.F. Lee (1975) The role of amino acids in the nitrogen cycle of Lake Mendota. *Limnology and Oceanography,* **20**: 379-388.

Gargas, E. (1971) "Sun-shade" adaptation in microbenthic algae from the Øresund. *Ophelia,* **9**: 107-112.

George, J.D. (1964) Organic matter available to the polychaete *Cirriformia tentaculata* (Montagu) living in an intertidal mud flat. *Limnology and Oceanography,* **9**: 453-455.

Gerlach, S. and M. Shrage (1972) Life cycles at low temperatures in some free-living marine nematodes. *Veröffentlichungen des Instituts für Meeresforschung in Bremerhaven,* **14**: 5-11.

Gersper, P.L., V. Alexander and S.A. Barkley (In press) The soils and their nutrients. In *An Arctic Ecosystem: The Coastal Tundra at Barrow, Alaska* (J. Brown, F.L. Bunnell, P.L. Miller and L.L. Tieszen, Eds.). Stroudsburg, Pa.: Dowden, Hutchinson and Ross, Chap. 3.

Ghassemi, M. and R.F. Christman (1968) Properties of the yellow organic acids of natural waters. *Limnology and Oceanography,* **13**: 583-597.

Glass, G.E. and J.E. Poldoski (1975) Interstitial water components and exchange across the water sediment interface of western Lake Superior. *Verhandlungen Internationale Vereinigung für Theoretische und Angewandte Limnologie,* **19**: 405-420.

Goering, J.J. and V.A. Dugdale (1966) Estimate of rates of denitrification in a subarctic lake. *Limnology and Oceanography,* **11**: 113-117.

Goldberg, E.D. (1965) Minor elements in sea water. In *Chemical Oceanography* (J.P. Riley and G. Skirrow, Eds.). Vol. 1. New York: Academic Press, pp. 163-196.

Goldman, C.R. (1963) The measurement of primary productivity and limiting factors in freshwater with ^{14}C. In *Proceedings of the Conference on Primary Productivity Measurement, Marine and Freshwater* (M.S. Doty, Ed.). U.S. Atomic Energy Commission, TID-7633, pp. 103-113.

Goldman, C.R., D.T. Mason and B.J.B. Wood (1963) Light injury and inhibition in antarctic freshwater phytoplankton. *Limnology and Oceanography,* **8**: 313-322.

Golterman, H.L. (Ed.) (1969) *Methods for Chemical Analysis of Fresh Waters.* International Biological Programme Handbook 8. Oxford: Blackwell, 172 pp.

Green, J. (1954) Size and reproduction in *Daphnia magna* (Crustacea: Cladocera). *Zoological Society of London Proceedings,* **124**: 535-545.

Green, J. (1956) Growth, size and reproduction in *Daphnia* (Crustacea: Cladocera). *Zoological Society of London Proceedings,* **126**: 173-204.

Grontved, J. (1962) Preliminary report on the productivity of microbenthos and phytoplankton in the Danish Wadden Sea. *Meddelelser fra Danmarks Fiskeri-og Havundersogelser, Ny Serie,* **3**: 347-378.

Gruendling, G.K. (1971) Ecology of the epipelic algal communities in Marion Lake, British Columbia. *Journal of Phycology,* **7**: 239-249.

Hall, D. (1964) An experimental approach to the dynamics of a natural population of *Daphnia galeata mendotae. Ecology,* **45**: 94-112.

Hall, K.J. (1972) Decomposition of leaves in sediment microcosms. In *Marion Lake Project Report, International Biological Program, Canada* (I.E. Efford, Ed.). Vancouver: Institute of Resource Ecology, University of British Columbia, pp. 50-52.

Hall, K.J., P.M. Kleiber and J. Yesaki (1972) Heterotrophic uptake of organic solutes by microorganisms in the sediment. *Memorie dell' Instituto Italiano di Idrobiologia,* **29**(Suppl.): 441-471.

Hanna, B.M., J.A. Hellebust and T.C. Hutchinson (1975) Field studies on the phytotoxicity of crude oil to subarctic aquatic vegetation. *Verhandlungen Internationale Vereinigung für Theoretische and Angewandte Limnologie,* **19**: 2165-2171.

Hansen, K. (1967) The general limnology of arctic lakes as illustrated by examples from Greenland. *Meddelelser om Grønland,* **178**: 1-79.

Harding, S.T. (1942) Evaporation from free water surfaces. In *Hydrology* (O.E. Meinzer, Ed.). New York: McGraw-Hill, pp. 56-82.

Hargrave, B.T. (1969) Epibenthic algal production and community respiration in the sediments of Marion Lake. *Journal of the Fisheries Research Board of Canada,* **26**: 2003-2026.

Hargrave, B.T. (1970) The effect of a deposit-feeding amphipod on the metabolism of benthic microflora. *Limnology and Oceanography,* **15**: 21-30.

Hargrave, B.T. (1971) An energy budget for a deposit-feeding amphipod. *Limnology and Oceanography,* **16**: 99-103.

Hargrave, B.T. (1972) Aerobic decomposition of sediment and detritus as a function of particle surface area and organic matter. *Limnology and Oceanography,* **17**: 583-596.

Hargrave, B.T. (1973) Coupling carbon flow through some pelagic and benthic communities. *Journal of the Fisheries Research Board of Canada,* **30**: 1317-1326.

Harner, R.F. and K.T. Harper (1973) Mineral composition of grassland species of the Eastern Great Basin in relation to stand productivity. *Canadian Journal of Botany,* **51**: 2037-2046.

Hart, L.T., A.D. Larson and C.S. McClesky (1965) Denitrification by *Corynebacterium nephridii. Journal of Bacteriology,* **89**: 1104-1108.

Harter, R.D. (1968) Adsorption of phosphorus by lake sediment. *Soil Science Society of America Proceedings,* **32**: 514-518.

Hauck, R.D., S.W. Melsted and P. Yankwich (1958) Use of N-isotope distribution in nitrogen gas in the study of denitrification. *Soil Science,* **86**: 287-291.

Hayes, F.R., J.A. McCarter, M.L. Cameron and D.A. Livingstone (1952) On the kinetics of phosphorus exchange in lakes. *Journal of Ecology,* **40**: 202-216.

Hicks, S.E. and F.G. Carey (1968) Glucose determination in natural waters. *Limnology and Oceanography,* **13**: 361-363.

Hilliard, D.K. and J.C. Tash (1966) Freshwater algae and zooplankton. In *Environment of Cape Thompson Region, Alaska* (N.J. Wilimovsky and J.N. Wolfe, Eds.). Oak Ridge, Tenn.: U.S. Atomic Energy Commission, pp. 363-413.

Hobbie, J.E. (1959) Limnological studies on Lakes Peters and Schrader, Alaska. M.S. thesis, University of California, Berkeley, 62 pp.

Hobbie, J.E. (1961) Summer temperatures in Lake Schrader, Alaska. *Limnology and Oceanography,* **6**: 326-329.

Hobbie, J.E. (1962) Limnological cycles and primary productivity of two lakes in the Alaskan Arctic, Ph.D. dissertation, Indiana University, 124 pp.

Hobbie, J.E. (1964) Carbon-14 measurements of primary production in two Alaskan lakes. *Verhandlungen Internationale Vereinigung für Theoretische und Angewandte Limnologie,* **15**: 360-364.

Hobbie, J.E. (1967) Glucose and acetate in freshwater: concentrations and turnover rates. In *Chemical Environment in the Aquatic Habitat* (H.L. Golterman and R.S. Clymo, Eds.). Amsterdam: Nord Hollandsche Uitgevers Maatschappij, pp. 245-251.

Hobbie, J.E. (1971) Heterotrophic bacteria in aquatic ecosystems: some results of studies with organic radioisotopes. In *The Structure and Function of Freshwater Microbial Communities* (J. Cairns, Jr., Ed.). Research Division Monograph 3. Blacksburg, Va.: Virginia Polytechnic Institute and State University, pp. 181-194.

Hobbie, J.E. (1973) Arctic limnology: a review. In *Alaskan Arctic Tundra* (M.E. Britton, Ed.). Arctic Institute of North America Technical Paper 25, pp. 127-168.

Hobbie, J.E. and C.C. Crawford (1969) Respiration corrections for bacterial uptake of dissolved organic compounds in natural waters. *Limnology and Oceanography,* **14**: 528-532.

Hobbie, J.E. and C.L. Lee (In press) Microbial production of extracellular material; importance in benthic ecology. In *Marine Benthic Dynamics* (K.R. Tenore and B.C. Coull, Eds.). Columbia, S.C.: University of South Carolina Press.

Hobbie, J.E. and G.E. Likens (1973) Output of phosphorus, dissolved organic carbon, and fine particulate carbon from Hubbard Brook watersheds. *Limnology and Oceanography,* **18**: 734-742.

Hobbie, J.E. and P. Rublee (1975) Bacterial production in an arctic pond. *Verhandlungen Internationale Vereinigung für Theoretische und Angewandte Limnologie,* **19**: 466-571.

Hobbie, J.E., R.J. Daley and S. Jasper (1977) Use of Nuclepore filters for counting bacteria by fluorescence microscopy. *Applied and Environmental Microbiology,* **33**: 1225-1228.

Holdren, G.C., Jr., D.E. Armstrong and R.F. Harris (1977) Interstitial inorganic phosphorus concentrations in Lakes Mendota and Wingra. *Water Research,* **11**: 1041-1047.

Holmes, R.T. and F.A. Pitelka (1968) Food overlap among coexisting sandpipers on northern Alaskan tundra. *Systematic Zoology,* **17**(3): 305-318.

Holmgren, S. (1968) Phytoplankton production in a lake north of the Arctic Circle. Fil. Lic. thesis, Uppsala University, 145 pp.

Holm-Hansen, O. (1972) The distribution and chemical composition of particulate material in marine and fresh water. *Memorie dell'Instituto Italiano di Idrobiologia,* **29**(Suppl.): 37-51.

Holmquist, C. (1959) *Problems on Marine-Glacial Relicts.* Lund: Berlingska Boktryckeriet, 270 pp.

Holmquist, C. (1973) Some arctic limnology and the hibernation of invertebrates and some fishes in sub-zero temperatures. *Archiv für Hydrobiologie,* **72**: 49-70.

Horowitz, A., A. Sexstone and R. Atlas (1978) Hydrocarbons and microbial activities in sediments of an arctic lake one year after contamination with leaded gasoline. *Arctic,* **31**: 180-191.

Howard, H.H. and G.W. Prescott (1973) Seasonal variation of chemical parameters in Alaskan tundra lakes. *American Midland Naturalist,* **90**: 154-164.

Hrbáček, T. (1962) Species composition and the amount of the zooplankton in relation to the fish stock. *Proceedings Czechoslovakian Academy of Sciences,* **72**(10): 1-116.

Hunding, C. (1971) Production of benthic microalgae in the littoral zone of a eutrophic lake. *Oikos,* **22**: 389-397.

Hutchinson, G.E. (1957) *A Treatise on Limnology.* Vol. 1. New York: Wiley, 1015 pp.

Hutchinson, G.E. and A. Wollack (1940) Studies on Connecticut lake sediments. II. Chemical analyses of a core from Linsley Pond, North Branford. *American Journal of Science,* **238**: 493-517.

Jackson, M.L. (1958) *Soil Chemical Analysis.* Englewood Cliffs, N.J.: Prentice-Hall, 498 pp.

Jassby, A.D. (1973) The ecology of bacteria in the hypolimnion of Castle Lake, California. Ph.D. dissertation, University of California, Davis, 198 pp.

Jobson, A., F.D. Cook and D.W.S. Westlake (1972) Microbial utilization of crude oil. *Applied Microbiology,* **23**: 1082-1089.

Johannes, R.E. (1964) Phosphorus excretion and body size in marine animals: microzooplankton and nutrient regeneration. *Science,* **146**: 923-924.

Johannes, R.E. (1965) Influence of marine protozoa on nutrient regeneration. *Limnology and Oceanography,* **10**: 434-442.

John, M.K. (1970) Colorimetric determination of phosphorus in soil and plant materials with ascorbic acid. *Soil Science,* **109**: 214-220.

Johnson, M.G. and R.O. Brinkhurst (1971) Production of benthic macroinvertebrates of Bay of Quinte and Lake Ontario. *Journal of the Fisheries Research Board of Canada,* **28**: 1699-1714.

Johnson, P.L. and L.L. Tieszen (1973) Vegetative research in arctic Alaska. In *Alaskan Arctic Tundra* (M.E. Britton, Ed.). Arctic Institute of North America Technical Paper 25, pp. 169-198.

Jonasson, P.M. (1972) Ecology and production of the profundal benthos in relation to phytoplankton in Lake Esrom. *Oikos* (Suppl.), **14**: 1-148.

Jones, R. (1975) Comparative studies of plant growth and distribution in relation to water logging. VIII. The uptake of phosphorus by dune and dune slack plants. *Journal of Ecology,* **63**: 109-116.

Jordan, M.J. and G.E. Likens (1975) An organic carbon budget for an oligotrophic lake in New Hampshire, U.S.A. *Verhandlungen Internationale Vereinigung für Theoretische und Angewandte Limnologie,* **19**: 994-1003.

Jordan, M.J., J.E. Hobbie and B.J. Peterson (1978) Effect of petroleum hydrocarbons on microbial populations in an arctic lake. *Arctic,* **31**: 170-179.

Juday, C. (1942) The summer standing crop of plants and animals in four Wisconsin lakes. *Transactions of the Wisconsin Academy of Sciences,* **34**: 103-135.

Junge, C.E. (1958) The distribution of ammonia and nitrate in rain water over the United States. *Transactions, American Geophysical Union,* **39**: 241-248.

Kajak, Z. and K. Dusoge (1970) Production efficiency of *Procladius choreus* MG (Chironomidae, Diptera) and its dependence on the trophic conditions. *Polskie Archiwum Hydrobiologii,* **17**: 217-224.

Kajak, Z. and J. Warda (1968) Feeding of benthic non-predatory *Chironomidae* in lakes. *Annales Zoologici Fennici,* **5**: 57-64.

Kalff, J. (1965) Primary production rates and the effect of some environmental factors on algal photosynthesis in small arctic tundra ponds. Ph.D. dissertation, Indiana University, 122 pp.

Kalff, J. (1967a) Phytoplankton abundance and primary production rates in two arctic ponds. *Ecology,* **48**: 558-565.

Kalff, J. (1967b) Phytoplankton dynamics in an arctic lake. *Journal of the Fisheries Research Board of Canada,* **24**: 1861-1871.

Kalff, J. (1968) Some physical and chemical characteristics of arctic fresh waters in Alaska and northwestern Canada. *Journal of the Fisheries Research Board of Canada,* **25**: 2575-2587.

Kalff, J. (1969) A diel periodicity in the optimum light intensity for maximum photosynthesis in natural phytoplankton populations. - *Journal of the Fisheries Research Board of Canada,* **26**: 463-468.

Kalff, J. (1970) Arctic lake ecosystems. In *Antarctic Ecology* (M.W. Holdgate, Ed.). London: Academic Press, pp. 651-663.

Kalff, J. (1971) Nutrient-limiting factors in an arctic tundra pond. *Ecology,* **52**: 655-659.

Kalff, J. and H.E. Welch (1974) Phytoplankton production in Char Lake, a natural polar lake, and in Meretta Lake, a polluted polar lake, Cornwallis Island, Northwest Territories. *Journal of the Fisheries Research Board of Canada,* **31**: 621-636.

Kalff, J., H.E. Welch and S.K. Holmgren (1972) Pigment cycles in two high-arctic Canadian lakes. *Verhandlungen Internationale Vereinigung für Theoretische und Angewandte Limnologie,* **18**: 250-256.

Kallendorf, R. (1974) Population dynamics and the ecological role of *Lepidurus arcticus* in coastal tundra ponds, Barrow, Alaska. M.S. thesis, University of Cincinnati, 107 pp.

Kangas, D.A. (1972) The ecology of some arctic tundra ponds. Ph.D. dissertation, University of Missouri, 343 pp.

Kanwisher, J. (1963) On the exchange of gases between the atmosphere and the sea. *Deep-Sea Research,* **10**: 195-207.

Kauffman, S.A. (1969) Metabolic stability and epigenesis in randomly constructed genetic nets. *Journal of Theoretical Biology,* **22**: 437-467.

Kaushik, N.K. and H.B.N. Hynes (1971) The fate of the dead leaves that fall into streams. *Archiv für Hydrobiologie,* **68**: 465-515.

Kelley, J.J. (1973) Micrometeorological investigations near the tundra surface. In *Alaskan Arctic Tundra* (M.E. Britton, Ed.). Arctic Institute of North America Technical Paper 25, pp. 109-126.

Kennedy, W.A. (1953) Growth, maturity, fecundity and mortality in the relatively unexploited whitefish, *Coregonus clupeaformis*, of Great Slave Lake. *Journal of the Fisheries Research Board of Canada,* **10**: 413-441.

Kibby, H.V. (1971) Effect of temperature on the feeding behavior of *Daphnia rosea. Limnology and Oceanography,* **16**: 580-581.

Kilham, P. (1971) A hypothesis concerning silica and freshwater planktonic diatoms. *Limnology and Oceanography,* **16**: 10-18.

Kilham, S.S. (1975) The kinetics of silicon limited growth in the freshwater diatom *Asterionella formosa. Journal of Phycology,* **11**: 396-399.

Kinney, P.J., D.M. Schell, V. Alexander, D.C. Burrell, R. Cooney and A.S. Naidu (1972) Baseline data study of the Alaskan arctic aquatic environment. *University of Alaska, Institute of Marine Science Report* R-72-3, 275 pp.

Klekowski, R.Z. (1970) Bioenergetic budgets and their application for estimation of production efficiency. *Polskie Archiwum Hydrobiologii,* **17**: 55-80.

Klir, G.J. (1969) *An Approach to General Systems Theory.* New York: Van Nostrand Reinhold, 323 pp.

Knight, A.W. and A.R. Gaufin (1966) Oxygen consumption of several species of stoneflies (Plecoptera). *Journal of Insect Physiology,* **12**: 347-355.

Knoechel, R. and J. Kalff (1976a) The applicability of grain density autoradiography to the quantitative determination of algal species production: A critique. *Limnology and Oceanography,* **21**(4): 583-590.

Knoechel, R. and J. Kalff (1976b) Track autoradiography: A method for the determination of phytoplankton species productivity. *Limnology and Oceanography,* **21**(4): 590-596.

Konstantinov, A.S. (1971) Ecological factors affecting respiration in Chironomid larvae. *Limnologica,* **8**: 127-134.

Korelyakova, I.L. (1968) The role of higher aquatic plants in the formation of organic matter in the shallow-water areas of Kiev Reservoir. *Second Conference on Problems of Matter and Energy Cycles in Lakes, Part II.* Limnological Institute, Siberian Division, U.S.S.R. Academy of Sciences, pp. 152-156.

Kudo, R.R. (1966) *Protozoology.* 5th edition. Springfield, Ill.: Thomas, 1174 pp.

Kuenzler, E.J. (1970) Dissolved organic phosphorus excretion by marine phytoplankton. *Journal of Phycology,* **6**: 7-13.

Kuzin, P.S. (1960) *Klassifikatsia Rek i Gidrologicheskoe Raionirvanie S.S.S.R. [Classification of rivers and hydrological regionalization of the U.S.S.R.]* Leningrad: Gidro-Meteorologicheskoe Izdat, 455 pp.

Lachenbruch, A.H. (1962) Mechanics of thermal contraction cracks and ice-wedge polygons in permafrost. Geological Society of America Special Paper No. 70, 69 pp.

Ladle, M. (1974) Aquatic crustacea. In *Biology of Plant Litter Decomposition* (C.H. Dickinson and G.J.F. Pugh, Eds.). Vol. 2. New York: Academic Press, pp. 593-608.

Lamar, W.L. (1966) Chemical character and sedimentation of the waters. In *Environment of the Cape Thompson Region, Alaska* (N.J. Wilimovsky and J.N. Wolfe, Eds.). Oak Ridge, Tenn.: U.S. Atomic Energy Commission, pp. 133-148.

Larow, E.J. and D.C. McNaught (1978) Systems and organismal aspects of phosphorus remineralization. *Hydrobiologia,* **59**: 151-154.

Lasenby, D.C. and R.R. Langford (1972) Growth, life history and respiration of *Mysis relicta* in an arctic and temperate lake. *Journal of the Fisheries Research Board of Canada,* **29**: 1701-1708.

Lean, D.R.S. (1973a) Phosphorus dynamics in lake water. *Science,* **179**: 678-679.

Lean, D.R.S. (1973b) Movements of phosphorus between its biologically important forms in lake water. *Journal of the Fisheries Research Board of Canada,* **30**: 1525-1536.

Lean, D.R.S. (1976) Phosphorus kinetics in lake water: influence of membrane filter pore size and low pressure filtration. *Journal of the Fisheries Research Board of Canada,* **33**: 2800-2804.

Lees, A.D. (1960) The role of photoperiod and temperature in the determination of parthenogenetic and sexual forms in the aphid *Megoura viciae* Buckton. II. The operation of the "interval timer" in young clones. *Journal of Insect Physiology,* **4**: 154-175.

Lees, A.D. (1965) Is there a circadian component in the megoura photoperiodic clock? In *Circadian Clocks* (J. Aschoff, Ed.). Amsterdam: North Holland Publishing Company, pp. 351-356.

Lehman, J.T., D.B. Botkin and G.E. Likens (1975) The assumptions and rationales of a computer model of phytoplankton population dynamics. *Limnology and Oceanography,* **20**: 343-364.

Lewellen, R.I. (1972) *Studies on the fluvial environment, arctic coastal plain province, northern Alaska.* Vol. 1. Published by the author, P.O. Box 1068, Littleton, Colorado 80120, 282 pp.

Li, W.C., D.E. Armstrong, J.D.H. Williams, R.F. Harris and J.K. Syers (1972) Rate and extent of inorganic phosphate exchange in lake sediments. *Soil Science Society of America, Proceedings,* **36**: 279-284.

Likens, G.E. and O.L. Loucks (1978) Analysis of five North American lake ecosystems. III. Sources, loading and fate of nitrogen and phosphorus. *Verhandlungen Internationale Vereinigung für Theoretische und Angewandte Limnologie,* **20**: 568-573.

Likes, E.H. (1966) Surface-water discharge of Ogotoruk Creek. In *Environment of the Cape Thompson Region, Alaska* (N.J. Wilimovsky and J.N. Wolfe, Ed.). Oak Ridge, Tenn.: U.S. Atomic Energy Commission, pp. 125-131.

Lindegaard, C. and P.M. Jonasson (1975) Life cycles of *Chironomus hyperboreus* Staeger and *Tanytarsus gracilentus* (Holmgren) (Chironomidae, Diptera) in Lake Mývatn, Northern Iceland. *Verhandlungen Internationale Vereinigung für Theoretische und Angewandte Limnologie,* **19**: 3155-3163.

Livingstone, D.A., K. Bryan, Jr. and R.C. Leahy (1958) Effects of an arctic environment on the origin and development of freshwater lakes. *Limnology and Oceanography,* **3**: 192-214.

Livingstone, D.A. (1963a) Alaska, Yukon, Northwest Territories, and Greenland. In *Limnology in North America* (D.G. Frey, Ed.). Madison: University of Wisconsin Press, pp. 559-574.

Livingstone, D.A. (1963b) Chemical composition of rivers and lakes. U.S. Geological Survey Professional Paper 440G, 64 pp.

Loden, M.S. (1974) Predation by chironomid (Diptera) larvae on oligochaetes. *Limnology and Oceanography,* **19**: 156-159.

Longhurst, A.R. (1955) The reproduction and cytology of the Notostraca (Crustacea, Phyllopoda). *Proceedings of the Zoological Society of London,* **125**: 671-680.

Lorenzen, C.J. (1967) Determination of chlorophyll and phaeopigments: spectrophotometric equations. *Limnology and Oceanography,* **12**: 343-346.

Loughman, B.C. (1968) The uptake of phosphate and its transport within the plant. In *Ecological Aspects of the Mineral Nutrition of Plants* (I.H. Rorison, Ed.). Oxford: Blackwell, pp. 309-322.

Lund, J.W.G. (1964) Primary production and periodicity of phytoplankton. *Verhandlungen Internationale Vereinigung für Theoretische und Angewandte Limnologie,* **15**: 37-56.

MacArthur, R.H. (1960) On the relative abundance of species. *American Naturalist,* **94**: 25-36.

Machan, R. and J. Ott (1972) Problems and methods of continuous *in situ* measurements of redox potential in marine sediments. *Limnology and Oceanography,* **17**: 622-626.

MacLean, S.F., Jr. (1974) Primary production, decomposition, and the activity of soil invertebrates in tundra ecosystems: A hypothesis. In *Soil Organisms and Decomposition in Tundra: Proceedings of the Microbiology, Decomposition and Invertebrate Working Group Meeting, Fairbanks, Alaska, August 1973* (A.J. Holding, O.W. Heal, S.F. MacLean, Jr. and P.W. Flanagan, Eds.). Stockholm: International Biological Programme, Tundra Biome Steering Committee, pp. 197-206.

MacLean, S.F., Jr., B.M. Fitzgerald and F.A. Pitelka (1974) Population cycles in arctic lemmings; Winter reproduction and predation by weasels. *Arctic and Alpine Research,* **6**: 1-12.

Marshall, S.M. (1973) Respiration and feeding in copepods. *Advances in Marine Biology,* **11**: 57-120.

Martin, J.H. (1968) Phytoplankton–zooplankton relationships in Narragansett Bay. III. Seasonal changes in zooplankton excretion rates in relation to phytoplankton abundance. *Limnology and Oceanography,* **13**: 63-71.

Mather, J.R. and C.W. Thornthwaite (1958) *Microclimate Investiga-*

tions at Point Barrow, Alaska, 1957-1958. Drexel Institute of Technology, Laboratory of Climatology, *Publications in Climatology,* **XI**(2), 239 pp.

Mathias, J.A. (1971) Energy flow and secondary production of the amphipods *Hyallela azteca* and *Crangonyx richmondensis occidentalis* in Marion Lake. *Journal of the Fisheries Research Board of Canada,* **28**: 711-726.

May, R.M. (1971) Stability in multispecies community models. *Mathematical Biosciences,* **12**: 59-79.

May, R.M. (1972) Will a large complex system be stable? *Nature,* **238**, 413-414.

May, R.M. (1973) *Stability and Complexity in Model Ecosystems.* Princeton, N.J.: Princeton University Press, 235 pp.

Maynard Smith, J. (1974) *Models in Ecology.* London: Cambridge University Press, 146 pp.

McAlpine, J.F. (1965) Insects and related terrestrial invertebrates of Ellef Ringnes Island. *Arctic,* **18**: 73-103.

McCart, P. and P. Craig (1971) Meristic differences between anadromous and freshwater-resident Arctic Char (*Salvelinus alpinus*) in the Sagavanirktok River drainage, Alaska. *Journal of the Fisheries Research Board of Canada,* **28**: 115-118.

McCart, P. and P. Craig (1973) Life history of two isolated populations of Arctic Char (*Salvelinus alpinus*) in spring-fed tributaries of the Canning River, Alaska. *Journal of the Fisheries Research Board of Canada,* **30**: 1215-1220.

McCown, B.H. (1978) The interaction of organic nutrients, soil nitrogen and soil temperature on plant growth and survival in the arctic environment. In *Vegetation and Production Ecology of an Alaskan Arctic Tundra* (L.L. Tieszen, Ed.). New York: Springer-Verlag, Inc., pp. 435-456.

McLaren, I.A. (1963) Effects of temperature on growth of zooplankton and the adaptive value of vertical migration. *Journal of the Fisheries Research Board of Canada,* **20**: 685-727.

McLaren, I.A. (1964) Zooplankton of Lake Hazen, Ellesmere Island, and a nearby pond, with special reference to the copepod *Cyclops scutifer* Sars. *Canadian Journal of Zoology,* **42**: 613-629.

McLaren, I.A. (1974) Demographic strategy of vertical migration by a marine copepod. *American Naturalist,* **108**: 91-102.

McMahon, J.W. and F.H. Rigler (1965) Feeding rate of *Daphnia magna* Straus in different foods labeled with radioactive phosphorus. *Limnology and Oceanography,* **10**: 105-113.

McMurtrie, R.E. (1975) Determinants of stability of large randomly connected systems. *Journal of Theoretical Biology,* **50**: 1-11.

McRoy, C.P. and V. Alexander (1975) Nitrogen kinetics in aquatic plants in arctic Alaska. *Aquatic Botany,* **1**: 3-10.

McRoy, C.P. and R.J. Barsdate (1970) Phosphate absorption in eelgrass. *Limnology and Oceanography,* **15**: 6-13.

McRoy, C.P., R.J. Barsdate and M. Nebert (1972) Phosphorus cycling in an eelgrass (*Zostera marina* L.) ecosystem. *Limnology and Oceanography,* **17**: 58-67.

Mendl, H. and K. Müller (1970) Die Laufaktivität von *Capnia atra* Morton (Plecoptera). *Oikos* (Suppl.), **13**: 75-79.

Menzel, D.W. and N. Corwin (1965) The measurement of total phosphorus in seawater based on the liberation of organically bound fractions by persulfate oxidation. *Limnology and Oceanography,* **10**: 280-282.

Menzel, D.W. and R.F. Vaccaro (1964) The measurement of dissolved organic and particulate carbon in seawater. *Limnology and Oceanography,* **9**: 138-142.

Meyer, J.L. (1979) The role of sediments and bryophytes in phosphorus dynamics in a headwater stream ecosystem. *Limnology and Oceanography,* **24**: 365-375.

Meynell, G.G. and E. Meynell (1965) *Theory and Practice in Experimental Bacteriology.* Cambridge University Press, 287 pp.

Milbrink, G. (1973) On the vertical distribution of oligochaetes in lake sediments. Institute of Freshwater Research, Drottningholm, Report 53, pp. 34-50.

Miller, M.C. (1972) The carbon cycle in the epilimnion of two Michigan lakes. Ph.D. dissertation, Michigan State University, 233 p.

Miller, M.C., V. Alexander and R.J. Barsdate (1978a) The effects of oil spills on phytoplankton in an arctic lake and ponds. *Arctic,* **31**: 192-218.

Miller, M.C., G.R. Hater and J.R. Vestal (1978b) Effect of Prudhoe crude oil on carbon assimilation by planktonic algae in an arctic pond. In *Environmental Chemistry and Cycling Processes: Proceedings of Symposium, Augusta, Georgia, 28 April-1 May 1976* (D.C. Adriano and I.L. Brisbin, Jr., Eds.). U.S. Department of Energy, CONF-760429, pp. 833-850.

Miller, M.C. and J.P. Reed (1975) Benthic metabolism of arctic coastal ponds, Barrow, Alaska. II. Lakes. 3. North America. *Verhandlungen Internationale Vereinigung für Theoretische und Angewandte Limnologie,* **19**: 459-465.

Miller, P.C. and L.L. Tieszen (1972) A preliminary model of processes affecting primary production in the arctic tundra. *Arctic and Alpine Research,* **4**: 1-18.

Morgan, K.C. and J. Kalff (1972) Bacterial dynamics in two high-arctic lakes. *Freshwater Biology,* **2**: 217-228.

Morris, J.T. (1978) The nitrogen uptake kinetics and growth response of *Spartina alterniflora.* Ph.D. dissertation, Yale University, 93 pp.

Moshiri, G.A., K.W. Cummins and R.R. Costa (1969) Respiratory energy expenditure by the predaceous zooplankter *Leptodora kindtii* (Focke) (Crustacea: Cladocera). *Limnology and Oceanography,* **14**: 475-484.

Moss, B. (1968) The chlorophyll *a* content of some benthic algal communities. *Archiv für Hydrobiologie,* **65**: 51-62.

Mozley, S.C. (1978) Effects of experimental oil spills on Chironomidae in Alaskan tundra ponds. *Verhandlungen Internationale Vereinigung für Theoretische und Angewandte Limnologie,* **20**, 1941-1945.

Mozley, S.C. and M.C. Butler (1978) Effects of crude oil on aquatic insects of tundra ponds. *Arctic,* **31**: 229-241.

Mugard, H. (1949) Contribution a l'étude des infusoires hymenostomes histiophages. *Annales des Sciences Naturelles,* (Ser. 11), **10**: 171-268.

Müller, K. (1970) Tages-und Jahresperiodik der Drift in Fliessgewassern in verschiedenen geographischen Breiten. *Oikos* (Suppl.), **13**: 21-44.

Müller-Haeckel, A. (1970) Wechsel der Aktivitätsphase bei einzelligen Algen fliessender Gewässer. *Oikos* (Suppl.), **13**: 134-138.

Müus, B. (1967) The fauna of Danish estuaries and lagoons. *Meddelelser fra Danmarks Fisheri-og Havundersogelser, N.S.,* **5**: 1-315.

Myers, J. (1970) Genetic and adaptive physiological characteristics observed in the Chlorellas. In *Prediction and Measurement of Photosynthetic Productivity (Proceedings Technical Meeting, Trebon).* Wageningen: Centre for Agricultural Publishing, pp. 447-454.

Myers, J. and J. Graham (1971) The photosynthetic unit of *Chlorella* measured by repetitive short flashes. *Plant Physiology,* **48**: 282-286.

Naumann, E. (1921) Spezielle Untersuchungen über die Ernährungsbiologie des tierischen Limnoplanktons. I. (Cladoceren). *Lunds Universitets Årsskrift (R.E.),* **17**: 3-26.

Nauwerck, A. (1959) Zur Bestimmung der Filterierrate limnischer Planktontiere. *Archiv für Hydrobiologie* (Suppl.), **25**: 83-101.

Nauwerck, A. (1963) Die Beziehungen zwischen Zooplankton und Phytoplankton im See Erken. *Symbolae Botanicae Upsalienses,* **17**(5): 1-163.

Nauwerck, A. (1968) Das phytoplankton des Latnajajaure, 1954–1965. *Schweizerische Zeitschrift für Hydrobiologie,* **30**: 188-216.

Newell, R. (1965) The role of detritus in the nutrition of two marine deposit feeders, the prosobranch *Hydrobia ulvae* and the bivalve *Macoma balthica. Proceedings of the Zoological Society of London,* **144**: 25-45.

Noland, L.E. and M. Gojdics (1967) Ecology of free-living protozoa. In *Research in Protozoology* (T.-T. Chen, Ed.). New York: Pergamon Press, pp. 215-266.

O'Brien, W.J. (1972) Limiting factors in phytoplankton algae: their meaning and measurement. *Science,* **178**: 616-617.

O'Brien, W.J. (1974) The dynamics of nutrient limitation of phytoplankton algae: a model reconsidered. *Ecology,* **55**: 135-141.

O'Brien, W.J. (1978) Toxicity of Prudhoe Bay crude oil to Alaskan arctic zooplankton. *Arctic,* **31**: 219-228.

Oliver, D.R. (1964) A limnological investigation of a large arctic lake, Nettilling Lake, Baffin Island. *Arctic,* **17**: 69-83.

Oliver, D.R. (1968) Adaptations of arctic Chironomidae. *Annales Zoologici Fennici,* **5**(1): 111-118.

Olofsson, O. (1918) Studien über die Süsswasserfauna Spitzbergens. *Zoologiska Bidrag fran Uppsala,* **6**: 183-646.

Otsuki, A. and T. Hanya (1972a) Production of dissolved organic matter from dead green algal cells. I. Aerobic microbial decomposition. *Limnology and Oceanography,* **17**: 248-257.

Otsuki, A. and T. Hanya (1972b) Production of dissolved organic matter from dead green algae cells. II. Anaerobic microbial decomposition. *Limnology and Oceanography,* **17**: 258-264.

Pamatmat, M.M. (1965) A continuous-flow apparatus for measuring metabolism of benthic communities. *Limnology and Oceanography,* **10**: 486-489.

Park, P.K., L.I. Gordon, S.W. Hager and M.C. Cissell (1969) Carbon dioxide partial pressure in the Columbia River. *Science,* **166**: 867-868.

Parsons, T.R. and J.D.H. Strickland (1962) On the production of particulate organic carbon by heterotrophic processes in sea water. *Deep-Sea Research,* **8**: 211-222.

Peters, R. and D.R.S. Lean (1973) The characterization of soluble phosphorus released by limnetic zooplankton. *Limnology and Oceanography,* **18**: 270-279.

Peterson, B.J., J.E. Hobbie and J.F. Haney (1978) *Daphnia* grazing on natural bacteria. *Limnology and Oceanography,* **23**: 1039-1044.

Picken, L.E.R. (1937) The structure of some protozoan communities. *Journal of Ecology,* **25**: 368-384.

Pitelka, F.A. (1973) Cyclic pattern in lemming populations near Barrow, Alaska. In *Alaskan Arctic Tundra* (M.E. Britton, Ed.). Arctic Institute of North America Technical Paper 25, pp. 199-215.

Pitelka, F.A., P.Q. Tomich and G.W. Treichel (1955) Ecological relations of jaegers and owls as lemming predators near Barrow, Alaska. *Ecological Monographs,* **25**: 85-117.

Platzer, I. (1967) Temperature adaptations in tropical chironomids. *Zeitschrift für Vergleichende Physiologie,* **54**: 58-74.

Pomeroy, L.R. and E.J. Kuenzler (1969) Phosphorus turnover by coral reef animals. In *Proceedings of Symposium on Radioecology (Se-*

cond National), Ann Arbor, Michigan, 1967. CONF-670503, pp. 474-482.

Pomeroy, L.R., L.R. Shenton, R.D.H. Jones and R.J. Reimold (1972) Nutrient flux in estuaries. In *Nutrients and Eutrophication: The Limiting-Nutrient Controversy* (G.E. Likens, Ed.). Special Symposia, Vol. 1. American Society of Limnology and Oceanography, pp. 274-293.

Porter, K.G. (1973) Selective grazing and differential digestion of algae by zooplankton. *Nature,* **244**: 179-180.

Poulsen, E.M. (1940) Biological remarks on *Lepidurus arcticus* Pallas and *Daphnia pulex* De Geer and *Chydorus sphaericus* O.F.M. in East Greenland. *Meddelelser om Grønland,* **131**: 1-50.

Prentki, R.T. (1976) Phosphorus cycling in tundra ponds. Ph.D. dissertation, University of Alaska, 275 pp.

Prentki, R.T., M.S. Adams, S.R. Carpenter, A. Gasith, C.S. Smith and P.R. Weiler (1979) The role of submersed weedbeds in internal loading and interception of allochthonous materials in Lake Wingra, Wisconsin, USA. *Archiv für Hydrobiologie Monographische Beiträge,* **57**: 221-250.

Prescott, G.W. (1953) Preliminary notes on the ecology of freshwater algae in the Arctic Slope, Alaska, with descriptions of some new species. *American Midland Naturalist,* **50**: 463-473.

Proper, G. and J.C. Carver (1966) Mass culture of the protozoan *Colpoda steinii. Biotechnology and Bioengineering,* **8**: 287-296.

Prosser, C.L. (Ed.) (1973) *Comparative Animal Physiology.* Third edition. Philadelphia: W.B. Saunders Co., 966 pp.

Rabinowitch, E. (1951) *Photosynthesis and Related Processes.* Volume 2, Part 1, pp. 603-1208. New York: Interscience.

Rawson, D.S. (1953) Limnology in the North American Arctic and Subarctic. *Arctic,* **6**: 198-204.

Rawson, D.S. (1960) A limnological comparison of twelve large lakes in Northern Saskatchewan. *Limnology and Oceanography,* **5**: 195-211.

Reed, E.B. (1962) Freshwater plankton Crustacea of the Colville River area, northern Alaska. *Arctic,* **15**: 27-50.

Reed, J.P. (1974) Benthic metabolism in arctic coastal ponds, Barrow, Alaska. M.S. thesis, University of Cincinnati, 118 pp.

Reimold, R.J. (1972) The movement of phosphorus through the salt marsh cord grass, *Spartina alterniflora* Loisel. *Limnology and Oceanography,* **17**: 606-611.

Reiss, F. and E.J. Fittkau (1971) Taxonomie und Okologie europäisch verbreiteter Tanytarsus-Arten (Chironomidae Diptera). *Archiv für Hydrobiologie* (Suppl.), **40**: 75-200.

Remmert, H. (1969) Tageszeitliche Verzahnung der Aktivität verschiedener Organismen. *Oecologia,* **3**: 214-226.

Rex, R.W. (1961) Hydrodynamic analysis of circulation and orientation of lakes in northern Alaska. In *Geology of the Arctic* (G.O. Raasch, Ed.). Vol. II. Toronto: University of Toronto Press, pp. 1021-1043.

Rhee, G.Y. (1972) Competition between an alga and a bacterium for phosphate. *Limnology and Oceanography,* **17**(4), 505-514.

Rhee, G.Y. (1973) A continuous culture study of phosphate uptake, growth rate and polyphosphate in *Scenedesmus* sp. *Journal of Phycology,* **9**: 495-506.

Rich, P.H. (1970) Utilization of benthic detritus in a marl lake. Ph.D. dissertation, Michigan State University, 89 pp.

Rickard, P.W. and F.C. Harmston (1972) Diptera and other arthropods of the Sukkertoppen Tasersiaq area, southwest Greenland. *Arctic,* **25**: 107-114.

Rigler, F.H. (1956) A tracer study of the phosphorus cycle in lake water. *Ecology,* **37**: 550-562.

Rigler, F.H. (1961) The uptake and release of inorganic phosphorus by *Daphnia magna* Straus. *Limnology and Oceanography,* **6**: 165-174.

Rigler, F.H. (1964) The phosphorus fractions and the turnover time of inorganic phosphorus in different types of lakes. *Limnology and Oceanography,* **9**: 511-518.

Rigler, F.H. (1966) Radiobiological analysis of inorganic phosphorus in lakewater. *Verhandlungen Internationale Vereinigung für Theoretische und Angewandte Limnologie,* **16**: 465-470.

Rigler, F.H. (1968) Further observations inconsistent with the hypothesis that the molybdenum blue method measures orthophosphate in lake water. *Limnology and Oceanography,* **13**: 7-13.

Rigler, F.H. (1972) The Char Lake project. A study of energy flow in a high arctic lake. In *Productivity Problems of Fresh Waters.* (Z. Kajak and A. Hillbricht-Ilkowska, Eds.). Warsaw: Polish Academy of Sciences, pp. 287-300.

Rigler, F.H. (1973) A dynamic view of the phosphorus cycle in lakes. In *Environmental Phosphorus Handbook* (E.J. Griffith, A. Becton, J.M. Spencer and D.T. Mitchell, Eds.). New York: Wiley, pp. 539-572.

Riley, J.P. and G. Skirrow (1975) *Chemical Oceanography.* 2nd edition. London: Academic Press, 564 pp.

Ringleberg, J. and H. Servass (1971) A circadian rhythm in *Daphnia magna. Oecologia,* **6**: 289-292.

Rodhe, W., R.A. Vollenweider and A. Nauwerck (1958) The primary production and standing crop of phytoplankton. In *Perspectives on Marine Biology* (A.A. Buzzati-Traverso, Ed.). Berkeley: University of California Press, pp. 299-322.

Roff, J.C. and J.C.H. Carter (1972) Life cycle and seasonal abundance of the copepod *Limnocalanus macrurus* Sars in a high arctic lake.

Limnology and Oceanography, **17**: 363-370.

Rogers, A.W. (1969) *Techniques of Autoradiography.* New York: Elsevier, 338 pp.

Romanenko, V.I. (1971) Total number of bacteria in the Rybinsk Reservoir. *Microbiology,* **40**: 617-622.

Rosen, R. (1970) *Dynamical System Theory in Biology.* Vol. I. *Stability Theory and Its Applications.* New York: Wiley-Interscience, 302 pp.

Round, F.E. (1964) The ecology of benthic algae. In *Algae and Man* (D.F. Jackson, Ed.). NATO Advanced Study Institute, Louisville, Kentucky, 1962. New York: Plenum Press, pp. 138-184.

Rublee, P.A. (1974) Production of bacteria in a pond. M.S. thesis, North Carolina State University, 39 pp.

Ryhänen, R. (1968) Die Bedeutung der Humussubstanzen im Stoffhaushalt der Gewässer Finnlands. *Mitteilungen Internationale Vereinigung für Theoretische und Angewandte Limnologie,* **14**: 168-178.

Ryther, J.H. and R.R.L. Guillard (1959) Enrichment experiments as a means of studying nutrients limiting to phytoplankton production. *Deep-Sea Research,* **6**: 65-69.

Ryther, J.H. and D.W. Menzel (1959) Light adaptation by marine phytoplankton. *Limnology and Oceanography,* **4**: 492-497.

Salonen, K. and A.-L. Kotimaa (1975) The determination of dissolved inorganic carbon, a possible source of error in determining the primary production of lake water phytoplankton. *Annales Botanici Fennici,* **12**: 187-189.

Sater, J.E. (Coordinator) (1969) *The Arctic Basin.* Revised edition. Washington, D.C.: Arctic Institute of North America, 337 pp.

Saunders, G.W. (1969) Some aspects of feeding in zooplankton. In *Eutrophication: Causes, Consequences, Corrections.* Washington, D.C.: National Academy of Sciences, pp. 556-573.

Saunders, G.W. (1972) The transformation of artificial detritus in lake water. *Memorie dell'Instituto Italiano di Idrobiologia,* **29** (Suppl.): 261-288.

Saunders, G.W. (1976) Decomposition in fresh water. In *The Role of Terrestrial and Aquatic Organisms in Decomposition Processes* (J.M. Anderson and A. Macfadyen, Eds.). Oxford: Blackwell Scientific Publications, pp. 341-374.

Schell, D.M. and V. Alexander (1970) Improved incubation and gas sampling techniques for nitrogen fixation studies. *Limnology and Oceanography,* **15**: 961-962.

Schindler, D.W. (1971) Carbon, nitrogen and phosphorus and the eutrophication of freshwater lakes. *Journal of Phycology,* **7**: 321-329.

Schindler, D.W. (1977) Evolution of phosphorus limitation in lakes. *Science,* **195**: 260-262.

Schindler, D.W., R.V. Schmidt and R.A. Reid (1972) Acidification and bubbling as an alternative to filtration in determining phytoplankton production by the ^{14}C method. *Journal of the Fisheries Research Board of Canada,* **29**: 1627-1631.

Schindler, D.W., H.E. Welch, J. Kalff, G.J. Brunskill and N. Kritsch (1974) Physical and chemical limnology of Char Lake, Cornwallis Island (75 °N lat.). *Journal of the Fisheries Research Board of Canada,* **31**: 585-607.

Scholander, P.F., W. Flagg, R.J. Hock and L. Irving (1953) Studies on the physiology of frozen plants and animals in the Arctic. *Journal of Cellular and Comparative Physiology,* **42**(1): 1-56.

Schultz, A.M. (1969) A study of an ecosystem: the arctic tundra. In *The Ecosystem Concept in Natural Resource Management* (G.M. Van Dyne, Ed.). New York: Academic Press, pp. 77-93.

Sellmann, P.V. and J. Brown (1965) Coring of frozen ground, Barrow, Alaska, spring 1964. *U.S. Army CRREL Special Report* 81, 8 p.

Sellmann, P.V. and J. Brown (1973) Stratigraphy and diagenesis of perennially frozen sediment in the Barrow, Alaska, region. In *Permafrost: North American Contribution to the Second International Conference.* Washington, D.C. National Academy of Sciences, pp. 171-181.

Sellmann, P.V., J. Brown, R.I. Lewellen, H. McKim and C. Merry (1975) The classification and geomorphic implications of thaw lakes on the Arctic Coastal Plain, Alaska. *U.S. Army CRREL Research Report* 344, 21 pp.

Shaver, G.R. and W.D. Billings (1975) Root production and root turnover in a wet tundra ecosystem, Barrow, Alaska. *Ecology,* **56**: 401-409.

Shaver, G.R., F.S. Chapin, III and W.D. Billings (1979) Ecotypic differentiation in *Carex aquatilis* on ice-wedge polygons in the Alaskan coastal tundra. *Journal of Ecology,* **67**: 1-21.

Sheldon, R.W. (1972) Size separation of marine seston by membrane and glass-fiber filters. *Limnology and Oceanography,* **17**: 494-498.

Sheldon, R.W. and T.R. Parsons (1967) A continuous size spectrum for particulate matter in the sea. *Journal of the Fisheries Research Board of Canada,* **24**: 909-915.

Shindler, D.B., B.F. Scott and D.B. Carlisle (1975) Effect of crude oil on populations of bacteria and algae in artificial ponds subject to winter weather and ice formation. *Verhandlungen Internationale Vereinigung für Theoretische und Angewandte Limnologie,* **19**: 2138-2144.

Sick, L.V. (1970) The nutritional effect of five species of marine algae on the growth, development and survival of the brine shrimp, *Artemia salina.* Ph.D. dissertation, North Carolina State University, 43 pp.

Siljak, D.D. (1974) Connective stability of complex ecosystems. *Nature,* **249**: 280.

Skirrow, G. (1965) The dissolved gasses—carbon dioxide. In *Chemical Oceanography* (J.P. Riley and G. Skirrow, Eds.). Vol. 1. London: Academic Press, pp. 227-322.

Skuja, H. (1964) Grundzüge der Algenflora und Algenvegetation der Fjeldgegenden um Abisko in Schwedisch-Lappland. *Nova Acta Regiae Societatis Scientiarum Upsaliensis,* Ser. IV, Vol. 18, No. 3, 462 pp.

Slobodkin, L.B. (1959) Energetics in *Daphnia pulex* populations. *Ecology,* **40**: 232-243.

Slobodkin, L.B. (1960) Ecological energy relationships at the population level. *American Naturalist,* **94**: 213-236.

Slobodkin, L.B. (1974) Prudent predation does not require group selection. *American Naturalist,* **108**: 665-678.

Smith, F.E. (1952) Experimental methods in population dynamics: a critique. *Ecology,* **33**: 441-450.

Smith, F.E. (1963) Population dynamics in *Daphnia magna* and a new model for population growth. *Ecology,* **44**: 651-663.

Snow, N.B. and D.M. Rosenberg (1975a) Experimental oil spills on Mackenzie Delta lakes. I. Effect of Norman Wells crude oil on Lake 4. *Fisheries Marine Service Technical Report* 548, 44 pp.

Snow, N.B. and D.M. Rosenberg (1975b) Experimental oil spills on Mackenzie Delta Lakes. II. Effects of two types of crude on Lakes 4C and B. *Fisheries Marine Service Technical Report* 549, 19 pp.

Snow, N.B. and B.F. Scott (1975) The effect and fate of crude oil spilt on two arctic lakes. In *Conference on Prevention and Control of Oil Pollution.* EPA-API-USCG, pp. 527-534.

Solórzano, L. (1969) Determination of ammonia in natural waters by the phenolhypochlorite method. *Limnology and Oceanography,* **14**: 799-801.

Sommers, L.E. and D.W. Nelson (1972) Determination of total phosphorus in soils: a rapid perchloric acid digestion procedure. *Soil Science Society of America, Proceedings,* **36**: 902-904.

Somorjai, R.L. and D.N. Goswami (1972) Relationship between stability and connectedness of non-linear systems. *Nature,* **236**: 466.

Sorokin, Y.I. and I.W. Konovalova (1973) Production and decomposition of organic matter in a bay of the Japan Sea during the winter diatom bloom. *Limnology and Oceanography,* **18**: 962-967.

Soto, C., J.A. Hellebust and T.C. Hutchinson (1975) The effects of aqueous extracts of crude oil and naphthalene on the physiology and morphology of a freshwater green alga. *Verhandlungen Internationale Vereinigung für Theoretische und Angewandte Limnologie,* **19**: 2145-2154.

Spindler, K.D. (1971a) Der Einfluss von Licht auf die Eiablage des Copepoden *Cyclops vicinus. Zeitschrift für Naturforschung,* **26**: 953-955.

Spindler, K.D. (1971b) Untersuchungen über den Einfluss äusserer Faktoren auf die Dauer der Embryonalentwicklung und den Häutungsrhythmus von *Cyclops vicinus. Oecologia,* **7**: 342-355.

Sprules, W.G. (1972) Effects of size-selective predation and food competition of high altitude zooplankton communities. *Ecology,* **53**: 375-386.

Stanley, D.W. (1974) Production ecology of epipelic algae in Alaskan tundra ponds. Ph.D. dissertation, North Carolina State University, 151 pp.

Stanley, D.W. (1976a) Productivity of epipelic algae in tundra ponds and a lake near Barrow, Alaska. *Ecology,* **57**: 1015-1024.

Stanley, D.W. (1976b) Carbon flow model of epipelic algal productivity in Alaskan tundra ponds. *Ecology,* **57**: 1034-1042.

Stanley, D.W. and R.J. Daley (1976) Environmental control of primary productivity in Alaskan tundra ponds. *Ecology,* **57**: 1025-1033.

Steele, J.H. (1962) Environmental control of photosynthesis in the sea. *Limnology and Oceanography,* **7**: 137-150.

Steemann Nielsen, E. and V.K. Hansen (1961) Influence of surface illumination on plankton photosynthesis in Danish waters (50°N) throughout the year. *Physiologia Plantarum,* **14**: 595-613.

Stewart, W.D.P., G.P. Fitzgerald and R.H. Burris (1967) *In situ* studies on nitrogen fixation using the acetylene reduction assay. *Proceedings of the National Academy of Sciences,* **58**: 2071-2078.

Strickland, J.D.H. (1960) Measuring the production of marine phytoplankton. *Bulletin of the Fisheries Research Board of Canada,* **122**: 172 pp.

Strickland, J.D.H. and T.R. Parsons (1965) *A Manual of Sea Water Analysis.* 2nd edition, revised. Ottawa: Fisheries Research Board of Canada, Bulletin 125, 203 pp.

Strickland, J.D.H. and T.R. Parsons (1968) *A Practical Handbook of Seawater Analysis.* Ottawa: Fisheries Research Board of Canada, Bulletin 167, 311 pp.

Strickler, J.R. and A.K. Bal (1973) Setae of the first antennae of the copepod *Cyclops scusider* Sars; their structure and importance. *Proceedings of the National Academy of Sciences,* **70**(9): 2656-2659.

Stross, R.G. (1969) Photoperiod control of diapause in *Daphnia*. II. Induction of winter diapause in the Arctic. *Biological Bulletin,* **136**: 264-273.

Stross, R.G. (1975) Cause of daily rhythms in photosynthetic rates of phytoplankton. II. Phosphate control of expression in tundra ponds. *Biological Bulletin,* **149**: 408-418.

Stross, R.G. and S.W. Chisholm (1975) Density stabilization in arctic populations of *Daphnia. Verhandlungen Internationale Vereinigung für Theoretische und Angewandte Limnologie,* **19**: 2879-2884.

Stross, R.G., S.W. Chisholm and T.A. Downing (1973) Causes of daily rhythms in photosynthetic rates of phytoplankton. *Biological Bulletin,* **145**: 200-209.

Stross, R.G. and D.A. Kangas (1969) The reproductive cycle of *Daphnia* in an arctic pool. *Ecology,* **50**: 457-460.

Stross, R.G., P.A. Nobbs and S.W. Chisholm (1979) SUNDAY, a simulation model of an arctic *Daphnia* population. *Oikos,* **32**:349-362.

Stross, R.G. and S.M. Pemrick (1973) Nutrient uptake kinetics in phytoplankton: a basis for niche separation. *Journal of Phycology,* **10**: 164-169.

Stumm, W. and J.J. Morgan (1970) *Aquatic Chemistry.* New York: Wiley, 583 pp.

Sutton, C.D. (1969) Effect of low soil temperature on phosphate nutrition of plants—a review. *Journal of Science and Food Agriculture,* **20**: 1-3.

Syers, J.K., R.F. Harris and D.E. Armstrong (1973) Phosphate chemistry in lake sediments. *Journal of Environmental Quality,* **2**: 1-14.

Szlauer, L. (1963) The resting stages of Cyclopoididae in Stary Dwor Lake. *Polskie Archiwum Hydrobiologii,* **11**: 385-394.

Taber, S. (1929) Frost heaving. *Journal of Geology,* **37**: 428-461.

Taber, S. (1930) The mechanics of frost heaving. *Journal of Geology,* **38**: 303-317.

Talling, J.F. (1957) Photosynthetic characteristics of some freshwater plankton diatoms in relation to underwater radiation. *New Phytologist,* **56**: 29-50.

Tash, J.C. and K.B. Armitage (1967) Ecology of zooplankton of the Cape Thompson area, Alaska. *Ecology,* **48**: 129-139.

Taylor, L.R. (1975) Longevity, fecundity, and size: control of reproductive potential in a polymorphic migrant, *Aphis fabae* Scop. *Journal of Animal Ecology,* **44**: 135-163.

Teal, J.M. (1957) Community metabolism in a temperate cold spring. *Ecological Monographs,* **27**: 283-302.

Teal, J.M. and J. Kanwisher (1966) The use of pCO_2 for the calculation of biological production with examples from waters off Massachusetts. *Journal of Marine Research,* **24**: 4-14.

Tenore, K.R. (1977) Utilization of aged detritus derived from different sources by the polychaete, *Capitella capitata. Marine Biology,* **44**: 51-55.

Tessenow, U. and Y. Baynes (1975) Redox-dependent accumulation of Fe and Mn in a littoral sediment supporting *Isoetes lacustris* L. *Naturwissenschaften,* **62**: 342-343.

Thomas, W.H. and A.N. Dodson (1972) On nitrogen deficiency in tropical Pacific oceanic phytoplankton. II. Photosynthetic and cellular characteristics of a chemostat-grown diatom. *Limnology and Oceanography,* **17**: 515-523.

Thompson, C.J., H.J. Coleman, J.P. Dooley and D.E. Hirsh (1971) Bumines analysis shows characteristics of Prudhoe Bay crude. *Oil and Gas Journal,* **69**: 112-120.

Tieszen, L.L. (1978a) Photosynthesis in the principal Barrow, Alaska, species: a summary of field and laboratory responses. In *Vegetation and Production Ecology of an Alaskan Arctic Tundra* (L.L. Tieszen, Ed.). New York: Springer-Verlag, Inc., pp. 241-268.

Tieszen, L.L. (1978b) Summary. In *Vegetation and Production Ecology of an Alaskan Arctic Tundra* (L.L. Tieszen, Ed.). New York: Springer-Verlag, Inc., pp. 621-649.

Tieszen, L.L. and D.A. Johnson (1975) Seasonal pattern of photosynthesis in individual grass leaves and other plant parts in arctic Alaska with a portable $^{14}CO_2$ system. *Botanical Gazette,* **136**: 99-105.

Tilzer, M. (1972) Bacterial productivity of a high-mountain lake. *Verhandlungen Internationale Vereinigung für Theoretische und Angewandte Limnologie,* **18**: 188-196.

Tiwari, J.L. and J.E. Hobbie (1976a) Random differential equations as models of ecosystems: Monte Carlo simulation approach. *Mathematical Biosciences,* **28**: 25-44.

Tiwari, J.L. and J.E. Hobbie (1976b) Random differential equations as models of ecosystems. II. Initial conditions and parameter specifications in terms of maximum entropy distributions. *Mathematical Biosciences,* **31**: 37-53.

Tiwari, J.L., J.E. Hobbie, J.P. Reed, D.W. Stanley and M.C. Miller (1978) Some stochastic differential equation models of an aquatic ecosystem. *Ecological Modeling,* **4**: 3-27.

Toerien, D.F., C.H. Huang, J. Radimsky, E.A. Pearson and J. Scherfig (1971) Final report. Provisional algal assay procedures. Water Quality Office, Environmental Protection Agency Project No. 16010 DQB, 211 pp.

Toetz, D.W. (1971) Diurnal uptake of NO_3 and NH_4 by a *Ceratophyllum*-periphyton community. *Limnology and Oceanography,* **16**: 819-822.

Toetz, D.W. (1973) The kinetics of NH_4 uptake by *Ceratophyllum. Hydrobiologia,* **41**: 275-290.

Trapnell, B.M.W. (1966) *Chemisorption.* London: Butterworth, 265 pp.

Twilley, R.R., M.M. Brinson and C.J. Davis (1977) Phosphorus absorption, translocation, and secretion in *Nuphar luteum. Limnology and Oceanography,* **22**: 1022-1032.

Ulrich, A. and P.L. Gersper (1978) Plant nutrient limitations of tundra plant growth. In *Vegetation and Production Ecology of an Alaskan Arctic Tundra* (L.L. Tieszen, Ed.). New York: Springer-Verlag, Inc., pp. 457-481.

Utermöhl, H. (1958) Zur Vervollkommnung der quantitativen Phytoplanktonmethodik. *Mitteilungen Internationale Vereinigung für Theoretische und Angewandte Limnologie,* **9**: 1-38.

Vaccaro, R.F., S.E. Hicks, H.W. Jannasch and F.G. Carey (1968) The occurrence and role of glucose in seawater. *Limnology and Oceanography,* **13**: 356-360.

Van Cleve, K. and L.A. Viereck (1972) Distribution of selected chemical elements in even-aged alder (*Alnus*) ecosystems near Fairbanks, Alaska. *Arctic and Alpine Research,* **4**: 239-255.

Vollenweider, R.A. (1965) Calculation models of photosynthesis–depth curves and some implications regarding day rate estimates in primary production measurements. *Memorie dell'Instituto Italiano di Idrobiologia,* **18** (Suppl.): 425-457.

Vollenweider, R.A. (1968) *Scientific Fundamentals of the Eutrophication of Lakes and Flowing Waters, with Particular Reference to Nitrogen and Phosphorus as Factors in Eutrophication.* Paris: Organization for Economic Co-operation and Development.

Waksman, S.A., C.L. Carey and H.W. Rueszer (1933) Marine bacteria and their role in the cycle of life in the sea. I. Decomposition of marine plant and animal residues by bacteria. *Biological Bulletin,* **65**: 57-79.

Waksman, S.A. and C.E. Renn (1936) Decomposition of organic matter in seawater by bacteria. III. Factors influencing the rate of decomposition. *Biological Bulletin,* **70**: 472-483.

Walker, H.J. (1973) Morphology of the North Slope. In *Alaskan Arctic Tundra* (M.E. Britton, Ed.). Arctic Institute of North America Technical Paper 25, p. 49-92.

Walsh-Maetz, B.M. (1953) Le Métabolisme de *Chironomus plumosus* dans des conditions naturelles. *Physiologia Comparaet Oecologia,* **3**: 135-154.

Waters, T.F. (1977) Secondary production in inland waters. *Advances in Ecological Research,* **10**: 91-164.

Watson, D.G., W.C. Hanson, J.J. Davis and C.E. Cushing (1966a) Limnology of tundra ponds and Ogotoruk Creek. In *Environment of the Cape Thompson Region, Alaska* (N.J. Wilimovsky and J.W. Wolfe, Eds.). Oak Ridge, Tenn.: U.S. Atomic Energy Commission, pp. 415-435.

Watson, D.G., J.J. Davis and W.C. Hanson (1966b) Terrestrial invertebrates. In *Environment of the Cape Thompson Region, Alaska* (N.J. Wilimovsky and J.W. Wolfe, Eds.). Oak Ridge, Tenn.: U.S. Atomic Energy Commission, pp. 565-584.

Watson, S.W., T.J. Novitsky, H.L. Quinby and F.W. Valois (1977) Determination of bacterial number and biomass in the marine environment. *Applied and Environmental Microbiology,* **33**: 940-946.

Watt, W.D. and F.R. Hayes (1963) Tracer study of the phosphorus cycle in seawater. *Limnology and Oceanography,* **8**: 276-285.

Webber, P.J. (1978) Spatial and temporal variation of the vegetation and its production, Barrow, Alaska. In *Vegetation and Production Ecology of an Alaskan Arctic Tundra* (L.L. Tieszen, Ed.). New York: Springer-Verlag, Inc., pp. 37-112.

Weers, E.T. and T.M. Zaret (1975) Grazing effects on nannoplankton in Gatun Lake, Panama. *Verhandlungen Internationale Vereinigung für Theoretische und Angewandte Limnologie,* **19**: 1480-1483.

Weiler, R.R. (1973) The interstitial water composition in the sediments of the Great Lakes. I. Western Lake Ontario. *Limnology and Oceanography,* **18**: 918-931.

Weiss, R. (1970) The solubility of nitrogen, oxygen and argon in water and sea water. *Deep-Sea Research,* **17**: 721-735.

Welch, H.E. (1973) Emergence of Chironomidae (Diptera) from Char Lake, Resolute, Northwest Territories. *Canadian Journal of Zoology,* **51**: 1113-1123.

Welch, H.E. (1974) Metabolic rates of arctic lakes. *Limnology and Oceanography,* **19**: 65-73.

Welch, H.E. (1976) Chironomidae in a polar lake. *Journal of the Fisheries Research Board of Canada,* **33**: 227-247.

Welch, H.E. and J. Kalff (1974) Benthic photosynthesis and respiration in Char Lake. *Journal of the Fisheries Research Board of Canada,* **31**: 609-620.

Weller, G. and B. Holmgren (1974) The microclimates of the arctic tundra. *Journal of Applied Meteorology,* **13**: 854-862.

Westlake, D.F. (1968) Methods used to determine the annual production of reed-swamp plants with extensive rhizomes. In *Methods of Productivity Studies in Root Systems and Rhizosphere Organisms.* Leningrad: Publishing House Nauka, pp. 226-234.

Wetzel, R.G. (1964) A comparative study of the primary productivity of higher aquatic plants, periphyton, and phytoplankton in a large shallow lake. *Internationale Revue der Gesamten Hydrobiologie,* **49**: 1-61.

Wetzel, R.G. (1969a) Excretion of dissolved organic compounds by aquatic macrophytes. *BioScience,* **19**: 539-540.

Wetzel, R.G. (1969b) Factors influencing photosynthesis and excretion of dissolved organic matter by aquatic macrophytes in hard-water lakes. *Verhandlungen Internationale Vereinigung für Theoretische und Angewandte Limnologie,* **17**: 72-85.

Wetzel, R.G. (1975) *Limnology.* Philadelphia: Saunders, 743 pp.

Wetzel, R.G. and B.A. Manny (1972) Secretion of dissolved organic carbon and nitrogen by aquatic macrophytes. *Verhandlungen Internationale Vereinigung für Theoretische und Angewandte Limnologie,* **18**: 162-170.

Wetzel, R.G., P.H. Rich, M.C. Miller and H.L. Allen (1972) Metabolism of dissolved and particulate detrital carbon in a temperate hardwater lake. *Memorie dell'Instituto Italiano Idrobiologia,* **29** (Suppl.): 185-243.

Wiggins, G.B. (1977) *Larvae of the North American Caddisfly Genera (Trichoptera).* Toronto: University of Toronto Press, 401 pp.

Wijler, J. and C.C. Delwiche (1954) Investigations on the denitrifying process in soil. *Plant and Soil,* **5**: 155-169.

Williams, J.D.H., J.K. Syers, D.E. Armstrong and R.F. Harris (1971a) Characterization of inorganic phosphate in noncalcareous lake sediments. *Soil Science Society of America, Proceedings,* **35**: 556-561.

Williams, J.D.H., J.K. Syers, R.F. Harris and D.E. Armstrong (1971b) Fractionation of inorganic phosphate in calcareous lake sediments. *Soil Science Society of America, Proceedings,* **35**: 250-255.

Williams, J.D.H., J.K. Syers, S.S. Shukla, R.F. Harris and D.E. Armstrong (1971c) Levels of inorganic and total phosphorus in lake sediments as related to other sediment parameters. *Environmental Science and Technology,* **5**: 1113-1120.

Winberg, G.G. (Ed.). (1971) *Methods for the Estimation of Production of Aquatic Animals.* (Translated from the Russian by A. Duncan.) London: Academic Press, 175 pp.

Wium-Andersen, S. and J.M. Andersen (1972) The influence of vegetation on the redox profile of the sediment of Grane Langsø, a Danish *Lobelia* lake. *Limnology and Oceanography,* **17**: 948-952.

Wohlschlag, D.E. (1953) Some characteristics of the fish populations in an arctic Alaskan Lake. *Stanford University Publications, University Series, Biological Sciences,* **11**: 19-29.

Wood, L.W. (1970) The role of estuarine sediment microorganisms in the uptake of organic solute under aerobic conditions. Ph.D. dissertation, North Carolina State University, 75 pp.

Wright, J.C. (1959) Limnology of Canyon Ferry Reservoir. II. Phytoplankton standing crop and primary production. *Limnology and Oceanography,* **4**: 235-245.

Wright, R.T. (1970) Glycolic acid uptake by planktonic bacteria. In *Organic Matter in Natural Waters* (D.W. Hood, Ed.). University of Alaska, Institute of Marine Science Publication 1, pp. 521-536.

Wright, R.T. and J.E. Hobbie (1966) Use of glucose and acetate by bacteria and algae in aquatic ecosystems. *Ecology,* **47**: 447-464.

Wülker, W. (1959) Drei neue Chironomiden-Arten (Diptera) und ihre Bedeutung für das Konvergenzproblem bie Imagines und Puppen. *Archiv für Hydrobiologie,* **25** (Suppl.): 44-64.

Wülker, W. and P. Götz (1968) Die Verwendung der Imaginalscheiben zur Bestimmung des Entwicklungszustandes von *Chironomus*-Larven (Dipt.). *Zeitschrift für Morphologie der Tiere,* **62**: 363-388.

Yentsch, C.S. and J.H. Ryther (1957) Short-term variations in phytoplankton chlorophyll and their significance. *Limnology and Oceanography,* **2**: 140-142.

Zadeh, L.A. and C.A. Desoer (1963) *Linear System Theory: The State Space Approach.* New York: McGraw-Hill, 628 pp.

Zadeh, L.A. and E. Polak (Eds.) (1969) *System Theory.* New York: McGraw-Hill, 521 pp.

Taxonomic Index

Acari, 298,299
Achromatium oxaliferum, 183
Agabus, 300, 302, 403
Alopex lagopus, 40
Amphibia, 14, 22
Amphidinium spp., 185, 194
A. cf. *lacustre,* 185
Amphipoda, 299
Anabaena lapponica, 183, 194
Ankistrodesmus, 193
A. acicularis , 184
A. falcatus, 184, 194
A. nannoselene, 184
A. spirale, 184, 194
Aphanocapsa sp., 194
Aphanotheca sp., 194
A. cf. *castagnei,* 183
A. clathrata, 183, 194
Aphanozomenon, 193
A. flos-aquae, 194
Arcella, 378
Arctophila fulva, 4, 102, 153, 224, 301
Artemia, 290
Asio flammius, 40
Astasia spp., 184
Asynarchus, 301, 403

Blepharisma lateritium, 377
Bodo, 374
Botryococcus braunii, 184
Bryophaenocladius sp., 301

Camptochironomus grandivalva, 300
Caenomorpha medusula, 373
Calanus, 291
Carex aquatilis, 4, 5, 9, 13, 15, 37, 41, 99, 102, 108, 134, 153, 154, 161, 224, 301, 303, 355, 357, 383, 401
Cercobodo spp., 185
Chironomidae, 298, 299
Chironomus, 15, 298, 299, 301, 305, 405
C. hyperboreus, 300, 306
C. pilicornis, 300, 301, 302, 305, 306, 310, 320
C. riparius, 300, 301, 306
Cymbella sp., 194
Chlamydomonas, 183, 193
C. caroleae, 183
C. frigida, 183, 194
C. lapponica, 194
C. liloeae, 184
C. reinhardtii, 266
C. sagittula, 183
C. sessila, 194
Chlorella, 291
Chlorobotrys regularis, 185
Chlorogonium spp., 183
Chromulina, 179, 180, 184, 187, 190, 221
C. diachloros, 184
Chroococcus prescotti, 183
C. turgidis, 194
C. turgidus var. *maximus,* 183
Chroomonas, 180
C. coerula, 184
C. nordstedtii, 185
Chrysococcus cf. *cordiformes,* 184
C. cystophorus, 184
C. rufescens, 184
Chrysomoron epherum, 184
Cladotanytarsus sp., 300, 305, 307
Closterium spp., 184, 193
C. aciculare, 194
C. kuetzingii, 184
C. moniliferum, 194
Coelosphaerium kuetzingianum, 183, 194
Coleps hirtus, 377

Coleoptera, 14, 299
Colpidium, 377
Constempellina sp., 300, 305
Corixidae, 22
Corynocera, 298, 299
Corynoneura, 298, 301, 302, 305, 306, 309, 315
Cosmarium botrytis, 194
C. granatum, 194
C. ornatum, 194
Cricotopus, 298, 302, 306, 315
Cricotopus sp. 3, 300
Cricotopus sp. 4, 300
C. perniger, 300
C. tibialis, 300, 305
Crucigenia tetrapedia, 184
Cryptochironomus, 300, 315
Cryptomonas, 180
Cryptomonas sp., 185, 194
C. borealis, 185
C. Marssonii, 185
C. obovata, 185
C. ovata, 185
Cryptogramma, 377
Cyclops, 4, 251, 403
C. languidoides, 263
C. strenuus, 252, 263
C. vernalis, 255, 257

Daphnia, 6, 14, 132, 133, 217, 251, 353, 403
D. longiremis, 251
D. pulex, 251, 261, 271, 272, 290
D. rosea, 267, 291
Derotanypus, 315
D. alaskensis, 300, 305, 307
Diaptomus, 251
D. alaskensis, 263
D. bacillifer, 256
D. glacialis, 256
Diflugia, 378
Dileptus anser, 377
Dinobryon sertularia, 184
D. sertularia var. *protuberans,* 184
D. sociale var. *americanum,* 184
Dictoteudipes lobiger, 300
Dictyosphaerium pulchellum, 184
D. simplex, 184
Dryadotanytarsus, 298
Dupontia fisheri, 38

Elakatothrix lacustris, 194
Equisetum, 229
Erkenia subaequiciliata (= Chrysochromulina parva?), 184
Eriphorum angustifolium, 38, 41
E. russeolum, 38
Escherichia coli, 291
Euastrum binale, 194
E. elegans, 194
Euglena gracilis, 184
E. cf. *intermedia,* 184
E. cf. *oxyuris,* 184
E. pisciformis, 184
E. cf. *viridis,* 184
Eunotia lunaris, 194
Euplotes patella, 377

Fragilaria sp., 194
F. virescens, 194
Frotonia acuminata, 377
F. leucas, 377

Glenodinium uliginosum, 185
Gloeococcus schroeteri, 184, 194
Gloeocystis planctonica, 184
Gomphonema, 193
Gomphospheria sp., 194
G. kuetzingianum, 194
G. lacustris, 183
G. robusta, 183
Gonium pectorale, 184
Gonatozygon monotaenium, 184
Gymnodinium spp., 185

G. cf. *lacustre,* 185
G. palustre, 185
G. triceratum, 185
G. uberrimum, 185
G. cf. *veris,* 185
Gyrinidae, 22

Hemidinium nasatum, 185
Heterocope, 4, 251, 256, 263, 403
Hyalotheca dissiliens, 184
Hydroporus sp. (larva), 302

Isoetes lacustris, 100

Kephyrion spp., 184
Kephyrion rubri-claustri, 184
Korshikoviella gracilipes, 184

Lacrymaria, 377
Lapposmittia sp., 301, 306
Lebertia, 300, 301, 302, 403
Lemmus sibericus, 39
Lepidurus arcticus, 132, 221, 298
302, 323, 371
Lepocinclis sp., 194
L. cf. *ovum,* 184
Limnocalanus macrurus, 23
Limnephilus sp., 302
Limnophyes sp., 301, 306
Lobelia, 99

Macromonas mobilis, 183
Mallomonas, 180, 190
M. akrokomos, 184
M. pumilio var. *canadensis*
n. var, 184
Menoideum spp., 184
Merismopedia glauca, 183
Mesosmittia sp., 301
Metopuses, 373
Metriocnemus, 301, 306, 307
Micrasema, 301, 302, 403
Microcystis, 187, 193
M. flos-aquae, 194

Monas, 374
Monomastix ophiostigma, 184
Mustela erminea, 40
M. nivalis, 40
Myriophyllum spicatum, 135
Mysis relicta, 299

Najas flexilis, 161
Nassula, 377
Navicula sp., 194
Nemoura, 300, 301, 302, 403
Nephroselmis discoides, 183
Nitzschia linearis, 194
Notonectidae, 22
Nuphar luteum, 135
Nyctea scandiaca, 40

Ochromonas, 180, 184
O. elsae n. sp., 184
Odonata, 14, 299
Oikomonas, 327, 374
Oligochaetae, 299, 301
Oncophorus wahlenbergii, 38
Oocystis arctica, 184
O. borgei, 184
O. gigas, 184
O. lacustris, 184, 194
O. submarina var. *variabilis,* 184
Ophryoglena, 377
Orthocladiinae, 301, 309, 319
Orthocladius (Eudactylocladius)
sp., 300
O. (Pogonocladius) sp., 300
Oscillatoria sp., 194
O. agardii, 194
O. cf. *borneti,* 183

Pandorina morun, 183
Parabodo attenuatus, 184
Parakiefferiella gracillima, 301
Paramecium, 14, 381
Paratanytarsus penicillatus, 300,
302, 305
Paulschulzia pseudovolvox, 184

Pediastrum boryanum, 184
P. duplex, 184
P. integrum, 184
P. kawraiskyi, 184
Pedinomonas minutissima, 183
Pelecypoda, 299
Peridinium cinctum, 185
P. inconspicum, 185, 194
P. palustre, 185
P. willei, 185
Phacus sp., 184
P. pyrum, 184
Phaenopsectra, 298
Phalaropus fulicaris, 317
Philodina, 378
Phragmites, 229
Physa, 300, 301, 302, 303, 403
Pinnularia sp., 194
P. mesolepta, 194
Planctonema lauterborni, 184
Poa arctica, 37
Polyartemiella, 256
Procladius, 298, 301, 315, 322
P. prolongatus, 300, 301, 305
P. versus, 300, 301, 302, 305
Propappus sp., 300, 302
Proroden teres, 377
Psectrocladius, 298, 305, 306, 307, 309
P. dilatatus gr. sp., 300
P. psilopterus gr. sp., 300
P. nigra, 301
Pseudonabaena spp., 183
Pseudokephyrion, 180
*P. enzii (=*184
P. undulatissimum, 194
Pseudoprorodon, 377
Pseudosmittia, 301, 306

Ranunculus pallasii, 224
Rhodomonas sp., 184
Rhodomonus minuta, 14, 179, 180, 185, 188, 190, 221, 222, 400
R. minuta var. *nannoplanctonica,* 184

Salix rotundifolia, 37
Saprodinium, 373
Scenedesmus armatus, 184
S. acuminatus, 184
Scirpus, 229
Scourfieldia cordiformis, 183
Sennia parvula, 185
Spartina alterniflora, 109, 135
Spathidioum, 377
Sphaerellopsis cf. *fluviatilis,* 183
Spirostomum teres, 377
Staurastrum spp., 184
S. gracile, 194
S. pachirhynchum, 184
S. polymorphum, 184
Stauroneis sp., 194
Stenokalyx monilifera, 184
Stentor, 377
Stercorarius pomarinus, 40
Stictochironomus sp., 300
S. rosenchoeldi, 305
Strombidium, 377
Stylonychia mytilus, 377
Synechocystis notatus, 194
Synura lapponica, 184
S. spagnicola, 184
S. uvella, 194

Tabellaria fenestrata, 194
Tanytarsini, 298, 301, 315, 405
Tanytarsus gregarius, 300, 301, 305, 306, 307, 311
T. inaequalis, 300, 301, 302, 305, 308, 311
Tetraedron minimum, 184
Tetrahymena, 377
T. pyriformis, 133, 380, 383
Tipula, 301
Trachelomonas spp., 184
T. hispida, 184
T. volvocina, 184

Tribonema spp., 185
Trichoptera, 14, 299
Trichotanypus, 301, 307, 315
T. alaskensis, 300, 302, 305, 307
Trimyema, 377
Tubifex sp., 300, 302
Turbellaria, 298, 302

Uroglena, 14, 188, 222, 400
U. americana, 184

Volvox aureus, 183
Vorticella, 377

Xanthidium antilop_eum, 184

Zostera marina, 135

Subject Index

Acarina. 41, 301
Aeration, oil spill, 403
Alaska, freshwater habitats, 23
Albedo, snow and ground, 32–33
Algae. *See also* Phytoplankton; Benthic algae
  benthic. *See* Benthic algae
  biomass compared to bacteria, 6
  biomass limitation, 449
  effect of temperature, 14, 201–202, 222–223
  factors controlling, 201
  in model, 434
  light inhibition, 206, 208
  nutrient limitation, 211
  P limitation, 212
  replacement of species, 14, 222
  seasonal succession, 180
  secrete DOC, 161
  uptake of N, 211
  uptake of P, 11
  zooplankton affect species, 14, 219
Alkalinity, seasonal cycle, 87
Ammonia
  effect of plants, 102
  in melt water, 102
  in rainfall, 106
  in water, 100
  interstitial water, 102, 104, 108
  supply rate, 13
  uptake, 13, 107
  uptake by roots, 109
Ammonification, 110, 211
Amphibia, 14, 22
Anaerobic metabolism, 353
Aquatic plants. *See Carex; Arctophila*
Arctic
  adaptions of plants, 38, 236
  adaptations of small organisms, 14, 381
  adaptations to, 14
  coastal plain, 25
  definition, 21
  effect on bacteria, 354
  effect on decomposition, 359
  effect on metazoans, 14, 381
  effect on P concentration, 151
  effect on plants, 242
  effect on ponds, 13
  effect on protozoans, 14
  foothills, 25
  lakes
    history of studies, 21
    and ponds, characteristics, 22
    experimental studies, 22
  limnology, results, 21
  ponds, unique properties of, 14
Arctic study, advantages of, 1
Arctophila fulva, 4, 102, 153, 224, 301
    life history, 228
    primary production, 231
Assimilation efficiency, zooplankton, 218
Autoradiography, phytoplankton, 188, 221

Bacteria.
  affected by grazing, 8, 382
  and exudates, 350
  and sediment grazers, 354
  attached, 340, 351

carbon content, 341
control by DOC supply, 350, 355
comparison with temperate zone, 14
control by grazing, 8, 354, 356, 379, 380, 382
control by substrate, 356
control by zooplankton, 354
effect of freezing, 354
effect of temperature changes, 355
grazing and activity, 383
grazing by protozoa, 379
growth rate, 344
heterotrophic activity, 345
in benthic model, 431
in model, 446
interactions with protozoa, 9
meltwater, 344
numbers, 340, 344
oil pond, 395
overwinter mortality, 344
P cycling, 9, 11 133–134
primary production relation, 346, 355
production, 351, 353
protection of attached forms, 351
respiration, 351
resuspension, 341
simulation of biomass, 450
soils, 41
specific types and oil, 395
temperature effect, 346
uptake of DOC, 160, 346, 353
Barrow, site description, 25
Benthic algae. *See also* Algae, 193-200, 219-221
biomass, 4, 193-194
burial, 208
chlorophyll, 193
control by burial, 17, 221
control by grazers, 219, 380
daily Ps patterns, 199
grazing by microbenthos, 380
in model, 431
light adaptation, 206
model, 414
N uptake, 108
oil pond, 397
photosynthetic capacity, 207
primary production, 196, 198–199
respiration, 352
seasonal Ps, 198
silica interaction, 223
simulation of Ps, 449
species, 193
vertical distribution, 195
Benthic animals.
community, 22
respiration, 352
Benthic microalgae. *See* Benthic algae
Benthic model, 413
Benthic respiration, 98, 363-372
Benthos. *See also* Midges, 303
Char Lake, 299
effect of temperature, 299
emergence, 301
habitats, 301
in arctic, 299
species at Barrow, 299
Bicarbonate, 80
Birds, 40
Bioassay for P limitation, 215
Biomass, control by nutrients, 217
Bioturbation.
chironomids, 297
in model, 431
tadpole shrimp, 333
Blue-green algae, 187
N fixation, 13, 112, 393
Bottom, movement, 69
Branchinecta, 256

Brooks Range, 25
Bryophytes, 38
Budget, nitrogen, 106

Caddisfly, 301
Calcium, 80
  immobilization, 41
  in *Carex*, 362
  in plants, 233
  interstitial water, 84
  immobilization, 41
Carbon, total inorganic, 88
  interstitial water, 96
Carbon dioxide
  comparison of losses, 367
  concentration, 91
  evasion rate, 95-96
  factors affecting, 94
  flux, 8
  gradients, 93
  in sediment respiration, 366
  partial pressure, 90
  release from soil 41, 97
  respiration effects, 93-94
  shallow lake, 93
  total loss, 352
  transfer rate, 93
Carbon flux
  general description, 5
  protozoa, 378
Carbon-14 dates,
  Barrow soils, 26
  lakes, 52
Carbon-nitrogen ratio
  algae, 108
  aquatic plants, 108
Carbohydrates in plants, 233, 235
Carex aquatilis, 4, 5, 9, 13, 15, 37, 41, 99, 102, 108, 134, 154, 161, 224, 301, 355, 357, 383, 401
  decomposition, 5, 41, 134, 355, 357, 362
  effect of arctic, 15, 242
  effect on redox potential, 99
  leaching, 359
  leaf color defined, 357
  leaves, 224
  life history, 224
  loss of P, 383
  nutrient content, 231, 362
  P uptake, 15
  primary production, 230
  releases DOC, 161
  releases P, 134
  resistance to decomposition, 355
  source of DOC, 154
  temperature adaptation, 15
  uptake of N, 13, 108
Carnivores, tundra, 40
Cell turnover, planktonic algae, 6
Char Lake (IBP) study, 22
Chemistry
  affected by oil, 392
  arctic lakes, 22
  effect of freezing, 80
  effect of ocean, 84
  pond water, 24, 79
Chironomidae, 15, 248, 299, 301 305, 405
  Chironomus pilicornis, 300–302, 305–306, 310, 320
  C. riparius, 301, 306
Chironomids. *See also* Midges
  burrows affect oxygen, 372
  consume microbenthos. 382
  effect on sediment respiration, 371
  food for birds, 40
  generation time, 15
  grazing, 219, 352
  in benthic model, 432
  mix sediment, 221, 297
  oil pond, 403, 405

simulation of biomass, 449
temperature effects, 15
Chloride, 80
Chlorophyll, seasonal cycle, 179
Chromulina, 179–180, 184, 187, 190, 221
Ciliates
distribution and abundance, 373, 375
feeding on algae, 220
functional types, 375, 377
temperate comparison, 380
Circadian rhythm, 273
Climate
description, 32
during IBP study, 53
Cold temperatures, effect on ecosystem processes, 14
Cold climate, effect on P cycling, 136
Coleoptera, 14, 299
Colloidal P, 116–117
Colloids, and humic compounds, 157
Color of water, 153, 155
Comparative natural history approach, 1
Competition, 243
in crustacea, 274
tadpole shrimp, 335
Connectance, 411
Condensation, 36
Consumers
in benthic model, 432
in planktonic model, 444
tundra, 39
Control of bacteria, 9
Copepods
abundance, 255
calanoid, 256
control of generation number, 276
cyclopoid, 255
egg production, 257
harpacticoid, 374, 378, 381
oil pond,
Corynoneura, 298, 301, 302, 305–306, 309, 319
Cricotopus, 298, 306–307, 315
Cyclops, 4, 251, 403
strenuus, 252, 263
vernalis, 255, 257
Currents
caused by wind, 3, 72
convective, 54
measurements, 71

Daphnia
assimilation efficiency, 269
brood size, 259–260, 278, 282
critical photoperiod, 253
density limits, 254
egg production, 259
fairyshrimp interactions, 284
feeding activity, 271
feeding rate, 265, 267
feeding rate controls, 267, 269
feeding rates in model, 430
food affects brood size, 260, 279
food and growth, 269
food controls, 278, 282
generation number control, 276
growth efficiency, 281
hatching synchrony, 271
Heterocope interactions, 285
in model, 444
life history, 257
middendorffiana, 6, 14, 132–133, 217, 251, 353, 403
P excretion, 133
pattern of control, 284
population control, 274
predation by *Heterocope*, 257
predation control, 285

production limits, 254
pulex, 251, 261, 271–272, 290
oil pond, 14, 403
reproduction, 252, 277, 279
reproduction control, 253
respiration rate, 269
simulation, 278
simulation of biomass, 449
size, 273
swarms, 254
switch in production, 253
synchrony, 271
temperature effects, 277, 280
Data bank, 408
Daylight, duration, 32
Dead storage, 65
Death rate, in model, 430
Decapoda, 299
Decomposition, 41
affects P supply, 151
Carex, 134, 355, 357
changes in chemistry, 362
coefficients, 360
control in tundra, 41
controls primary production, 43
detritus, 351
effect of low temperature, 151
effect of slow rate, 22
hydrolysis, 360
immersion effect, 234
in benthic model, 431
in planktonic model, 446
leaching, 359
Lepidurus affects, 371
litter bags, 153, 357
N cycle, 115
photo-oxidation, 163
stimulation by grazing, 9
three phases, 359
time course, 361
trituration, 359
Denitrification, sediments, 104
Deoxygenation, 98
Description of ponds, 2
Deterministic model, 15, 409
Detritus
as biological buffer, 290
decomposition, 351
dilutes zooplankton food, 271, 288
food chain, 4, 7–8
in model, 433
in plankton model, 446
quantity, 8
Diatoms, 187
control, 223
silica needs, 84
Discharge. *See* Runoff, 35
Dissolved organic carbon
affects light, 70
algal secretion, 161
and bacteria, 350
composition, 151
control, 159, 163
control by bacteria, 159
effect of rainfall, 155
formation by bacteria, 353
from *Carex*, 154
from plants, 161
from zooplankton, 163
humic compounds, 155
in model, 433
in planktonic model, 446
in rainwater, 155
input, 154,163
labile compounds, 159
leachates from plants, 153
leaches from sediment, 153
percent humics, 158
origin, 151
refractory, 362
resistant to breakdown, 154
re-solution from sediments, 153
seasonal cycle, 152
sources, 8, 153, 161
sources of labile fraction, 161

turnover, 160
Dissolved organic nitrogen
precipitation, 106
sediments, 102
water, 102
Dissolved reactive phosphorus
equilibrium concentration, 142
errors in analysis, 115, 119
interstitial water, 121
relation to sorbed P, 142
temporal variations, 116
variations among ponds, 145
Dissolved total phosphorus, interstitial water, 121
Dissolved unreactive phosphorus, 116,132
diel and seasonal variations, 117
from sediments, 132
in snow, 116
interstitial water, 121
Diversity, 1
DOC. *See* Dissolved organic carbon
DON. *See* Dissolved organic nitrogen
DRP. *See* Dissolved reactive phosphorus
DTP. *See* Dissolved total phosphorus
DUP. *See* Dissolved unreactive phosphorus
Dytiscid beetles, 22, 299, 301

Eh. *See* Redox potential, 98
Emergence, benthos, 301
Energy cycling, tundra, 42
Energy, input, 42
Endogenous rhythm, photosynthesis, 210
Ephemeroptera, 14, 299
Epipelic algae. *See* Benthic algae
Ephippial eggs, 252
Eriophorum, 38, 41, 229
Esatkuat Creek, 64
Evaporation
factors affecting, 36
measurement techniques, 67
pan, 67
rates, 36, 67
Evasion coefficient, carbon dioxide, 96
Excretion, in model, 430

Fairyshrimp
control of generation number, 276
egg production, 259
excretion, 133
food limitation, 282
life history, 14, 256
oil pond, 403
reproductive groups, 254
Feces, lemming, 362
Fertilization, 110
artificial ponds, 115
photosynthesis effects, 212
plants, 237
with P, 212
Filtering rates, *Daphnia*, 218
Food chain
grazing, 8
protozoan and micrometazoan, 9
Fish, 14, 22
Food quality, zooplankton, 288, 292
Food supply, temperature effect, 13
Freeze-thaw description, 3, 60
Freezing, effect on ion concentration, 80
Freshwater habitats, 23
Fungi
compared to bacteria, 397
in ponds, 356
in soils, 41

Gastropoda, 299
Gastrotrichs, 378
Geology of Barrow area, 26
Grazing
affects algae and bacteria, 218–219, 382
affects bacterial activity, 356
affects benthic algae, 219
affects P turnover, 134
chironomid, 352
enhances algae activity, 207, 218
food chain, 8
in model, 445
microbenthos, 352, 379
protozoa, 354
zooplankton, 188, 217, 353
Grazing rates, protozoans and micrometazoans, 9

Half-saturation, light, 205, 206
Harpacticoid copepods, 374, 378, 381
Hatching, synchrony in *Daphnia*, 272
Heat balance, seasonal, 33
Hemiptera, 14, 299
Heterocope, 4, 251, 256, 263, 403
infant killer, 288
predator of *Daphnia*, 285
Herbivores, 4
Hirudinea, 299
Histophagous ciliates, 377
History, pond site, 51
Humic compounds
composition, 155
concentration, 157
light extinction changes, 70
percent of DOC, 158
water budget effects, 158
Hydrology, description, 33
Hydrolysis, 360

Ice
effect on plants, 243
formation time, 22, 95
melt time, 33, 34
Ice cover, lakes and ponds, 22
Ice wedges, 28, 52
Ikroavik Lake, 20, 24, 25
ammonification and nitrification, 110
benthic algal productivity, 199
DRP, 119
interstitial water, 85
light extinction, 71
N concentrations, 100, 102
nutrient uptake, 108
oxygen, 98
phytoplankton productivity, 199
PP, 120
temperature, 57
zooplankton, 251
Imikpuk Lake, 24, 25
temperature in sediments, 58
thaw dates, 61
Infant killer, *Heterocope*, 288
Ingestion, in model, 429, 432
Insects. *See* Chapter 7
oil pond, 403
soils, 42
International Biological Program, description, 19
Interstitial water
ammonia, 108
inorganic carbon, 96
inorganic ions, 85
N concentration, 104
P concentration, 121
Invertebrates, factors affecting in soil, 41, 42
Iron
dissolved in water, 82
extractable, 141
in *Carex*, 362

interstitial water, 84
mobilization, 138
relation to P, 121, 137, 139
sediment enrichment, 138
sorbs P, 11, 121
source in plant zone, 138
Isopoda, 299

Lakes
northern Alaska, 23, 24
origin, 24
P concentration, 119
Leaching
Carex, 82, 359
lemming feces, 82
potassium, 82
Lead, 86
Leaves, color defined, 153
Lemmings, 39
affect plants, 38–39, 234
chemisty of feces, 362
control of cycles, 39, 40
nutrient effects, 45
Lepidurus, 132, 221, 298, 302, 323, 371
P excretion, 133
sediment respiration effects, 371
Libertia, 300–302, 403
Lichens, 38
Light
extinction in sediments, 71, 198
extinction in water, 70
half-saturation, 205–206
in model, 414
inhibition in algae, 208
limits plants, 235
photosynthesis effects, 205
Litter, from plants, 153
Low temperature, biological effects, 14

Magnesium, 80
Carex, 362
interstitial water, 84
Manganese, 85
Carex, 362
Marsh, surrounds ponds, 51
Megaloptera, 14, 299
Meltwater
bacteria concentration, 344
DOC, 152
N concentration, 107
POC, 164
Metazoans, adaptations to arctic, 14, 15
Methane, 351, 353
Michaelis–Menten functions, in model, 429
Microbenthos
carbon flux, 383
distribution and composition, 373
feeding rates, 379
grazing, 352, 375
particle size effects, 373
predation control, 382
respiration, 352
Microbes, control soil P, 239
Microcosm experiment, 383
Micrometazoa, 9
Micrometazoans, grazers, 219
Midges
adaptations to Arctic, 309, 311, 317
bird predation, 317
cohort distinction, 309
control of production, 321–322
determination of species, 306
detritivores, 315, 319
emergence, 307–308
feeding rate, 318
food availability, 322
habitats, 303
life cycle, 309–311
living food limitation, 318

mating behavior, 311
mortality, 316, 321
overwintering, 311
parasitism, 307
predators, 315
production, 318–322
production variations, 321
recruitment, 321
respiration, 311, 313
selection, 311
species, 15, 307
synchronous emergence, 307
temperature acclimation, 312
temperature and growth, 313
trophic structure, 315
vertical distribution, 315
Mites, 41, 301
Model
abiotic input variables, 414
as management tool, 410
benthic, 413
benthic algae photosythesis, 15
block diagram, 414
change in algae biomass, 414
comparison of deterministic and stochastic, 453
deterministic, 16, 409
deterministic framework, 447
general formulation, 412
Monte Carlo simulation, 452
notational convention, 412
planktonic, 413
planktonic carbon flow, 434
positive aspects, 453, 455
processes, 413
requirements, 408
restrictions, 454
simulation, 449
state variables, 413
stochastic, 17
stochastic framework, 451
test of hypotheses, 17
tuning and calibration, 451
value to project, 410
variability effects, 17
Modeling
advantages, 15, 16
and scientists, 409
assumption, 15
conclusions, 15, 17
general description, 410
steps, 407
Modeling process, steps, 412
Molting in *Daphnia*, rhythm, 273
Monovoltine, 252
Monte Carlo simulation, 452
Mosquitos, 299
Moss, dominates ponds, 158

Nannoplankton. *See also* Phytoplankton
dominates algae, 22
numbers, 182
relation to net plankton, 180
Naval Arctic Research Laboratory, 19, 25
Nematodes, 373–375, 378, 381
Nemoura, 300–302, 403
Nitrate
interstitial water, 104
sediments, 102
uptake, 107
water, 100
Nitrification, 110
Nitrite, 102
Nitrogen
algal uptake, 211
annual budget, 13, 106
budget for tundra, 46
Carex, 362
exchangable, 109
fertilization, 110, 111
forms, 13
input, 13, 106
interstitial water, 13
limits plants, 241

meltwater, 107
oil pond, 392
particulate, 110
plants, 233
primary production effects, 13
rainfall, 106
relative to C, 13
relative to P, 115
root absorption, 240–242
sediment, 13
soil, 43
turnover in water, 13
uptake, 13, 107–108
Nitrogen fixation, 13, 104
oil pond, 393
P effects, 13, 112
Nitrogen gas, 105
Nitrogenase activity. *See* N fixation, 104
Nitrous oxide, 105
Nutrients
bioassy and uptake, 24, 215, 217
biomass control, 217
control photosynthesis, 24, 217
limit algae, 211
limit plants, 45
regeneration, 188
retention within plants, 43
soil organic content, 43
uptake in model, 429

Oil
aeration and toxicity, 403
algae, 14, 219
bacterial effects, 395
benthic algae effects, 397
chemistry, 392
composition changes, 394
insects affected, 403
loss rate, 390, 393
N and P content, 392
N fixation, 393
oxygen changed, 392
phytoplankton affected, 187, 222, 397
rooted plants affected, 401
sediment respiration effects, 395
seston effects, 292
spill of 1970, 388
spill of 1975, 391
temperature effects, 391
volatization and degradation, 393
water soluble fraction, 395, 399, 403
zooplankton affected, 14, 188, 218, 403
Oligochaetes
as grazers, 219
consume microbenthos, 382
Organic matter
accumulates in soil, 43
sediments, 8, 78
Oriented lakes, 24, 28
Ostracods, 344, 378
Oxalate extraction, 12, 137, 145
Oxygen
arctic lakes, 22, 98
oil pond, 392
ponds, 97

Particulate organic carbon. *See also* Seston, 290
composition, 271
control, 164, 168
defined, 163
living percentage, 164, 271
meltwater, 164
sedimentation, 166–168
zooplankton grazing, 168, 289
Particulate organic nitrogen, water, 102
Particulate phosphorus, 117,119
Permafrost
Barrow area, 26

biological effects, 28
pond freezing, 61
pH
seasonal cycle, 87
sediments, 98
Phalaropes, 283
Phosphatase, 117, 132
Phosphate sorption index (PSI), 137, 144, 149
relation to algae, 149
Phosphorus
addition, 111, 212
algae limited, 212
algal bloom, 213
algal cycling, 216–217
available, 131
bacteria, 133
bacterial cycling, 9
budget, 124
buffering, 143–144
Carex, 134–135, 362
chemical fractions, 120
colloidal, 130–131
concentration prediction, 12
controlled by sediments, 213
cycling rates, 11, 127, 130
decomposition losses, 41
distribution, 145
equilibrium concentration, 143
exchange, sediment-water, 136
excretion, 130, 132, 133
flux, 126
fractionation, 120–121, 139
in model, 431, 444
input, 10, 13
interstitial water, 11, 124
iron relationship, 11, 137, 139
iron sorption, 11, 121
leaching, 125
luxury storage, 129
meltwater, 125
microcosm cycling, 383
N fixation stimulation, 112
oil pond, 392
organic transformation, 130–131
output, 10
oxalate extractable, 137, 145
particulate, 132
photosynthesis control, 12
plants, 233
plant cycling, 43
plants limited, 45
primary production affected, 11
protozoa, 9, 133
rainfall, 10, 13, 126
release by *Carex*, 134–135
resin exchangeable, 141
root uptake, 136, 240
sediment buffering, 150
sediment control, 146, 150, 213
sediments, 10, 120, 124, 136, 138, 141
sediments of trough, 124
soil, 43, 141, 239
soil exchange, 126
soil microbes, 239
soil water, 126
sorption, 11, 135–136, 138
sorption controls, 141
sorption isotherms, 137
supply rates, 150
total, sediment, 124
total, water, 124
turnover times, 11, 130, 135
uptake by biota, 11
water, control, 150
XP (low molecular weight pool), 130, 217
zooplankton cycling, 132–133, 286
Photic zone, sediment, 71
Photoinhibition, temperature effect, 210
Photo-oxidation, phosphorus,

117, 132, 163
Photosynthesis
benthic algae, 6, 207
biomass related, 189, 235
efficiency, 210, 236
fertilization effects, 111, 212
in model, 429, 444
inorganic carbon changes, 87, 89
light inhibition, 14
light limitation, 205, 235
nutrient limitation, 112, 217
P effects, 111
P cycling, 131
per cell, 207
temperature effect, 15, 235
water column, 6
Photosynthetic capacity, phytoplankton, 207
Physa, 300–303, 403
Phytoplankton
arctic vs. temperate, 14
autoradiography, 189
biomass, 8, 180, 182, 189
blue-green algae, 187
control of species, 222
daily changes, 182
description, 180
diatoms, 187
endogenous rhythm, 210
oil pond, 187, 397
P excretion controls, 286
P excretion, 130
primary production, 188, 189
photosynthetic capacity, 207
silica interactions, 223
species number, 14
species replacement, 400
spring bloom, 213
zooplankton, control, 400
Planktonic model, 413
Plants, *See also* Rooted plants, 108
arctic adaptations, 38, 242
biomass, 228, 235
carbohydrate, 233, 235
carbon flux, 5
competition, 243
controls on productivity, 238
DOC secretion, 161
DOC source, 153
genetic differences, 243
growth rate, 235
habitat description, 224
ice effects, 243
in model, 434
iron in beds, 138
light conditions, 235
N uptake, 240
nutrient analyses, 231
nutrient limitations, 237, 241
oil effects, 401
P uptake, 239
photosynthesis control, 235
ponds, favorable, 234
reproduction, 224
retranslocation, 233
root respiration, 96
roots and rhizomes, 228–229
sun angle effects, 235
temperature control, 236
variability, 243
Plecoptera, oil pond, 403
Polygonal ground, 28, 52
Ponds
bulk sediment density, 77
description, 51
dimensions, 51
dry up, 36
evaporation affects chemistry, 81
formation, 52
history of the site, 51
ice melt, 34
location of study site, 25
northern Alaska, 23–24
origins, 29
phosphorus, 119

polygon and trough, 53
sediment description, 76
temperature, 54
water chemistry, 79
Pond studies, description, 20
Pond water, N, 100
Potassium
Carex, 362
decomposition loss, 41
interstitial water, 84
PP. *See* Particulate phosphorus
Precipitation. *See* Rainfall, 32, 36, 53, 63
Predators, 4
Predation
bacteria, 352
controls *Daphnia*, 285
controls microfauna, 382
controls zooplankton, 264
fish, 251
invertebrate, 274
midges, 322
tadpole shrimp, 332
tundra, 40
zooplankton, 263
Primary production. *See also* Photosynthesis
Arctophila
benthic and planktonic, 200
by species, 190
calculation, 190
Carex, 230
controls, 149, 180
distribution, 6
freshwater, 22
general, 4
N uptake, 108
P interaction, 131
ponds, 24, 188–189, 192
roots and rhizomes, 38
seasonal effect, 13
sediment control, 213
terrestrial, 38, 230
Procladius, 298, 301, 315, 322
Production
bacteria, 351, 353
midges, 319, 322
tadpole shrimp, 328
zooplankton, 218, 264
Protozoa, 9
arctic adaptation, 14, 380
bacteria interactions, 9
carbon flux, 378
feeding rates, 379
food control, 381
generation time, 381
grazing, 9, 354, 379
P cycling, 9, 133, 383
predation control, 382
resistant stages, 14
yield, 379
Ps. *See* Photosynthesis
Psectrocladius, 298, 305–307, 309
PSI. *See* Phosphorus sorption index

Q10
algal Ps, 202
in model, 429

Rainfall
annual, 32, 53, 63
chemistry, 82
DOC affected, 155
N concentration, 13, 106
P concentration, 10, 13, 126
Redox potential, plant beds, 99
Replacement, phytoplankton species, 400
Resource limitation, *Daphnia* and fairyshrimp, 278
Respiration
affects inorganic carbon, 87, 89
affects oxygen, 97
bacteria, 351

benthic algae, 352
benthic animals, 352
in model, 434
lake vs. pond, 94
midges, 31
roots, 96, 352
sediment, 100, 364
soils, 41
temperature effects, 14
Resuspension, bacteria, 341
Retranslocation, 233
Rhodomonas minuta, 14, 179–180, 185, 188, 190, 221–222, 400
Rhizopods, 377
Rivers, northern Alaska, 23–24
Rooted plants. *See* Plants
Roots
ammonia uptake, 109
biomass, 38
Carex, 228–229
oxygen affects, 109
P uptake, 136, 239
production, 38
respiration, 352
Rotifers, 251, 375
Runoff
Barrow watersheds, 36
description, 35
measurement, 63
ponds, 36
seasonal pattern, 65

Secretion, in model, 430
Sediments
algal control, 13, 213
biomass, 8
bioturbation, 221, 297
buffer P, 150
bulk density, 77
control, 373
control P, 13, 213
control photosynthesis, 13, 213
description, 76
disturbance by animals, 221, 297
freezing, 61
light extinction, 71, 198
midge distribution, 315
midge habitat, 303
N concentrations, 102, 115, 238
nutrients in plant beds, 238
organic matter, 78, 362
oxygen diffusion, 371
P concentrations, 136, 141
particle size, 76, 373
protozoan carbon flux, 378
release DUP, 132
respiration, oil pond, 395
source of DOC, 153
temperature, 57
thaw depth, 61
thaw-lake cycle, 37
Sediment respiration
animal effect, 369
compared to evasion, 367
nutrient effect, 369
particle size effect, 372
rate, 364
seasonal pattern, 366
temperature effect, 364, 367
transect, 364
Sedimentation rate, 166
Seston
composition, 290
zooplankton effects, 290
Shredders, 301
Silica
controls diatoms, 223
dissolved, 82, 84
Snails. *See Physa*, 303
Snow
affects ice-melt, 60
amount, 32, 33
extinction coefficient, 33
Snowmelt, description, 33

Snowcover, duration, 32
Snowfall, Barrow, 62
Sodium
  interstitial water, 84
  water, 80
Soils
  Barrow area, 37
  chemistry, 37
  moisture content, 37
  P concentrations, 141
  temperature, 37
Solar radiation, 32–33, 54
Sorption, controls P concentration, 141
Spartina alterniflora, 109, 135
Sponges, 14
Springs, northern Alaska, 23, 25
Springtails (Collembola), 41
Stability
  in stochastic model, 456
  of system, 411
State variable
  defined, 410
  of model, 413
Stochastic model, 17
  results, 451
Streams
  Barrow area, 25
  northern Alaska, 23–25
Sulfate, 84
SUNDAY, simulation, 278
Synchrony, midge emergence, 307

Tadpole shrimp
  abundance, 323, 325
  as predators, 332
  bioturbation, 333
  egg production, 323, 334
  factors affecting density, 334–335
  food for nauplii, 330
  growth efficiency, 330
  length and temperature, 326
  life history, 323
  mix sediment, 221, 333
  mortality, 334
  production, 328
  reproductive strategy, 334
  respiration, 328
Tanytarsini, 298, 301, 315, 405
Tanytarsus inaequalis, 300–302, 305, 308, 311
Tanytarsus gregarius, 300–301, 305–307, 311
Tardigrades, 378
Temperature
  air, 32, 53, 56
  and algae, 223
  effect on algae, 201–202, 223
  effect on *Daphnia*, 277
  effect on photoinhibition, 210
  Ikroavik Lake, 57
  in model, 414
  lakes and ponds, 22, 56
  midge growth, 314
  oil pond, 391
  P uptake, 240
  ponds and lakes, 57, 391
  seasonal pattern, 56
  sediment, 57
  soils, 37
  water and air, 54, 56
Temkin isotherm, 142
Thaw
  depth in soil, 26
  factors affecting, 60
  sediments, 61
Thaw-lakes, 61
Thaw-lake cycle, 28, 37, 43, 52
Trace metals
  sediment, 87
  water, 86
Trajectory, of vectors, 411, 413
Trichoptera, 14, 229
  oil pond, 403
Trichotanypus, 300–302, 305,

307, 315
Trituration, 359
Trough ponds, 77
DOC, 157
DRP, 119
origin, 29
Tundra Biome study, 19
Turbellarians, 374, 378

Ultraviolet absorbance, as DOC measure, 157–158
Uroglena, 14, 188, 222, 400

Vegetation. *See also* Plants
Barrow area, 37
growth on land, 38

Water
budget, 69
level, 63, 67
movement, 3, 72
runoff, 33
temperature, 3
Wind
affects currents, 72
causes resuspension, 341
effect on ponds, 32
speed and direction, 32, 72
Whole system models, 409
Worms, Enchytraeidae, 41

XP (a phosphorus fraction), 116–117

Zinc
in *Carex*, 362
in water, 86
Zooflagellates, 374
Zooplankton
assimilation efficiency, 218
C content, 261
control, 288, 292
control algae, 14, 219, 284, 400
detritus feeding, 289
distribution, 251, 254
effect of arctic environment, 15
egg production, 257
energy limited, 290
grazing, 6, 168, 188, 217 353
growth rates, 262, 287
interaction with seston, 168, 289, 291–292
life cycle, 22
mortality, 252
nutrient cycling, 188, 284, 293
oil effects, 188, 403
overwintering strategy, 15
P excretion, 132, 286-287
predation, 263–264
production, 8, 15, 218, 264, 288–293
production and seston, 292
release DOC, 163
reproduction, 252
size in Arctic, 22
species in ponds and lakes, 24, 251
standing crop, 8, 254